Magmatism and Geodynamics

Magmatism and Geodynamics Terrestrial Magmatism Throughout the Earth's History

Edited by O.A. Bogatikov

Institute of Ore Deposits Geology, Petrology, Mineralogy and Geochemistry (IGEM) Russian Academy of Sciences, Moscow, Russia

Translated from the Russian by

R.F. Fursenko, G.V. Lazareva, E.A. Miloradovskaya, A.Ya. Minevich and R.E. Sorkina

CRC Press
Taylor & Francis Group
Boca Raton London New York

CRC Press is an imprint of the
Taylor & Francis Group, an **informa** business

First published 2000 by Gordon and Breach Science Publishers

Published 2019 by CRC Press
Taylor & Francis Group
6000 Broken Sound Parkway NW, Suite 300
Boca Raton, FL 33487-2742

First issued in paperback 2019

No claim to original U.S. Government works

ISBN 13: 978-0-367-44743-4 (pbk)
ISBN 13: 978-90-5699-168-5 (hbk)

Visit the Taylor & Francis Web site at
http://www.taylorandfrancis.com

and the CRC Press Web site at
http://www.crcpress.com

British Library Cataloguing in Publication Data

Magmatism and geodynamics : terrestrial magmatism
  throughout the Earth's history
  1. Magmatism   2. Geodynamics
  I. Bogatikov, O. A.
  551.1'3

Cover illustration: the eruption of the Tolbachik volcano, Kamchatka Peninsula, 1976. Photograph by A.P. Chrenov.

Contents

List of Contributors

O.A. Bogatikov
Academician, Head of the Laboratory of Petrology
Institute of Ore Deposits Geology, Petrology, Mineralogy and Geochemistry
(IGEM), Russian Academy of Sciences
35 Staromonetny
Moscow 109107
Russia

V.I. Kovalenko
Corresponding Member of RAS, Head of Laboratory of Rare-metal Magmatism
Institute of Ore Deposits Geology, Petrology, Mineralogy and Geochemistry
(IGEM), Russian Academy of Sciences
35 Staromonetny
Moscow 109107
Russia

E.V. Sharkov
Professor, Leading research worker
Laboratory of Petrology
Institute of Ore Deposits Geology, Petrology, Mineralogy and Geochemistry
(IGEM), Russian Academy of Sciences
35 Staromonetny
Moscow 109107
Russia

V.V. Yarmolyuk
Dz.Sc., Leading research worker
Laboratory of Rare-metal Magmatism
Institute of Ore Deposits Geology, Petrology, Mineralogy and Geochemistry
(IGEM), Russian Academy of Sciences
35 Staromonetny
Moscow 109107
Russia

Acknowledgements

We would like to express our gratitude to our colleagues at the Laboratories of Petrology and Rare-metal Magmatism of the Institute of Ore Deposits Geology, Petrology, Mineralogy and Geochemistry (IGEM) (Russian Academy of Sciences) for consultation and discussion of salient topics during the preparation of this book. Our thanks also go to T.Yu. Malishevskaya, M.V. Girnis and N.N. Lavrenova for technical assistance.

We are most grateful to Academician M.A. Semikhatov, whose help and initiative enabled us to establish contacts with our publishers, and also to the Directorate of IGEM RAS for support while this book was being written.

We gratefully acknowledge also the financial support provided by the Russian Foundation of Basic Researches and the Federal Programme of Support of Leading Scientific Schools of the Russian Federation.

Introduction

The way in which the Earth has evolved through geological time is one of the fundamental concerns confronting geologists today. Judging from the available geological data, the Earth has experienced several phases of evolution over its approximately 4.5 Ga life span, which have affected the internal workings of our planet and the nature of the geological events taking place on its surface. As a result of these processes the Earth has changed irreversibly. Obviously, these changes are ongoing, thereby making it harder to reconstruct past processes. The best method is to study the magmatic events responsible for the continuous supply of molten material to the surface from the depth. Studying igneous rocks provides a unique opportunity to obtain useful data on the composition of melt substrates for the whole of the Earth's history. Over the past few years extensive field studies and the application of highly efficient and precise analytical techniques, i.e. geochemical and isotopic studies, have made it possible to obtain new data, which could be used to infer the spatial and temporal distribution of various types of igneous process and identify the nature of the melts.

It has become more and more obvious that the areas of melting in the Earth's outer layers are related in some way to mantle geodynamics and to the ascent of heated asthenospheric material to the surface through the cooler and more viscous lithosphere. This ascent was irregular and occurred in the form of diapirs (plumes) of different shapes and sizes, and therefore the interaction of these plumes with the lithosphere led to the complex tectonic settings observed at the Earth's surface. Present-day magmatism and geodynamics are undoubtedly interrelated and there is no reason to believe that they operated separately in the past. Therefore, they are two aspects of the same process of planetary evolution. Thus, it is important to consider magmatic and geodynamic processes together to understand better the principal features in the evolution of the Earth as a self-developing system.

However, it is still unclear whether comparisons with present-day tectono-magmatic processes are valid when interpreting textures and structures formed in the Archaean, for example; at present, opinions differ greatly on this subject. This is not unexpected as orogenic belts, especially Precambrian ones, contain rocks that have been strongly folded and intensely altered by metamorphism. We will focus mainly on the evolution of igneous rocks, as far as this provides evidence for long-term changes in geological processes deep in the lithosphere. Data from the study of igneous rocks can be used to reconstruct the major phases in the evolution of our planet.

This book provides a systematic discussion of the main igneous environments, ranging from young tectonic settings to the fold belts of the Phanerozoic, Proterozoic and Archaean. The major phases in the Earth's evolution are also identified and compared. The final chapters deal with the origin of magmas at different stages in the Earth's development, and with the reasons for the irreversible trends in the regional and global evolution of magmatic processes. For comparison, Chapter 8 contains data on lunar magmatism. This is followed by a brief discussion of the main principles of the petrological analysis of igneous rocks adopted in this book. In summary, this volume outlines the general trends in igneous evolution throughout the Earth's history; analyses the timing of the major phases of igneous activity in different regions; and summarizes the characteristics and paths of igneous evolution along with their possible causes.

The discussion presented in this book is based mainly on data from Russia, the states of the former USSR and neighbouring countries, and these data have not been available to non-Russian readers. They were collected during preparation for the publication of *The Magmatic Rocks* (1984–88), a six-volume work, under the supervision of Academician O.A. Bogatikov. The wide range of topics discussed in this book requires a comprehensive reference list. The list for this book refers both to fundamental studies and to recent publications, which in turn will carry references to earlier publications. We have attempted to critically assess the published data and to interpret them according to our personal viewpoint. Where a topic is controversial we have tried to emphasize this.

Magmatic evolution has always been a subject of great interest to Russian geologists. Their data have enabled us to conclude that differences between the Archaean and Phanerozoic types of deep-seated petrogenesis are reflected both in the composition of the initial magmatic melts and in their geodynamic environments. It is also noteworthy that critical changes took place in the Early Proterozoic, which means that the present-day tectonomagmatic regime can only be extrapolated back to 2 Ga ago.

Abbreviations

A — alkaline
ACM — active continental margin
CA — calc-alkaline
CFB — continental flood basalts
CHUR — chondrite uniform reservoir
DM — depleted mantle
DSDP — Deep Sea Drilling Project
EM — enriched mantle
EPR — East Pacific Rise
Ga — thousand million years
GGST — granite–greenstone terrane
GGT — granulite–gneiss terrane
HA — high-alkaline
ILE — incompatible lithophile elements
LILE — large-ion lithophile elements
Ma — million years
MAR — Mid-Atlantic Ridge
MORB — mid-oceanic ridge basalt
REE — rare earth element (LREE = light REE, HREE = heavy REE)
SMOW — standard mean oceanic water
T — temperature (°C)
TH — tholeitte
TTG — tonalite–trondhjemite–granodiorite
$^{87}Sr/^{86}Sr$ (o) — initial isotopic ratio
$\varepsilon_{Nd(T)}$ — initial ε_{Nd}
$\varepsilon_{Sr(T)}$ — initial ε_{Sr}

Minerals:
Cpx — clinopyroxene
CrSp — chromespinelide
Gr — garnet
Hbl — hornblende
Mgt — magnetite
Ol — olivine
Opx — orthopyroxene
Ort — orthoclase
Pig — pigeonite
Pig–Aug — pigeonite–augite
Pl — plagioclase
Py — pyroxene
Qtz — quartz

CHAPTER 1
PRINCIPLES OF PETROLOGICAL
ANALYSIS OF IGNEOUS ROCKS

E.V. SHARKOV and O.A. BOGATIKOV

Igneous processes are responsible for heat and mass exchange in the outer layers of the terrestrial-group planets. Igneous melts originating at depth bring magma to the surface to form planetary crust. The constant loss of low-melting-point components from the deeper layers leads to differentiation of the planets' bodies. Thus it can be predicted that the igneous melts changed continuously over time, and this can provide important information about the nature and rate of planetary evolution. It is also evident that the melt composition will differ under certain geodynamic conditions, depending on the distribution of ascending and descending convection cells, whether or not differentiation has already occurred, and the melt's involvement in later reworking. This explains the close relationship between certain types of igneous activity and tectonic settings.

Certainly, magmatism is a complex phenomenon, and therefore the interpretation of particular types of igneous activity and their significance in the context of planetary evolution (in this case that of the Earth) requires the application of geophysical, petrological and isotope-geochemical methods.

Undoubtedly, the history of terrestrial magmatic processes should be analysed from the standpoint of both intrusive and volcanic rocks. It is therefore necessary to discuss briefly the interrelated issues relevant to the paragenesis of igneous rocks, and how the intrusive equivalents of volcanic rocks can be recognized.

1.1 THE IMPORTANCE OF SOLIDIFICATION PROCESSES IN THE PETROGENESIS OF IGNEOUS ROCKS

1.1.1 The Differences Between Volcanic and Intrusive Rocks

Lavas and intrusions are a result of crystallization of the same magmas. However, their emplacement takes place under quite different conditions, leading to different textures and structures. Volcanic rocks are generally poorly crystallized; differentiation does not usually occur except for rare cases of superheated Early Precambrian komatiite or picrite melts. Intrusive igneous rocks are generated under quite different conditions, especially within large intrusive chambers, accompanied by intense fractional crystallization.

Layered intrusions are ideal for studying the solidification and differentiation processes, because their original structures and textures are easily detected in the field. As has been discussed repeatedly (Jackson, 1961; Wager and Brown, 1968; Sharkov, 1980; etc.) the solidification of large intrusions takes place during the upward movement of a thin (only metres thick) crystallization zone. This process owes its character to the fact that the adiabatic PT gradient in silicate melts is steeper than that of the melting point related to the barometric pressure (Jeffries, 1929; Jackson, 1961). Therefore, at any specific moment in time, crystallization within the magma chamber occurs only at its base, because most of the melt is superheated with respect of liquidus (Fig. 1.1). This relationship is of paramount importance for the solidification of large volumes of magmatic melt. According to Jeffries (1929) it must have controlled the solidification of the molten planets during the formation of the Solar system.

However, the picture is different in the marginal parts of intrusions. There, solidification proceeds from the margins to the core due to the marked thermal gradient at the contact with the cold country rocks (Fig. 1.2). This pattern of emplacement affects the overall structure, i.e. conformable with the contacts of the marginal series with the country rocks and autonomous with respect to the contacts of the major central series (Fig. 1.3), whose formation mainly controls the geochemical evolution of melts in the body of the intrusion (Wager and Brown, 1968; Sharkov, 1980).

The upper edge of the crystallization zone (Fig. 1.4) at the base of the magma chamber corresponds with the liquidus of the main melt volume, i.e. the liquidus solid phases precipitate here. They form about 70–75% of the crystallization zone with residual liquid among them. The lower edge of the zone

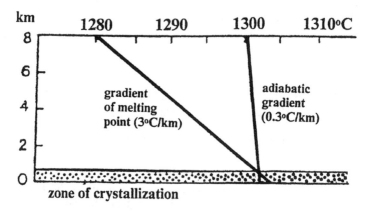

FIGURE 1.1 The relationship between adiabatic and melting point gradients, which determines the direction of solidification of large volumes of magma. After Jackson (1961).

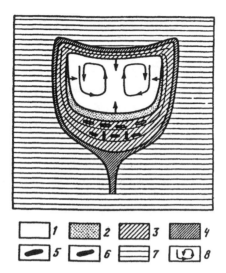

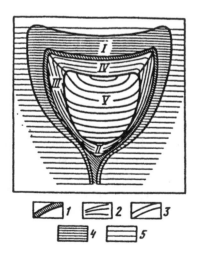

FIGURE 1.2 The pattern of intrusion solidification (Sharkov, 1980). 1 — the main melt volume; 2 — crystallization zone; 3 — solidified parts of the intrusion; 4 — chilled zone; 5 — batches of residual melt; 6 — veins; 7 — country rocks; 8 — convection currents; straight arrows show the direction in which the solidification front moves.

FIGURE 1.3 Schematic diagram of a layered intrusion (Sharkov, 1980). 1 — chilled zone; 2 — marginal series; 3 — central series; 4, 5 — country rocks: 4 — metamorphic aureole of the intrusion, 5 — unaltered country rocks; I–V — main structural units: I — zone of contact metamorphism, II–IV — marginal series (II — lower, III — lateral, IV — upper), V — central (layered) series.

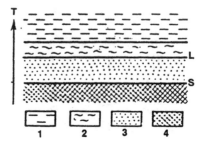

FIGURE 1.4 Schematic diagram of the solidification front of an intrusion (Sharkov, 1980). 1 — main melt volume; 2 — diffusion zone melt; 3 — crystallization zone; 4 — completely solidified part of the intrusion. Arrow shows the direction in which the solidification front moves; T — temperature; L — liquidus (front of beginning of crystallization); S — solidus (crystallization termination front).

(the crystallization termination front) corresponds with the solidus isotherm of residual melt in it. Hence, two types of crystals occur in intrusive rocks (cumulates). The euhedral crystals of the liquidus phases are known as cumulus minerals and the anhedral solidus minerals between them are called intercumulus minerals (Wager and Brown, 1968). The crystallization zone moves continuously upward when most of the melt is being cooled and crystals which settled out earlier are covered by later accumulations. The constant loss of high-temperature phases changes the composition of the melt and leads to the appearance of new, lower-temperature mineral assemblages as the liquidus phases, which are responsible for the occurrence of primary layering.

Another consequence of this directed solidification is a rhythmic layering represented by the regular alternation of different cumulates, gabbro and pyroxenites, for example. Most investigators suggest that they derived as a result of gravity differentiation. However, a special granulometric study of rhythmic layered rocks has shown that their cumulus phases are not equivalent in their hydraulic properties and could not have such an origin (Jackson, 1961; Sharkov, 1980). In fact, they are defined by alternation of high- and low-temperature cumulates — a type of temperature-concentration waves, characteristic of a dissipative system (Glensdorf and Prigozhin, 1973), which consolidated intrusions are. These waves are formed by the removal of heat from the system which occurs faster than the assimilation of the residual liquid ejected from the crystallization zone, by the main volume of the melt. As a result the liquid is concentrated ahead of the solidification front in the form of a diffusion zone (Fig. 1.4), leading to the concentration supercooling of the front of the beginning of crystallization, to its instability of the melt and the formation of a rhythmic pattern (Sharkov, 1980).

In lavas, in contrast to intrusions, the liquidus phases form phenocrysts, subliquidus phases form microphenocrysts and the residual melt forms a groundmass recrystallized to differing degrees, all of which can be seen in a single thin section. In lavas, these phases correspond with different stages of crystallization, whereas in intrusions they form separate layers of rocks similar in structure and texture and tend to be aligned upward. Naturally, each cumulate layer differs in composition from that of the melt because the layers are formed by either liquidus or subliquidus phases, and so on, up to the very low-temperature assemblages, i.e. the original melt has been fractionated.

Therefore, despite the close genetic relationship between intrusive and volcanic rocks they differ greatly in both their textures and structures. Moreover, unlike the solidification of lavas, the crystallization of intrusions may occur at different depths, which influences the way in which mineral assemblages segregate, even for the same composition of initial melts. For these reasons, the reconstruction of the intrusive equivalents of volcanic rocks poses a particular challenge.

In terms of magmatic series, the study of intrusions, especially layered types, provides important additional information about the true extent of a series, because all the rock types occur in the single cross-section. Therefore, in intrusive series the differentiation process can be observed as a whole, unlike in volcanic series, where each lava flow is an individual body, and so the question arises as to whether or not the associated rocks derived from the same magmatic source.

Until recently, intrusives were considered to be closed systems, i.e. they were believed to result from a single intrusion of magma into the chamber. However, detailed geological and isotopic data from large layered intrusions (Bushveld, Stillwater and Monchetundra, etc.; see Chapter 5) suggest that they represented open systems during the crystallization process and that they resulted from multiple influxes of new batches of the melt into the magma chamber of solidified intrusions. Therefore, the composition of magma in these chambers can change not only due to fractional crystallization, but also due to the replenishment of new batches of fresh melt. Obviously, the fresh melt could be derived from both the magma generation zone and the intermediate crustal chambers.

Apparently, intermediate chambers were represented by large, solidifying intrusions. As a result of convection, heat is mainly lost through the top of the magma chamber. Therefore the country rocks at the top of the chamber could be melted by high-temperature primary melts; a newly formed melt can become involved in convection and become mixed in with the main volume of melt in the magma chamber. Thus, magma systems display very complex structures, and hence the origin of a magmatic series may not be as simple as has been thought previously.

Based on available geological and geophysical data, it is thought that lava is mainly ejected from solidified intermediate magma chambers, i.e. it has already been segregated from the main volume of the melt. This melt could well be in equilibrium with segregating solid phases (cumulates), whose fragments often occur as cognate xenoliths. Therefore, lava phenocrysts could reflect the composition of cumulates at the time of separation of a given batch from the main melt. Hence, intrusives can provide important information on the specific mechanism responsible for the formation of volcanic assemblages, the dynamics and features of development of both magma chambers and magma generation zones and, in some cases, even about the mechanism of melt formation of the whole magmatic series. Sometimes, the series can have a heterogeneous origin, due to the involvement of intermediate magma chambers in melt formation. Apart from the solidification of purely petrological problems, the study of both volcanic and intrusive rocks allows the determination of the distribution pattern of certain types of magmatism independently of the erosion level in the area. This is particularly important in the case of a fold belt, where rocks formed at different depths have been uplifted to the present erosion level by tectonic processes.

1.2   ASSEMBLAGES AND SERIES OF IGNEOUS ROCKS

The results of field and laboratory studies of igneous rocks imply a relationship between certain igneous rock series and the geodynamic conditions and evolutionary phases of the lithosphere. This problem has been studied for more than a century, using as an example fold belts that have passed through all phases of development: from oceanic crust to orogenic belts. Each phase was accompanied by particular assemblages of igneous rocks, from ophiolites to bimodal basalt–rhyolite and granodiorite–granite assemblages, which according to Russian nomenclature are known as "magmatic formations" (Kuznetsov, 1964).

Two types of common natural assemblages of igneous rocks have been recognized:

1. assemblages not actually related to a particular region or structure that tend to recur during the geological evolution of different fold belts and
2. assemblages formed specifically over a particular period of time (corresponding with a geological period or stage) within a separate region and reflecting a specific stage in the development of a particular fold system.

The recognition of true igneous assemblages is based on the following criteria:

1. similarity in the petrographic and geochemical properties of the rocks;
2. development within a single geological structure;
3. relationship with a single stage of igneous activity within a particular structure.

It should be noted that igneous assemblages are recognized on the basis of their petrographic characteristics and that their name reflects that of the contained igneous rocks. The study of these assemblages can establish the relationship between igneous activity, magma composition and the structure of the Earth's crust and of the lithosphere as a whole, and the geodynamic conditions characteristic of a particular tectonic settings.

On the other hand, the establishment of the type of igneous activity, irrespective of the type and stage of development of geological structures where it occurs, is very important, especially when igneous processes are analysed on a global scale. It is even more important when events from the geological past are studied, since geodynamic processes, especially those in the Early Precambrian, could be different in character from those prevailing at the present day. For this purpose we can use the concept of igneous rock series, as based on geochemical data.

As long ago as the 1930s, petrologists discovered that basalts form two assemblages. It was suggested that they were formed by differentiation processes (N. Bowen, F. Fenner, V.S. Sobolev, etc.) and therefore they were termed igneous rock series. Since then, many petrologists have studied changes in the mineralogy of igneous series, among them L. Wager, U. Deer, O. Rittman, A. Osborn, D. Dickenson, J. Gill, etc. There are also the Russian publications of A.N. Zavaritsky and A.A. Marakushev and of the authors of the six-volume monograph Magmatic Rocks, edited by O.A. Bogatikov, published in 1984–1988.

Unfortunately, igneous rock series have been described in many different ways. Therefore, here we will provide a systematic classification of various igneous series, encompassing all types of igneous rocks. An igneous series is a generalized assemblage of igneous rocks forming a petrographic series (for example, from basic to acid in series with normal alkalinity). Each series is characterized by the particular mineralogical and geochemical features of its rocks, usually spatially related. Series are recognized on the basis of petrochemical (i.e. major elements of geochemistry) data; each series is distinguished by the evolutionary trend of its melts, suggesting a genetic relationship between the rocks within it (Bogatikov et al., 1987; Sharkov and Tsvetkov, 1990).

Within igneous rock assemblages we can recognize abstract (series type) and true series. Abstract series can be recognized only by petrochemical (i.e. major elements) means, and so are known as petrochemical series. This is an assemblage of rocks in general resulting from differentiation of certain types of parental magma, and as such shows an evolutionary trend corresponding with that of naturally occurring magmatic melts. The trend itself is determined by general physicochemical regularities and the number of trends is defined by physicochemical constraints. Physicochemical regularities play a role similar to that of DNA in living organisms, predetermining the development of individual magmatic melts irrespective of their origin. During crystallization it should occur within a general sequence, expressed as a petrochemical series. On this basis we will try to bring together data on igneous rocks from different regions of the world and for the Earth as a whole.

Petrographic (or true) series, whose definition incorporates igneous rock names (i.e. shoshonite, boninite, etc.) should be considered as special cases of realization of the abstract series in nature.

As an example, abstract tholeiite series should consist of high-Mg (komatiite) and three moderate-Mg series; the calc-alkaline series should comprise high-Mg rocks such as boninite and low-Mg rocks such as andesites (Fig. 1.5). In fact, a specific series is an assemblage of rocks derived from similar sources and possessing similar physicochemical properties. They also follow similar differentiation trends, but have their own peculiar features associated with specific conditions of magma generation. The latter, as a rule, are determined by geodynamic settings: this enables data on igneous activity to be used to indicate geodynamic conditions. However, it should be taken into account that throughout geological time the same types of rock could form under different geodynamic conditions. In this respect, the most typical is the siliceous high-Mg boninite series, which in the Phanerozoic is associated only with active oceanic and continental margins, but in the early Palaeoproterozoic was formed under intra-plate conditions.

In the ideal case, all the rocks of a series should have constant isotope ratios (Sr, Nd, Pb, etc.) and their incompatible elements, and the concentration of the latter should vary within the limits determined by paragenetic relationships. However, sometimes this rule is broken, even within a single intrusion or volcano (Wilson, 1989; Lambert et al., 1995). This means that processes of contamination and magmatic mixing, which strongly affect trace elements, continue to play a role in the petrogenesis, while not changing the overall pattern, which is governed by the evolution of the major elements. Nevertheless, in each particular case we should try to establish the paragenetic relationships of rocks in particular series, or at least this should be anticipated. In this book we will try to utilize fully the criteria for series recognition.

1.2.1 Petrochemical Criteria for the Recognition of Igneous Series

Volcanic Series

At present the igneous series are studied mainly for volcanic rocks and most criteria were worked out specifically for them. The identification of volcanic series in this book is as follows:

1. the recognition of trends of alkalinity (e.g. highly and moderately alkaline, on the one hand, and normal alkalinity on the other);
2. the subdivision of rocks of normal alkalinity into tholeiite and calc-alkaline, and the subdivision of alkali rocks into moderately alkaline (subalkaline) and alkaline (i.e. containing modal feldspathoids);
3. the subdivision of alkaline rocks using their K_2O/Na_2O ratio.

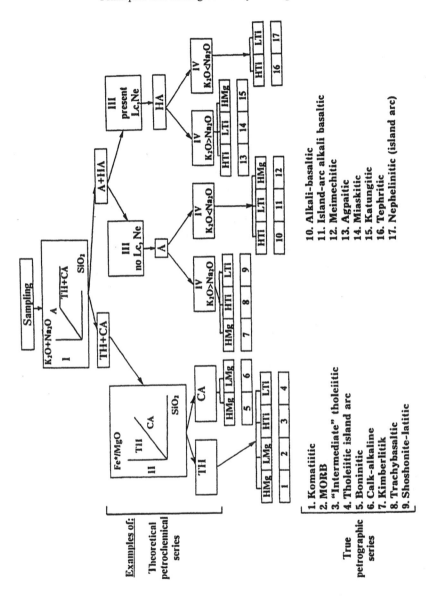

FIGURE 1.5 Igneous rock series.

In all, six main (abstract or petrochemical) series have been recognized: tholeiitic (TH), calc-alkaline (CA), potassium–sodium and moderately alkaline potassic (K–Na and K–A), and high-alkaline potassic and sodic (K–Na HA and K HA).

The analysis of the mineral composition of the rocks making up igneous series tends to bring out their distinctive features.

The tholeiite series contains picrobasalt, basalt, ferrous andesite (icelandite), dacite and rhyolite. Andesite tends to occur in subordinate quantities, which results in a bimodal distribution of rocks, with the acid varieties also being subordinate. The SiO_2 content generally amounts to 48–63%. The groundmass contains augite and sometimes pigeonite. A trend of enrichment in iron and chondritic REE is also characteristic. Igneous rocks of the tholeiite group contain at least four varieties: high-Mg (komatiite), and three moderate-Mg: low-K (MORB), low-Ti (island-arc tholeiites) and high-Ti (intermediate tholeiites).

The calc-alkaline series contains mainly andesite, dacite and rhyolite and usually subordinate amounts of high-Al basalt and andesite–basalt. The SiO_2 content varies from 52% to 70%; the groundmass contains hyperstene and pigeonite. There is no trend of enrichment in iron and low Ti content. Cerium-group LREE are more numerous as compared with those in the tholeiite series, and the total REE content is also high. At present, at least two specific series — predominantly low-Mg andesite and less commonly high-Mg boninite — can be recognized within the calc-alkaline series. The boninite series has high chromium and nickel content. The composition of the rock-forming minerals is distinctive, with the concomitant occurrence of four pyroxenes: bronzite, augite, pigeonite and pigeonite–augite; clinoenstatite is rare. The feature of the rocks is the andesitic composition of the volcanic glass in the basic and ultrabasic members of the series.

The K–Na moderately alkaline series (alkali basalt) consists of high-Ti and low-Ti varieties. The first one is composed of moderately alkaline picrobasalt (ankaramite), olivine alkali basalt, alkali basalt, hawaiite, trachyte and moderately alkaline and alkaline dacite and rhyolite. Rocks of andesitic composition (trachyandesite, benmoreite) are subordinate, as in the tholeiitic series, resulting in a bimodality characteristic of the volcanic activity of ocean islands and continental rift zones. The low-Ti series is characteristic of active plate margins.

The potassic moderate alkaline series consists of rocks from potassic picrobasalts and basalts to andesites and potassic rhyolites. There are also two varieties of potassic moderately alkaline series: high-Ti (trachybasalt) and low-Ti (shoshonite–latite). In the latter, basic rocks, intermediate rocks and dacites amount respectively to 50%, 40% and 10%. As in the calc-alkaline series, the rocks themselves, and the clinopyroxenes they contain, are relatively poor in iron. Typical features are low titanium content and lantanoids rich in LREE. The kimberlite assemblage is an usual variety of high-Mg moderately alkaline potassic rocks.

The potassic–sodic and potassic high alkaline series contain feldspathoid-bearing rocks (nepheline in the K–Na series and/or leucite in the K series). Geochemically, they differ from the moderately alkali series in their high concentrations of alkali and their low concentration of silica. The potassic high-alkaline series has both high-Ti and low-Ti varieties.

Intrusive Series

All of the above-mentioned volcanic series have intrusive equivalents, represented by small subvolcanic bodies or large complex intrusions. As a rule, they have a specific mineral composition and chemistry, allowing us to accurately identify to which series they belong.

Basic and ultrabasic layered intrusions of the Skaerggard type, for example Rhum Island, Dovyrensky, Lysogorsky, etc. (Wager and Brown, 1968; Sharkov, 1980; Laz'ko and Sharkov, 1988d), along with the layered gabbroic complexes of ophiolitic assemblages, provide typical examples of the intrusive equivalents of the tholeiitic series. Figure 1.6 shows typical cumulate sections.

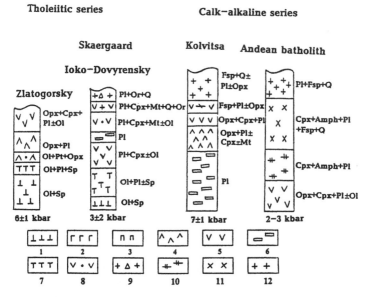

FIGURE 1.6 Succession of cumulates in intrusions emplaced during the crystallization of tholeiitic and calc-alkaline melts. 1 — dunite; 2 — harzburgite; 3 — orthopyroxenite; 4 — norite; 5 — gabbro and olivine gabbro; 6 — gabbro-anorthosite; 7 — troctolite; 8 — ferrogabbro; 9 — granophyre; 10 — diorite and tonalite; 11 — granodiorite; 12 — granite. Pressure during the emplacement of an intrusion is measured in kbar.

The intrusive equivalents of the calc-alkaline series form the gabbronorite–diorite–granite batholithic complexes of the Circum-Pacific Belt, including the island arc and the active continental margins. The Andian batholiths, represented by a number of poliphase intrusions exhibiting primary igneous layering, are well known (Pitcher, 1974). In terms of composition, they range from olivine gabbro and gabbronorite to adamelite and leucogranite, with diorite, granodiorite and tonalite being predominant. The Palaeoproterozoic intrusive equivalents of the calc-alkaline series are fairly usual, being characterized by the presence of large harzburgite–norite–anorthosite intrusions and norite–gabbronorite dyke swarms, emplaced as a result of the crystallization of specific siliceous high-Mg (boninite-like) melts (see Chapter 5). The Proterozoic was marked by the development of anorthosite–mangerite intrusions, which were confined to ancient collision structures (see Chapter 5). Their sections begin with plagioclase cumulates (anorthosites), overlain by intermediate rocks (monzonites, mangerites, pyroxene–diorites), and are terminated by charnockites.

The intrusive equivalents of the titanium potassic-sodic moderately alkaline series are represented by syenite–gabbro layered intrusions, which are very common in fold belts. They can be exemplified by the Proterozoic Gremyakha–Vyrmes and Elet'-Ozero massifs of the Baltic Shield (Sharkov, 1980) and the Palaeozoic Kizir, Bolshoi Taskyl and Bolshoi Kul'-Taiga massifs in the Altai–Sayan Fold belt (Bogatikov, 1966). Figure 1.7 shows that they can be subdivided into two groups, which differ in their mineral assemblages. The Precambrian massifs' sections begin with ultrabasic cumulates: namely, olivinites, wehrlites, and clinopyroxenites. The rocks are dominated by gabbroids, including gabbro–anorthosite and ferrogabbro. The accumulation of iron and alkalies resulted in the appearance of rocks characterized by high concentrations of titanomagnetite hortonolite olivinites, gabbroids and clinopyroxenites, and syenite and nepheline syenite, at the final stages of differentiation.

The syenite–gabbro layered intrusions of the Altai–Sayan are also characterized by a high concentration of titanomagnetite in the middle to upper part of the section crowned by syenites; nepheline-bearing varieties have not been reported. There is also a fairly drastic transition from titanium ferrogabbro to syenites (the equivalent of trachytes) which often represent a later, separate phase of intrusions. The rare, if any, presence of intermediate varieties suggests a bimodal pattern typical of the volcanics of this series.

The intrusive equivalents of the potassic moderately alkaline (shoshonite–latite) series are represented by subalkaline and gabbro–monzonite intrusions, recorded from the Kurile Islands and Japan (Tsvetkov, 1984). They make up the thick, sill-like bodies of the southern group of the Kuriles (Tanfilieva and Zeleny, etc.) and the Nemura Peninsula of Japan. They exhibit some layering features: from orthoclase–olivine gabbroids at the base, to monzonite, monzodiorite and syenite in the upper parts of the sills. Monzonite is the dominant rock type. Melt crystallization took place under near-surface conditions, as suggested by the spatial association with shoshonite lavas and pyroclastics.

The well-known massifs of the Khibinas and Lovozero on the Kola Peninsula provide an example of the intrusive equivalent of the potassic–sodic alkaline series that resulted from phonolite crystallization (Kogarko *et al.*, 1995). The lower part of the section of the layered Lovozero Intrusion is dominated by urtites, the middle part by foyaites and the upper part by lujavrites (nepheline–feldspar–aegizine cumulate), reflecting the sequence of segregation of corresponding mineral phases during the crystallization of phonolites (Sharkov, 1980).

Synnyr-type massifs are the intrusive equivalents of the potassic alkaline series (Orlova *et al.*, 1993). The cross-section and horizontal slices through the centre of Synnyr intrusion shows that nepheline syenites (nepheline–feldspar cumulates) give way to pseudoleucite syenites (nepheline–leucite cumulates) and then to pseudoleucites or synnyrites (leucite cumulate). It is an example of low-Ti K high-alkaline series. The alkaline igneous complex at Magnet Cove in Arkansas, USA (Erickson and Blade, 1963) provides an example of a massif derived from high-Ti alkaline potassic melt where pseudoleucite syenites are

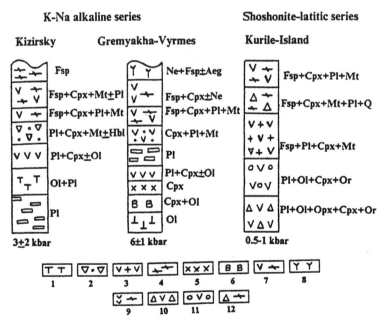

FIGURE 1.7 Rock succession (cumulates) in intrusions emplaced during crystallization of potassium medium-alkaline melts. 1 — troctolire; 2 — subalkaline ferrogabbro; 3 — monzonite; 4 — nordmarkite; 5 — clinopyroxenite; 6 — wehrlite; 7 — syenite; 8 — nepheline syenite; 9 — essexite; 10 — orthoclase–olivine gabbro; 11 — monzodiorite; 12 — quartz monzonite; The remaining symbols are the same as those defined in the caption of Fig. 1.6.

related to the marginal zone of the intrusion, and ijolites (nepheline–feldspar cumulates) with its central part. Apart from this, there are also jacupirangites and sphene pyroxenites and the youngest differentiates are represented by carbonatites. Volcanic equivalents of these rocks are recorded within the western branch of the East African rift system north of Lake Kivu (Tanganyka Rift).

We have already discussed some of the best-known intrusive equivalents of the volcanic series. In reality, they are much more diverse but they have not been completely identified as yet.

When summarizing the classification of the intrusive igneous rock series, it should be noted that they are much more diverse than their volcanic equivalents. They differ in their wider range of rock composition, with some varieties having no direct volcanic equivalents. This concerns mainly dunites, pyroxenites, anorthosites and pseudoleucitites. Probably the main reason for this situation is the difference in the way lavas and intrusions solidify (see above): unlike intrusions, where all liquidus phases could form their own separate rocks (cumulates), in lavas they could present themselves only as phenocrysts. These monomineralic assemblages of cumulates, like the more common polymineralic associations, mark certain stages in the evolution of melts in intermediate volcanic chamber (or intrusions). In this case, the cumulates remained in the form of solidified parts of the magma chambers and the melt could have been erupted at the surface by volcanoes. The primary melt of particular intrusion could be in turn a differentiate from a deeper magma chamber, therefore not all the members of a series had to be present in any particular case. Table 1.1 compares some volcanic series and their intrusive equivalents.

It should be noted that statements on igneous rock series are based mainly on data on the mineral composition and geochemistry of major elements; the trace element content and the isotope ratio are not of primary importance. The physico-chemical parameters that determine the evolutionary course of the melts (independent of their earlier history) play the leading role in the formation of a true series. The stability of the main parameters of an igneous rock series suggests several general trends in the evolution of melts. Depending on the specific features of mineral composition, a melt tends to fall on one of the general trends and follows it to form a certain sequence of igneous rocks. These series can be used as an independent source of data about deep-seated processes operating under different geodynamic conditions at different stages in the Earth's history.

1.3 SUMMARY

1. Igneous activity is a process responsible for heat and mass exchange in the outer layers of the terrestrial-group planets. It has influenced their evolution

Table 1.1 Comparison of volcanic and intrusive series.

Series	Volcanic assemblage	Intrusive assemblage
Tholeiite	MORB Basalt assemblages of Hawaii; British–Arctic Province	Layered complexes of ophiolitic assemblages; Mesozoic and Cenozoic layered intrusions of the Skaergard and Rhum Island types
Calc-alkaline	Andesite assemblages of the Andes, the Kuriles and Aleutian Islands, Japan, recent volcanic zones of the Caucasus; Early Proterozoic boninite-like series of the Baltic shield	Mesozoic–cainozoic batholiths of the Andes, Kuriles and Aleutian Islands; Proterozoic anorthosite–mangerite intrusions of the Kolvitsky and Kalarsky types; Early Proterozoic basic and ultrabasic layered intrusions
Titanium potassic–sodic medium alkaline	Alkali-basalt assemblages of continental rift zones	Layered syenite–gabbro intrusions of the Gremyakha–Vyrmesky, Kizirsky and Patynsky types
Low-Ti potassic medium-alkaline (shoshonite)	Shoshonite assemblage of Yellowstone Park, USA	Gabbro–monzonite–syenite intrusions of the Massif Central, France, the Kuriles and Japan
Potassic–sodic high-alkaline	Nephelinite assemblage of the Kenya Rift	Layered intrusions of Lovozero and Khibiny types
potassic–sodic potassium	Leucite–tephrite assemblage of Mediterranean island arcs	Layered intrusions of pseudo-leucite syenite of the Symnyrsky type

by controlling differentiation of planetary material, and by being the main agent of crust formation.

2. The nature of newly formed igneous melts is to a greater degree determined by specific geodynamic settings in the outer layers of planets controlled by mantle convection. This in turn defines the type of material to be melted, the conditions of magma generation, storage, the fluid regime, etc. Therefore, igneous formations can be considered as a very sensitive indicator of geodynamic regimes.

3. A special feature of igneous activity is the major role of solidification laws that govern the formation of igneous rock assemblages that are related by a common origin.

4. Igneous rock series can be recognized on the basis of their petrochemistry, independent of the type and phase of structural evolution. The recognition of abstract series (e.g. tholeiitic, calc-alkaline) is based on petrochemical data. Each occurrence of a particular rock type (true or natural series) within a petrochemical series will reflect the geological settings under which it was generated. Series reflect differentiation trends of natural magmatic melts.

5. Assemblages and series of igneous rocks can be recognized among both volcanic and intrusive rocks that otherwise differ greatly in their texture and structures. In a lava, the entire sequence of mineral parageneses, from liquidus to solidus phases, can occur in a single sample or thin section; in an intrusion the sequence is extended vertically, i.e. liquidus, intermediate and solidus parageneses can form separate rocks, i.e. cumulates. Therefore, despite the superficial petrochemical similarity of intrusive and volcanic series, they differ significantly in detail. This should be taken into account when comparing intrusive and volcanic formations. Moreover, a single melt, subjected to different pressure conditions, can fractionally crystallize different mineral phases, which will make it difficult to identify these formations.

6. Fortunately, often some shallow intrusions represent an intermediate stage between deeper intrusions and volcanic activity. Therefore, when these are studied together they allow us to understand better the mechanism and dynamics of magma evolution processes responsible for the generation of certain assemblages and series of igneous rocks, as well as to specify the extent of a particular type of igneous activity.

CHAPTER 2
MAGMATISM IN THE CONTEXT OF THE PRESENT-DAY TECTONIC SETTINGS

O.A. BOGATIKOV, V.I. KOVALENKO, E.V. SHARKOV and V.V. YARMOLYUK

The study of present-day magmatism is of primary importance in clarifying the relationship between magmatism and geodynamics. Here we will consider processes operating at the present day, which have their own dynamics, topographic expression, seismicity, geophysical fields, etc. and can provide the maximum data about their mechanism and character. Present-day geodynamic environments and their related igneous activity are known to be continuations of geological processes initiated in the Mesozoic but associated mainly with the Late Cenozoic. They are essentially a snapshot of those processes. It is obvious that existing tectonomagmatic systems are at different stages of development, thereby providing an overview of their main evolutionary trends during the Phanerozoic.

Present-day igneous activity can be subdivided into two major types:

1. at lithospheric plate boundaries; these include constructive (divergent) and destructive (convergent) boundaries, each of which shows different types of magmatism, and
2. intra-plate, often known as hot-spots (Fig. 2.1). The igneous activity of present-day geodynamic environments has been discussed in many publications, in particular: *Evolution of magmatism in the history of the Earth* (Kovalenko, 1987) and *Igneous petrogenesis* (Wilson, 1989).

2.1 MAGMATISM AT LITHOSPHERIC PLATE BOUNDARIES

In both the diversity of its rocks and its extent, magmatism at lithospheric plate boundaries vastly exceeds within-plate magmatism, accounting for about 90% of all magmatic activity (Kovalenko, 1987). Plate boundaries are recognized mainly on the basis of their higher seismicity. In the oceans, the base of a lithospheric plate coincides with the asthenosphere, exhibiting low viscosity. Convection in the upper mantle apparently gives rise to lithospheric plate motion. Based on different estimates, including the Sm and Nd isotope balance

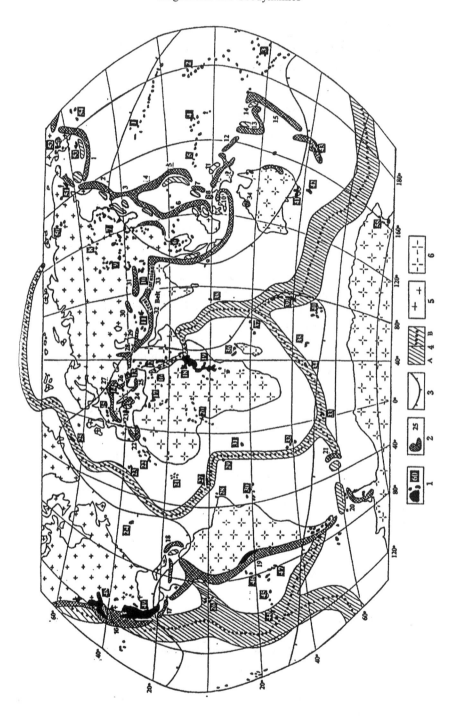

FIGURE 2.1 Distribution of late Cenozoic magmatism. Modified from Sharkov and Bogatikov (1987). Key: 1 — main areas of intra-plate magmatism (Fe–Ti picrite–basalt–tholeiite, K–Na and K moderate- and high-alkaline series (numbers are ringed). 1 — Hawaii; 2 — Line Islands; 3 — Tuamotu–Réunion–Tabaui islands; 4 — Marshall Islands; 5 — Caroline Islands; 6 — Indochina; 7 — Far East; 8 — Baikal; 9 — Mongolia; 10 — Tibet–Nan Shan; 11 — Iran–Afganistan; 12 — Arabia; 13 — Asia Minor; 14 — Pannonia; 15 — Central Europe; 16 — western Red Sea; 17 — North Africa (Ahaggar, Tibesti); 18 — Ethiopia; 19 — Kenyan; 20 — cameroon; 21 — Cape Verde; 22 — Canary; 23 — Azores; 24 — Bermuda; 25 — Iceland; 26 — Galapagos; 27 — San Paulu; 28 — Fernando de Noronha; 29 — Ascension; 30 — Trinidad; 31 — St. Helens; 32 — Tristan da Cunha and Gough; 33 — Bouvet; 34 — Prince Edward Island; 35 — Crozet; 36 — Comores–Madagascar; 37 — Mascarene (Reunion Mauritius); 38 — Maldives; 39 — Amsterdam and St Paul; 40 — Kerguelen; 41 — South Australia (Victoria); 42 — Tasman; 43 — South Island of New Zealand; 44 — Easter Island; 45 — Sala i Gomez; 46 — Nasca; 47 — Juan Fernandez; 48 — Western USA; 49 — Basin and Ranges Province; 50 — Indigirka; 51 — Chukchi; 52 — Alaska; 53 — Pribylov; 54 — Gulf of Alaska; 55 — West Antarctic. 2 — main areas (arcs) of andesite–latite magmatism of active ocean margins, continents and microplates (calc-alkaline series, low-titanium and K–Na moderate- and high-alkaline series (numbers are not ringed). Arcs: 1 — Aleutian–Alaska; 2 — Kurile–Kamchatka; 3 — Japan; 4 — Izu–Bonin; 5 — Marianas; 6 — Philippine–Sulawesi; 7 — Caroline Islands; 8 — Molucca; 9 — Indonesia–Burma; 10 — Banda; 11 — New Guinea and New Britain; 12 — Solomon Islands; 13 — New Gebrides; 14 — Fiji; 15 — Tonga–Kermadec; 16 — western North America; 17 — Transmexican; 18 — Lesser Antilles; 19 — western South America; 20 — Antarctic Peninsula; 21 — South Sandwich Islands; 22 — Alboran; 23 — Sardinia; 24 — southern Italy; 25 — Aegean; 26 — Balkan; 27 — Carpathians; 28 — Caucasus–Anatolian; 29 — Elburs; 30 — Pamir–Tien Shan; 31 — Kuenlin; 32 — South African; 33 — Himalayan. Key: 3 — Solid lines: geoidal height (Le Pichon and Huchon, 1984); 4 — diagonal shading: MORB (a — back-arc basins, b — mid-oceanic ridges); 5 — crosses: fragments of Laurasia; 6 — open crosses: fragments of Gondwanaland.

in the crust and in the Earth as a whole, the depth of convection is taken to be about 700 km (McKenzie, 1985).

Generally the convection scheme is as follows. Heated mantle material comes to the surface at a mid-oceanic ridge, where it undergoes partial melting as a result of adiabatic decompression and newly formed basalts are added to the Earth's crust (this is known as a constructive plate margin) and solidified material returns into the mantle via subduction zones (destructive plate margins). The age of the lithosphere increases away from the mid-oceanic ridge axis, as inferred from the symmetrical pattern of banded magnetic anomalies.

On continents the plate-tectonic setting is more complex. According to seismic tomography (Dziewonski and Anderson, 1983; Anderson et al., 1992; Zhang and Tanimoto, 1993) the base of the continental lithosphere occurs at a depth of about 400 km or more. However, both the scope and character of magmatism is similar to that at the active plate boundaries of continents and oceans. This similarity is not dependent on the composition of the interacting plates (Fig. 2.1) (Sharkov and Svalova, 1991). This is exemplified by the Alpine–Himalayan Belt, which is a destructive plate boundary. Molnar and Atwater (1971) were the first to suggest the generation of subduction zones in the Himalayas. This could be ascribed to the layering of the lithosphere, namely, the continental lithosphere is divided into layers, whose motion and subduction was christened "two-storey plate tectonics" by Lobkovsky (1989).

2.2 MAGMATISM AT DIVERGENT PLATE BOUNDARIES

Oceanic rift zones where sea-floor spreading and accretion of the oceanic crust take place are assigned to constructive geodynamic environments. These zones are known as mid-oceanic ridges (MOR), and they form a system extending for more than 60,000 km and girdling the whole of the Earth's surface (Fig. 2.1). They are associated with medial regions of the World Ocean and sometimes continue on to continents where they are expressed in a complex structural pattern, for example in western North America. Spreading systems have been reported from many back-arc (marginal) basins (the Philippine Sea, the Japan Sea and the Bering Sea, among others).

Morphologically, the MOR is a linear high with gentle slopes and a total width of 200–500 km. The height does not exceed 2,000–3,000 m above the sea-floor. Median valleys (axial rifts) occur on the slow-spreading Mid-Atlantic and Arabian–Indian (Western Indian) ridges but they are less obvious on the East Pacific Rise, which has a fast spreading rate of about 10–12 mm per year (Nicolas, 1989). Mid-oceanic ridges are characterized by high heat flow, suggesting the presence of ascending convective flows of asthenospheric material.

An unusual feature of the mid-oceanic ridge is the symmetry of its deep structure and morphology on either side of the ridge axis. The lithosphere shows a three-layered structure, as inferred from geophysical and geological data. The uppermost layer is a sedimentary cover up to 0.5 km thick. The second layer consists of interbedded basalts, often pillow lavas, underlain by a vertical dyke complex. The thickness of this layer may reach 2 km. The lowermost layer is composed of gabbroic rocks with a thickness of up to 5 km. Mantle ultramafics underly the crust (Nicolas, 1989).

In the course of sea-floor spreading, partial melting of rising asthenospheric material (10–15% of the volume) occurs due to a decrease in the total pressure. New oceanic crust is generated in the central part of the spreading zone, due to injection of magmatic material from below. The magma chamber is located beneath the spreading axis, and cumulate gabbro and peridotite form on the floor of the chamber. Based on seismic data, the chamber is actually quite small in size, for example, on the East Pacific Rise (EPR) it is 1–2 km in width and less than 1 km in height, although its length can reach several tens of kilometres (Detrick, 1991). The cumulate gabbros from the western Indian Ridge are well known: there, on the transform fault Atlantis II, DSDP Hole 735 exposed a section of layered gabbroids containing troctolites, olivine gabbros, gabbros and Fe–Ti-rich gabbros (Kempton *et al.*, 1991a). The hole was drilled in thin (35–50 cm) basaltic and trondhjemitic dykes. The plutonic rocks follow the tholeiitic fractionation trends, to highly evolved melts from which two-pyroxene and Fe–Ti-rich gabbroids and possibly trondhjemites crystallized. In addition, there are samples with mg [molar $(Mg^{2+} + Fe^{2+})/Mg^{2+}$] c. 80–82 suggesting a more or less regular supply of primitive melt to the magma chamber. According to seismic tomography, the thickness of the wedge of heated material beneath most spreading ridges does not exceed 100 km and only locally in the Pacific and Indian ocean is it thicker than 300 km (Anderson *et al.*, 1992). The lowest seismic velocity was recorded in the southern- and northernmost parts of the Mid-Atlantic Ridge (MAR), on the EPR, in the SE part of the Indian Ridge and at triple junctions and hot-spots. It should be noted that mid-oceanic ridges are mobile structures and can move relative to the jots of heated mantle, as shown by the Mid-Atlantic and Indian Ridges, which drift in a westerly direction.

At present, two types of magma series: tholeiitic and moderately potassic–sodic alkaline, can be recognized within mid-oceanic ridges (Basaltic volcanism, 1981). Ridges are made up mainly of basalts of the tholeiitic series, while alkaline volcanics occur on islands located on ridge axes (Iceland) or on their flanks (St. Helens', Gough and others) and on some transform faults.

The first most comprehensive petrographic review was presented by Engel *et al.* (1965). Later studies of bedrocks obtained during the Deep Sea Drilling Project (DSDP) and dredged from the sea-floor show that mid-oceanic ridge basalts (MORB) are not as homogeneous as had been thought previously. Tholeiitic basalts, which make up most of the ridges, differ slightly from the basalts of submarine highs or plateaus and are associated with oceanic islands located on ridge axes (Basaltic volcanism, 1981). Hence the recognition of two types of tholeiitic basalts: N (normal) and E (enriched), whose chemistry is shown in Table 2.1. There is also a T-type basalt, which is transitional between the N and E types. Even the early study of mid-oceanic ridge basalts has shown that as compared with tholeiitic basalts from Hawaiian-type oceanic islands and continental traps, the former are depleted in Cs, Rb, K, Ba and Sr. Therefore, the K/Rb, K/Ba, Sr/Rb ratios are higher in MORB than in their petrographic equivalents occurring in other plate-tectonic settings. It has been established that this depletion is mainly characteristic of N-type basalts, whereas E-type basalts resemble the tholeiitic basalts of oceanic islands. Basalts of fast- and slow-spreading ridges also differ in their geochemistry, the latter containing less titanium and zirconium (respectively 1.20–1.77, 0.80–0.87 wt.% TiO_2 and 64–110, 40–42 ppm Zr).

Primitive melts included in MORB olivines have low H_2O content (mean at 0.12 wt.% for N-MORB, 0.17 wt.% for T-MORB and 0.51 wt.% for E-MORB). A strong coupling between H_2O and K_2O has been found in some MORB primary melts which might well be explained by the presence of an H_2O-bearing CO_2-rich fluid (Sobolev and Chaussidon, 1996).

Transform faults greatly influence MORB geochemically. Basalts occurring near faults are marked by higher abundances of REE, Ti and Na; therefore, in their geochemistry they are closer to the tholeiitic basalts of oceanic islands. The transform-fault zones are characterized by weaker igneous activity as compared with the series of ridges in between (Pushcharovsky and Melanholina, 1989; Kogarko et al., 1993).

Gabbroids from the lower oceanic crust are very similar to MORB in their geochemistry and isotopic composition, as evidenced by samples from Site 735 (Kempton et al., 1991a). However, locally these rocks are intensely deformed and altered and their isotope ratio implies that metamorphic and hydrothermal processes have not strongly affected the Sr-, Nd- and Pb-isotopes. These rocks have higher $^{143}Nd/^{144}Nd$ (0.51301–0.50319) and lower $^{206}Pb/^{204}Pb$ (17.35–17.67) ratios as compared with those of MORB. The $^{87}Sr/^{86}Sr$ ratio (0.7025–0.7030) is higher than that of MORB from the West Indian Ridge and the Rodriguez triple junction. Strong variations in these ratios suggest that the chamber is fed by fresh batches of melt. The secondary processes affect only the O-isotopes; the pattern of the process suggests that the reaction took place at high temperatures and low sea-water content.

Spreading ocean zones are sometimes complicated by islands. In their pattern of development and their geochemistry they are similar to intra-plate formations and will be discussed in a separate section.

This suggests that a section of the ancient oceanic lithosphere is represented in ophiolite sequences, which were obducted onto one of the colliding continental forelands during the final stages of ocean closure and formation of fold belts. Their lower part is usually formed by the mantle ultramafites (harzburgites, lherzolites, dunites, etc.); the crustal section starts with a layered complex

Table 2.1 Mid-oceanic-ridge basalts (MORB). Average major and trace element compositions.

Components	Tholeiitic series		
	N-MORB	T-MORB	E-MORB
SiO_2	49.02	50.81	47.25
TiO_2	1.16	1.22	3.49
Al_2O_3	16.39	14.14	12.61
Fe_2O_3	10.27	10.95	17.48
MnO	0.16	0.16	0.20
MgO	7.55	9.89	5.20
CaO	13.27	10.58	10.03
Na_2O	1.91	1.91	2.90
K_2O	0.15	0.18	0.49
P_2O_5	0.13	0.14	0.35
Total	100.00	100.00	100.00
LOI	1.00	0.87	2.45
Mg*	59.30	64.16	37.00
Cr	266	496	31
Ni	97	262	38
Co	41	48	—
Sc	36	33	—
V	302	209	391
Rb	3	4	10
Cs	0.13	0.06	—
Ba	16	40	125
Sr	98	129	310
Ta	0.17	0.50	1.89
Nb	2.3	8.7	28.5
Nf	1.85	1.97	5.46
Zr	72	84	222
Y	32	27	45
Th	0.13	0.54	1.88
U	0.04	0.15	0.69
La	0.04	0.15	0.69
Ce	7.66	15.98	62.82
Nd	6.15	8.90	32.58
Sm	2.75	2.82	7.72
Eu	0.99	1.04	2.66
Tb	0.86	0.63	1.39
Dy	—	3.14	7.74
Yb	3.17	2.26	4.02
Lu	0.42	0.30	0.51
$^{87}Sr/^{86}Sr$	0.703169	0.703392	0.704141
$^{143}Nd/^{144}Nd$	0.513139	0.512999	0.512979

Hereafter: LOI — loss of ignition. Major elements in wt.%, total iron as Fe_2O_3; rare elements in ppm; $Mg* = Mg/Fe^{3+}$. After Boungault, H. and Cande, S.C. *et al.* (1985) Init. Rep. DSDP, **82**, 395–530.

(dunites, wherlites, clinopyroxenites, gabbro, etc.); a further dyke-sheeted complex occurs and the upper part is formed by pillow lavas and deep-sea sediments (Coleman, 1977; Nicolas, 1989). Detailed isotope studies of Voykar ophiolite, from the Polar Urals, have shown that oceanic crustal rocks and the mantle harzburgites are complementary and come from a MORB source (Sharma *et al.*, 1995).

2.3 MAGMATISM AT CONVERGENT PLATE BOUNDARIES

Island arcs, active continental margins and continental plate collision zones are assigned to destructive or convergent geodynamic environments. All, except

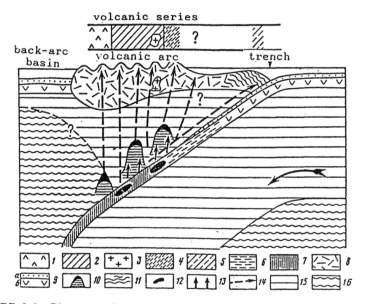

FIGURE 2.2 Diagrammatic representation of the subduction zone in an island-arc area. After Kovalenko (1987). Key: 1–5 — igneous rock series within the arc: 1 — a series with elevated alkalinity, including high alkaline; 2,3 — calc-alkaline, including granitoids (3); 4 — arc tholeiite; 5 — igneous rocks transitional between the calc-alkaline and the arc tholeiite series; 6 — rocks of the amphibolite facies in the oceanic crust; 7 — rocks of the eclogite facies in oceanic crust; 8 — island-arc crust; 9 — oceanic crust (a — sediments, b — tholeiitic basalts, ultrabasic rocks, gabbroids and other igneous rocks); 10 — mantle diapirs: peridotitic and pyroxenitic in composition, with zones of magma generation; 11 — zone of intense compression, folding and thrust formation; 12 — sites of magma generation; 13 — zone of metasomatism: containing fluids rich in water, silica, alkalis and incompatible elements; 14 — probable routes of magma movement to form various volcanic series; 15 — lithospheric mantle; 16 — asthenospheric mantle.

some collision zones, are characterized by the presence of a zone of seismic foci with a depth of 650–700 km. In the 1940s Wadatu, Zavaritsky and Benioff independently studied this phenomenon.

In terms of plate tectonics the Benioff zone corresponds with a lithospheric slab (80–100 km thick) which is being subducted into the mantle. At a certain depth the subducting plate material dewaters and melts by injection of the melt and volatiles into the mantle wedge, resulting in partial melting, the intrusion of magma into the upper crust, and volcanism.

Many geologists believe that the magmas responsible for the formation of most of the continental crust were generated in subduction zones over long periods of geological time.

In the western Pacific, zones of seismic foci occur beneath island arcs and marginal seas, while in the eastern Pacific they occur beneath active continental margins (of Andean type and other types). These zones are also recorded from continental collision zones (from the Alps through Turkey and Iran to the Himalayas and South-East Asia). It should be noted that island arcs, continental margins and collision zones form a single global system of destructive plate boundaries (margins) equal in extent to the global system of oceanic ridges (Fig. 2.1).

Newly available geophysical and geological data imply the structural complexity of subduction zones. Figure 2.2 shows the simplified structure of a typical island arc, with its Benioff zone.

2.3.1 Island-Arc Magmatism

Island arcs and active continental margins (ACM) are characterized by the following features (Kovalenko, 1987; Wilson, 1989):

1. Linear volcanic belts extending for several thousand kilometres and with a width of 200–300 km.
2. The presence of deep-sea trenches, often with a depth of over 7 km, in the ocean and back-arc basins near the continent.
3. Active volcanism in a relatively narrow zone (volcanic front), parallel to the trench, at a distance of 150–250 km. The intensity of volcanism decreases away from the trench. There is no direct correlation between the thickness of the lithosphere and the position of volcanoes.
4. Active seismicity, including shallow, intermediate and deep earthquakes concentrated in the Benioff zone along a particular surface from the trench toward the back-arc basin or continent.
5. Negative heat-flow anomalies in the trench areas and positive anomalies above arcs, back-arc basins and active continental margins. The high heat flow in the volcanic fronts and back-arc basins is actually related to mass transport of heated material during mantle convection or magma ascent.

6. Some island arcs (see below) consist of volcanic rocks that display transverse zoning (Fig. 2.2). Tholeiitic island-arc basalts are erupted mainly near the volcanic front, more distant volcanoes erupt andesites and the rear part of the arc and the ACM are characterized by alkali-rich magmas of shoshonite, latite and moderate alkaline basalt types. This zoning is shown in a higher potassium content at a given SiO_2 concentration.

7. The Subduction zone is complex both in its morphology and its stress pattern. In some cases it shows a two-layered structure; the upper layer is locally dominated by compressional stresses and the lower layer by tensinal stressed (Kosygin, 1991) (Fig. 2.3).

8. Some arcs and ACM have a variable slope (35 to 90°) and a different subduction rate (0.9 cm/yr for the Calabrian Arc of the Mediterranean, and 10 cm/yr in Peru, Chile and New Hebrides) and length of time taken for subduction (5 to 200 Ma).

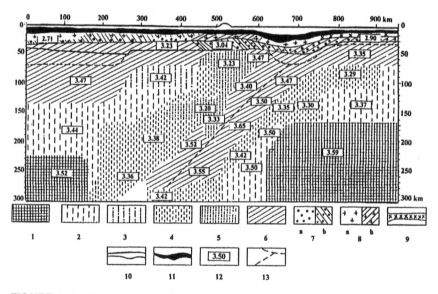

FIGURE 2.3 Density model of the tectonosphere along the Sakhalin–Pacific traverse. After Kosygin (1991). Key: 1 — dense parts of the subasthenospheric layer; 2 — asthenospheric areas, as inferred from low density gradients; 3 — areas of partial melting in the asthenosphere; 4 — asthenospheric areas inferred from density inversions; 5 — inferred centres of magma generation; 6 — mantle part of the lithosphere; 7 — basalt layer (a), dense zones of the basaltic layer (b); 8 — granitic layer (a), dense zones of the granitic layer (b); 9 — transitional layer of the oceanic crust; 10 — unconsolidated sedimentary layer; 11 — water layer; 12 — density (g/cm³); 13 — inferred boundaries (discontinuities in the lithosphere).

The stress regime changes across the strike of the island-arc system (i.e. from the fore-arc to the arc to the back-arc basin). According to Fedotov *et al.* (1985) the deep-sea trench is fronted by a stress zone about 200 km wide, manifested by a shallow (up to 50 km) earthquake belt (Fig. 2.4). The earthquake depth shows a regular increase, up to 700 km deep away from the trench toward the continent. The density of earthquake foci decreases drastically at depths of 140–180 km at the site of intersection of the subducted plate by the bases of the volcanoes. Back-arc basins are characterized by a tensional regime, accompanied by sea-floor spreading and submarine basaltic volcanism.

Sharapov *et al.* (1984) analysed the main axial stress tensors beneath the Kurile–Kamchatka Arc to depths of 0–35, 40–120 and 120–130 km, and showed by the predominance of compressional stresses that there are also sites character-ized by tensile stresses. These sites shift from the edges of the arc toward the centre as the depth increases.

An important feature of the magmatism of destructive environments is its average andesite composition and the wide variations in SiO_2 (and iron content and alkalinity) (Fig. 2.5). The geochemical characteristics of the igneous assem-blages of island arcs are presented in Tables 2.2 and 2.3. All formations from every petrochemical series can be seen there. However, these series differ greatly from formations of the same name but from different geodynamic regimes. In

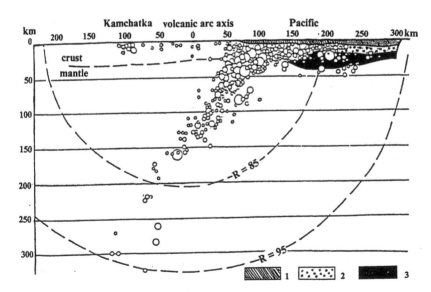

FIGURE 2.4 Projections of epicentres of Kamchatka earthquakes for 1968–1969 on to a plane perpendicular to the Kamchatka Arc. After Fedotov *et al.* (1974). Key: Horizontal distances are measured from axis of the volcanic arc. 1 — water layer; 2 — granitic layer; 3 — basaltic layer.

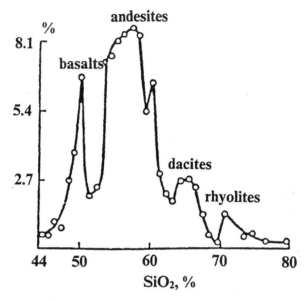

FIGURE 2.5 Variation of silica content in the volcanics of the Kurile–Kamchatka Arc. After Markhinin (1967).

addition, there is a difference between igneous assemblages generated by convergent environments, i.e. of ACM, continental plate collision zones and island arcs.

The latter are in turn subdivided into ensimatic, i.e. emplaced on oceanic crust, and ensialic, emplaced on continental crust. Therefore tholeiite series formations and boninite assemblages have been recorded only in island-arc environments. Ensimatic island arcs do not contain high-potassic andesites, shoshonites are uncommon and high-alkali series are absent.

The Kurila–Kamchatka Arc, located in the NW Pacific, provides a typical example of the island-arc system (Fig. 2.1). In the south it joins the Japan Arc and almost at a right angle abuts the Aleutian Arc in the North. The Klyuchevskoi volcanic centre, which is currently very active, occurs at the site of their junction. The Klyuchevshaya Sopka is the highest (4,850 m) active volcano in Eurasia. It is at this volcanic centre that the largest catastrophic eruptions of the twentieth century took place: the 1956 eruption of the Benzymyanny Volcano and the 1975 eruption of the Tolbachik Volcano, which has been studied in great detail (Fedotov *et al.*, 1975).

Several zones can be recognized within the Kurile Island segment (Fig. 2.6):

1. the trench slope;
2. the frontal non-volcanic zone (the Lesser Kurile Ridge and submarine Vityaz Ridge);
3. the axial volcanic zone (Greater Kurile Ridge); and
4. the back-arc zone (Okhotsk Sea arc slope).

Table 2.2 Igneous series of ensimatic island arcs. Average major and trace element composition of rocks in the Bonin and Marianas island arcs South-East Pacific.

Components	Tholeiite series			
	Basalt	Andesite–basalt	Andesite	Dacite
SiO_2	50.09	55.20	62.80	70.83
TiO_2	0.84	1.49	0.94	0.49
Al_2O_3	15.72	15.09	15.00	13.42
FeO	12.06	12.74	8.80	5.62
MnO	0.27	0.22	0.21	0.19
MgO	7.34	3.58	1.83	0.72
CaO	11.98	8.23	6.10	3.97
Na_2O	1.38	2.75	3.48	3.97
K_2O	0.24	0.55	0.64	0.78
P_2O_5	0.09	0.15	0.20	0.11
Total	100.01	100.00	100.00	100.00
Mg*	52.03	33.37	27.04	18.59
Cr	76	74	7	—
Ni	19	16	2	—
Co	36	40	10	—
Sc	45	52	10	—
V	—	—	—	—
Cu	—	—	—	—
Rb	5	5	17	—
Ba	86	160	242	382
Sr	182	225	240	181
Ta	0.01	0.04	0.17	—
Hf	0.66	1.63	4.38	—
Zr	24	67	138	—
Y	12	22	—	—
Th	0.12	0.31	1.28	—
U	0.04	0.13	0.50	—
La	2.39	5.90	2.98	—
Ce	6.35	13.80	7.13	—
Nd	5.67	16.20	6.87	—
Sm	2.08	5.60	2.20	—
Eu	0.80	1.72	0.95	—
Gd	—	—	—	—
Tb	0.53	—	0.66	—
Dy	—	—	—	—
Er	—	—	—	—

Table **2.2** *contd.*

Components	Tholeiite series			
	Basalt	Andesite–basalt	Andesite	Dacite
Yb	2.30	5.45	2.76	—
Lu	0.38	0.82	0.43	—
$^{87}Sr/^{86}Sr$	0.70320 – 0.70363			
$^{143}Nd/^{144}Nd$	0.513056 – 0.51073			
$^{206}Pb/^{204}Pb$	18.390			
$^{207}Pb/^{204}Pb$	15.52			
$^{208}Pb/^{204}Pb$	38.45			

Table 2.2 *contd.*

Components	Boninite series					
	Boninite	Magnesian andesite	Andesite	Dacite	Quenched glass	Rhyolite
SiO_2	57.94	58.09	63.90	71.08	65.43	80.53
TiO_2	0.17	0.32	0.26	0.35	0.19	0.18
Al_2O_3	11.51	16.02	14.26	13.31	17.64	10.89
FeO	8.41	7.62	6.72	4.28	5.53	1.38
MnO	0.18	0.15	0.12	0.07	0.10	0.04
MgO	12.02	7.69	4.33	1.24	1.22	0.23
CaO	7.76	7.03	6.68	5.44	6.35	1.32
Na_2O	1.46	2.54	2.75	3.17	2.77	3.92
K_2O	0.50	0.48	0.93	1.01	0.77	1.47
P_2O_5	0.04	0.06	0.06	0.06	—	0.05
Total	99.99	100.0	100.01	100.01	100.0	100.01
Mg*	71.81	64.26	53.45	34.05	28.22	22.90
Cr	786	420	187	5	—	4
Ni	192	180	—	11	—	1
Co	44	37	25	27	—	—
Sc	42	36	28	22	—	—
V	—	—	—	—	—	—
Cu	—	—	—	—	—	—
Rb	10	10	—	20	—	18
Ba	27	29	46	56	—	155
Sr	71	77	—	114	—	83
Ta	0.04	0.05	0.05	—	—	—
Hf	0.48	0.70	1.60	1.34	—	—
Zr	19	28	—	52	—	125
Y	4	8	—	7	—	—
Th	0.09	0.12	0.16	—	—	1.60
U	0.07	0.06	—	—	—	0.85
La	0.90	1.02	2.51	1.82	—	5.50
Ce	1.99	2.04	6.39	3.96	—	12.50
Nd	1.29	1.67	—	2.69	—	8.50
Sm	0.35	0.52	1.34	0.77	—	2.24

Table 2.2 *contd.*

Components	Boninite series					
	Boninite	Magnesian andesite	Andesite	Dacite	Quenched glass	Rhyolite
Eu	0.13	0.19	0.38	0.27	—	0.59
Gd	—	—	—	—	—	—
Tb	0.09	0.13	0.29	0.19	—	—
Dy	—	—	—	—	—	—
Er	—	—	—	—	—	—
Yb	0.55	0.77	1.24	1.08	—	2.15
Lu	0.10	0.12	0.20	0.19	—	0.43
$^{87}Sr/^{86}Sr$			0.70330–0.70383			
$^{143}Nd/^{144}Nd$			0.512948–0.512989			
$^{206}Pb/^{204}Pb$			18.37–18.74			
$^{207}Pb/^{204}Pb$			15.47–15.53			
$^{208}Pb/^{204}Pb$			38.02–38.32			

After: Fryer *et al.*, 1992: Sharaskin, 1992; Bloomer *et al.* (1989) *Geophys. Res.*, **B4**, 4469–4496.

Table 2.3 Igneous series of ensialic island arcs. Average major and trace element composition of Kurile–Kamchatka and Aleutian island arcs.

Components	Tholeiitic series				
	Basalt	Andesite–basalt	Andesite	Dacite	Rhyolite
SiO_2	50.43	55.38	60.78	67.09	72.57
TiO_2	0.87	0.89	0.77	0.53	0.28
Al_2O_3	19.08	17.95	17.33	16.16	14.66
Fe_2O_3	4.14	3.93	2.74	1.81	1.26
FeO	6.09	5.21	4.34	2.55	1.22
MnO	0.19	0.16	0.17	0.14	0.06
MgO	5.65	4.13	2.28	1.33	0.50
CaO	10.58	8.47	6.48	4.31	2.93
Na_2O	2.37	2.91	3.87	4.45	4.43
K_2O	0.45	0.73	1.06	1.48	1.98
P_2O_5	0.15	0.24	0.19	0.14	0.11
Total	100.00	100.00	100.00	100.00	100.00
H_2O	0.25	0.14	0.14	0.32	0.41
LOI	0.35	0.42	0.40	1.13	1.49
Mg*	50.63	45.71	37.41	36.11	27.38
Cr	—	—	—	—	—
Ni	29	42	19	7	8

Table 2.3 *contd.*

Components	Tholeiitic series				
	Basalt	Andesite–basalt	Andesite	Dacite	Rhyolite
Co	32	28	16	8	3
V	315	217	148	101	21
K	3768	6022	—	—	—
Rb	15	9	13	22	28
Ba	224	321	388	443	567
Sr	410	373	352	310	342
Ta	0.50	—	1.01	—	—
Nb	0.6	—	1.1	—	—
Hf	0.50	—	3.33	—	—
Zr	18	—	106	—	—
Ti	5200	5315	—	—	—
La	2.62	6.15	—	8.72	—
Ce	6.76	12.59	—	20.38	—
Nd	5.35	12.09	—	16.02	—
Sm	1.90	5.04	—	6.13	—
Eu	0.69	1.39	—	1.60	—
Gd	2.52	6.25	—	7.10	—
Tb	0.46	1.23	—	1.33	—
Dy	—	—	—	—	—
Ho	—	—	—	—	—
Er	—	—	—	—	—
Tm	0.31	—	—	—	—
Yb	2.06	3.49	—	4.55	—
Lu	0.29	0.61	—	0.72	—
F	329.8	358.65	379.53	343.67	193.80
B	55.47	45.34	55.52	62.85	65.28
Be	0.48	0.49	0.50	0.58	0.55
$^{87}Sr/^{86}Sr$	0.7030 – 0.7035				

Table 2.3 *contd.*

Components	Calc-alkaline series				
	Basalt	Andesite–basalt	Andesite	Dacite	Rhyolite
SiO_2	51.53	55.12	60.96	66.81	72.73
TiO_2	1.06	0.99	0.77	0.61	0.36
Al_2O_3	18.09	17.59	16.85	15.87	14.39
Fe_2O_3	3.86	3.37	3.08	2.11	0.90
FeO	5.90	5.27	3.41	2.17	1.28
MnO	0.18	0.16	0.15	0.11	0.10
MgO	5.69	4.44	2.93	1.55	0.66
CaO	9.44	8.02	6.01	3.95	2.02

Table 2.3 *contd.*

Components	Calc-alkaline series				
	Basalt	Andesite–basalt	Andesite	Dacite	Rhyolite
Na_2O	3.00	3.48	3.88	4.32	4.52
K_2O	0.99	1.31	1.75	2.32	2.92
P_2O_5	0.25	0.25	0.21	0.18	0.12
Total	100.00	100.00	100.00	100.00	100.00
H_2O	0.26	0.26	0.31	0.34	0.15
LOI	0.35	0.36	0.43	0.53	0.59
Mg*	51.99	48.80	45.76	40.35	35.87
Cr	109	73	62	—	11
Ni	58	34	29	10	6
Co	41	26	19	8	4
V	345	197	133	79	20
K	8,200	10,876	—	—	—
Rb	15	19	26	36	58
Ba	614	455	773	788	1,294
Sr	406	464	423	317	200
Ta	1.01	0.91	1.01	0.25	0.25
Nb	2.1	2.2	1.9	2.8	2.4
Hf	3.53	4.73	3.83	3.13	—
Zr	135	246	138	101	—
Ti	6,345	5,916	—	—	—
La	8.97	11.88	13.72	18.80	16.56
Ce	18.75	26.08	25.72	33.97	36.35
Nd	13.20	19.13	17.25	19.21	20.09
Sm	3.44	4.43	4.64	4.65	6.16
Eu	1.07	1.25	1.12	1.11	1.08
Gd	3.93	5.94	5.68	4.45	5.45
Tb	0.66	0.54	0.72	0.69	1.00
Dy	—	—	—	—	—
Ho	—	—	—	—	—
Er	—	—	—	—	—
Tm	0.33	—	—	—	—
Yb	2.30	1.92	2.45	2.11	3.94
Lu	0.31	0.30	0.40	0.32	0.65
F	383.03	373.58	341.94	370.03	367.53
B	22.18	25.17	19.16	32.35	19.18
Be	0.62	0.76	0.73	0.76	0.77
$^{87}Sr/^{86}Sr$	0.7033–0.7040				

Table 2.3 *contd.*

Components	Shoshonite–Latite series			
	Absarokite	Shoshonite	Latite	Trachyte
SiO_2	50.72	56.12	60.58	68.08
TiO_2	1.12	0.95	0.92	0.61
Al_2O_3	17.92	17.13	16.88	15.65
Fe_2O_3	4.17	3.43	3.11	1.60
FeO	5.73	4.07	2.77	1.42
MnO	0.19	0.10	0.13	0.11
MgO	5.14	3.93	2.41	0.67
CaO	9.78	6.60	4.56	1.87
Na_2O	2.87	3.62	4.06	4.69
K_2O	2.00	3.58	4.19	5.17
P_2O_5	0.35	0.46	0.38	0.12
Total	100.00	100.00	100.00	100.00
H_2O	0.61	0.87	0.46	0.52
LOI	0.18	0.85	0.58	0.16
Mg*	49.12	49.45	43.53	29.28
Cr	65	24	18	9
Ni	30	14	8	8
Co	29	17	11	3
V	216	159	102	22
K	16641	—	—	—
Rb	39	52	66	93
Ba	557	1456	1438	1421
Sr	670	806	692	317
Ta	0.30	—	0.86	0.35
Nb	3.8	5.5	5.9	7.2
Hf	5.2	5.92	6.58	7.36
Zr	193	206	278	345
Ti	6703	—	—	—
La	11.79	19.39	17.21	15.12
Ce	25.18	46.44	35.44	28.22
Nd	16.02	29.60	20.45	18.14
Sm	4.37	8.37	5.57	4.13
Eu	1.41	2.45	1.65	0.86
Gd	4.94	6.63	4.46	3.43
Tb	0.82	—	—	—
Dy	—	5.6	3.65	3.43
Ho	—	1.22	0.81	0.77
Er	—	3.27	2.10	1.71

Table 2.3 *contd.*

Components	Shoshonite–Latite series			
	Absarokite	Shoshonite	Latite	Trachyte
Tm	0.43	—	—	—
Yb	2.68	3.37	2.33	2.32
Lu	0.34	0.34	0.32	0.30
F	482.52	1075.73	711.83	797.30
B	12.09	12.25	32.40	22.58
Be	2.12	2.55	3.65	3.33
$^{87}Sr/^{86}Sr$		0.7032–0.7035		

Table 2.3 *contd.*

Components	Alkaline K–Na series			Alkaline high–K series		
	Trachy-andesite	Trachyte	Comendite	Feldspar leucitite	Leucite phonolite	
SiO_2	56.09	62.09	72.97	51.43	56.54	61.07
TiO_2	1.46	0.93	0.21	1.15	1.54	0.90
Al_2O_3	17.90	17.64	13.29	15.51	16.40	18.56
Fe_2O_3	4.60	2.85	1.47	3.86	2.85	1.91
FeO	3.41	2.40	0.85	4.70	3.58	1.96
MnO	0.15	0.14	0.09	0.14	0.07	0.05
MgO	2.65	1.44	0.45	8.28	4.27	2.16
CaO	5.51	3.14	0.95	7.10	4.23	2.09
Na_2O	4.84	5.06	5.05	1.94	2.17	2.51
K_2O	2.80	3.98	4.64	5.15	7.92	8.48
P_2O_5	0.60	0.32	0.03	0.74	0.43	0.30
Total	100.00	100.00	100.00	100.00	100.00	100.00
H_2O	0.78	0.50	0.19	0.58	0.26	—
LOI	0.85	0.79	0.36	3.18	3.40	2.26
Mg*	38.554	34.11	27.13	64.35	55.30	51.20
Cr	40	45	15	127	86	53
Ni	35	10	6	63	31	18
Co	23	7	1	24	23	13
V	94	48	6	244	312	148
K	—	—	—	42756	—	—
Rb	32	53	117	171	221	254
Ba	722	1165	113	1836	2555	—
Sr	755	369	41	464	564	—
Ta	—	1.63	5.54	0.52	0.62	—
Nb	31.9	37.4	95.6	6.4	10.3	—
Hf	7.02	9.24	10.17	12.41	12.49	—

Table 2.3 *contd.*

Components	Alkaline K–Na series			Alkaline high–K series		
	Trachy-andesite	Trachyte	Comendite	Feldspar leucitite	Leucite phonolite	
Zr	361	589	489	417	495	—
Ti	—	—	—	6876	—	—
La	—	38.58	27.68	56.51	23.11	—
Ce	—	70.56	47.82	111.56	52.06	—
Nd	—	30.46	20.64	43.06	21.86	—
Sm	—	7.97	5.03	5.84	4.38	—
Eu	—	2.28	0.31	1.92	1.42	—
Gd	—	5.58	3.88	4.90	4.27	—
Tb	—	0.93	—	0.77	0.62	—
Dy	—	5.28	5.64	—	—	—
Ho	—	1.12	1.16	—	—	—
Er	—	3.15	3.83	—	—	—
Tm	—	—	—	0.34	0.21	—
Yb	—	3.20	3.23	2.11	1.35	—
Lu	—	0.36	0.35	0.23	0.19	—
F	223.69	590.86	251.66	1355.41	2030.19	1220.6
B	16.78	33.50	112.74	14.91	24.99	38.98
Be	2.54	2.60	5.64	3.65	4.69	3.80
$^{87}Sr/^{86}Sr$	0.7031–0.7062			0.7035–0.7040		

After: Bogatikov and Tsvetkov (1988); Tsvetkov, 1990; Frolova *et al.* (1989).

The crustal structure of the Kurile Arc is fairly heterogeneous. The thickness of the crust varies from 8 to 10 km beneath the Central Kuriles, and beneath the North and South Kuriles it is up to 20–25 km. Beneath the Lesser Kuriles the crustal thickness is about 20 km. Therefore, the thickness of the Earth's crust beneath the arc flanks and in the centre is close to that of the continental and oceanic crust, respectively. The typical granite–metamorphic layer does not occur in the central part of the arc, unlike the southern and northern flanks where xenoliths are represented by granite–gneiss and plagioclase–hyperstene schists. Acid plutonic and metamorphic rocks, similar to those recorded in the South Kuriles, are common among metamorphic rocks of the Hidaka Belt of Hokkaido, northern Japan, and therefore, flanks of the Kurile Arc can be classified as ensialic (Bogatikov and Tsvetkov, 1988a).

Based on the above data, four assemblages of differing ages — from the Late Cretaceous to the Palaeogene — alternating in succession, are recognized in the frontal zone (the Lesser Kurile Ridge):

1. basaltic (100–80 Ma);
2. basalt–andesite–basaltic (80–70 Ma);

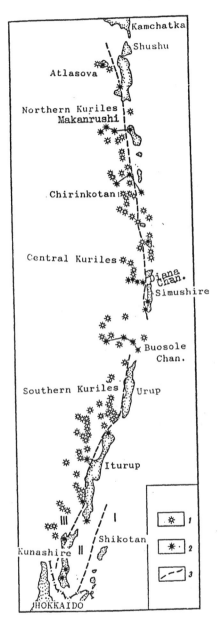

FIGURE 2.6 Structural zonation of the Kurile island arc. After Bogatikov and Tsvetkov (1988). Key: 1 — submarine volcanoes; 2 — volcanoes dated using Nd and Sr isotopes; 3 — profiles; I–III — structural zones: I — fore-arc, II — axial, III — back-arc; the boundary between II and III is based on isotopic data.

3. shoshonite (70–60 Ma) and
4. gabbro–dacite–basaltic (60–40 Ma).

Based on their petrography, these assemblages belong to the calc-alkaline series and the potassic moderate-alkaline series (low-Ti subseries).

The axial zone extends for 1200 km, from Kamchatka to Hokkaido. It is divided by the Bussol and Kruzenstern straits into three segments, each differing in its crustal structure. Moreover, the eastern and western submeridional zones were recognized within the Greater Kurile Ridge. The western zone, which consists of islands (Alaid, Shirinki, Naramrushi and Chirinkotan), was traced only in the northern Kuriles, with the remaining part being built up by submarine volcanoes. Most of the Quaternary volcanoes (64%) occur in the axial zone of the Kurile island arc. Compositionally the rocks are fairly diverse (ranging from basalts to rhyolites) and formed over a long period of time (Neogene–Quaternary). The history of the formation of rocks in the back-arc zone was not as long, they are all of similar composition, being dominated by andesites, and acid rocks are absent. The rocks of the back-arc zone are rich in K, Rb, Sr, Ba, Th, U and LREE and depleted in Fe, V, HREE, Y; they also have lower $^{87}Sr/^{86}Sr$.

The eruption products of the Quaternary volcanoes in the axial zone of the Kurile Arc can be placed in the calc-alkaline and tholeiite series.

Bogatikov and Tsvetkov (1988a) grouped the Cenozoic igneous rocks of the Greater Kurile Ridge into five igneous assemblages of different ages: basalt–rhyolite, gabbro–plagiogranite, basalt–andesite, andesite–rhyolite, and basalt–rhyolite.

The basalt–rhyolite assemblage, together with the subordinate andesites (N_1), is the oldest on the Kurile Ridge. The rocks of this assemblage, especially in the lower part of the section, where acid tuffs predominate, are often strongly propylitized, and therefore some geologists describe them as green tuffs.

The gabbro–plagiogranite association (N_1–N_2) is made up of structurally complex bodies of olivine and olivine-free gabbro, gabbronorites, gabbrodiorites, quartz diorites, granodiorites, and plagiogranites. Most massifs are stock- and sill-like in shape and have a multiphase structure; their area is about several tens of square kilometres. In Japan, the so-called Tertiary granites, represented by plagiogranites and quartz diorites, are the petrological equivalents of these formations. Compositionally similar coeval intrusives were recorded from Kamchatka, Komandorskie and the Aleutian Islands (Tsvetkov, 1990; Kay and Kay, 1990).

The rocks of the Middle-Upper Miocene basalt–andesite assemblages consist of Al-rich basalts and andesites. Their intrusive equivalents — leucogabbro, gabbro–anorthosites and gabbronorites — were reported from Paramushir Island.

The Upper Miocene–Lower Pliocene andesite–rhyolite assemblage was actually found on all of the islands of the ridge. The lavas and pyroclastics are dominated by andesitic-dacites, dacites and rhyolites. Rocks of the Pliocene–Quaternary basalt–rhyolite assemblage, where the lavas are dominated by basalts and two-pyroxene andesites, and pyroclastics of mainly dacitic and rhyolitic composition are common on the North and South Kurile Islands. The latter two assemblages occur in the central Kuriles. Despite the fact that the islands are underlain mainly by suboceanic crust, the lavas are generally acid in composition.

The western (back-arc) zone of the Greater Kuriles contains a Quaternary basalt–andesite assemblage and a moderately alkaline low-Ti basalt assemblage. The andesitic lavas of this zone are characterized by the presence of lherzolite nodules (Tsvetkov and Avdeiko, 1982). These nodules differ from the typical spinel lherzolites of intra-plate alkali basalts in the amount of chrome spinel and low-Al clinopyroxene they contain, resembling the depleted lherzolites of ophiolite complexes.

The least-differentiated rocks from the volcanoes studied were found to be rich in lithophile elements (of the LILE group: i.e. K, Rb, Cs and Li) and Sr and Th, while being depleted in the high-field-strength elements (Ti, Zr, Ta and Nb) and to a lesser degree in Ce with respect to other incompatible elements. The primary magmas of the Kurile Arc were assumed to have been generated by the mixing of two sources: the altered oceanic lithosphere, similar in composition to MORB, and depleted mantle peridotite source. The former produces LILE and H_2O (the wet component) and the latter, refractory elements (the dry component) (Bogatikov and Tsvetkov, 1988a).

Stationary high-temperature fumaroles ($T = 800°C$), which eject diverse minerals, including a new ReS mineral, were observed on a volcano in the Kuriles (Kudryavy volcano, Iturup Island) (Korzhinsky et al., 1994).

A similar succession of Cenozoic igneous rocks is a distinctive feature of Kamchatka. It is underlain by continental crust, as inferred from geophysical data. The system of outer non-volcanic and inner volcanic arcs of the Kuriles within Kamchatka gives way to an échelon system of horsts and grabens. The earliest igneous activity ended there at the Cretaceous–Palaeocene boundary and has much in common with that of the Kurile Ridge, including the presence of basalt–rhyolite and shoshonite assemblages and Neogene gabbro and dolerite bodies. In the Cenozoic, the Palaeogene–Neogene evolution of the magmatism was similar in nature to that of the Kuriles. The products of Quaternary volcanism have been reported only from the Central and East Kamchatka belts. However, the East Kamchatka and the outer Kurile arcs contain high-Al formations of calc-alkaline and island-arc tholeiitic series.

Rocks of the K–Na and K moderately alkaline series and potassic high-alkaline series occur mainly in Central Kamchatka (Median Ridge), i.e. the same transverse zoning is characteric of both Kamchatka and Kuriles.

The geodynamic parameters and island-arc magmatism vary strongly, Miyashiro (1974) and Gill (1981) being the first to mention this phenomenon. Later, further development of the concept allowed Bogatikov and Tsvetkov (1988a) to subdivide island arcs into young, developed and mature types based on differences in the Earth's crustal thickness and in the degree of similarity to continental crust (Fig. 2.4). The evolutionary maturity of ensimatic island arcs can be clearly determined by the relative position of igneous assemblages of the same type in a stratigraphic section. The island-arc tholeiite series that give way to calc-alkaline and then alkaline series are the oldest magmatic series of the arcs. This was used as the basis for the recognition of these three evolutionary types. The young type encompasses only intra-oceanic arcs (Tonga–Kermadec, Mariana, South Sandwich, etc.) underlain by thin crust which does not have a granite–metamorphic layer. The developed type is mainly peripheral-oceanic arcs, underlain by thicker crust (30–40 km), locally by subcontinental and continental type (Kurile–Kamchatka, Aleutians, etc.). Mature arcs are extensive, often polycyclic, island-arc systems underlain by thick continental crust mainly formed during the previous stages; the system was separated from the continent in the course of back-arc spreading (Japan, Philippines and Indonesia, etc.). The degree of arc maturity does not always match its geological age, but rather the basement on which it developed (Kovalenko et al., 1987a).

The replacement of some igneous series by others, which is easily discernible in young and developed arcs, does not occur in mature island arcs. Mature arcs contain rocks of practically all known magmatic series.

The active margins and collision zones are characterized by the widespread distribution of acid volcanics and highly alkaline formations. In general, igneous rocks at a destructive plate margin have a low titanium content (TiO_2 of 1.3 wt.% and below). The Al_2O_3 content reaches 18 wt.% in most of the rocks discussed. Low-Al andesites (14–16 wt.% Al_2O_3) are assigned to the tholeiitic series. The K_2O content in rocks of potassic, highly alkaline series accounts for 8–9 wt.% at a low TiO_2 content (e.g. the lamproites of southern Spain, the leucitites of the southern Italian and the Indonesian arc).

In contrast to mid-oceanic ridges, primitive melts of subduction zones basalts are very rich in H_2O (between 1.0 and 2.9 wt.% for boninites and between 1.2 and 2.5 for island-arc tholeiites); most of these melts have high H_2O/K_2O ratios which are consistent with a transfer of H_2O as a fluid phase from the subducted slab to the mantle wedge (Sobolev an Chaussidon, 1996). Based on different estimates, most of the andesite and boninite magmas prior to eruption contain 1–5 wt.% of H_2O.

The content of large-ion lithophilic elements (LILE) depends directly on the K_2O and SiO_2 content of the rocks, reaching a maximum in the andesites, and are very high in formations of the shoshonite–latite series. The K/Rb ratio decreases with K_2O and SiO_2. The Rb/Sr, Rb/Ba and Ba/Sr ratios increase with that of SiO_2 (Kovalenko, 1987; Wilson, 1989).

Additionally, H_2O/CO_2, Cl/K, Cl/S and probably Cl/H_2O values are higher in the andesites of convergent margins, as compared with those of igneous rocks of other environments. As a rule, andesites are enriched in elements of the K- and Th groups as compared with REE, and depleted in elements of the Ti group (Nb and Ta etc.), therefore Ba/La > 15, La/Nb > 2, La/Th < 7. They are rich in Rb, Cs, Ba, Sr, Th and Pb and depleted in Nb, Ta and to lesser degree in Ti, Zr and Hf. A high Ba/Ta ratio (above 450) is the most reliable diagnostic character for island-arc and ACM igneous rocks.

The rocks in destructive plate margins are marked by a chondrite/rock flat spectrum of REE, an enrichment or depletion in LREE and often an enrichment in HREE (5–20 times as compared with chondrites). Rocks of the low-potassic tholeiitic series often show a depleted or flat REE spectrum, their content being the lowest among all the known terrestial volcanics.

The abundance of thorium-group elements (Th, U and Pb) and their ratio increase from tholeiite to calc-alkaline and alkaline series. There is a positive correlation between the titanium-group elements (Ti, Zr, Hf, Nb and Ta) in SiO_2-rich magmas. The compatible elements (Ni, Co, Cu, V and Sc) are concentrated in Mg–Fe minerals, and their abundance decreases with an increase in SiO_2. Andesites of ACM and collision zones have a higher Ni content than those of island arcs with the SiO_2 and MgO content being equal (Kovalenko et al., 1987).

Although Quaternary basalts and andesites of island arcs often do not differ from MORB in isotopic composition, their $^{87}Sr/^{86}Sr$ ratios are higher, with $\varepsilon Nd < 9$ and $\varepsilon Nd > 9$ in andesites and MORB, respectively; $^{207}Pb/^{204}Pb$ and locally $^{208}Pb/^{204}Pb$ are often higher than the $^{206}Pb/^{204}Pb$ ratio. The presence of ^{10}Be in arc magmas ($> 10^6 g^-$) (Brown et al., 1982; Tsvetkov, 1990) may imply subduction during magma generation, with the involvement of material that has recently been present at the Earth's surface.

Boninite-series rocks are fairly unusual. Compositionally, they are a mixture of the highly depleted mantle material, with a high Mg, Cr, Ni and PGE content and crust-derived material, as shown by high SiO_2, alkalies and "crustal" specificity of Sr, Nd and Pb isotopes (Crawford, 1989; Sharaskin, 1992). The distribution pattern of REE in boninites is very similar to that in island-arc tholeiites with the boninites differing only in their lower REE abundances (Fig. 2.7).

2.3.2 Magmatism of Active Continental Margins

The igneous assemblages of the western coast of America are typical representatives of ACM magmatism (Zeil and Pichler, 1967; Thorpe et al., 1982). The main magma series recognized in island arcs (calc-alkaline, shoshonite–latite, high-K) occur here. The most conspicuous difference between island-arc and ACM magmas is greater abundance of more silica-rich varieties (dacites, rhyolites). Large granitic batholiths took place here; for example, granitoids are exposed in nearly forty per cent of Chile. The commonest rock types, regardless of age, are granodiorite and tonalite, but the compositional range is broad, and diorite, granite, quartz diorite, and gabbro are common (Aguirre, 1983).

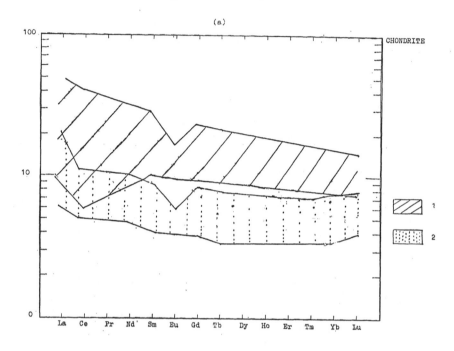

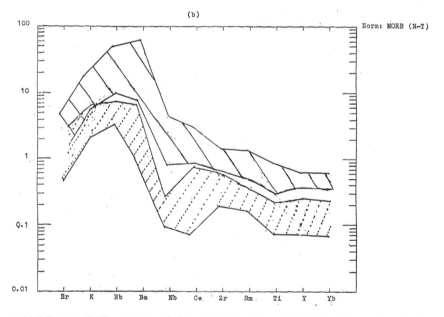

FIGURE 2.7 REE spectra of the tholeiitic (1) and boninitic (2) series of the Mariana Arc normalized to (a) chondrite C1 and (b) MORB. After Fryer *et al.* (1992).

The most typical occurrences of ACM magmas is in South America, in the Andes. The continental basement beneath the Andes was formed by tectonic and igneous processes operating during the whole of the Phanerozoic. In the Andes of Colombia and Ecuador the basement is composed of pre-Mesozoic and Mesozoic igneous rocks and in the central Andes it is made up of Precambrian and Hercynian rocks. The basement of the southern Andes is formed by Mesozoic–Cenozoic folded rocks.

The belt is discontinuous and consists of three parts: the northern Andes (between 5° and 3°S), the central Andes (16–28°S), and the southern Andes (32–52°S). Magmatism is considered to appear only above those segments of subduction zones where the angle of subduction is fairly steep (25–30°). Both intrusive and extrusive Cenozoic formations are presented.

The volcanic rocks of the central and southern Andes are the best known. They differ mainly in the presence of acid volcanic products of the rhyolite and andesite assemblages in the central and southern Andes respectively (Fig. 2.8). However, both areas have a similar petrochemical zoning across the strike of the marginal belt, mainly due to an increase in K_2O content away from the deep-sea trench. The increase is regular and is correlated with the depth to the Benioff zone. The calc-alkaline volcanics dominating the western and central parts of the volcanic areas given way to the shoshonite–latite assemblage of the eastern Altiplano graben and some volcanoes of the South Andes (Deruelle, 1982). The latter forms a belt about 100 km wide and is controlled by parts of the subduction zone at the depth of about 300 km from the surface.

For detailed characterization of the rocks of rhyolite assemblage the reader is referred to the publications by Ziel and Pichler (1967) and Kussmaul *et al.* (1977). The assemblage comprises pumice and ash deposits, ignimbrites, tuffs, lavas and subvolcanic formations, varying in composition from quartz andesites and latites to trachyrhyolites dominated by moderately acid and acid rocks. Typical compositions of these rocks are presented in Table 2.4.

According to Zeil and Pichler (1967) the petrography, along with unusually high copper (1,880 ppm on average) and aluminium (up to 8.7 normative corundum) content suggests that these rocks were generated by crystal anatexis. The strontium isotopic data agree with the above statement. The assemblage is characterized by a high initial $^{87}Sr/^{86}Sr$ ratio (0.7052–0.7091 for northern Chile; 0.7125 for Bolivia). In addition, they are much higher than those in rocks of the andesite assemblage. The ignimbrites are enriched in LREE and have a distinct europium anomaly.

Acid volcanic rocks of unusual composition — the equivalents of S-type granites or rare-metal Sn-bearing granites — were reported from the central zone of south-east Peru (Noble *et al.*, 1984). They have extremely high alumina and high phosphorus content. A very high $^{87}Sr/^{86}Sr(0)$ (0.7158, 0.7226 and 0.7216) in the volcanic rocks suggests a high degree of melting of ancient crystalline rocks, with a high Rb/Sr ratio and a moderate Rb content, during the formation

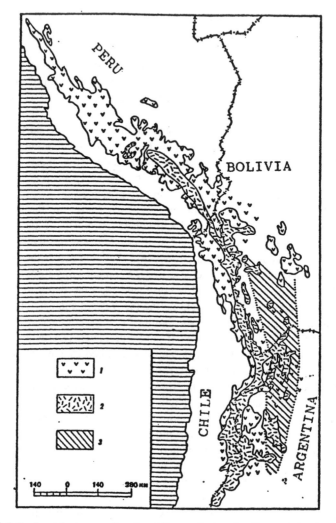

FIGURE 2.8 Location map of rocks of the andesite, rhyolite and shoshonite–latite assemblages of the Central Andes (Klerx *et al.*, 1997; Deruelle, 1982). Key: 1, 2 — rocks of assemblages of: 1 — rhyolite, 2 — andesite; 3 — shoshonite and latite.

of these magmas. High ^{18}O ($+11.2–10.2‰$) also implies that these volcanics are similar to S-type granites. The rocks formed at about 775°C and were named macusanites.

The andesite assemblage consists of rocks varying in their silica content (50–70 wt.%), their alkali content and especially their K_2O content. More acid varieties are common in the Central Andes, being dominated by quartz latites,

Table 2.4 Igneous series of active continental margins. Average major and trace element composition of Andean volcanic rocks.

Components	Calc-alkaline series			
	Basalts	Andesite–basalt	Andesite	Rhyolite
SiO_2	51.72	53.17	56.95	70.75
TiO_2	1.14	1.12	1.30	0.38
Al_2O_3	16.87	17.47	15.86	14.07
Fe_2O_3	10.02	9.33	10.08	4.24
FeO	—	—	—	—
MnO	0.16	0.15	0.18	0.10
MgO	6.62	5.11	3.58	0.45
CaO	9.76	9.92	7.23	1.93
Na_2O	2.91	2.90	3.53	5.13
K_2O	0.61	0.64	1.03	2.86
P_2O_5	0.21	0.20	0.27	0.09
Total	100.00	100.00	100.00	100.00
Mg*	56.69	52.02	41.28	17.44
Cr	248	135	20	81
Ni	85	42	19	10
Co	37	25	26	3
Sc	33	33	33	11
V	246	247	267	13
Zn	87	90	108	101
Rb	15	15	25	75
Cs	1.60	1.63	2.23	5.72
Ba	191	201	286	660
Sr	428	439	404	161
Ga	18	19	19	18
Ta	—	—	—	—
Nb	2.4	2.6	3.2	7.3
Hf	2.50	2.31	3.21	8.63
Zr	92	96	125	348
Y	22	22	28	51
Th	1.50	1.71	2.21	6.63
U	—	—	—	—
La	7.79	7.78	1065	25.10
Ce	19.77	19.61	27.93	59.83
Nd	12.18	13.27	17.08	32.02
Sm	3.29	3.52	4.44	7.54
Eu	1.15	1.08	1.32	1.35
Tb	0.68	0.66	0.77	1.26
Yb	2.35	2.34	3.06	5.86
Lu	0.36	0.33	0.42	0.88
$^{87}Sr/^{86}Sr$	0.704030	0.704010	0.704100	0.703980
$^{143}Nd/^{144}Nd$	—	0.512878	0.512866	0.512873

Table 2.4 *contd.*

Components	Calc-alkaline series, enriched in K			
	Basalt	Andesite	Dacite	Rhyolite
SiO_2	51.23	62.85	64.46	71.27
TiO_2	1.14	1.09	0.36	—
Al_2O_3	17.96	15.23	15.41	13.85
Fe_2O_3	10.97	7.77	5.57	4.43
FeO	—	—	—	—
MnO	0.18	0.22	0.14	0.15
MgO	4.68	1.20	1.25	0.15
CaO	8.60	3.21	3.28	1.78
Na_2O	3.43	5.21	5.47	5.11
K_2O	1.25	2.73	2.97	2.82
P_2O_5	0.40	0.43	0.36	0.08
Total	100.00	100.00	100.00	100.00
LOI	—	—	—	—
Mg*	45.78	23.44	30.81	6.30
Cr	23	6	3	2
Ni	24	9	10	7
Co	32	6	7	1
Sc	28	18	15	14
V	218	17	62	3
Zn	97	100	74	121
Rb	34	67	80	71
Cs	1.89	3.04	4.52	5.67
Ba	383	803	757	786
Sr	624	335	298	144
Ga	19	20	18	19
Ta	—	—	—	0.55
Nb	6.6	13.3	12.3	10.6
Hf	3.69	6.71	7.72	8.47
Zr	157	312	361	389
Y	27	46	41	59
Th	5.68	9.92	9.72	8.87
U	—	—	—	—
La	21.53	40.77	34.07	34.80
Ce	51.14	89.26	76.45	87.76
Nd	25.82	45.38	37.37	41.19
Sm	5.73	9.47	7.78	9.37
Eu	1.62	2.49	1.74	1.83
Tb	0.81	1.20	1.31	1.50

Table 2.4 *contd.*

Components	Calc-alkaline series, enriched in K			
	Basalt	Andesite	Dacite	Rhyolite
Yb	2.84	4.58	4.37	6.64
Lu	0.41	0.68	0.64	1.00
$^{87}Sr/^{86}Sr$	0.703970	0.703960	0.704020	0.704090
$^{143}Nd/^{144}Nd$	0.512820	0.512843	0.512899	
$^{206}Pb/^{204}Pb$	—	—		18.583000
$^{207}Pb/^{204}Pb$	—	—		15.599000
$^{206}Pb/^{204}Pb$				38.487999

	Shoshonite–Latite series			
	Shoshonite	Latite	Trachyte	Rhyolite
SiO_2	53.95	60.09	54.97	69.67
TiO_2	1.75	1.02	1.04	0.50
Al_2O_3	15.05	16.22	16.06	15.60
Fe_2O_3	2.15	4.30	3.40	—
FeO	5.01	0.97	1.08	3.08
MnO	0.14	0.10	0.04	0.02
MgO	7.56	3.95	1.43	1.09
CaO	6.36	5.99	2.80	2.39
Na_2O	2.82	2.93	4.41	3.78
K_2O	4.34	3.89	4.30	3.88
P_2O_5	0.89	0.52	0.48	—
Total	100.00	100.00	100.00	100.00
H_2O	—	—	0.91	0.20
LOI	0.26	1.25	0.32	—
Mg*	67.0	59.3	38.2	38.7
Cr	284	—	—	—
Ni	139	9	—	—
Co	25	13	—	—
Sc	13	10	—	—
V	156	98	—	—
Zn	96	82	—	—
Rb	139	158	—	183
Cs	4.0	11.56	—	—
Ba	1,206	688	—	745
Sr	728	633	—	358
Ga	25	26	—	—
Ta	3.00	2.03	—	—

Table 2.4 *contd.*

	Shoshonite–Latite series			
	Shoshonite	Latite	Trachyte	Rhyolite
Nb	35.0	24.3	—	—
Zr	343	216	—	—
Y	21	17	—	—
Tb	15.02	17.24	—	—
U	6.01	6.08	—	—
La	68.08	63.77	64.34	—
Ce	144.17	104.63	131.15	—
Nd	61.97	49.07	—	—
Sm	11.61	8.01	8.11	—
Eu	2.76	1.83	2.31	—
Tb	0.97	0.81	0.55	—
Yb	1.70	1.01	0.98	—
Lu	0.28	0.22	0.16	—

After: Hickey-Vargas *et al.* (1989) *Contrib. Mineral. Petrol.*, **103**, 361–386; Gerlach *et al.* (1988) *J. Petrol.*, **29**(2), 333–382; Deruelle, B.J. (1978) *Volc. & Geochem. Res.*, **3**, 281–298; Dostal *et al.* (1977) *Lithos*, **10**, 173–183.

quartz latite–andesites, dacites, rhyodacites and scarce andesites. The southern Andes is a true andesite province, where andesites make up volcanic fields and volcanoes. It should be noted that petrochemically, the rocks may be assigned to two series, i.e. calc-alkaline (MgO-poor high alumina basalts, andesitic basalts and andesites) and shoshonite–latites (shoshonites, latites and quartz latites) series. The most common are rocks of the calc-alkaline series, which make up the main volcanic regions, while rocks of the shoshonite–latite series occur in the easternmost parts of the volcanic region. The SiO_2 content varies from 50 to 64 wt.% and the K, Sr, Ba, incompatible element and LREE content is higher than those of the calc-alkaline series.

As mentioned above, a feature of these rocks is their progressive enrichment in alkalies from the outer, Pacific, structural zones of the Andes to the interior of the continent. The dominant calc-alkaline rocks give way farther east to volcanic rocks of the subalkaline (shoshonite–latite) series. Moderate- and high-alkaline basalts and basanites, representing back-arc-related magmatism (the basalts of Patagonia and similar rocks) occur in the back-arc region. In general, calc-alkaline lavas of the southern and central Andes have some features in common with the rocks of the Pacific Ring of Fire (high Al_2O_3 content, constant Na_2O content and increase in K_2O with differentiation) and some differences (Deruelle, 1982). For example, the andesites of the central Andes have a lower TiO_2 and iron content, and their REE show a higher degree of fractionation and higher Sr content than the andesites of the southern Andes. In addition, the andesites of the central Andes are fairly inhomogeneous in their $^{87}Sr/^{86}Sr$ ratio, which shows a tendency to increase from west to east, further into the continental part of the volcanic region. In the volcanics of northern Chile it reaches 0.7051–0.7071, in NW Argentina: 0.7055–0.7088 and in SW Bolivia: 0.7059–0.7133. The andesites of the southern Andes have a lower and more stable $^{87}Sr/^{86}Sr$ ratio (0.7039–0.7043).

The longitudinal variations in rock composition in igneous regions are mainly due to the structural features of the underlying basement. In areas underlain by Mesozoic basic-composition folded basement, e.g. the northern and southern Andes, southern Central America, the coastal ranges of North America, basic and intermediate rocks are very common, whereas the areas underlain by older sialic basement (the Central Andes, Guatemala and El Salvador) are dominated by formations of acidic composition.

Granitiod plutonic rocks make up large batholiths (for example, the Coastal Batholith of Peru, which is over 1,600 km long and 60 km wide) whose emplacement began in the Late Palaeozoic, being associated with crustal evolution beneath the Andes. According to Aguirre (1983), the Mesozoic and Cenozoic Andean granitoids cut rocks ranging from Palaeozoic to Late Tertiary. The Coastal Batholith mentioned above comprises more than 1,000 plutons emplaced over a 60 Ma timespan from 100 to 37 Ma. Their transgressive contact relations indicate a post-kinematic character, and thermal aureoles are present in several regions. Mineral deposits, especially copper, iron, molybdenum, and zinc, are genetically related to these granitoids. The Tertiary granitoids are richer in alkali than those of the Mesozoic, and some are directly associated with porphyry copper deposits. A genetic relation between the Tertiary granitoids and Pliocene andesitic flows has been suggested on the basis of initial $^{87}Sr/^{86}Sr$ ratios. The age of the Andean granitoids decreased from west to east, and the initial $^{87}Sr/^{86}Sr$ ratio in granitoids, with ages ranging from mid-Cretaceous to Quaternary, shows a systematic west to east increase from 0.7022 to 0.7077, which indicates a change in the locus melting. Magmatism within batholiths was distinctly episodic, with quiescent periods often longer than 15 Ma between intrusive phases (Beckinsale et al., 1985).

2.3.3 Magmatism of Collision Zones

This type of magmatism is similar to that of island arcs and, especially, ACM (see Table 2.4). Magmatism of arc–arc collision (NW Indonesia) and arc-passive continental margin (East Indonesia) does not differ from common types of magmatism in these environments.

The most interesting case is a collision of continental plates which occurs within the Alpine–Himalayan Belt. The latter was initiated in the Late Cretaceous–Early Palaeogene by the closure of the Tethys, and its development still continues (Fig. 2.9). At that time, the southern edge of Eurasia was an active margin and the pattern of deep-seated processes was to a large extent inherited by the later collision of continental plates. As a result an extensive Cenozoic andesite–latite volcanic belt was formed, extending across the whole of Eurasia, from the Alboran Sea to the Indonesian–Burma Arc (the Great Andesite Belt of Eurasia: Fig. 2.1) (Milanovsky and Koronovsky, 1974; Borsuk, 1979; Pearce *et al.*, 1990; Arnaud *et al.*, 1992; etc.). It consists of the eastern or Aegian–Caucasus and western or Alpine–Mediterranean segments. Although their structures are of different morphology, they have a number of features in common. Thrust slices, stacked to form an accretionary prism, form arcuate mountain ridges around the periphery. They tend to follow the suture zones of the structures. The thrust sheets often contain fragments of deep-water Tethyan

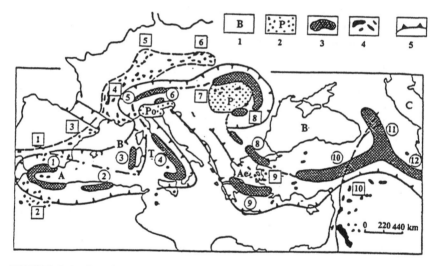

FIGURE 2.9 Development of late Cenozoic igneous rocks within the Alpine terrane. After Sharkov and Svalova (1993). Key: 1 — back-arc seas (A — Alboran, T — Tyrrhenian, Ae — Aegean, B — Black, C — Caspian); 2 — back-arc sedimentary basins (P — Pannonian, Po — Po Valley); 3 — andesite–latite arcs. Circled numbers: 1 — Alboran, 2 — Tell–Kabil, 3 — Sardinia, 4 — southern Italy, 5 — Drava–Insubria, 6 — Eugancy, 7 — Carpathian, 8 — Balkans, 9 — Aegeas, 10–12 — Anatolia–Caucasus: 11 — area of young volcanism in the Caucasus, 12 — Caucasus–Elburs. Areas of basalt volcanism marked by numbers in squares (4 on key): 1 — South Spain, 2 — Atlas, 3 — East Spain, 4 — Central France, 5 — Rhine Graben, 6 — Polabia–Silesia, 8 — Pannonian, 9 — Asia Minor, 10 — northern Arabia; 5 — suture zones of major thrust structures.

sediments, ophiolitic rocks and locally blocks from the lower crust and even the upper mantle (Milanovsky and Koronovsky, 1974; Chopin, 1987; Quick *et al.*, 1993). The rocks are usually metamorphosed to form schists.

Volcanic arcs, composed of igneous rocks of calc-alkaline, potassic, moderately alkaline (shoshonite–latite) and locally potassic low-Ti high alkaline series, occupy the rear part of the structures and follow their outline (Fig. 2.9). Depressions underlain by transitional or thinned oceanic crust, often marked by intense basaltic volcanism, are located in the back-arc areas (for example, the Tyrrhenian Sea). In general, the setting is similar to that at active continental or ocean margins, although in Alpine belts the lithospheric plates mainly consist of continental crust.

Within the Caucasus segment the plate-tectonic setting is controlled by underthrusting of the continental Arabian and the oceanic East Mediterranean plates beneath the Eurasian Plate. The underthrusting suture zone is marked by ophiolites of the Peri-Arabian Arc. North of this zone there is the large, structurally complex andesite–latite Anatolian–Elburz volcanic arc. It is formed by Anatolian–Caucasus and Caucasus–Elburz arcs, which are joined together in the Transverse Zone of the Recent Caucasian volcanism. The Black and Caspian seas are underlain by oceanic crust, and occur in the back-arc zone of the structures. The Caucasus segment is a zone of the Arabian syntaxis (Burtman, 1989), where the Arabian Plate is subducted at an angle under the Eurasian Plate, hence the morphology of the Anatolian–Elburz Arc. The emergence of the northern part of the Arabian Plate began in the early Miocene, with the initiation of high-Ti K–Na moderately alkaline basaltic volcanism and the opening of Red Sea Rift in the south-west, which separated this plate from Africa.

In terms of geophysics the Caucasian segments possess two strong positively isostatic anomalies, implying excess mass beneath these areas (Fig. 2.10). The first of these is related to the Aegean Sea, and the second to a zone of recent Caucasian volcanism, especially the Lesser Caucasus and to the zone of Arabian syntaxis.

Unlike the Caucasian segment, the Alpine segment contains major complex structures, created by interaction of the African and Eurasian plates (Recou *et al.*, 1986). Andesite–latite volcanic arcs are easily discernible there, partly enclosing back-arc basins underlain by the thinned crust and characterized by basaltic volcanism (see Fig. 2.8).

An unusual feature of the Alpine segment is a far-reaching extension zone (fore-arc rift) stretching across the whole of Central and Western Europe from Silesia to Portugal. Rift structures are common there, for example the Rhona, Liman, the Rhine graben, Polabian graben, etc. (Milanovsky and Koronovsky, 1974) as well as moderately or high alkaline high-Ti K–Na basaltic volcanism and even carbonatites (as at Kaiserstühl in Germany). To the south the Alboran Arc is marked by a zone of a powerful basaltic volcanism in the Atlas Mountains. The basaltic volcanoes of Sicily (including Etna), Pantelleria, Lemos and the volcanoes of seamounts of the Tunisian underwater uplift lie in the south-western part of the Alpine segment.

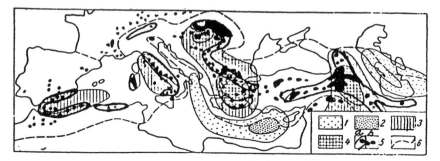

FIGURE 2.10 Distribution of main regional isostatic anomalies and areas of orogenic volcanism in the Alpine Belt. After Milanovsky and Koronovsky (1974). Key: 1 — regional lows of average intensity; 2 — of high intensity; 3 — regional highs of average intensity; 4 — intense; 5 — volcanic rocks: a — andesite–latite series, b — basalt series; 6 — boundaries of the Alpine Belt.

Geophysically, the Alpine segment is a region of reduced crustal thickness, due to omission of the lower low-velocity layer characteristic of the Eastern European Platform and the higher heat-flow density (Gize and Pavlenkova, 1989). The thrust structures that are now mountain ridges have immensely thickened crust beneath them (for example, up to 200 km in the Alps: Laubscher, 1989).

The seas of the Western Mediterranean, which formed in the Late Cenozoic, are floored by oceanic crust generated by the thinning and rupture of the continental lithosphere. The latter is preserved around the coasts of these seas, in particular the Tyrrhenian Sea, where the crustal thickness decreases from 20–25 to 6–8 km from the coast to the centre of the sea, as inferred from seismic data (Rehault *et al.*, 1987). In the Pannonian Basin, the decrease in crustal thickness shows the same pattern: it decreases from 30 to 18–20 km, mainly at the expense of the "basaltic" layer (Nikolaev, 1986).

Areas of average (Alboran and Tyrrhenian seas) and highly intense (Pannonian) maximum isostatic anomalies are related to the back-arc basins of the Alpine segment and to Aegean Sea (Fig. 2.10). This may suggest a non-compensated mass excess beneath these structures due to the presence of asthenospheric diapirs. These data, along with evidence of magmatism, imply mantle-derived geological processes, which are confirmed by intermediate earthquakes, whose epicentres have depths of 100 to 500 km, within the Alpine Belt (Gize and Pavlenkova, 1988).

The above-discussed structures are mainly Late Cenozoic in age. This means that after the closure of the Tethys during the late Cretaceous–early Palaeogene, the Alpine Belt became the site of very active geological processes which, as shown by current intense seismic and volcanic activity, continue to operate. This could be attributed to the presence of a huge mantle plume beneath it; in the Mesozoic the plume was responsible for the existence of the Western Tethys and the geological processes within it during the Cenozoic. A head of this plume must have had a complex topographic expression, and back-arc basins are located above its regional highs. Intense Cenozoic andesite–latite volcanism is typical of the Carpatians, Turkey (Anatolia), the Caucasus, Iran and Tibet (Fig. 2.1, Table 2.4). Volcanic rocks are represented by high-alkaline andesites, dacites and rhyolites with a $^{87}Sr/^{86}Sr$ ratio of 0.705; their potassium content increases from the south to the north, i.e. from the collision zone suture deep into the Eurasian Plate. Alkaline K–Na and K volcanism is estimated to have occurred 5 Ma ago.

Based on geochemical and isotopic data, these formations show similarities with volcanic rocks at active ocean and continental margins (Pearce *et al.*, 1984). Moderately alkaline basalts occur within the continental arc; however, back-arc tholeiites and boninites are absent.

Studies of isotopic systematics of Sr and Nd and the distribution pattern of rare elements suggest that magma generation occurs at the expense of material from the subducted plate. For example, the lithosphere of the Arabian Plate provided material for the formation of the Anatolian Arc (Pearce *et al.*, 1990). These authors point to the important role of mantle metasomatism in the petrogenesis of potassic rocks.

The presence of intrusive granitoids is an important feature of collision zones. Two types of massif: the Himalayan and Alpine, or syn-collisional and post-collisional, have been established (Pearce, 1986). The first group includes ultraacid undifferentiated leucogranites of subalkaline series of Neogene granite type (12–30 Ma old) occurring in the Himalayas. They form laccoliths, sills and other large intrusions in the metamorphic basement. The $^{87}Sr/^{86}Sr$ ratio of 0.712 and O of 11.38–12.55‰ suggest a crustal source, and a strong Eu anomaly implies a fairly shallow depth for the magma generation. Their origin is assigned to crustal melting *in situ* due to its thickening by the collision and the rise in temperature.

Post-collisional granites show a marked similarity with those of ACM. They are often formed by the rock suite from gabbronorites to granites, where the latter dominate. They display a calc-alkaline trend and are the equivalents of the type 1 granites of Chappel and White (1974). The Cenozoic (29–19 Ma) granitoids of the Alps provide a typical example (the Drava–Insubrian Complex: Milanovsky and Koronovsky, 1974; Laubscher, 1988). Recent intrusive rocks in the Caucasus (2.5–1.9 Ma old), including the youngest Eldjurtinsky massif (Borsuk, 1979), can be assigned to the same type of granitoids. All of these granitoids show low Sr-isotope ratios (0.707–0.708) and O values (8.0–8.420‰) and weak, if any, Eu anomaly. Table 2.5 displays the chemical characteristics of the collisional granites of the Caucasus.

2.3.4 Magmatism of Back-Arc Basins

Back-arc basins are part of a destructive plate margin. They are areas of back-arc spreading where generation of new crust takes place, in a similar way to that beneath mid-oceanic ridges. Based on three types of destructive environment (island-arc, ACM and continental plate collision zone), three types of back-arc basins are recognized:

1. back-arc seas, mainly along the western periphery of the Pacific (the Sea of Okhotsk, the Japan Sea and Lau, etc.) and to a lesser degree in the Atlantic (the Caribbean and the Scotia Sea);
2. back-arc sedimentary basins at active continental margins marked by active basalt volcanism as in western USA and South America (marginal rifts) and
3. the back-arc basins of collision zones in the Alpine–Himalayan Belt, where they form back-arc seas (Western Mediterranean, Aegean, Black and Caspian Seas) and sedimentary basins (the Pannonian Basin and the Po Valley). In the Pannonian Basin, the eruption of moderately alkaline basalts, containing spinel lherzolite xenoliths, took place during the Pliocene and Pleistocene.

In back-arc basins the crust is generally structurally complex. In most cases, there are fragments of thinned (20–30 km thick) continental crust (so-called "transitional" crust), along with newly formed oceanic crust (Fig. 2.1). The extent of the latter varies from relatively small in area (the Kurile Basin in Sea of Okhotsk, and the Okinawa Trough in the Southern China Sea) to about half of the area beneath the Japan Sea (Fig. 2.11) and the Bering Sea, and actually occupies the entire area beneath the back-arc basins of the western Mediterranean, the Philippine Sea and the Lau Sea, etc. As mentioned above, this enables two types of back-arc basins to be recognized: ensimatic, i.e. emplaced on oceanic crust (the Philippine and the Lau Sea), and ensialic, i.e. emplaced on continental crust (the sea of Okhotsk and the Japan Sea).

Rifts of the back of active continental margins also contain basalts similar to MORB; these are the Late Cenozoic basalts of the Columbia Plateau and the Snake River plain which are often described as "traps" (Wilson, 1989), although some authors attribute their generation to back-arc spreading (Presvic and Goles, 1985).

The study of ensialic seas, particularly the Japan Sea, shows that the oldest igneous formations are dominated by acid varieties, mainly rhyolites (Frolova et al., 1989; Ishizuka et al., 1990), which are similar to the initial ignimbrites of many continental rifts. Later they give way to moderately alkaline titanium K–Na basalts and then to basalts similar to MORB. Unlike mid-oceanic ridges, a multiaxial pattern is present (also known as scattered spreading). Table 2.5 gives analyses of typical rocks from this environment.

Basaltic magmatism of back-arc basins is generally similar to that of the oceans: the deep parts of the sea floor consist mainly of MORB (Ueda and Kanamori, 1979; Bogdanov, 1988), and are overlain by volcanic islands and seamounts made up high-Ti tholeiites and alkali–olivine basalts typical of intra-plate magmatism: for example, Mt. Vavilov in the Tyrrhenian Sea

Table 2.5 Igneous series of collision zones. Average major and trace element composition of rocks of Western Anatolia and the Caucasus.

Components	Calc-alkaline series	Calc-alkaline series, enriched in K			
	Rhyolite	Andesite–basalt	Andesite	Dacite	Rhyolite
SiO_2	76.48	56.74	61.27	65.12	74.60
TiO_2	0.04	0.76	0.77	0.50	0.23
Al_2O_3	13.36	15.75	16.53	16.40	13.18
Fe_2O_3	1.05	6.87	5.33	4.37	1.76
FeO	0.14	—	—	—	—
MnO	0.01	0.12	0.08	0.08	0.04
MgO	0.41	6.30	3.30	1.90	0.55
CaO	0.87	7.78	5.81	3.77	1.79
Na_2O	2.67	3.48	3.33	3.85	3.18
K_2O	4.93	1.90	3.28	3.81	4.57
P_2O_5	0.04	0.29	0.30	0.20	0.09
Total	100.00	100.00	100.00	100.00	100.00
H_2O	2.10	—	—	—	—
LOI	—	3.52	1.81\$; LOI	3.76	0.66
Mg*	40.26	64.50	55.11	46.26	38.36
Cr	19	224	188	23	6
Ni	6	118	75	11	6
Co	9	26	12	9	3
V	—	146	108	58	24
Cu	23	41	26	12	4
Pb	—	28	21	30	30
Zn	53	62	47	45	17
Rb	401	54	162	141	264
Ba	39	1340	1041	1480	360
Sr	15	1067	538	349	269
Nb	—	11.3	19.4	13.5	23.2
Zr	63	152	215	184	163
Y	75	25	28	24	19
La	8.21	—	—	—	—
Ce	55.41	—	—	—	—
Nd	—	33.89	27.56	52.23	23.70
Sm	—	5.89	5.00	8.77	3.72
$^{87}Sr/^{86}Sr$	0.710–0.720	0.706416	0.707695	0.708139	0.709139
$^{143}Nd/^{144}Nd$		0.512532	0.512430	0.512393	0.512362

Magmatism and Geodynamics

Table 2.5 *contd.*

Components	Shoshonite–Latite series		
	Shoshonite	Latite	Dacite
SiO_2	53.69	59.69	63.13
TiO_2	1.16	0.62	0.55
Al_2O_3	12.18	15.98	17.04
Fe_2O_3	5.72	6.22	4.69
FeO	—	—	—
MnO	0.10	0.13	0.07
MgO	6.93	3.54	1.64
CaO	11.02	4.66	4.48
Na_2O	2.68	5.07	3.87
K_2O	5.18	3.78	4.26
P_2O_5	1.33	0.32	0.26
Total	100.00	100.00	100.00
H_2O	—	—	—
LOI	2.41	6.22	0.52
Mg*	70.59	52.97	40.98
Cr	280	33	16
Ni	82	14	5
Co	25	11	9
V	124	119	85
Cu	38	27	16
Pb	23	11	43
Zn	67	13	39
Rb	122	130	152
Ba	2202	1150	1249
Sr	3077	416	976
Nb	67.8	20.3	18.2
Zr	379	248	217
Y	23	27	25
La	—	—	—
Ce	—	—	—
Nd	114.73	25.31	41.89
Sm	16.36	4.75	6.92
$^{87}Sr/^{86}Sr$	0.703527	0.707158	0.707391
$^{143}Nd/^{144}Nd$		0.512533	0.512519

After: N. Gulec (1991) *Geol. Mag.*, **128**(5), 417–435; J. Keller and L. Villary (1993) *Bull. Volcanol.*, **36**(2), 342–358; Y. Yilmaz (1990) *J. Volcan. Geotherm. Res.*, **44**, 69–87; data on the Caucasus are from the unpublished material of M.M. Bogina.

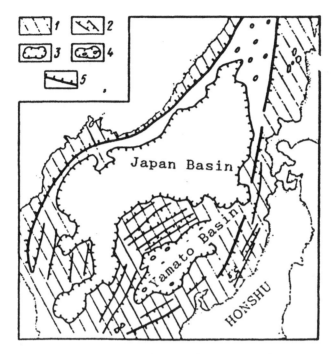

FIGURE 2.11 Structural–geological map of the Japan Sea. After Bogdanov (1988). 1 — zones of stacked up nappes; 2 — faults on the sea-floor. Outlines of deep basins: 3 — underlain by oceanic crust; 4 — underlain by suboceanic crust; 5 — boundaries of extensional zones.

(Bertrand *et al.*, 1990) among typical MORB, which form the floor of this sea (Kastens *et al.*, 1990). Basalts of back-arc seas, especially at sites adjacent to island arcs, as evidenced by the Philippine, Lau, Sulu and Celebes seas, are enriched in silica compared with MORB, are depleted in TiO_2 and have lower K/Rb ratio, i.e. they are enriched in crustal components (Spadea *et al.*, 1992).

Studies of rocks from the base of the crustal section in the Philippine Sea (Parese Vela Basin: Laz'ko and Gladkov, 1991) revealed two types: ultramafic rocks (mainly harzburgites with subordinate clinopyroxene), which are typical mantle tectonites, and gabbroids of the lower oceanic crust (troctolites, olivine gabbros, gabbro and gabbronorites (?) exhibiting characterisitic cumulate textures). Studies of the ultramafic rocks show that they are typical restites. The P–T data suggest the separation of the melt at T = 1258–1284°C and P of 11–12 kbar. The gabbroids show a slight shift of the tholeiitic trend towards the calc-alkaline trend, which can be attributed to an increase in the H_2O content in the parent melt, as a result of water supply from the subduction zone into the area of magma generation.

Based on geophysical data (the high density of heat flow, positive gravity and isostatic anomalies), widespread basaltic magmatism and the presence of xenoliths of mantle spinel lherzolites, etc., back-arc basins are sites of intense asthenospheric diapirism. Additionally, these basins show a tendency, with time, toward one-way spreading, accompanied by the arc movement. This spreading can take place gradually, as in the Pannonian Basin (Royden, 1989) or spasmodically, as in the Philippine Sea (Karig, 1971; Bogdanov, 1988). Therefore, back-arc basins are very important centres of endogenic activity, which to a great degree govern the processes operating at active plate margins. It is no coincidence that, as inferred from seismic tomography, many Pacific back-arc basins, including seas in the West Pacific, Australasia, the Tasman Sea and the western coasts of North and South America, are underlain by heated upper mantle at a depth of more than 400 km (Anderson *et al.*, 1992).

2.4 INTRA-PLATE MAGMATISM

Areas of intra-plate magmatism are characterized by its local occurrence during a specific, fairly limited (10–20 Ma) period of time. High-Ti picrites and basalts of normal, moderate and high alkalinity, as well as diverse K–Na and K alkaline rocks, represent this type of magmatism. Areas of intra-plate magmatism are often related to domes up to 200–300 km across, and the activity is accompanied by gravity and thermal anomalies, due to the presence of asthenospheric diapirs beneath these regions (Kovalenko, 1987; Wilson, 1989). Therefore, the intra-plate magmatism is also known as hot-spot-related magmatism.

The current hypothesis is that hot-spots are mantle plumes whose surface expression is represented by small-scale igneous regions often associated with domes. A peculiar feature of these areas is long-term igneous activity and specific, mostly moderately alkaline magmas. They range up to tens of thousands of square kilometres in size. At some stages of their development they may have a triple junction, due to the accumulation of the products of tectonomagmatic activity (grabens and volcanic fields) in linear zones radiating outward at the same angle from the hot-spot centre.

For example, the South Baikal (Fig. 2.12) has now been recognized as a hot-spot in the western part of the Baikal Rift (see below) (Yarmolyuk *et al.*, 1992). Its development began in the Oligocene–Early Miocene (34–18 Ma ago), with the emplacement of a slow-growing arch at the southern extremity of Lake Baikal and the eruption of Fe–Ti moderately and highly alkaline basalts. Basaltic eruption continued up to the end of Miocene. In the Pliocene–Quaternary valley eruptions become predominant. The most important feature of the recent structure of a volcanic area initiated in the Pliocene is large graben such as Tunka,

Khubsugul and East Tuva (Oka), forming a triple junction some 200 km west of the southern extremity of Lake Baikal. The pattern is very similar to that in north-east Africa, where there is a triple junction of Red Sea, the Gulf of Aden and the East African Rift.

Figure 2.12 shows the outline of the anomalous mantle in this region, at a depth of less than 50 km. A high heat flow corresponds with the 1200°C isotherm to the depth of 40 km, while along the rift zone bordering this area the isotherms occur at depths of 150–180 km (Zorin *et al.*, 1989).

According to Yarmolyuk *et al.*, (1992) the hot-spot developed as follows. Its inception, in the Oligocene, was followed by the formation of a triple junction. One of its branches extended towards the South Khangai hot-spot, which has existed in Southern Mongolia since the Late Mesozoic. Their interaction resulted in the emplacement of asthenospheric diapirs, which later merged into a single asthenospheric protrusion (Fig. 2.13). These diapirs in the present-day

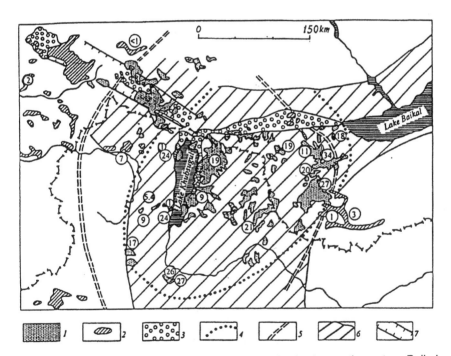

FIGURE 2.12 Location map of Cenozoic volcanics in the south-western Baikal Rift area. After Yarmolyuk *et al.* (1992). Key: 1–2 — basalts: 1 — summit, 2 — valley; 3 — Pliocene–Holocene molasse; 4 — distribution of Pliocene–Holocene (valley) volcanism; 5 — outlines of anomalous mantle at depths above 50 km; 6 — anomalous mantle at depths below 50 km; 7 — Eastern Tuva Zone of Cenozoic volcanism and intermontane lows. Circled numbers indicate the radiometric age of the basalts (Ma).

lithosphere are expressed as protruding linear zones of anomalous mantle. The ascent of the latter from a depth of 120 km (the thickness of the lithosphere prior to the formation of the rift area) to the base of the crust took place over a period of 30 Ma. Since that time, the asthenosphere was able to directly affect processes operating in the crust; in particular, it caused the formation of grabens and shaped the margins of microplates (Fig. 2.13).

At the present time, more than 120 hot-spots that were active in the Late Cenozoic have been established (Zonenshain *et al.*, 1991). The roots of hot-spots occur at much greater depth than that of the base of lithospheric plates and they are almost fixed in their position. Therefore, as the lithospheric plates move over these hot-spots the magmatism associated with the plume forms a clear trace on the plate surface, leaving a record of the plate's movements. Due to this fact, the trace of the hot-spot is often used for reconstructing this absolute rather than relative plate motion (Morgan, 1972; Le Pichon and Huchon, 1984; Wilson, 1989). Aseismic submarine ridges such as the Walvis Ridge in the Atlantic or the Ninety-East Ridge in the Indian Ocean could be of the same type (Morgan, 1972; Kovalenko, 1987; Wilson, 1989).

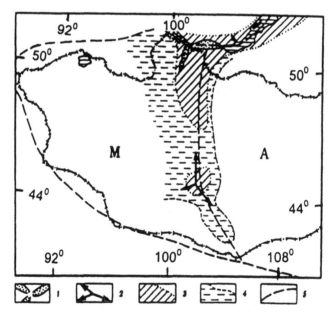

FIGURE 2.13 Location map of geodynamic settings in the south-western Baikal area and Mongolia. After Yarmolyuk *et al.* (1992). Key: 1, 2 — mantle hot-spots: 1 — South Baikal, 2 — South Khangai; 3, 4 — depth to the top of anomalous mantle: 3 — below 50 km, 4 — below 70 km; 5 — microplate boundaries: A — Amur Plate, M — Mongolian Plate.

It should be noted that, based on satellite gravity data, positive gravity anoma-lies have been reported from hot-spots. The hot-spots tend to differ in size and depth of occurrence. Mass transport of material within the hot-spot is not continu-ous flow but tends to occur in the form of isolated batches; the asthenospheric material of a hot-spot spreads along the base of the lithosphere like a mushroom cap. Hot-spots are created, evolve and disappear within a period of about 100 Ma.

From the evidence of seismic tomography, hot-spots do not actually form tubular bodies piercing the entire mantle. Instead they bend to provide routes for the passage of mantle material along extensional and fracture zones within the lithosphere, above large segments of hot mantle (Anderson et al., 1992). Many hot-spots occur at depths below 200–300 km and are underlain by cold mantle. The St. Helens, Tristan da Cunha, Iceland, Bouvet and Yellowstone hot-spots are examples of this type. Deeper hot-spots about 400 km deep include the Azores, Ascension Island, the Galapagos, Marquesas, Crozet and Carolines. Most of these hot-spots are located close to triple junctions, fault zones or rifts. Hot-spots beneath the Atlantic or Indian Ocean are related to mid-oceanic ridges, and with an allowance being made for ridge migration, this regular feature seems to be quite important. Hot-spots unrelated to mid-oceanic ridges (Hawaii, Reunion, Canaries, Cape Verde, Kerguelen and Tasmania) have the deepest roots above extensive high-temperature regions of the upper mantle.

There are two main types of intra-plate magmatism, spatially related to areas of continental rifting and oceanic floor environment. In general, there is not a major difference in the manifestation of intra-plate magmatism within the litho-sphere of a single type or in a certain geological environment, and the intra-plate magmatism of oceans and continents differs slightly: acid members of the mag-matic series are more common in continental regions' lithosphere. The petro-graphic and geochemical characteristics of igneous assemblages produced by oceanic and continental intra-plate magmatism are presented in Tables 2.6 and 2.7.

2.4.1 Structure and Magmatism of Continental Rifts

Locally continental rifts are linked, along strike, with oceanic rifts by transitional structures. Here these structures are considered to be located at plate margins. The transitional zone of the Arabian–Indian oceanic ridge and the Red Sea oceanic spreading zone into the continental rift zones of East Africa, incorporating the Ethiopian and Kenyan continental rifts, provide an example of the connection between continental and oceanic rifts (Fig. 2.14). However, continental rifts are often intra-plate in nature and not related to plate margins as inferred from geo-physical data (Western Europe, Eastern Australia, Central and East Asia, etc.).

The Baikal Rift Zone (BRZ) is a typical example of this type of rift zone (Logachev, 1977; Grachev, 1987; Yarmolyuk et al., 1992). It lies within the Sayan–Baikal Dome, extending south-eastward for 1,500 km and locally widening

Table 2.6 Magmatic series of continental rifts. Average major and trace element composition of volcanics of Afar, West and East African rifts.

Components	Tholeiite series (Afar rift)					Alkaline K–Na series			
	Basalt	Ferrobasalt	Alkaline–olivine basalt	Hawaiite	Trachyte	Comendite	Pantellerite	Trachyte	Trachyte
SiO_2	47.80	47.64	47.59	55.56	58.10	70.69	73.27	58.87	63.78
TiO_2	2.35	3.10	1.97	1.64	1.80	0.43	0.39	1.58	0.94
Al_2O_3	14.24	14.36	15.96	14.50	14.20	13.29	9.50	16.27	14.15
FeO	11.20	13.31	11.14	8.89	10.23	4.12	5.96	6.25	7.85
MnO	0.19	0.25	0.20	0.16	0.30	0.14	0.21	0.21	0.27
MgO	9.97	5.45	7.40	4.78	2.18	0.01	0.01	2.15	0.04
CaO	10.31	10.87	12.37	8.98	5.15	0.72	0.35	4.64	1.31
Na_2O	2.76	3.08	2.64	3.56	5.11	6.02	5.83	5.57	6.35
K_2O	0.82	1.29	0.48	1.68	2.20	4.57	4.47	4.07	5.23
P_2O_5	0.36	0.64	0.24	0.25	0.74	0.01	0.01	0.39	0.07
Total	100.00	100.00	100.00	100.00	100.00	100.00	100.00	100.00	100.00
H_2O	0.33	—	0.12	0.04	0.37	—	0.10	—	—
LOI	0.78	—	—	—	0.95	0.62	1.26	—	—
Mg*	61.32	42.17	54.21	48.94	27.51	0.44	0.30	38.04	0.90
Cr	378	97	—	—	6	—	78	20	—
Ni	77	67	—	—	6	2	8	—	—
Co	51	44	30	30	8	—	3	14	1
Sc	24	30	—	—	15	—	0	14	6
Rb	18	30	9	71	50	112	149	67	115
Ba	256	512	256	278	417	659	10	1047	160
Sr	391	416	385	282	368	24	3	339	10

	(1)	(2)	(3)	(4)	(5)	(6)	(7)	(8)	(9)
Ta	—	2.61	—	70.1	—	—	—	5.23	11.12
Nb	—	55.2	23.2	—	—	—	—	113.7	207.4
Hf	4.09	4.72	—	—	14.32	—	32.51	9.6	17.74
Zr	124	184	72	183	414	933	1189	261	766
Y	—	45	19	43	—	—	—	44	93
Th	—	5.02	—	—	4.81	—	15.14	10.47	19.44
U	0.66	—	—	—	1.69	—	3.56	—	—
La	20.04	45.35	17.04	42.26	61.36	—	162.06	77.21	152.31
Ce	36.70	—	36.60	77.81	117.10	—	292.02	128.85	185.37
Nd	22.29	51.17	18.25	36.26	72.31	—	131.27	57.38	91.18
Sm	4.60	9.45	4.14	7.82	16.98	—	27.43	10.90	17.94
Eu	1.74	2.70	1.65	1.70	5.42	—	3.35	3.41	3.18
Gd	4.40	—	4.54	8.33	17.80	—	25.40	—	—
Tb	0.72	1.28	—	—	2.35	—	3.96	1.33	2.40
Dy	4.19	—	4.08	9.29	14.22	—	26.42	—	—
Ho	1.02	—	—	—	3.17	—	6.50	—	—
Er	—	—	2.27	5.52	—	—	—	—	—
Yb	1.94	3.41	2.15	5.40	7.16	—	14.33	3.62	8.92
Lu	0.20	0.61	—	—	1.02	—	2.03	0.58	1.64
$^{87}Sr/^{86}Sr$	0.704600	0.704800	0.704579	0.704816	0.705200	0.705900	0.705675	—	—
$^{143}Nd/^{144}Nd$	—	—	0.512763	0.512628	—	—	0.512519	—	—
$^{206}Pb/^{204}Pb$	—	—	19.778999	19.787001	—	—	19.805000	—	—
$^{207}Pb/^{204}Pb$	—	—	15.720000	15.790000	—	—	15.809000	—	—
$^{208}Pb/^{204}Pb$	—	—	39.608002	39.695009	—	—	39.729998	—	—

Table 2.6 *contd.*

Components	High-alkaline K–Na series				High-alkaline K series		
	Basanite	Phonolite	Alkaline phonolite	Nephelinite	Leucitite	Katunzite	Leucite trachybasalt
SiO_2	44.09	52.10	56.93	36.16	44.30	39.92	44.26
TiO_2	2.30	0.94	0.16	2.78	4.64	5.97	3.81
Al_2O_3	13.45	20.05	21.42	10.12	10.23	7.81	12.56
FeO	11.96	8.61	3.90	13.41	11.57	13.08	12.07
MnO	0.18	0.28	0.21	0.22	0.17	0.21	0.21
MgO	9.97	1.48	0.09	14.06	9.02	9.12	8.76
CaO	12.64	4.56	2.04	14.84	11.85	17.88	12.13
Na_2O	3.45	7.86	9.87	5.99	2.76	1.23	2.08
K_2O	1.46	3.72	5.34	1.57	4.89	3.71	3.49
P_2O_5	0.42	0.39	0.04	0.84	0.56	1.07	0.64
Total	100.00	100.00	100.00	100.00	100.00	100.00	100.00
LOI	0.47	0.95	0.92	1.03	—	—	—
Mg*	59.78	23.47	4.00	65.13	58.14	55.41	56.39
Cr	455	12	14	213	476	183	379
Ni	140	13	15	158	127	64	76
V	—	—	—	—	359	476	—
Rb	38	96	143	49	118	73	130
Ba	498	1095	1143	801	1410	1585	1148

	577	853	788	947	1446	2385	1031
Sr	577	853	788	947	1446	2385	1031
Ta	—	—	—	—	—	—	9.33
Nb	—	—	—	—	214.2	284.9	110.8
Hf	—	—	—	—	—	—	10.67
Zr	144	314	391	277	287	345	314
Y	—	—	—	—	15	15	30
Th	—	—	—	—	—	—	11.59
U	—	—	—	—	—	—	2.36
La	20.09	74.13	107.19	59.72	145.34	181.13	83.09
Ce	36.16	100.53	141.57	100.21	284.02	346.36	168.22
Nd	—	—	—	—	108.24	135.58	65.65
Sm	—	—	—	—	14.35	18.32	10.98
Eu	—	—	—	—	3.59	4.77	3.08
Gd	—	—	—	—	8.30	8.47	8.31
Dy	—	—	—	—	5.02	5.19	—
Er	—	—	—	—	1.64	1.69	—
Yb	—	—	—	—	1.23	1.17	2.15

After: Barbery et al. (1975); Baker et al. (1977) Contrib. Miner. Petrol., **64**, 303–332; Davies and Macdonald (1987) J. Petrolgy, **28**(6), 1009–1031; Davis and Lloid (1988) Proc. 4th Int. Kimberl. Conf., Perth, W. Australia. Oxford, Blackwell Scientific; Price et al. (1985) Contr. Miner. Petrol., **85**, 394–409; Thompson et al. (1984) Phil. Trans. R. Soc. London, **A310**, 549–590.

Table 2.7 Oceanic islands. Average major and trace element composition of volcanics of Hawaii and Tristan da Kunha islands.

Components	Hawaii							
	Tholeiite series				Alkaline basalt	Moderate alkaline K–Na series		
	Picrite	Tholeiite	Fe–Ti basalt	Ancaramite		Hawaiite	Mugearite	Trachyte
SiO_2	44.03	48.12	46.18	46.60	46.58	49.26	54.49	58.75
TiO_2	2.42	3.04	4.83	2.82	3.17	2.89	1.61	1.02
Al_2O_3	9.62	13.47	14.61	10.63	13.20	17.06	17.21	18.12
Fe_2O_3	13.85	13.74	15.74	14.36	13.80	11.71	9.64	7.09
MnO	0.19	0.18	0.21	0.20	0.19	0.20	0.22	0.24
MgO	17.94	6.63	5.15	12.54	8.55	4.24	2.65	1.23
CaO	9.96	11.80	8.57	9.60	9.92	7.16	5.03	2.78
Na_2O	1.21	2.06	2.90	2.19	3.17	4.81	5.51	7.15
K_2O	0.47	0.59	1.20	0.71	0.92	1.82	2.57	2.91
P_2O_5	0.30	0.35	0.61	0.34	0.48	0.86	1.08	0.70
Total	100.00	100.00	100.00	100.00	99.98	100.01	100.00	100.00
H_2O	0.41	0.26	1.56	0.30	0.32	—	0.14	0.30
CO_2	—	—	—	—	—	—	—	0.07
Mg	71.96	48.88	39.32	63.35	55.10	41.76	35.22	25.65
Cr	1146	169	61	916	338	12	18	9
Ni	664	87	60	343	190	16	21	—
Co	—	—	44	68	54	—	11	3
Sc	28	33	30	28	27	11	6	3
V	257	296	338	239	227	101	17	—
Rb	6	10	23	13	8	33	57	78

Ba	266	192	422	258	307	286	797	934
Sr	442	457	510	451	505	1286	10	1331
Ga	16	22	27	20	23	21	24	—
Ta	1.71	1.51	2.47	1.38	1.53	—	4.63	5.02
Nb	26.4	21.9	40.2	23.4	29.0	57.0	76.5	—
Hf	3.92	5.03	8.10	4.79	6.90	9	15.69	13.85
Zr	180	216	357	215	313	430	735	—
Y	16	28	43	25	37	44	50	—
Th	1.21	1.11	3.00	0.90	1.50	3.50	6.04	6.63
La	19.71	18.52	34.90	19.15	25.90	45.10	75.55	72.48
Ce	45.25	46.59	82.00	44.89	63.30	111.00	185.09	158.60
Nd	25.04	28.38	48.00	25.34	36.70	67.00	88.52	706.69
Sm	5.35	6.81	11.00	6.23	9.22	13.90	17.00	14.15
Eu	1.84	2.49	3.73	2.26	3.15	4.42	5.05	4.20
Tb	0.72	0.99	1.68	0.97	1.54	1.62	1.90	1.62
Ho	0.60	1.11	1.50	0.80	1.10	—	2.21	2.11
Yb	1.24	2.19	3.05	1.89	2.86	3.11	4.20	4.16
Lu	0.16	0.30	0.44	0.28	0.37	0.42	0.58	0.63
$^{87}Sr/^{86}Sr$	0.703360	0.703500	0.703610	0.703510	0.703600	0.703500	0.703430	0.703470
$^{143}Nd/^{144}Nd$	0.513055	0.513044	0.513019	0.513076	0.513026	0.513004	0.513043	0.513034
$^{206}Pb/^{204}Pb$	18.438000	18.430000	18.358999	18.403000	18.413000	18.358000	18.513000	18.492001
$^{207}Pb/^{204}Pb$	15.486000	15.501000	15.441000	15.490000	15.458000	15.478000	15.494000	15.476000
$^{208}Pb/^{204}Pb$	37.994999	38.015999	37.887001	38.018002	37.992001	37.952000	38.049999	38.007999

Table 2.7 *contd.*

Components	Hawaii				Tristan da Kunha			
	High-alkaline K–Na series				High-alkaline K series			
	Alkaline Ol basalt	Basanite	Nephelenite	Melilitite	Basanite	Phonotephrite	Tephrophon.	Phonolite
SiO_2	45.45	45.49	40.37	36.34	46.40	50.97	55.59	59.14
TiO_2	1.98	2.27	1.91	2.84	3.26	2.31	1.61	1.15
Al_2O_3	13.46	11.90	12.08	11.26	16.93	19.67	19.37	20.17
Fe_2O_3	3.88	3.55	5.83	10.05	—	—	—	—
FeO	9.06	9.03	7.24	7.29	10.88	5.44	5.44	3.74
MnO	0.20	0.20	0.23	0.26	0.19	0.21	0.21	0.13
MgO	11.54	12.68	13.34	11.72	4.46	1.45	1.45	0.77
CaO	10.46	10.70	12.80	12.45	9.24	5.55	5.55	2.95
Na_2O	2.96	2.64	4.15	4.95	4.58	5.56	5.56	5.78
K_2O	0.65	0.52	1.07	1.76	3.06	4.95	4.95	5.96
P_2O_5	0.35	0.60	0.99	1.07	0.99	0.37	0.37	0.20
Total	100.00	100.00	100.00	100.00	100.00	100.00	100.00	100.00
H_2O	0.19	0.94	0.14	0.58	0.17	0.17	0.11	0.48
CO_2	0.05	0.01	0.03	0.26	—	—	—	—
LOI	0.15	0.81	0.14	0.63	2.25	0.57	0.62	1.18
Mg	62.10	64.90	65.57	56.12	42.19	38.42	32.18	26.85
Cr	507	529	578	334	2	2	2	22
Ni	322	360	312	235	3	2	5	1
Co	70	64	67	77	29	11	4	4
Sc	24	23	27	14	13	5	3	2
V	297	284	282	235	197	87	65	33
Cu	—	—	—	—	12	3	4	1
Pb	—	—	—	—	6	5	9	—
Zn	—	—	—	—	114	92	105	81
Rb	16	19	28	52	76	121	131	153

Ba	388	490	719	962	819	1309	1311	1232
Sr	601	709	1161	1586	1311	1725	1524	787
Ga	—	—	18	—	—	—	—	—
Ta	1.21	2.88	2.61	4.71	102.0	126.8	143.8	141.9
Nb	—	—	42.2	—	—	—	—	—
Hf	2.52	3.56	2.82	5.42	—	—	—	—
Zr	111	154	134	246	388	420	519	599
Y	—	—	24	—	35	32	38	25
Th	2.11	4.04	6.64	8.08	—	—	—	—
U	—	—	—	—	2.37	2.82	3.37	—
La	18.12	44.22	74.40	74.68	65.00	—	91.00	160.00
Ce	41.28	77.86	125.68	149.36	110.00	—	79.00	85.00
Pr	—	—	—	—	12.00	—	8.60	8.10
Nd	20.14	36.53	60.33	67.52	50.00	—	35.00	29.00
Sm	5.04	8.46	13.17	15.75	13.60	—	9.60	6.00
Eu	1.85	2.49	3.74	4.81	3.76	—	2.80	1.77
Gd	—	—	—	—	11.00	—	7.40	5.60
Tb	0.70	0.96	1.31	1.53	1.60	—	1.20	0.86
Dy	—	—	—	—	9.00	—	5.70	4.80
Ho	—	—	—	—	1.50	—	1.20	1.00
Tm	—	—	—	—	4.10	—	3.10	2.80
Er	—	—	—	—	0.57	—	0.41	0.38
Yb	1.51	1.63	2.01	1.23	3.30	—	2.10	2.10
Lu	0.2	0.27	0.26	0.14	0.56	—	0.30	0.30
F	—	—	0.07	—	—	—	—	—
Cl	—	—	0.03	—	—	—	—	—
$^{87}Sr/^{86}Sr$	0.703340	0.703260	0.703450	0.703370	0.705040	0.705130	0.705070	0.705110
$^{143}Nd/^{144}Nd$	0.513024	0.513018	0.513029	0.513056	0.512290	0.512550	0.512530	0.512510
$^{206}Pb/^{204}Pb$	—	—	—	—	18.556000	18.513000	18.497999	—
$^{207}Pb/^{204}Pb$	—	—	—	—	15.529000	15.543000	15.507000	—
$^{208}Pb/^{204}Pb$	—	—	—	—	39.006001	39.009998	38.918999	—

After: *Basaltic volcanism* (1981); le Roex et al. (1990) *J. Petrology*, **31**(4), 779–812; Kogarko et al. (1984) *Geochemistry Int.*, **5**, 639–652; Frey et al. (1990) *J. Geophys. Res.*, **95** (B2), 1271–1300, **96**(B9), 14347–14375; Kennedy et al. (1991) *Earth Planet Sci. Lett.*, **103**, 339–353; Decker et al. (1984); Roden et al. (1984) *Earth Planet. Sci. Lett.*, **69**, 141–158.

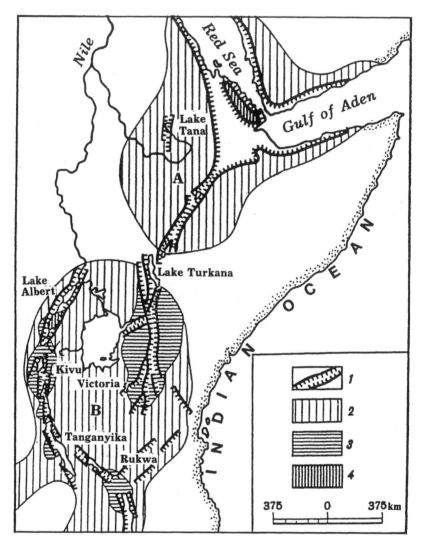

FIGURE 2.14 Structure of the East African Rift zone. After Logatchev (1977). Key: 1 — rift valleys; 2 — main domes: A — Ethiopian, B — East African; 3 — domes and semi-arched uplifts of second and higher orders; 4 — large internal horsts in rift valleys.

to 300 km and narrowing to 150 km (Fig. 2.15). Its central and eastern parts are related to the southern edge of the ancient Siberian Platform and the western part extends into the Caledonian fold belt. The BRZ includes a network of often asymmetrical graben-like depressions bounded by faults and separated by horst

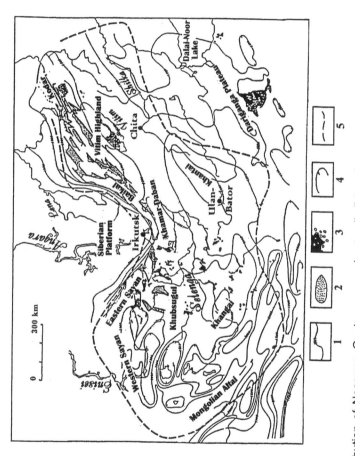

FIGURE 2.15 Distribution of Neogene–Quaternary moderate-alkali basalts in southern East Siberia and in Mongolia. After Kiselev *et al.* (1979). Key: 1 — generalized stratum contours of domes; 2 — depressions in the Baikal rift area; 3 — basalt lava flows and volcanoes; 4 — the main faults forming rifts; 5 — distribution of anomalous mantle.

blocks. The largest Baikal depression consists of two enchelon basins, separated by submarine Academi Ridge. The BRZ evolved in two major phases. An extensive dome was formed there during the first phase (35 to 3–5 Ma ago; pre-rift stage). Only during the second rift stage proper did the rate of tectonic movement increase drastically and give rise to the modern topography. Based on geodesic data and the high level of seismicity, rifting still continues. Seismic surveys revealed tensile stresses across the strike of the depressions, in the brand earthquake foci, thus suggesting crustal extension (Grachev, 1987).

Figure 2.15 shows that volcanism within the BRZ does not depend directly on the development of rift structures. The pre-rift stage was characterized by the appearance of a large number of basaltic domains scattered over a large area; fissure eruptions tend to predominate. The base of the lava plateaux is made up by pyroclastic and volcano-sedimentary rocks, formed by a thick succession (600–800 m) of basalt flows locally interbedded with sedimentary rocks and cut by numerous dykes 3–4 m thick. Lava and pyroclastic of similar origin were emplaced at the final stage. An unusual feature of the volcanic rocks is the widespread occurrence of diverse ultramafic mantle xenoliths and megacrysts, which can be used for reconstruction of the mantle substrate and the conditions of magma generation (Grachev, 1987; Dobretsov et al., 1989).

Volcanic rocks are dominated by moderately alkaline Fe–Ti basalts of the K–Na series associated with hawaiites and mugearites; intermediate tholeiites and trachybasalts are subordinate. Even less common are formations of the K–Na high-alkaline series (melanephelinites and tephrites) and potassic–leucite basanites. Potassic igneous rocks occur mainly within the Khubsugul Rift. All of these rocks show a high titanium content, thus differing from formations at active plate margins.

At the rift stage, volcanic rocks show a tendency to accumulate within rift valleys (Tunka, Khubsugul, etc.). Unlike the fissure eruption described for the previous stage, central-type volcanoes and the development of differentiated series, containing benmoreites and trachytes as well as basalts (Udokan Range), are more common there. The pattern of evolution of the subcrustal material was studied in detail for the Vitim Plateau (Fig. 2.16). According to Dobretsov et al. (1989), the deep-seated xenoliths in the oldest picrobasalts (47–27 Ma) tend to be garnet lherzolites. They are depleted in LREE, and, on the basis of their isotope parameters ($^{87}Sr/^{86}Sr = 0.7019$–0.7023 and $\varepsilon_{Nd(T)} = +15$ to $+27$), they were derived from strongly depleted mantle and are fragments of the lithosphere brought up from depths of 60–110 km.

Basaltic volcanic activity reached its maximum in Miocene (17–9 Ma ago) when the Vitim and other lava plateaux were formed. Xenoliths consist mainly of spinel lherzolites, which are fragments derived from the asthenosphere. Dobretsov et al. (1989) suggest that the top of the diapir finally reached the base of the crust. This has resulted in a thermal anomaly about 100 km across, with a temperature of 1000°C, which successively decreases to 850°C towards the

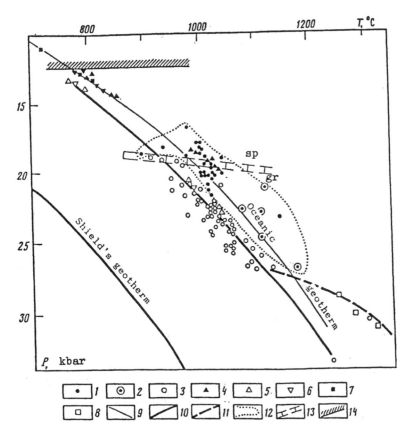

FIGURE 2.16 Thermal evolution of the mantle under the Vitim Highlands (white symbols — xenoliths in picrobasalts, black symbols — from hawaiites). After Dobretsov *et al.* (1989). Key: 1, 3 — garnet lherzolites; 2 — the same with elevated iron contents; 4, 5 — spinel lherzolites; 6 — depleted spinel peridotites (in assemblages 3–6 the pressure has not been measured, temperatures are plotted using geo-isotherm); 7, 8 — cumulates of the basaltic melts; 9 — standard geo-isoterms; 10 — geo-isoterm from xenoliths within picrobasalts of stage 1; 11 — geo-isoterm (10) from xenoliths of high-temperature websterites; 12 — P–T field evaluated on xenoliths from hawaiites of stage III; 13 — garnet–spinel transition boundary; 14 — position of Moho discontinuity.

periphery; this anomaly occurs beneath the Vitim Plateau in the centre of the structure. The igneous activity became less intense in the Pliocene (5–2 Ma), which coincided with the initiation of the rifting stage. This time period is characterized by the presence of spinel and spinel–garnet ultramafic rocks, probably implying a lower level of magma generation at a depth of 90–70 km, apparently caused by diapiric extension (Grachev, 1987).

Numerous geophysical surveys of the BRZ have revealed an extensive asthenospheric diapir (Fig. 2.15). The greatest elevation of the asthenosphere can be found beneath the rift valleys. For example, the Moho discontinuity occurs at a depth of 34 km beneath the South Baikal Depression, while beneath the sides of the rift, the crustal thickness increases to 42–46 km. Moreover, narrow local heat anomalies, interpreted by Zorin et al. (1989) as large dykes, often occur beneath the depression. A typical dyke beneath the South Baikal Depression is 7–10 km wide and occurs at a depth of 6 km or more. However, these structures can also be interpreted as local asthenospheric diapirs (cf. the Zabargad Island in the Red Sea or St. Paul Rocks in the Atlantic: Bonatti et al., 1986; Roden et al., 1984). Additionally, a special study of xenolith-bearing basalts from the BRZ showed that they derived from shallow (18–20 km) mantle reservoirs (Sharkov and Bindeman, 1991). Therefore, these thermal structures require further elucidation.

According to Grachev (1987) and Logachev and Zorin (1992), the presence of an asthenospheric "lens" beneath the BRZ suggests that stretching was due to the spreading out of the upper part of the mantle diapir. This means that the BRZ is an active rather than a passive structure. The thickness of the asthenospheric lens beneath the BRZ, as inferred from seismic tomography, cannot exceed 200 km. The intrusion of the asthenosphere first (at the pre-rift stage) led to the uplift of the Earth's crust, whose topographic expression was the Sayan–Baikal Dome. The subsequent spreading out of the top of the diapir (rift stage) caused crustal extension across the strike of the dome.

Other continental rifts exhibit similar structural features and unusual features in their development (Grachev, 1989; Wilson, 1989). As in case of the BRZ, their emplacement is related to a mantle diapir rising from the asthenosphere and associated doming, and splitting of the Earth's crust along which the lithospheric blocks were pulled apart. Continental rifting is not related to plate margins and is actually an intra-plate phenomenon. However, the Red Sea Rift shows that in some cases, steady inflow of the asthenospheric material can cause a continental rift zone to develop into an oceanic spreading zone, i.e. to be transformed from an inter-continental to an intra-continental zone.

Based on seismic tomography (Anderson et al., 1992), in most cases (the Baikal Rift, the Rhine Graben, the Kenyan Rift, etc.) the thickness of the asthenospheric "pillow" beneath the rift areas does not exceed 200 km, and at a greater depth of 330 km the mantle beneath these regions is fairly cold. The asthenospheric mantle beneath most of the Red Sea Rift, except for its southern part which joins the triple junction with the East African and Aden Rift, also occurs at a shallow depth. At this site, the basement lies at great depths and can be traced down to 400 km or more. This could have caused the transformation of the continental rift into oceanic one.

It is possible to suggest that the asthenosphere generated at a depth of above 400 km could, like the magmatic melt, intrude into the overlying colder lithosphere. This would have resulted in the emplacement of huge intrusion-like bodies

150–200 km thick and thousands of kilometres long. Apparently, these lenticular asthenospheric "intrusions" occur beneath most of the rift areas.

Present-day continental rift zones, which show similarities in terms of structure and morphology, differ greatly in their igneous rock composition. Some rift zones (for example, the BRZ) contain no volcanic products within extensive segments, and generally are characterized by poorly developed igneous assemblages. The other rift zone, the East African system, provides the best example, and is characterized by intense and diverse volcanic activity.

Table 2.6 presents the geochemical characteristics of igneous assemblages from continental rifts. These are igneous rocks of the tholeiitic, moderately alkaline and high-alkaline series, as well as of the K–Na and K series. They are represented by different Fe–Ti basalts and picrobasalts of varying alkalinity and to a lesser degree by various alkaline rocks, including intermediate and acid rocks. The magmatism is often bimodal, being most typical of the late rift stage of the system, while weakly differentiated alkali–basaltic assemblages, locally giving way to tholeiite series, are characteristic of the pre-rift stage. Diverse alkaline rhyolites (comendites and pantellerites), often represented by ignimbrites, lie at the base of the rift series section.

2.4.2   Intra-plate Magmatism of Oceanic Islands

Oceanic spreading zones, especially in the Atlantic and Indian Oceans, are known to be complicated by islands. At present, these volcanic islands are subdivided into two groups:

1. at the mid-oceanic ridge axis, and
2. on the flanks or slopes of the ridges.

There is a definite relationship between the nature of the magmatism and its association with major structures on the sea-floor. Therefore, islands at the mid-oceanic ridge axis (e.g. Iceland, Ascension Island and the Bouvet Islands) are composed mainly of rocks of the tholeiitic series (the basalt–icelandite–rhyolite assemblage) while those on the ridge slopes (St. Helens, Azores, Tristan da Cunha and Gough) are made of K–Na moderately and high-alkaline series (alkali basalt–trachyte–trachyrhyolite, comendite–pantellerite, phonolite and alkali–trachyte assemblages). Moderately alkaline basalts of islands at MOR show a fairly low SiO_2 content, a high alkali content, especially Na_2O, and rare K_2O (1.45– 2.00 wt.%), and TiO_2 which is high as compared with MORB (up to 4%) and Al_2O_3 (up to 18%) and low abundance CaO and MgO (see Table 2.7). Geochemically, moderately alkaline basalts, unlike tholeiites, are enriched in Ba, Nb, Sr, Rb and Zr. Their isotopic ratio is lower: $^{87}Sr/^{86}Sr$ (0.7028–0.7031) and $^{143}Nd/^{144}Nd$ (0.51322– 0.51318) (Basaltic Volcanism, 1981). It suggests that intra-plate oceanic volcanism is about 10% of the magmatic products erupted on mid-oceanic ridges.

Iceland is of special interest because it has the most concentrated volcanic activity on Earth. In contrast to the statement that the tholeiitic basalts of Iceland and those of the Mid-Atlantic Ridge are identical (Imsland, 1983), they do actually show clear differences: the former are enriched in K, Na, Ba, Sr, Rb, Ti, P, La and depleted in Al_2O_3 (Table 2.7). Compared with oceanic rocks, the rocks of Iceland show low Na/K, K/Rb and Sm/La and have a higher $^{87}Sr/^{86}Sr$ (0.70291–0.70357) ratio, as compared to basalts of adjacent Reykjanes Ridge (0.7027). The tholeiitic basalts of Iceland are most similar in their petrological and geochemical characteristics to those of aseismic oceanic ridges (e.g. Ninety East Ridge, Walvis, Maldives–Chagos, etc.) which, as can be seen from Table 2.8, differs from MORB in its higher Fe, Ti, K, Ba, Sr, V and Zr content.

The magmatism of hot-spots from inner parts of oceanic plates is shown on volcanic islands and seamounts far from mid-oceanic ridges and island arcs. It is characterized by intense eruptions over a certain time, per unit area, being higher than the supply of igneous material to the mid-oceanic ridges. For example, for Hawaii this ratio varies from 0.01 to 0.02 km/yr (Decker et al., 1987).

Trinidadi, the Canary and Cape Verde Islands in the Atlantic, Reunion, Rodriges and Mauritius in the Indian Ocean, and Hawaii, Line, Galapagos and Tahiti in the Pacific are examples of these intra-plate magmatic environments. This magmatism is Fe–Ti tholeiitic, moderately and high alkaline. Rocks of the K–Na series are dominant, although potassic varieties have also been reported (Cape Verde Islands). Seamount, which can make up 25% of the entire sea-floor, are composed of similar rocks (Batiza, 1982). The ratio of tholeiites, moderately alkaline and high alkaline basic igneous rocks varies according to region. So, on some islands (e.g. Hawaii and Reunion) rocks of the tholeiite series predominate over others (Canaries and Tahiti) — these are moderate- and high-alkaline rocks.

On the whole, the igneous assemblages of oceanic islands are similar to those of continental intra-plate environments, although the latter often contain various acid igneous rocks. The Cameroon Line, a volcanic chain made up of moderately alkaline basalts and high-alkaline volcanics, begins on the sea-floor and continues till it reaches the African continent. The fundamental composition of the igneous assemblages actually does not change, except on land, where pantellerites occur together with moderately alkaline basalts (Liotard et al., 1982).

Tholeiitic basalts, andesites (icelandites) and rhyolites from hot-spots are hypersteene-normative, and the acid rocks often contain modal quartz. They differ greatly from MORB in their high TiO_2, iron and K_2O content and their low Al_2O_3 content. These differences in petrochemistry are not related to fractionation effects at low pressure and could be attributed to different magma sources for the two groups of basalts. In their K_2O content and REE distribution pattern, the tholeiitic basalts of volcanic islands and continental rifts are close to moderately alkaline basalts, therefore they are sometimes classified with the intermediate tholeiites.

Table 2.8 Magmatic series of back-arc basins. Average major and trace element composition of volcanics of Philippine Sea and Sea of Japan.

Components	Tholeiitic series	
	N-MORB	T-MORB
SiO_2	49.47	50.29
TiO_2	0.86	1.44
Al_2O_3	15.85	15.39
Fe_2O_3	1.18	1.10
FeO	7.79	7.26
MnO	0.16	0.17
MgO	8.45	7.62
CaO	13.06	13.15
Na_2O	2.72	3.24
K_2O	0.39	0.23
P_2O_5	0.06	0.11
Total	100.00	100.00
Mg	62.97	62.22
Cr	335	257
Ni	118	112
Co	41	50
Sc	35	35
Zn	72	71
Rb	5	2
Cs	0.20	0.07
Ba	6	13
Sr	80	135
Ga	18	18
Ta	0.06	0.15
Nb	1.0	1.0
Hf	1.18	2.26
Zr	43	90
Y	22	31
Th	0.07	0.15
U	0.06	0.10
La	1.82	3.12
Ce	5.05	7.74
Eu	0.75	1.16
Tb	0.48	0.69
$^{87}Sr/^{86}Sr$	0.70288	0.70232
$^{143}Nd/^{144}Nd$	0.51317	0.513154
$^{206}Pb/^{204}Pb$	17.91	
$^{207}Pb/^{204}Pb$	15.44	
$^{208}Pb/^{204}Pb$	37.62	

After: Initial Rep. of DSDP (1980), **58**, 789–914; (1981) **59**, 625–800; Cohen and O'Nions (1982) *J. Petrology*, **23**(3), 299–324; Sharaskin (1992).

Isotopic data show that oceanic island basalts (OIB) in different locations are different from each other and could be modelled by mixing depleted mantle and ancient (1–2 Ga) oceanic crust and sediments (White and McKenzie, 1989). However, high titanium, iron and alkali content in OIB shows that it is not enough to explain the situation in general, and the involvement of a specific material, non-typical of crust and upper mantle material, in the magma generation process seems to be obvious: mantle plumes could not originate only from the delamination of subcontinental and suboceanic lithosphere.

Rocks of the moderate- and high-alkaline series are members of hot-spot igneous assemblages; however, they occur in widely differing proportions in different areas. Their composition also varies widely, with the more acid products of the series incorporating trachytes, comendites, pantellerites and phonolites. In the character of their igneous assemblages, the alkaline rocks of oceanic hot-spots are similar to those of continental intra-plate areas, for example the East African Rift, including the occurrence of carbonatites (Cape Verde: Mazarovich et al., 1990).

It is also important to note that in both intra-plate continental and oceanic volcanic areas, moderate- and high-alkaline magmatism shows a bimodal compositional distribution for the intermediate members of the oceanic basalt–trachyte assemblages (Chayes, 1963).

Figure 2.17(a)–(d) presents the REE composition for volcanic rocks from different tectonic settings.

2.5 GENERAL FEATURES OF RECENT MAGMATISM

A correlation can be made between present-day magmatism and the geodynamic environment. However, the data summarized in Tables 2.1–2.7 show a wide variety of magmatism, even within a single geodynamic environment, together with a wide variation in the geodynamic environments. Therefore, it is important to consider the distinctive features of magmatism from a real environment, which has different geodynamic parameters as well as specific rocks of a single type.

Present-day igneous rocks include all known types, and even variations, with only a few exceptions (komatiites and kimberlites, etc.). MORB show the least variation in chemical composition. However, compositional variations in igneous rocks are wider for islands on oceanic ridges, where, apart from basalts, andesites (icelandites), dacites and plagiorhyolites also occur. Continental rift rocks display even wider variations in composition (see Table 2.6). The most diverse magmatism occurs on active plate margins (Table 2.4).

The distribution of magmatism in different present-day geodynamic environments enables them to be subdivided into elementary (or simple) and complex

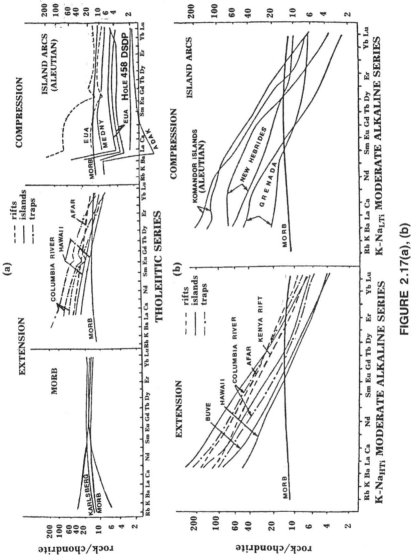

FIGURE 2.17(a), (b)

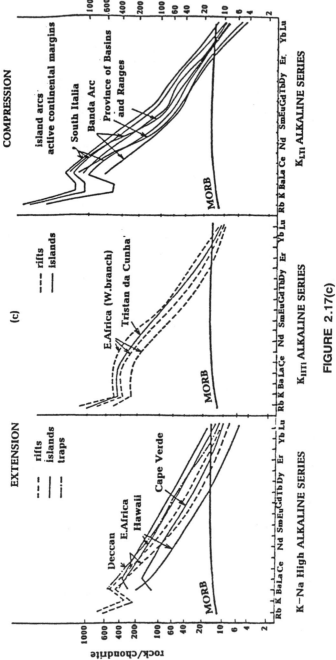

FIGURE 2.17(c)

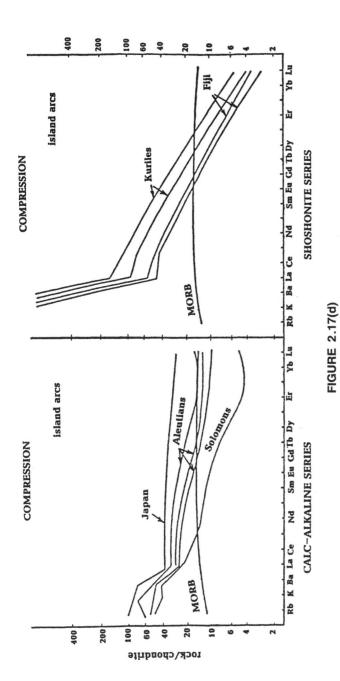

FIGURE 2.17(d)

FIGURE 2.17(a)–(d) REE distribution in igneous rocks from different geodynamic settings. After E.V. Sharkov and A.A. Tsvetkov (1985). Key: 1 — rifts; 2 — islands; 3 — trap basalts; all diagrams show the REE distribution in tholeiitic basalts of the mid-oceanic Gorda Ridge.
(a) REE distribution in rocks of the tholeiitic series
(b) in rocks of K–Na moderately alkaline series
(c) in rocks of K–Na and K alkaline series
(d) in rocks of the calc-alkaline and shoshonite–dacite series

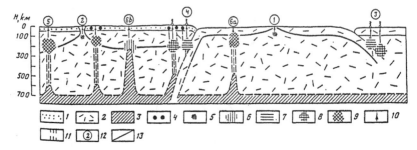

FIGURE 2.18 Correlation of magmatism and the simplest geodynamic environments. After Kovalenko *et al.* (1987). Key: 1–3 — sources of magma: 1 — continental crust, 2 — depleted upper mantle, 3 — non-depleted lower mantle; 4–9 — magma types: 4 — anatectic sialic (continental), 5 — tholeiite basaltic (rift), 6 — tholeiite basaltic intra-continental (trap), 7 — calc-alkaline of normal alkalinity (andesite), 8 — calc-alkaline of moderate alkalinity (shoshonite-latite), 9 — alkaline, kimberlite and carbonatite; 10 — direction of magma movement; 11 — metasomatic mantle reworking zones (enriched mantle); 12 — geodynamic environments (1 — oceanic rifts, 2 — continental rifts, 3 — intra-oceanic island arcs, 4 — active Andean-type continental margins with or without island arcs, 5 — collision of continents, 6a — intra-plate oceanic, 6b — intra-plate continental); 13 — lower boundary of lithospheric plates.

ones according to plate-tectonic environments (Bogatikov *et al.*, 1987). "Simple" environments cannot be further subdivided. They are formed when lithospheric plates are pulled apart, or by their collision, and also within plates. "Complex" environments result from the spatial superimposition of two or more simple geodynamic environments of the same age.

Generally these environments are as shown in Fig. 2.18.

2.5.1 Formulae for the Characteristic Magmatism of Young Tectonic Settings

Characterizing of the indicative magmatism of each geodynamic environment can be done using schematic formulae, including petrochemical assemblages such as the tholeiite (T), calc-alkaline (CA), moderately alkaline (A) and high-alkaline (HA) series. A more detailed description of the components of these series is as follows (see chapter 1): tholeiite series — T_{MORB}, T_{LTi} (low-titanium) and T_{HTi} (high-titanium) varieties; calc-alkaline series — CA_{HMg} (high-magnesian) and CA_{LMg} (low-magnesian); K–Na moderately alkaline series — $A_{K-NaLTi}$ (low-titanium) and $A_{K-NaHTi}$ (high-titanium); K moderately alkaline series — A_{KLTi} and A_{KHTi} consequence; K–Na high-alkaline series — $HA_{K-NaLTi}$ (low-titanium) and $HA_{K-NaHTi}$ (high-titanium); and K high-alkaline series — HA_{KLTi} and HA_{KHTi}.

a) Simple Geodynamic Environments

The concept of plate tectonics suggests the existence of the following simple geodynamic environments with characteristic types of magmatism.

1. Oceanic rift zones, comprising igneous rocks of the mid-oceanic ridges (MOR), series represented by tholeiitic basalts, gabbro and ultrabasic rocks (Fig. 2.19) that make up the oceanic crust; in fold belts the ophiolites are the equivalents of ancient oceanic crust.
2. Continental rift zones, characterized by basaltic and basalt–rhyolite (locally with comendite) volcanism, with an elevated titanium content and high alkalinity, alkali–ultramafic complexes containing carbonatites (see Fig. 2.19: environment 2). They have much in common with intra-plate magmatism and are often considered as a variety of the latter.
3. Oceanic plate-collision zones resulted in the generation of oceanic (young) island arcs (see Fig. 2.19: environment 3). This geodynamic environment is mainly characterized by igneous formations of the tholeiite and calc-alkaline series containing boninites. Developed and mature arcs are dominated by calc-alkaline igneous series; the low-Ti moderately alkaline and alkaline rocks, mainly of the potassic series, are subordinate.
4. Oceanic and continental plate-collision zones resulted in the generation of either "developed" island arcs underlain by continental crust or active Andean-type continental margins (see Fig. 2.19: environment 4). The magmatism is mainly of the calc-alkaline and potassic moderately alkaline series, but is very rarely from tholeiitic island-arc and high-alkaline series. Acid igneous rocks, including granitoid batholiths, are common at active Andean-type ACM. Basalts of the tholeiitic series occur in back-arc basins.
5. Continental plate-collision zones are characterized by magmatism that is very similar to that of the Andean margin, where calc-alkaline and potassic alkaline series are found; granitoid batholiths are also very common (see Fig. 2.19: environment 5).
6. Intra-plate environments (see Fig. 2.19: environment 6) including domes and extensive positive thermal and gravity anomalies (hot-spots). Their magmatism is represented by titanium tholeiitic, moderate- and high-alkaline series mainly of K–Na, less-common K groups, alkali–ultramafic rocks, kimberlites and, locally, acid rocks. In general, this type of magmatism is very similar to that of continental rifts (environments 2).

b) Complex Geodynamic Environments

The relationships between indicative magmatic series and geodynamic environments described above are superimposed on each other. In this case, complex geodynamic environments display mixed-type magmatism. Magmatism at the

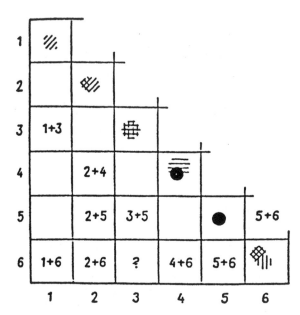

FIGURE 2.19 Correlation of magmatism of simple geodynamic environments and known mixed geodynamic environments. After Kovalenko *et al.* (1987). Key: Simple (1–6) and mixed (combination of simple, for example 1 plus 3) geodynamic environments. Symbols as in Fig. 2.18.

MOR, which combines tholeiitic mid-oceanic ridge basalts and the intra-plate magmatism of oceanic hot-spots, is an example of this type of environment. Thus this appears to involve the superposition of an intra-plate environment on a constructive plate boundary (Fig. 2.19). This superposition is confirmed in the case of Iceland, where the trace of the hot-spot on the plate moving over it can be detected as far away as Ireland (Fig. 2.20).

Island-arc and lack-arc basin systems (environments 1 and 3, Figs. 2.21 and 2.22), which contain back-arc spreading zones, with igneous rocks similar to MORB, can be overprinted by typical arc igneous rocks. These systems should be assigned to complex geodynamic environments. Volcanic islands and seamounts, which have basaltic intra-plate volcanism (for example, Mt. Vavilov in the Tyrrhenian Sea, Ullyndo Island in the Japan Sea, etc.), occur within back-arc seas. Often these volcanoes are located within island arcs, especially the mature Japan-type arcs, for example Mt. Fuji, Itinomegata and others.

The geodynamic equivalent characteristic of western North America in Late Cenozoic (California type, see Fig. 2.19, environments 2, 4 and 6, Fig. 2.23) is also considered to be complex. There we can see a spatial combination of calc-alkaline and locally tholeiitic magmatism of the active Andean-type continental

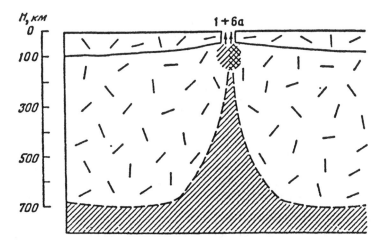

FIGURE 2.20 Combination of oceanic-rift environments and intra-plate environments with mixed magmatism (as exemplified by Iceland). Symbols as in Fig. 2.18.

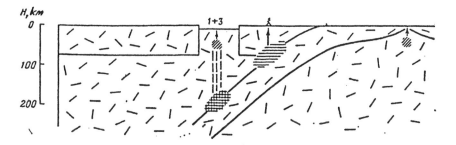

FIGURE 2.21 Combination of intra-oceanic island-arc and oceanic rift (inter-arc basin) environments with mixed magmatism. Symbols as in Fig. 2.18.

margin of the Cascade Range and other regions, with the continental-rift magmatism of the Columbia Plateau.

Undoubtedly, complex environments are very common in continental plate-collision zones, in particular in the Alpine part of the Alpine–Himalayan collision zone. There we can find a combination of andesite–latite collision arcs with typical intra-plate magmatism of fore-arc zones (e.g. the Massif Central (France) and Rhine Graben, among others). Mid-oceanic ridge basalts (the Tyrrhenian Sea) or titanium alkaline basalts in the Pannonian Basin have been recorded within back-arc areas here.

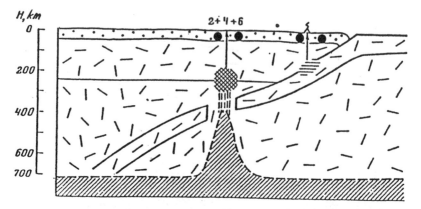

FIGURE 2.22 Combination of an active continental margin of Andean type, an oceanic rift and an intra-plate environment with mixed magmatism. Symbols as in Fig. 2.18.

c) Magmatic Formulae

1) Magmatism of Constructive Plate Margins

MOR magmatism. This is dominated by low-K (0.25 wt.% K_2O for N-type and 0.30% for E-type) tholeiitic basalts (Table 2.1). These basalts are relatively low in titanium (0.80–1.20 wt.% TiO_2), and depleted in potassium-group elements (Rb, Sr and Ba) and radiogenic strontium. MOR environments differ in the following indicative features. MOR with high spreading rates contain tholeiitic basalts close to N-type, enriched in Ti and Zr as compared with similar basalts of low spreading-rate ridges. E-MORB close to the composition of volcanic island tholeiites occur within submarine plateaux and hot-spots.

2) Magmatism of Intra-plate Environment

$T_{MORB\ (N,\ T,\ E)}$ is magmatic formula for MOR. Oceanic intra-plate magmatism, including islands on the slopes of mid-oceanic ridges in addition to intermediate Fe–Ti basalts, is characterized by rhyolites (the basalt–rhyolite bimodal assemblage), icelandites (basalt–thachyte–rhyolite assemblage) and also rocks of the Fe–Ti moderately alkaline series of the K–Na group (moderately alkaline basalt–trachyte assemblage, e.g. St. Helens and Gouth) to high-alkaline (nephelinite–tephrite assemblage, locally with carbonatites, e.g. the Cape Verde Islands) and rocks of the K group: trachybasalts, trachytes, phonolites, pantel- lerites and comendites (e.g. Tristan da Cunha and Kerguelen). $T_{HTi} + A_{K-NaHTi} + HA_{K-NaHTi}$ and A_{KHTi} is formula.

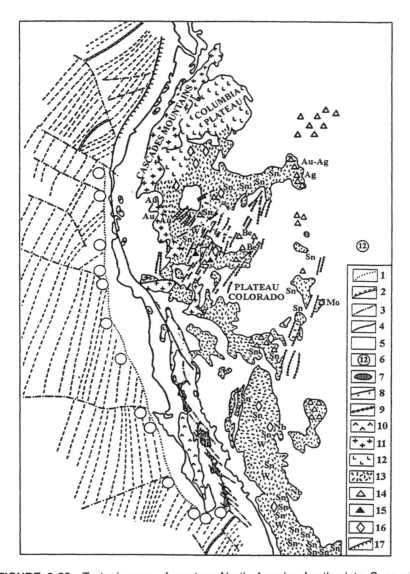

FIGURE 2.23 Tectonic map of western North America for the late Cenozoic (Kovalenko *et al.*, 1987). Key: 1 — boundary of continental slope (boundary of continent); 2 — spreading axes; 3 — transform faults; 4 — San Andreas Fault; 5 — banded magnetic anomalies in the ocean; 6 — age of the ocean floor (Ma); 7 — deep sedimentary trough of the Gulf of California; 8 — boundary of the Salton Trough; 9 — faults in the Basin and Ringe Province; 10–16 — igneous rocks: 10 — tholeiitic, 11 — calc-alkaline, 12 — alkaline basalt, 13 — contrasting basalt–rhyolite, 14 — alkaline, acid and intermediate (alkaline granites, comendites, syenites, latites and phonolites), 15 — carbonite, 16 — acid plumasite lithium–fluoric (topaz-bearing rhyolites and ongonites), 17 — subduction zone.

The magmatism of an aseismic ridge is similar to that of E-MORB: the basalts are sightly enriched in K (up to 0.82% K_2O) and lithophilic elements, more iron-rich and depleted in chromium. The paragenesis of the igneous rocks shows a certain similarity to those of Iceland, although there are some differences.

Continental-rift magmatism is fairly diverse (Table 2.6). In this geodynamic environment, various igneous rocks of moderate and high alkalinity K–Na and K groups, including ultramafic alkaline rocks and carbonatites, are common, along with tholeiite-series formations (Fe–Ti basalts). Continental rifts are characterized by bimodal basalt–rhyolite, moderately alkaline basalt–rhyolite–comendite, alkaline basalt–trachyte, trachybasalt–trachyte–phonolite, and other igneous suites. The general magmatic formula for this geodynamic environment is:

$$T_{HTi} + A_{K-NaHTi, KHTi} + HA_{K-NaHTi, KHTi}.$$

The similarity between the magmatism of this environment and that of oceanic islands is not surprising, because both are products of intra-plate magmatism.

3) Magmatism at Destructive Plate Margins

The main feature of this magmatism is the predominance of low-Ti rocks of the calc-alkaline and potassic moderately alkaline series. Rocks of the low-Ti tholeiitic, K–Na moderately alkaline and high-alkaline series also occur, but in different volumes. Rocks of the boninite series are unique to this environment. The shoshonite–latite series mainly characterize mature arcs, active continental Andean-type margins and collision zones. In the latter two cases, dacites and rhyolites are more common as compared with island arcs.

Igneous rocks of convergent plate margins differ from those of other environments, including back-arc basins, in:

(i)   their predominance of intermediate rocks (mainly andesites and latites);
(ii) the low degree of enrichment in iron and especially in titanium;
(iii) the importance of water and volatiles;
(iv) their specific REE composition; and
(v)   their Sr and Pb isotopic composition, suggesting the importance of crustal material in the petrogenesis of these rocks.

The general formula for magmatism of island arc, ACM and collision zones is:

$$T_{LTi} + CA_{HMg, LMg} + A_{KLTi, K-NaLTi} + HA_{KLTi, K-NaLTi}.$$

As mentioned above, island arcs are subdivided into young, developed and mature types based on their degree of similarity to continental crust (Fig. 2.24)

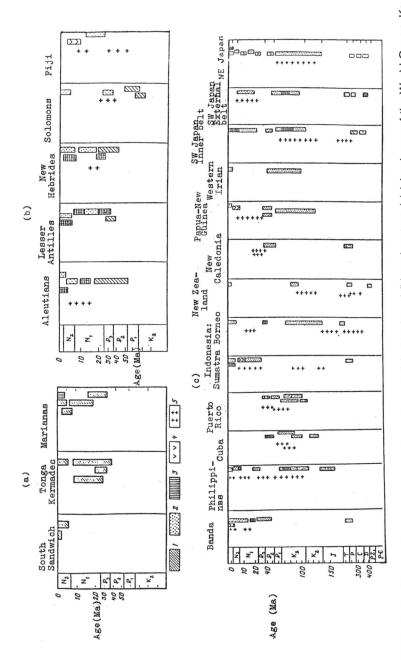

FIGURE 2.24 Schematic geological sections of young (a), developed (b) and mature (c) island arcs of the World Ocean. Key: 1–4 — igneous series: 1 — tholeiitic, 2 — calc-alkaline, 3 — K–Na subalkaline and alkaline, 4 — high-K subalkaline; 5 — periods of intrusive activity.

The formulae for classifying each of the island-arc groups are presented below:

Young island arcs:

$$T_{LTi} + CA_{HMg, LMg} + A_{K-NaLTi};$$

Developed island arcs:

$$T_{LTi} + CA_{LMg} + A_{K-NaLTi, KLTi};$$

Mature island arcs:

$$CA_{LTi} + A_{KLTi, K-NaLTi} + HA_{KLTi, K-NaLTi}.$$

Plate-collision zone magmatism is also dominated by rocks of the calc-alkaline series. These are mainly andesites, dacites, rhyolites and comagmatic granitiods, often with high $^{87}Sr/^{86}Sr$ ratios.

Unlike island arcs, moderate- and high-alkaline melts of the low-Ti K-group, from the shoshonite–latite assemblage to diverse leucite-bearing formations, including lamproites and phonolites, are fairly common there. This applies to both the continental arcs of the Anatolian–Caucasus–Elbursian type and to the island arcs, such as the Calabrian and the Aegean arcs in the west and the arcs of South-East Asia in the east of the Alpine–Himalayan Belt. Moderately alkaline volcanics of the K–Na series are less common and usually give way to intermediate and acid rocks.

The general formula of collision-zone magmatism is:

$$T_{LTi} + CA_{LTI} + A_{KLTi, K-NaLTi} + HA_{KLTi}.$$

4) Back-Arc Magmatism

Unlike island arcs, back-arc or marginal seas are characterized by MORB-type basaltic magmatism. The indicative formula of this magmatism is T_{MORB}; locally it is enriched in crustal material.

Table 2.6 shows the overall distribution of magmatic series in different geodynamic environments.

5) Magmatism of Complex Californian Geodynamic Environments

Despite the controversy surrounding the geodynamic environment, which resulted from the overriding of the East Pacific Rise by the active margin of the North American continental plate, the characteristic feature there is a combination of all types of magmatism characteristic of the above-mentioned environments: taking account of the Gulf of California, they are tholeiitic MORB assemblages, locally containing icelandites (Tortuga Island); calc-alkaline basalt–andesite–dacite–rhyolite assemblages; moderately alkaline assemblages (trachybasalts, trachyandesites, latites, rhyolite and topaz rhyolites); alkaline assemblages (basanites, trachytes, comendites, kimberlite- and carbonatite-like rocks, leucite-bearing rocks and phonolites) and K–Na moderately alkali basalt and tholeiite assemblages.

This magmatism shows an unusual zoning: within the continent, tholeiitic magmatism gives way to calc-alkaline magmatism, followed by basaltic (some researchers attribute it to back-arc spreading) and then by bimodal basalt–rhyolite magmatism. This sequence can be disrupted by local rift structures of the Rio Grande type, which have their own tectonomagmatic zoning. The magmatic formula for this environment is:

$$T_{MORB} + CA_{LMg} + A_{K, K-Na} + HA_{K-Na, K}.$$

2.5.2 Magmatic Zoning

Magmatic zoning within active structures can be both across and along the strike of plate boundaries.

a) Island Arcs

Transverse zoning of island arcs is represented by a gradual increase in the concentration of incompatible elements, especially alkalies, from the volcanic front towards the back-arc areas. A large body of data (Gill, 1981; Thorpe, 1982; Kay and Kay, 1990; etc.) shows that, as a rule, there was a lateral change from the tholeiitic series to the clac-alkaline series, then alkaline series (mainly potassic, rarely K–Na) till MORB in back-arc basins occur here. The abundance of K, Rb, La, Th and U, and the La/Yb and Rb/Sr ratios, always increase from the deep-sea trench towards the back-arc zone, whereas K/Rb, Ba/La and Zr/Nb decrease. There is also isotopic zoning of magmatism and zoning of ore deposits associated with igneous rocks according to the scheme: Fe — (Cu, Mo and Au) — (Cu, Pb, Zn and Ag) — (Sn, W, Ag and Bi) (Mitchell and Garson, 1981).

The early models of island-arc magmatism, based on data of the Japan Arc (Kuno, Dickenson and Hatherton, Sigimura), suggested that an increase in the magmas' alkalinity was linked to the depth of the subduction zone. However, in the Lesser Antilles, volcanoes are most depleted in potassium near the northern edge of the volcanic front, four highly potassic volcanoes (including Granada) lie on its southern edge, and the remaining volcanoes of the arc are intermediate in composition and their location provides an example of longitudinal (N–S) zoning (at an approximately constant depth of 100 km to the Benioff zone). The primary [87]Sr/[86]Sr ratio increases and varies southwards (0.7040 between

Guadeloupe and Dominica) while the Nd content is close to that of MORB in the north and decreases to the south. The transition from suboceanic to continental mantle takes place in the same direction (Gill, 1981).

Within the Kurile Arc, the K, Rb, Ba and Sr abundances in basalts, andesites and dacites from the northern part of the arc are 20–59% higher than those in the central part (Bogatikov and Tsvetkov, 1988).

Data available for island arcs and ACM enable four types of transverse (isotopic Nd and Sr) zoning to be recognized:

1. an increase in $^{143}Nd/^{144}Nd$ and a decrease in $^{87}Sr/^{86}Sr$ ratios, from the front towards the back-arc zone in northern Honshu (Nohda and Wasserburg, 1986);
2. an increase of $^{87}Sr/^{86}Sr$ and a constant $^{143}Nd/^{144}Nd$ for the Aleutian Arc (Tsvetkov, 1990);
3. no difference in the isotopic composition between individual structural zones in New Britain (De Paolo and Johnson, 1979);
4. a decrease in $^{143}Nd/^{144}Nd$ and $^{87}Sr/^{86}Sr$ ratios from the volcanic front toward the back-arc basin of the Kurile Arc (Zhuravlev et al., 1985).

The above examples apparently suggest a correlation not only between isotopic parameters and distance to convergent plate margins, but also with crustal thickness, the age of the crust and the upper-mantle composition. Data on Pb-isotopic zoning are less common; in general this zoning is similar to that of Sr. Therefore in northern Honshu the $^{206}Pb/^{204}Pb$ ratio decreases away from the fore-arc zone, not only for Quaternary but also for Miocene volcanics (Gill, 1981).

b) Zoning in Structures of Other Types

Longitudinal and transverse zoning of magmatism was also recorded in other geodynamic environments. For example, in Iceland, located on a dome of the Mid-Atlantic Ridge, basalts in the interior of the island are very similar to intraplate basalts and only gradually become similar to MORB along the axial rift zone oceanwards (Basaltic Volcanism, 1981).

Continental rifts are characterized by a distinct longitudinal and transverse zoning. The distribution pattern of igneous rocks, as reflected in the Ethiopian–Kenyan rift system, provides the best example of longitudinal igneous zoning probably associated with a change of rift type. From north to south there is a sequence of igneous assemblages, from tholeiitic basalt with poorly developed alkaline comendite in the Afar rift valley, through moderately alkaline olivine basalts containing pantellerites in the main Ethiopian valley and alkaline basalts with trachytes in the Kenyan Rift, to high-alkaline basanite–nephelinite lavas in the southern Kenyan Rift (Almond, 1986). The transverse zoning in this rift zone is represented by a transition from high-alkaline assemblages of the plateau occurring on the rift flanks, to less alkaline in the rift valleys themselves. A certain type of transverse zoning may occur within the entire rift system. For example, in East Africa, the eastern branch (Kenyan Rift) and western branch (Tanganyika Rift) are characterized respectively by K–Na and K alkaline magmatism; in the Baikal Rift Zone potassic magmatism is typical of the Khubsugul Rift, while its remaining part is dominated by K–Na igneous assemblages.

2.5.3 The Evolution of Young Magmatism

The evolutionary trend in the change in igneous rock composition can be easily found in all young tectonic settings, except for mid-oceanic ridges. For example, on islands such as Iceland, one can trace the evolution in basalt composition from Miocene tholeiitic flood basalt eruptions to recent rift complexes. During the Quaternary, the compositionally homogeneous Miocene flood basalts gave way to basaltic melts with a wide range of composition, from N-MORB to moderately alkaline basalts (Imsland, 1983; Gibson and Gibbs, 1987).

As shown above, two stages can be recognized in the development of continental rifts (Grachev, 1987). The early pre-rift stage is characterized by an overall uplift of the area, without a clear relationship between volcanism and rift structure. Eruptions of weakly differentiated lava series, often represented by homogeneous lava flows of moderately alkaline basalts or tholeiites (e.g. the Baikal Rift and the Ethiopian Rift) and in other cases by phonolites (Kenyan Rift), are typical for this stage. As a whole, it resembles a continental flood basalt province. The second stage of continental-rift evolution is marked by more intensive tectonic activity, resulting in the formation of rift basins and mountainous topography. Volcanic activity is limited mainly to the rift basins or takes place within them. A wide spectrum of magma composition characterized this stage, often under the domination of alkaline basalts. Fissured eruptions of weakly differentiated alkaline volcanics were accompanied by the formation of central-type volcanoes responsible for the generation of differentiated rocks.

Age trends in the pattern of magmatism are clearly reflected in the vertical succession of volcanic products. For example, in the eastern branch of the East African rift system, this trend is reflected in time-dependent changes in the composition of the Eocene–Oligocene tholeiitic assemblages towards moderately alkaline basalts. There is a correlation between the change in composition and change in nature of the eruptions, from fissured to central type. Within the Red Sea Rift, there is a spatial and temporal change in composition (from the edge inwards) from mainly alkaline–basaltic to tholeiite basaltic volcanism of the transitional type (Afar). Tholeiitic basalts of MORB type were recorded only in the axial part of the rift (Almukhamedov et al., 1985).

If the above-mentioned variations in composition reflect general trends in the development of rift magmatism in time and space, the pattern of evolution of the separate magma chambers is displayed in the structure of differentiated igneous series related mainly to central-type volcanoes. The general sequence of compositional variations in these volcanoes is a replacement of basic by younger acid rocks. Different rift zones can provide examples of such series; their evolutionary trend actually is not affected by the composition of the initial melt. In Afar, in the Erta Ale Ridge and in the Boina volcanic centre, we can see the following sequence: moderately alkaline basalt (hawaiite) — moderately alkaline ferrobasalt–mugearite–trachyte–comendite–pantellerite with the abundance of rocks decreasing in the same direction (Barberi et al., 1975).

The Main Ethiopian Rift zone is dominated by mildly alkaline basalt–pantellerite–comendite volcanic rocks also containing transitional rocks (mugearite, benmoreite and trachyte), i.e. they occupy a position between earlier alkaline basalts and pantellerites; they are limited in volume, hence the distinct bimodal pattern of assemblages (Brotzu et al., 1974). The geochemistry of the assemblages fits the fractionation model, and their bimodal structure is due to the melting of the crustal material around the intermediate reservoirs by which magma was brought to the surface (Yarmolyuk and Kovalenko, 1991). A similar structure typifies the Kenyan Rift volcanoes, which erupted alkaline basalts and trachytes. Their final differentiates are represented by small comendite stocks.

A different differentiation trend is characteristic of volcanic centres composed of alkaline rocks and carbonatites. They were recorded in the Rhine Graben, but are more common in the Kenyan Rift. The stage of main cone formation was dominated by the eruption of nephelinites and of phonolites accompanied by agglomerates and tuffs of the same composition. Nephelinite and carbonatite flows (Oldonyo Lengai Volcano) and carbonatite agglomerates, tuff and ash were generated at the final stage.

Structures of hot-spot type, not related to rifting areas, show a similar pattern of evolution. The most typical is the compositional variability, due apparently to intra-chamber differentiation. This is reflected in a sequence of eruptions from older, moderately alkaline basalts to younger sialic and alkaline products. In Africa (Ahaggar and Djos plateaux), there are phonolite and trachyte stocks and domes capping basaltic plateaux (Tibesty), and in eastern Australia there are trachytes, trachyrhyolites, comendites and pantellerites generated during the final stages of the development of large shield volcanoes. In the Hawaiian Islands, incipient magmatism is alkaline-basaltic, and there are tholeiite basalts (shield volcano stage, the largest in volume), which, during their evolution, give way to alkaline and high alkaline rocks (Decker et al., 1987). Continental rift and hot-spot moderately alkaline basalts and nephelinites often contain mantle xenoliths: mainly spinel lherzolite and less commonly garnet lherzolite and websterite, which could be fragments of asthenospheric material.

The evolution of magmatism at destructive margins has been discussed (Gill, 1981; Thorpe, 1982; Bogatikov *et al.*, 1987; etc.). The best known is the evolution of the calc-alkaline series, within the scope of a single volcano, an individual island arc or part of an arc. Eruptions of different individual volcanoes over a period of tens of years may show a different evolutionary trend:

1. from more basic to acid rocks;
2. from more acid to basic rocks;
3. synchronous eruption of rocks of various composition in the form of physical mixtures (in tephra) of unrelated magma types;
4. persistent (constant) composition.

Longer time intervals in magmatic evolution are reflected in stratovolcanoes (hundred of thousands of years). A first and second trend and a cyclic trend are reflected in each andesite cycle. The standard pattern of magmatic evolution has not yet been established for all of the andesitic volcanoes.

The most common trend of changes in the melt composition for long time intervals is from basic to acid varieties. It could well be attributed to fractional crystallization in transitional magma chambers beneath volcanoes, and probably to partial assimilation of crustal material by deep-seated magmas. It is notable that there can be several such transitional (intermediate) reservoirs. For example, in Japan, such reservoirs occur at depths of 12–17 km and 23–37 km beneath Mt. Morieshi (Hori and Hasegawa, 1991); beneath Krakatoa in Indonesia the edge of the upper reservoir occurs at a depth of 9 km and the lower reservoir occurs at the depth of 22–30 km (Harjono *et al.*, 1989). Often near-surface magma chambers occur at depths of 1–5 km: they are probably responsible for catastrophic eruptions (Slezin, 1991; Sharkov, 1992).

Owing to the difference in total pressure, the evolution of magma in each chamber follows its own specific pattern, complicated by irregular refilling from the lower chambers and from the area of magma generation. The magma supply from these reservoirs and the evolution in each reservoir have an irregular pattern; the complex evolution pattern of melts brought to the surface is also apparent. Island arcs as a whole, and their separate parts, also evolve over time. The evolutionary maturity of ensimatic island arcs can be clearly determined by the relative position of igneous assemblages of the calc-alkaline series, occurring stratigraphically above rocks of the initial tholeiite series, similarly to those of young arcs. The presence of ultrabasic xenoliths, whose petrological–geochemical characteristics are similar to those of ophiolitic assemblage rocks, in basalts and andesites of "developed" arcs suggests that they are primary ensimatic in nature. The developed arcs are composed of calc-alkaline and shoshonite–latite assemblage rocks. The formation of the latter is often attributed to the existence or emplacement of a newly generated granite–metamorphic layer of the Earth's crust, which greatly affects the composition of mantle magma. Later, shoshonite and alkaline melts were subject to differentiation in near-surface magma chambers.

It is common for low-titanium, K–Na, moderately alkaline basaltic magmas to be generated during the late stages of evolution of developed arcs; however, sometimes they precede the formation of the calc-alkaline series (Tsvetkov, 1990). Therefore, we can recognize standard and non-standard types of igneous evolution; respectively they will be consistent and inconsistent with the evolution scheme of Miyashiro (1974).

The replacement of some igneous series by others, which is easily discernible in young and developed arcs, does not occur in mature island arcs. Mature arcs contain rocks of all known magmatic series, but their proportions vary considerably. Igneous rocks of mature arcs consist mainly of calc-alkaline series. The proportion of plutonic assemblages of tonalite–granodiorite–granite type and high-alkaline rocks, such as syenites, monzonites and their comagmatic acid volcanics of the shoshonite series, together with volcanics from the potassic alkaline series, tends to be higher. The evolution of island arcs, on the one hand, led to the accretion of sialic crust, due to the supply of juvenile igneous material, and, on the other hand, differentiation continued, including anatectic melts and remobilization accompanied by an increase in metamorphic rocks. The growth of the crust as whole, and of the separate "granitic" and "basaltic" layers, was actually continuous. Table 2.9 shows the characteristic igneous assemblages from the whole range of geodynamic environments.

2.5.4 Magmatic Evolution during the Emplacement of Major Geological Structures

Magmatic evolution occurs at two levels: in magmatic centres and within a regional framework. In the first case, its development follows a known, regular physicochemical trend, i.e. it is a sequence of eruptions from less-differentiated magmas, similar to the primordial magmas, to extreme differentiates whose composition is determined by that of the initial melt (and probably by a degree of contaminated material). For tholeiitic and calc-alkaline magmas, these are rhyolites; for the moderately alkaline magmas of the K–Na group — trachytes, comendites, pantellerites and ongonites, and of the K-group — latites, trachydacites and trachyrhyolites; and for alkaline magmas — nepheline and leucite phonolites. This general sequence can be disrupted when particular members of differentiation series do not fully manifest themselves, or due to repeated eruptions of differentiated melts injected into the magma chamber from the magma generation zone, or in the case of magmatic mixing; however, for the general case, this trend is pretty persistent.

In the regional framework, the situation is different. To a greater extent it is determined by the type of magmatism. On the one hand, it varies within the

Table 2.9 Series of igneous rocks of main young tectonic settings.

Igneous rock series	At lithospheric plate boundaries			Intraplate	
	Constructive	Destructive		Continental rifts	Oceanic rifts
		Islands arc, ACM, collision	Back-arc seas		
Tholeiite					
MORB	+ +	—	+ +	—	—
low-Ti (T_{LTi})	—	+ +	+	—	—
high-Ti (T_{HTi})	+	—	+	+ +	+ +
Calc-alkalic					
high Mg (CA_{HMg})	—		—	—	—
low Mg (CA_{LMg})	—		—	—	—
K–Na moderately alkaline					
low Ti ($A_{K–NaLTi}$)		+ +	+ +	—	—
high Ti ($A_{K–NaHTi}$)	+	+ +	+ +	+ +	+ +
K-moderately alkaline					
low Ti (A_{KLTi})	—	+ +	+ +	—	—
high Ti (A_{KHTi})	+	—	+	—	+
K–Na high-alkaline					
low Ti ($HA_{K–NaLTi}$)	—	+	—	—	—
high Ti ($HA_{K–NaHTi}$)	—	—	—	+ +	+
K-high-alkaline					
low Ti (HA_{LTi})	—	+ +	—	—	
high Ti (HA_{HTi})	—	—	—	+ +	

+ + — most common series types indicative of a given setting

+ — series rare if any in a given settting

basic series related to spreading zones and areas of intra-plate magmatism, and on the other hand, to divergent margins. The example of the Red Sea Rift shows that magmatic evolution in time and space proceeds from high-titanium K–Na, moderately alkaline series to the tholeiite series of MORB type, characteristic of newly formed sea-floor spreading zones. Comendite–pantellerite ignimbrites are very common at the initial stage of structural development.

A similar sequence of magmatic evolution is found during the formation of back-arc basins, for example the Japan Sea (Tamaki, 1988; Frolova *et al.*, 1989). Its development began about 30 Ma ago, in the form of a continental-margin rift in the rear part of the newly formed active margin of Eurasia (Amur microplate). As in the case of the Red Sea Rift, subaerial K–Na moderately alkaline basalts and ignimbrites ("green tuffs") were the predominant rocks. The relics of this continental crust were preserved beneath submarine rises such as the Yamato, Oki and Korean rises. The submergence of the region was accompanied by extension, starting in the early Miocene and intensifying during the middle and late Miocene, with the formation of a marine basin and the eruption of tholeiitic basalts, including MORB, and generation of oceanic crust took place in the newly formed deep-sea basins, for example in the Yamato Basin. Thus, the evolution of deep-seated melts, from Fe–Ti moderately alkaline to tholeiite with the generation of MORB in the more developed regions, takes place by the opening of both new oceans and back-arc basins. This tendency is exhibited on some large oceanic islands, for example in the early stages of the development of Hawaii, although MORB have not been generated there. This type of evolution could be attributed to melting of the ultrabasic substrate of upper-mantle asthenospheric plumes; the general evolution trend is towards oceanization of the crust.

For destructive plate margins, magmatic evolution follows quite a different trend. The emplacement of island arcs begins with the eruption of island-arc tholeiitic basalts (and in some cases, boninites), which later give way to rocks of the calc-alkaline and low-Ti moderately alkaline (K and K–Na series) and even the potassic high-alkaline series. The latter commonly terminate the development of this type of magmatism. Within ensialic arcs developed on continental crust, such as the Japan and Kurile–Kamchatka arcs, magmatism, as in back-arc basins, was initiated with the generation of acid ignimbrites ("green tuffs"). The general trend of evolution is towards continentalization of the crust.

Available data suggest the presence of vast tectonomagmatic systems, which are at different stages of their development. This give us an idea about the processes operating there and about the character of the associated igneous activity. Therefore, in a single area, we can observe the initiation of ocean opening (the Red Sea Rift), recently opened young oceans (Atlantic and Indian oceans) and the active collision zone occupying the site of the Tethys, which closed in the late Cretaceous–early Palaeogene (the Alpine–Himalayan orogenic belt), where stacking of nappes and intense crustal reworking took place.

Therefore, in the Atlantic–Eurasian segment, we can recognize three main stages in the evolution of the Earth's crust, from ocean opening to ocean closure: initial, oceanic–continental and continental, where the simple and complex geodynamic environments are the most diverse (Bogatikov et al., 1987b).

In the Pacific segment the situation is more complex. Available geological data show that this ocean existed during the whole of the Phanerozoic and probably in the Late Precambrian (Pushcharovsky and Melanholina, 1989). The Earth's crust there, unlike the Atlantic–Eurasian segment, developed by suturing of island-arc formations to continental during the collapse of back-arc basins. The evolution of island arcs proceeds from young through developed to mature, and the composition of the rocks approaches that of continental crust. However, the ocean continues to exist and there are no collision structures of Alpine–Himalayan type emplaced during ocean closure.

2.6 MAGMATISM AND GEOLOGICAL CATASTROPHES

Catastrophe (Greek for "turn" or "destruction") is a term denoting sudden events that necessitate an abrupt change in, or the destruction of disappearance of, some inanimate or animate object. Historically, this concept was applied to events connected with human beings and their environment. The scope of its usage subsequently expanded: references were made to catastrophes unrelated to human activity and to catastrophes affecting mechanisms and constructions, as well as to cosmic, geological, social and other catastrophes.

Over the past decade, catastrophic events have attracted the attention of mathematicians. A mathematical theory of catastrophes was created by R. Tom, V.I. Arnold, T. Poston, I. Stewart and others. Essentially, the mathematical approach identifies variations of parameters, within which a given event (continuous function) experiences an interruption and shifts from one stable state to another. In this case, the function is the object of the catastrophe, and the values of the parameter corresponding with its disturbance represent a catastrophic event. The special place that they hold among other natural phenomena is due, above all, to their shock effect, which eventually replaces some natural state (or system) with another.

2.6.1 The role of catastrophes in geology

The concept of the object of a catastrophic impact is of key importance in the application of catastrophe theory to the study of natural phenomena. This concept stems from the identification of dynamically stable systems with certain spatial parameters and with a life span greatly in excess of the duration of the transition between them. In studying geological processes, including catastrophic ones, researchers must always remember the relative nature of geological time and the low accuracy with which the time and place of a particular event can be assessed in the geological past. For example, when analysing ongoing geological processes, researchers think in terms of years, months, hours and even seconds (in volcanology and seismology) and pinpoint the spatial manifestations of these processes.

When one deals with the earliest stages of geological history, the available dating techniques make it possible to evaluate events with a margin of accuracy of tens and even hundreds of millions of years, and the significance and non-accidental nature of such events are determined by their connection with major geological structures. Therefore, geological methods can be used to define a broad range of geologically stable states of systems (from strictly local ones with a duration recorded by a conventional time scale, to global ones lasting for tens and even hundreds of millions of years), which may have been the object of catastrophic transformations. Widely different geological structures, whose parameters change little as these structures form, provide a good example. They can comprise a particular volcano, mountain range, island arc, ocean, continent, and even complexes of geological formations inherent in a particular geodynamic environment, stage in the development of the Earth's crust, etc. Actually any geological object that a researcher singles out on the basis of the stability of its inner characteristics can be regarded as such a system. The above remarks make it possible to view geological catastrophes as relatively short-term processes or events (in terms of geology) that temporally divide long-term and stable geological states. The different components of such events, including magmatism, are seen as geological catastrophe factors.

Like any natural process, magmatism *per se* cannot be a catastrophe. Its manifestations are catastrophic only for certain natural systems. This viewpoint governs approaches to the identification of catastrophic magmatism in geological history. On the one hand, magmatism operates as an agent of catastrophe in relation to a particular natural process or stable natural state, by sharply changing this process or state (or forming part of a more general cause). On the other hand, magmatism can be interpreted as a dynamic system, whose drastic disturbances of its structural stability reflect catastrophes in the system's existence. In particular, catastrophes may be caused by local igneous processes.

Let us outline the range of objects whose state changed drastically under the exclusive or partial impact of magmatism in the geological past. They include the biosphere, the atmosphere and atmosphere-related climatic conditions, and the lithospheric outer layers of the planet. Let us try to evaluate the states of these environments whose changes, to varying degrees, are connected with magmatism and can be regarded as catastrophic.

In mathematical theory, natural catastrophes are mostly described by the geometrical form of a fold, while the transformation from one state into another can be easily pictured as a leap from one step of a staircase to another step. Therefore, for the sake of graphical convenience, stable natural states will be represented by horizontal planes.

2.6.2 Biospheric catastrophes

Boundaries at which massive extinctions of living organisms occurred have long been known in the geological record (Fig. 2.25). The Cretaceous–Palaeogene boundary is the best studied of these; it marks the extinction of many organisms that thrived throughout the Mesozoic. Geological correlation shows that a short-lived (about 500,000 yr) outbreak of basalt magmatism took place at this boundary (Courtillot et al., 1988; McCartney et al., 1990). Specifically, it resulted in the formation of the Deccan traps, totalling some two million cubic kilometres of volcanic rocks.

There are ample grounds to believe that other manifestations of trap volcanism may also have caused catastrophic extinctions of living organisms. This can be inferred from the usual temporal coincidence of trap eruptions with extinction events and from the absence of other large-scale manifestations of igneous activity at relevant stages of geological history. Specifically, the enormous volumes of trap volcanics of the Siberian Platform were established to have formed in the northern Tungusska syncline over a period of about one million years, 247 Ma ago (Renne and Basu, 1991), i.e. at the Permian–Triassic boundary, marking the catastrophic replacement of the Palaeozoic fauna with Mesozoic forms (see Fig. 2.25).

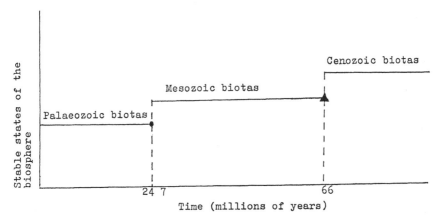

FIGURE 2.25 Changes in the biosphere at geological boundaries. Age position of the Siberian (circle) and Deccan (triangle) traps.

However, the accidental nature of these events gives no indication of the factors that caused the drastic environmental change. Trap volcanism, in the form of lava outpourings, is an extremely large-scale, yet spatially isolated, event. Aside from any assessment of its role, it can be described as only one of the parameters that radically and globally changed the balance of a dynamically stable ecosystem. The release of a large volume of gases with a high sulphur content in the course of trap eruptions is assumed to be of great importance (McCartney *et al.*, 1990). These gases were poisonous to living organisms and, apparently, affected the optical properties of the atmosphere and, through them, the climate. In the final analysis, these gases could have been the cause of global extinction.

In application of catastrophe theory to the analysis of environmental variability produces an unexpected conclusion: viewed from a formal angle, the peak flourishing of organic life that corresponds with a drastic change in the previous state of the biosphere should also be regarded as a natural catastrophe. The cause of such catastrophes may also include magmatism. Particularly telling in this regard was the emergence and rapid expansion of hard-shelled invertebrates which took place at the late Riphean–Phanerozoic boundary (Fig. 2.26). A major

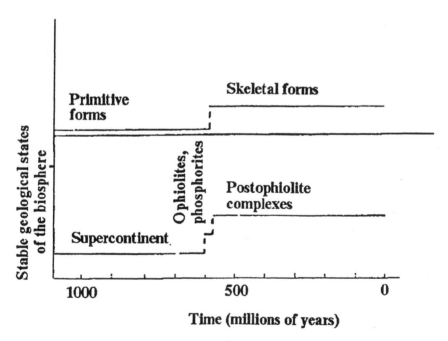

FIGURE 2.26 Correlation of a sequence of geologically stable states, with changes in the species composition of the biosphere at the Precambrian–Phanerozoic boundary.

geological event in the Earth's history — the break-up of the late Precambrian supercontinent and the formation, between its fragments, of numerous basins filled with igneous rocks of the ophiolite complex — may have been a causative factor in this explosive adaptive radiation.

The peak of ophiolite magmatism occurred somewhere between 600 and 540 Ma. Correlated with these ophiolites are phosphorites, for which this episode was among the most important in geological history (Bykhover, 1984). Phosphorites accumulated on the continental shelves that bordered basins underlain by ophiolitic basement (Byamba, 1991). The initial melts of lavas in ophiolite complexes are basic in composition and are usually rich in phosphorus. Therefore, like Academician N.S. Shatsky, we can assume that it was the source of this phosphorus. Phosphorus is a component of skeletal material and of chitin and it may have stimulated the appearance of hard-shelled organisms. Therefore, magmatism might have been a causal factor behind the rapid development of new life forms at the beginning of the Phanerozoic. Magmatism initiated one of the greatest catastrophes (in terms of catastrophe theory) in the history of the Earth's biosphere: the advent of life forms that came to dominate the subsequent history of the planet.

2.6.3 Catastrophic changes in the Earth's atmospheric composition

Changes in the Earth's atmosphere that occurred in the geological past can be traced only indirectly because we have yet to develop techniques that would enable us to reconstruct its quantitative composition accurately. Such disturbances are borne out by global changes in living material, whose stability depends primarily on the composition of the atmosphere, and on the climatic conditions that this composition creates. Volcanic activity is the only geological process that can radically change the composition of the atmosphere. Let us take the example of the 1883 eruption of Krakatoa. In the course of the eruption, fine dust reached the stratosphere and enveloped the planet, causing unusually bright twilight and sunsets throughout the world. It was several years before this fine dust filtered down to the surface from the upper layers of the atmosphere. Over large areas of the earth, partially screened solar radiation led to a decrease in the annual average temperature (Bullard, 1980).

Apart from ash, volcanic eruptions produce enormous volumes of gases (H_2O, CO_2, CO, H_2S, HCl, HF, NH_3, etc.), whose content accounts for several per cent of the solid products of the eruption. Climatic changes depend on the composition and volume of liberated gases. Let us remember that nitrogen dioxide impairs thermal radiation of the Earth; hence, an increase in its concentration in the atmosphere can cause the planetary temperature to rise. Conversely, aerosols containing sulphur dioxide impair solar radiation and lead to a decline in the atmospheric temperature.

At this point, let us refer to a relatively small-scale (about $0.6 \, km^3$ of ash) eruption of El Chichon in April 1982 (Rampino and Self, 1984). About 3.3 million tonnes of sulphur dioxide were discharged into the stratosphere to heights of 20 to 26 km. Within a month, this gas cloud formed an elongated strip 20° wide (5°–30°N). Consequently, the average monthly temperature at 17°N fell by 0.2°C as early as June 1982. The change in the vertical temperature profile disrupted atmospheric circulation.

Mitchell (1970) has studied the impact of volcanic activity on planetary temperatures. Judging by his results, it can be concluded that volcanic dust and gases from major eruptions spread virtually throughout the atmosphere, causing stable temperature changes on the Earth's surface (Fig. 2.27).

There is a good reason to believe that the latest Ice Age, which began 73,000 years ago, was initiated by the Toba Volcano of Sumatra: the eruption took place 73,500 years ago. As a result of this natural catastrophe, some $2.8 \times 10^3 \, km^3$ of volcanic pumice and tuff were deposited on the Earth's surface. Volcanic ash has been reported from marine sediments and Antarctic glaciers, and this has enabled Rampino and Self (1992) to link the eruption with other natural events (Fig. 2.28). Specifically, it was during this period that sudden change in the isotopic composition of oxygen, a fine-tuned indicator of sea-water temperature, was recorded in marine biogenic carbonates. For that same time interval, the bottom sediments of the North Atlantic showed a marked increase in ice-rafted material, implying an increase in the ice cover of the planet. At that time, ice fields in Canada spread south down to 50°N. Rampino and Self (1992) ascribe these climatic changes to the consequences of the Toba eruption, i.e. to the formation, at a height of 27–37 km in the stratosphere, of a cloud containing about $10^{16} g$ of H_2SO_4 aerosol and approximately $2 \times 10^{16} g$ of very fine dust. This aerosol–dust cloud dramatically changed the optical characteristics of the atmosphere, and average yearly temperature fell by three to five degrees: effectively a "volcanic" winter set in.

Eruptions on a similar and even much larger scale have occurred many times throughout the Earth's history. Thick layers of acid tuff and ignimbrites are very common in the geological sections of ancient volcanic belts. As a rule, ignimbrite volcanism is related to caldera volcanoes such as Toba and Krakatoa. The large volume of these beds ($10,000–100,000 \, km^3$) covers vast areas without any signs of internal unconformities, suggesting extremely intensive volcanic activity of explosion type. Under certain circumstances, the cumulative effect of such eruptions may have caused substantial climatic changes, with catastrophic consequences for the biosphere.

The aggregate effect of volcanic gases on atmospheric composition and on climate depends on the intensity of their discharge into the atmosphere. If they are discharged at a rate higher than that of their assimilation by geological, hydrogeological and biological systems, these gases will accumulate in the atmosphere causing abrupt (catastrophic) changes in the environment and climate.

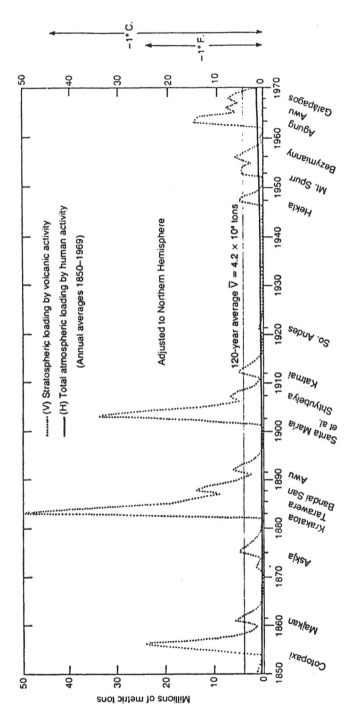

FIGURE 2.27 Fluctuation of average yearly northern Hemisphere temperatures due to atmospheric emissions of volcanic (dotted line) and man-made (solid line) ash and dust.

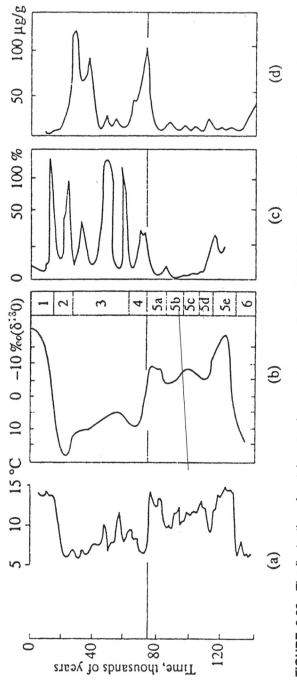

FIGURE 2.28 The fluctuation of certain natural parameters over the past 120,000 years. Key: (a) summer sea-surface temperature; (b) isotopic composition of oxygen in pelagic carbonates, the stages of oxygen scale are in the right-hand column; (c) percentage of ice-rafted debris in North Atlantic Sediments (data from the Me 69-17 deep-sea borehole); (d) aluminium content in Antarctic ice (Vostok Station area), registering discharges of atmospheric dust. The horizontal line marks the time of the Toba eruption.

Many researchers believe that the accumulation of volcanic gases in the atmosphere produced marked effects during the period of trap volcanism at the Palaeozoic–Mesozoic and Mesozoic–Cenozoic boundaries; both were major stages in the development of organic life.

The late Palaeozoic period of carbon formation, when most coals were formed (Bykhover, 1984), led to the extraction from the atmosphere of an enormous volume of carbon, while not changing the environment in which the coal-forming plants were growing. The source of all this atmospheric carbon is debatable. Probably continental magmatism provided such a source. During the Earth's history, the scale of continental magmatism was greatest during the late Palaeozoic, when extensive volcanic belts were formed. The largest of them ran through Eurasia from the British Isles to the Pacific coast, a distance of more than 14,000 km (Yarmolyuk and Kovalenko, 1990). The volume of erupted products within the area totalled on fewer than several million cubic kilometres. In the Central Asian section of the belt alone, stretching for about 2,000 km, more than 0.5 million km^3 of volcanic rocks were formed. Volcanic gases amounted to about 1% of all overall volume of eruption products. The composition of the gases was dominated by H_2O and CO_2 and their weight ratio varied widely. Even if we proceed from deliberately low estimates ($1 \times 10^6 \, km^3$) of volcanic products, and CO_2 comprising 10% of the gases (Giggenbach and Le Guern, 1976), it follows that more than $2,500 \times 10^9 \, t$ of CO_2, or about $700 \times 10^9 \, t$ of carbon, were discharged into the atmosphere. This figure is close to that of the late Palaeozoic coal reserves of the planet. Therefore, continental volcanism could have been responsible for the discharge of large amounts of gaseous carbon compounds into the atmosphere, resulting in the flourishing of plant species.

2.6.4 Catastrophic changes in the Earth's lithosphere

The lithosphere reaches a crisis point, above all, when new lithospheric plate boundaries are formed. In the interval between such successive events, the overall state of the lithosphere can be regarded as structurally stable. The transition from one stable state to another, provided this transition is relatively brief, fits the definition of a catastrophe in catastrophe theory. Often such a transition is accompanied by igneous processes localized in the rift zones of continental plates. There are some zones where plates diverge and others where they converge.

Divergent boundaries usually form within continental plates which are gradually eroded until they are fragmented and new (oceanic) crust is formed. The break-up of Gondwanaland provides an example of this type of boundary.

In the early Jurassic, Gondwanaland was divided into an eastern and western part, with what is still the Indian Ocean between them. In the area that linked

Africa and Antarctica, the break-up (at 170 Ma) occurred immediately following a period of intense trap volcanism of the Karoo system and bimodal rift volcanism of the Lembombo flexure (Forster, 1975; Powell *et al.*, 1988). The subsequent fragmentation of Gondwanaland took place mainly during the Cretaceous. Some 130 Ma ago, the separation of the African and South American continents and the formation of the ocean basin of the South Atlantic began. This coincided with the emplacement of the Parana and South African traps (with an average age of 131 ± 9 Ma) (Rocha-Campos *et al.*, 1988). The break-up of East Gondwanaland into India, Australia and Antarctica began at about the same time. The process lasted for 34 Ma and was accompanied by the development of extensive troughs along the belts of plate fragmentation. It should be emphasized that there was no significant synchronous igneous activity spatially related to the break-up. Furthermore, let us stress the separation of a continental block, marked by the Seychelles, from India, which occurred at the Cretacous–Palaeogene boundary (66 Ma ago). This process was preceded by an outburst of trap magmatism, which resulted in the formation of the Deccan traps and the plateau immediately east of the Seychelles (Courtilot *et al.*, 1988; McKenzie and Bickle, 1988). The main volume of lava flows was erupted over a period of approximately 0.5 Ma (McCartney *et al.*, 1990). Trap magmatism ceased after the separation of continental blocks and the formation of an oceanic basin with a spreading ridge.

Therefore, there is undoubtedly a relationship between the salient features of the break-up of continental plates, and unusual intra-plate (trap, continental and rift) magmatism (Fig. 2.29). Relatively fast (catastrophic: over a time scale of few million years) break-up began with this type of magmatism in a zone of subsequent plate fragmentation and was superseded by magmatism of spreading ridges. In such cases, intra-continental magmatism can be regarded as catastrophic.

While the formation of a new lithospheric plate boundary is a long-term process, large-scale igneous events are not centred around this process. Their absence implies that deep-seated sublithospheric processes which, in particular, gave rise to magmatism, did not initiate the lithospheric break-up (or did not really stimulate it). It resulted from the interaction of lithospheric plates and occurred in accordance with evolutionary processes that fit into the overall system of stable states of lithospheric blocks.

It is only at the initial stage of their development that convergent plate boundaries (subduction and collision) cause changes in the dynamic balance of the immediately preceding ensemble of plates. Once they have formed, such boundaries are, on the whole, stable in time, and processes within them are more or less stationary. Along convergent boundaries, magmatism begins to develop as they take shape, and forms volcanic arcs and belts above subduction zones (or zones of convergence). This magmatism constitutes an important parameter, reflecting the stationary development of these boundaries over tens of millions of years.

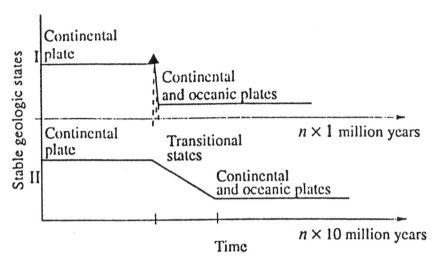

FIGURE 2.29 Two types of changes of stable geological states occurring during the break-up of the continental lithosphere: I — with intra-plate magmatism accompanying the break-up (triangle); II — without accompanying magmatism.

According to catastrophe theory, only a spasmodic transition from one stationary state to another fits the notion of a catastrophe. In the case of convergent boundaries, it coincides with their inception. Certain information about the nature of the transformation of one stable geological state into another (with convergent boundaries) can be inferred from igneous rocks formed at the time of such transformations. However, magmatism related to convergent boundaries is an attribute of dynamically stable systems and has nothing to do with geological catastrophes.

The manifestations of intra-plate magmatism should be discussed separately. They include the traps already mentioned and igneous rocks of intra-continental rifts and eruptions formed by hot-spots in the mantle. In the past, they often stimulated the appearance of rift structures (including those of the global rift system) which played the principal role in the fragmentation of lithospheric plates (White and McKenzie, 1989). Products of the igneous activity of hot-spots (some researchers assign them to traps) usually precede the break-up. Apart from large-scale lithospheric destruction, hot-spots manifest themselves in other ways in geological structures. In such cases, they can be correlated with local and often elongated areas that were formed over tens and even hundreds of millions of years. In such areas, igneous activity are discrete in both time and space. For example, the activity of the South Khangai hot-spot in Mongolia can be traced over the past 160 Ma (Yarmolyuk et al., 1991). Several successive short-term pulses spatially displaced with respect to each other have been identified in its development. Each pulse was accompanied by local fractures in the Earth's crust, and each was reflected in faults wedging out along strike, rift valleys and grabens. Viewed individually, each of these pulses displays characteristics of a geological catastrophe that causes disturbance of the dynamic balance of the plate, including short-term outbursts of igneous activity, and leads to breaks in and collapse of crustal blocks. The above examples show that intra-plate magmatism (that of traps, continental rifts and hot-spots) is an integral element of geological catastrophes due to lithospheric destruction and hence can be classed as catastrophic.

2.6.5 Magmatism as an agent of catastrophic change

Individual volcanoes, groups of volcanoes, volcanic strata, igneous areas, and the like, up to the evolution of magmatism as a whole in Earth history, can be seen as stable states or systems as far as magmatism is concerned. The parameters of their stability determine gradual (or evolutionary) changes in the composition of products with time. Those changes in the trends in the development of magmatism that are expressed in abrupt disturbances of its stable states, in terms of catastrophe theory, can be classed as proper igneous catastrophes. Based on their scale, the latter are divided into local, regional and global.

The succession of volcanic events in the development of caldera volcanoes of the Krakatoa type is a dramatic example of local catastrophes. In the history of Krakatoa, Van Bemmelen (1957) identified a number of cycles, including quiescent and paroxysmal phases of volcanic activity. The quiescent phase is characterized by eruptions of small volumes of andesite–basalt and andesites; this phase lasts for hundreds of years. During an extremely short paroxysmal phase (lasting several days) huge volumes (up to $18 \, \text{km}^3$) of ash and acid tuff were ejected. This was followed immediately by formation of basic and intermediate volcanics, thus marking the onset of a new cycle in the development of the volcano.

Let us trace the changes in the composition and volume of the ejected material in the history of the Krakatoa Volcano (Fig. 2.30). Segments of dynamic stability and areas of drastic disruption in such stability can be recognized in its volcanic activity. The former correspond with the smooth development of volcanic activity, with a successive increase in the silica content of the volcanic products. The latter correspond with marked changes in the volcanic products and their volumes, as a result of which the system returns to its original state. Such paroxysmal phases can be classified as catastrophes. They are often accompanied by the formation of large collapse structures or calderas; therefore, they can be seen as integrated (on a set of parameters) catastrophes.

Changes in the composition and environments of formation of igneous rock assemblages in the marginal volcanic–plutonic belts involved in the development

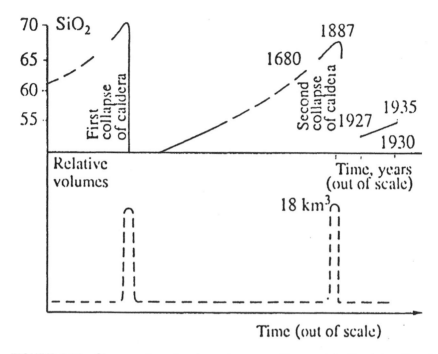

FIGURE 2.30 Changes in volcanic rock composition and in the intensity of volcanism (volume of products per time unit) in the development of the Krakatoa Volcano.

of California-type active continental margins provide an example of regional igneous catastrophes. In the structure of such belts, age-related trends of changes in magmatism usually result from the successive replacement of basic and intermediate rocks by moderately acid and acid types. The latter belong to the calc-alkaline series and are grouped into a differentiated volcanic complex. At the final stages of formation of margins of this type, the above sequence is abruptly broken by the appearance of bimodal assemblages of quite different composition, which are composed of alkaline basalts and alkali rhyolites. In terms of geological time, localized manifestations of bimodal magmatism are extremely brief. Therefore, its outbursts can be described as catastrophes (in terms of catastrophe theory).

Let us illustrate the above statement, using as an example the events that occurred along the late Palaeozoic active margin of Central Asia (see Chapter 3). There are actually two parallel terranes — the South Mongolian and the Mongolian–Transbaikalian. In the South Mongolian terrane, rocks of the differentiated complex had been forming since Visean times (early Carboniferous) to early late Carboniferous. Igneous activity of this type ceased abruptly after the formation of narrow rift zones infilled by products of bimodal magmatism, i.e. subalkaline basalts, trachyrhyolites and comendites.

Magmatism in the Mongolian–Transbaikal terrane developed according to the same scenario, albeit at a different time. There, the igneous products of the differentiated complex were formed mainly during the early Permian. This process terminated due to outbursts of bimodal magmatism that occurred in the Central Mongolian segment during the late early–early late Permian. After the bimodal igneous activity the area entered an amagmatic stage of evolution. On the whole, the formation of each igneous terrane took 20 to 30 Ma. The duration of bimodal magmatism during this time interval can be inferred from the following data. Comendites and trachyrhyolites occur throughout the entire vertical section of volcanic rocks, in correlation with particular stratovolcanoes as evidenced by the structure of the bimodal volcanic assemblages. This allows us to determine the time of formation of the bimodal assemblages, proceeding from the time that the volcano started to erupt. Based on volcanological studies, even for slowly developing volcanoes, this does not exceed several thousand years. Against the overall background of the evolution of marginal belt magmatism, this is an extremely brief time interval (Fig. 2.31). Nevertheless, it was a turning point in the development of the continental margin type of igneous activity.

It is notable that catastrophe theory can be applied to the spasmodic migration of volcanic centres — events that over the course of time change the configuration of igneous areas. For example, centres of volcanic activity abruptly shifted 100 km inland in the Andes during the late Miocene, whereas during the late Palaeozoic, at the Upper Carboniferous–lower Permian boundary, igneous centres in the igneous region of Central Asia shifted dramatically from southern to central and northern Mongolia. Such spatial and temporal shifts undoubtedly should be regarded as catastrophic in relation to igneous activity. Information on the nature of the processes operating at that time could have included the earliest rock complexes formed after such a shift. It is widely accepted that magmatism has been a permanent factor; it is inherent in the dynamics of development of the upper shells of the Earth throughout its history. Therefore, terrestrial magmatism as such can be viewed as a functionally evolving system. An abrupt change in its state, primarily in its real parameters and intensity characteristics, can be classed as a global igneous catastrophe. However, the problem has only been stated and we shall discuss some of the changes in the evolutionary tree relevant to catastrophe theory. Rocks of the tholeiitic and calc-alkaline series are known to occur throughout the geological record. Other rock types either disappear or appear at certain stages in geological history. For example, peridotite komatiites hardly cross the Archaean boundary, whereas the unusual siliceous high-magnesian series occur mainly during the early Proterozoic. Autonomous anorthosites occur throughout the Proterozoic. Iron–titanium picrites and basalts and associated alkaline rocks were formed about 2 Ga ago. As mentioned above, they are typical of Phanerozoic intra-plate magmatism. The rapakivi granites were formed during a brief period in the late Early Proterozoic. This is approximately when lithium–fluorine granites and ongonites are recorded in magmatism.

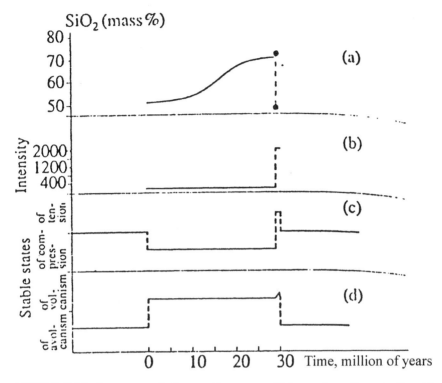

FIGURE 2.31 Igneous activity on the upper Palaeozoic active continental margin off Central Asia: (a) the change in composition of igneous rock types; (b) change in the volcanic activity, as inferred from the thickness (metres) of layers per square kilometre over a period of 1 Ma; (c) replacement of moderate compression by extension, leading to the formation of trans-lithospheric rifts and rift grabens; (d) episodes of bimodal volcanism.

Large-scale trap development has been recorded only since the late Proterozoic. Time intervals in geological history, when particular types of igneous rocks appear or disappear, are similar to the boundaries at which new faunal groups appear or become extinct. These points in time reflect changes in the conditions of magma generation at deep levels in the Earth. They also mark deep-seated catastrophic transformations in the Earth's mantle.

In this connection, apart from the above-listed catastrophic events in the lithosphere we should mention global catastrophes affecting the progress of processes operating in the Earth's core. The formation of the liquid core at the Archaean–Proterozoic boundary and the change in the geodynamic regime from plume tectonics to plate tectonics about 2 Ga ago could at present be regarded as such a catastrophe (Hall, 1987). It was this event that was accompanied by the

Table 2.10 Classification of catastrophic igneous activity.

Agent causing the catastrophe	Nature of the catastrophe	Catastrophic igneous activity
Fauna and flora	Extinction: Global Regional Local	 Traps Traps and large tuff fields Individual eruptions
Atmosphere and climate	Change in the composition and physical properties of the atmosphere due to emission of gases (H_2S, CO_2 and N_2O) and ash	Traps, large-scale eruptions
The Earth's crust	Lithospheric rifts of varying scales: Divergent plate boundaries Convergent plate boundaries	 Traps, continental-rift magmatism of hot-spots and deep rift zones. Magmatism of the initial stages of island arcs, active continental margins and collision zones.
Evolution of magmatism		
	Global	
Of Earth history	End of the formation of rock types with a particular composition (e.g. komatiitic peridotites). Formation of rock types of specific compositions (e.g. alkaline) Spasmodic appearance of rock types	Changes in the dynamics of in-depth magma generation The same Autonomous anorthosites
	Regional	
Of marginal belts	Disruption of magmatic activity of continental margin type	Bimodal magmatism of axial rifts on active continental margins
Of magmatic areas	Spasmodic changes in the geometry of volcanic eruption zones	Magmatism at the initial stages of newly formed eruption zones
	Local	
Of isolated volcanoes	Abrupt changes in the composition and/or type of eruption (e.g. in Krakatoa-type caldera volcanoes)	Ignimbrite volcanic activity of the caldera stage

appearance of a new type of igneous activity, i.e. the appearance of Fe–Ti picrites and basalts (see Chapter 5). Analysis of catastrophic igneous episodes has allowed them to be classified schematically (Table 2.10). This classification is based on the recognition of the object of the catastrophe, the nature of the catastrophic event and the igneous activity involved in catastrophic transformations.

Many episodes of igneous activity act as agents of catastrophe, concurrently with respect to several dynamically stable natural systems. Traps are particularly relevant: they give rise to (or accompany) catastrophes in the structure of the lithosphere, in the composition of the atmosphere in the biosphere. Multicomponent catastrophes also accompany rift magmatism. They occur when lithospheric plates break up, and in some complex geodynamic environments (at active continental margins of Californian type), they can lead to severe disruption in the development of the magmatism that is typical of these geodynamic environments. Igneous activity related to rapid changes in the geometry of the area that is generating the magma, for example the formation of new centres of magmatism within developing igneous areas, can be considered as a special case. Here the changes in the geological setting depend on the time of inception of such centres. This type of catastrophic magmatism can only be traced if we identify the initial stages in the formation of igneous assemblages when new centres are formed, because they mark the transition from one stable geological state to another.

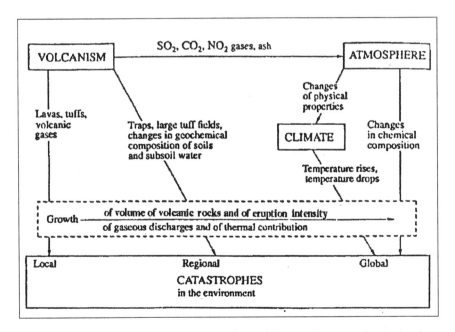

FIGURE 2.32 Volcanic activity: factors determining environmental catastrophes of varying scales.

Volcanism is dominant in catastrophes affecting the state of the environment (the atmosphere, hydrosphere, soil and climate). Based on the scale of the impact, catastrophes can be classed as local, regional or global (Fig. 2.32). Local catastrophes are connected with the activity of particular volcanoes, whose eruption products destroy all living things in the areas where these products accumulate.

2.7 SUMMARY

1. The present-day (late Cenozoic) environments allowed a relationship to be established between the extent and character of magmatic activity and geodynamics. The most extensive igneous activity (60–65% of the entire volume) occurs at constructive margins, i.e. axes of sea-floor spreading. About 20–30% of the volume of igneous activity is associated with destructive margins or zones of active plate interaction (island arcs, active continental margins and collision zones). Intra-plate magmatism accounts for 10–15% of the volume. Examples of this type of magmatism occur everywhere: both in intra-plate environments and at plate margins, resulting in the combination of different types of igneous activity.
2. A specific magma assemblage of sea-floor spreading zones is composed of MORB-type tholeiitic basalts. Similar basalts also occur in back-arc basins at the active margins of the lithospheric plates.
3. Magma series at active margins are represented by calc-alkaline and potassic moderately (shoshonite–latite) and high-alkaline rocks. Island arcs are dominated by rocks of the calc-alkaline series. Potassic moderate- and high-alkaline rocks occur at active plate margins and in collision zones. At active margins rocks of the low-Ti (island-arc) tholeiitic and K–Na moderately alkaline series are subordinate. The boninite series was recorded only at island arcs.
4. Intra-plate (hot-spot) magmatism is represented by formations of the Fe–Ti, K–Na and potassic normal-, moderate- and high-alkaline series. Intra-plate magmatism is determined by the type of Earth's crust: on the oceanic crust it occurs in the form of volcanic islands, seamounts and aseismic ridges, while on continents it is common in rift zones and forms lava plateaux.
5. As well as the simple geodynamic environments there are complex environments resulting from the combination of two or more environments. Among these are:

   a) oceanic islands at mid-oceanic ridges and in back-arc basins;
   b) the overriding of mid-oceanic ridges by active (California-type) margins;
   c) collision zones (Alpine or Mediterranean type).

6. There are certain trends in distribution within particular geodynamic environments. Therefore, active plate interaction zones show a transverse zoning

reflected in the following sequence: low-Ti tholeiite-series rocks in the fore-arc zone and calc-alkaline to low-Ti moderate- and high-alkaline series, mainly of K-group and MORB, in the back-arc zone. There is also a longitudinal zoning, probably caused by changes in the composition of the subducted slab involved in melting.

7. Transverse zoning is characteristic of large rift areas such as the Ethiopian Rift, with tholeiite basalts within the rift valley and moderately alkaline basalts on the flanks; within the Red Sea Rift the eruption of MORB takes place in the axial zone, MORB are followed by intermediate tholeiites, and moderately alkaline basalts occur on the rift flanks (in Egypt) and in adjacent parts of the Arabian plate. Locally, this zoning is asymmetrical: for example, the eruption of high-alkaline K–Na rocks takes place in the Kenyan Rift and the eruption of potassic rocks on its western branch in the Tanganyika Rift. In parallel with the transverse zoning there is longitudinal zoning, so within the same East African Rift, from south to north, the tholeiitic basalts of the Afar Triangle are replaced by the moderately alkaline basalts of the Ethiopian Rift and the high-alkaline volcanics of the Kenyan Rift.

8. The evolution of young tectonic settings magmatism can take place in individual magmatic centres and can be regional in scope. In the former case, the general trend follows that of the physicochemical development of magmas and is apparently subject to partial magmatic assimilation of crustal material. In the second case, it most often follows the melting trend of the subcrustal basement. For example, within the Red Sea Rift, there is a transition in time and space, from moderately alkaline basalts to MORB; the general trend of evolution is towards oceanization of the crust.

Evolution of another type is typical of active zones of plate interaction, where there is a spatial and temporal transition from rocks of the tholeiitic series to calc-alkaline and then to the moderately alkaline and alkaline series (mainly of the K-group). At active margins of continents and oceans, the transition is from young arcs, through developed arcs, to mature arcs differing in the thickness of their underlying crust and their similarity to continental crust.

9. On Earth, there are tectonomagmatic systems that are at different stages of development. This gives an idea of the main evolutionary trends of these systems during the Cenozoic. Two main types of evolution can be recognized: Atlantic–Eurasian and Pacific. The former is characterized by three main phases in the evolution of the Earth's crust, from the inception of an ocean to its closure. The situation on the Pacific segment is more complex. The development of the Earth's crust took place by the suturing of island-arc structures to continents in the course of the collapse of back-arc basins; however, the ocean itself continued to exist.

10. Magmatism is an important agent of natural catastrophes, which has played an essential role in the changing of geoecological systems and the evolution the biosphere.

CHAPTER 3
MAGMATISM OF PHANEROZOIC FOLD BELTS

V.V. YARMOLYUK and V.I. KOVALENKO

Phanerozoic fold belts are an important element of continental plate structure. Their study enabled the construction of a geological paradigm based on the geosynclinal hypothesis. This was in use for about 100 years and only in the 1960 and 1970s was it replaced by plate tectonics.

Fold belts are linearly elongated and have the following characteristics:

1. structural zonation;
2. a set of unusual assemblages of sedimentary, igneous and metamorphic rocks formed in a particular historical sequences and correlated with the different developmental stages of these areas;
3. the presence of different types of folding.

Rocks of fold belts form fairly regular age series, whose individual portions occur in the same repetitive sequence in the different fold belts. For example, in many cases, the early formations of fold belts are represented by assemblages typical of ocean basins. They include tholeiitic basalts, ultrabasic rocks and siliceous sedimentary rocks, grouped into ophiolitic assemblages. Their study theoretically allows elucidation of the deep structure of oceanic crust or back-arc seas which cannot yet be sampled by direct drilling. Granitoids, which imply more mature underlying crust, are very common at the final stages of the development of these fold belts.

Many researchers identify fold belts as structural regions of the Earth, with the transformation of oceanic crust into the continental crust during their formation. According to these viewpoints, a number of stages can be recognized during the formation of the continental crust underlying fold belts, each differing in their type of crustal structure and inherent in a particular past geological or tectonic environment (Peive et al., 1980).
These are:

1. an oceanic stage, with oceanic crust and palaeogeographical environments similar to present-day oceans and some back-arc seas;
2. transitional, with transitional-type crust, a fragmentary granite–metamorphic layer and palaeogeographical environments similar to present-day marginal seas and island arcs;
3. continental, with continental crust and the characteristic ubiquitous occurrence of a granite–metamorphic layer and continental sedimentary environments.

For the first, oceanic, stage, earlier identified as a geosynclinal stage, a number of assemblages can be recognized. The ophiolitic assemblage is undoubtedly the most important. It consists of a mantle ultrabasic basement complex and crustal layered complex, sheet complex and volcanic complex, represented by basaltic pillow lavas (Coleman, 1977; Nicolas, 1989). Rocks of the latter complex are most common in fold belts: in these cases, they are identified as spilite–diabase, spilite–siliceous and other assemblages. The formations of this stage also include an early basalt–andesite assemblage composed mainly of rocks of the calc-alkaline series and island-arc tholeiites, as well as a plagiogranite–plagiorhyolite assemblage consisting of plagiorhyolites, plagiorhyolite–dacites and Na-dacites; their intrusive equivalents are plagiogranites and tonalites (Borsuk et al., 1987). These rocks are often associated with pillow lavas, forming the bimodal assemblages also known as spilite–keratophyre assemblages.

The second, transitional, stage is characterized by the presence of diverse andesite assemblages dominated by rocks of the calc-alkaline (basalt–andesite–dacite–rhyolite) and potassic subalkaline (shoshonite–latite–alkali rhyolite) series. The diverse gabbro–granite assemblages are their intrusive equivalents. Less common and of local occurrence are the bimodal basalt–rhyolite assemblages and their intrusive equivalents: the gabbro–plagiogranite assemblages (Borsuk et al., 1987). Unusual dunite–clinopyroxenite–gabbro complexes, similar to those of the Platinonosny (platinum-bearing) Belt of the Urals and the massifs of Alaska, were recorded locally (Irvine, 1974; Laz'ko, 1988b). And, finally, the third, continental stage (often classed as orogenic) includes a wide range of igneous rocks. The most common are andesite–dacite–rhyolite, dacite–rhyolite and their intrusive equivalents: the diverse granodiorite–granite series. Apart from this, high-titanium basic rocks of normal, moderate and high alkalinity, usually of the K–Na series, are also common; the acid rocks contain trachytes and alkaline rhyolites.

The regular features of magmatism and the development of fold belts such as the Central Asian belt will now be discussed and compared with the other fold belts of Eurasia.

3.1 MAGMATISM AND THE FORMATION OF THE CENTRAL ASIAN FOLD BELT

The Central Asian Fold Belt is a mosaic-like system of fold structures separated by stable massifs located between the Siberian and Chinese platform (Kovalenko et al., 1995b) (Fig. 3.1). Three major stages: the early Caledonian, Hercynian and late Hercynian, can be recognized in the development of the fold structures within the belt. Each of these can be correlated with complexes of

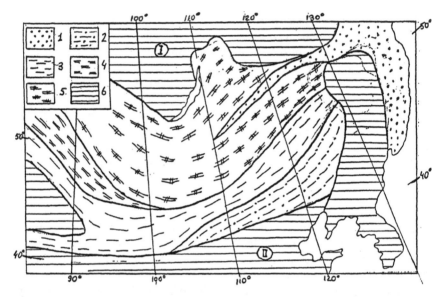

FIGURE 3.1 Position of the Central Asian Fold Belt in the structural system of Asia. Key: 1 — Mesozoic; 2–5 — fold zones of the Central Asian Belt: 2 — late Hercynian, 3 — Hercynian, 4 — Caledonian, 5 — early Caledonian; 6 — complexes formed prior to Pre-Riphean stabilization complexes. Circled figures: I — Siberian Platform; II — Sino-Korean Platform.

geological formations forming wide linear belts, successively accreted on to the margin of the Siberian Platform from the south. The stabilization of the fold belt structure ended in the late Permian, due to collision with the Sino-Korean Platform.

During its development, the Central Asian Fold Belt passed through the stages of continental crust formation: oceanic, transitional and continental. Most ocean- and transitional-stage complexes occur in the same structures, and it is difficult to differentiate them. Therefore, the magmatism of these stages will be discussed together.

3.1.1 Oceanic- and Transitional-Stage Magmatism

Within the Central Asian Fold Belt, there are three ages of formation of oceanic and transitional crust (Kovalenko *et al.*, 1996b): the early Caledonian (late Riphean–early Cambrian), Hercynian (Ordovician–Silurian) and late Hercynian (late Carboniferous–early Permian). Structures from each period are spatially isolated, but generally their distribution corresponds with successive accretion of additional younger crust from the south, on to the Siberian Palaeocontinent (Fig. 3.2).

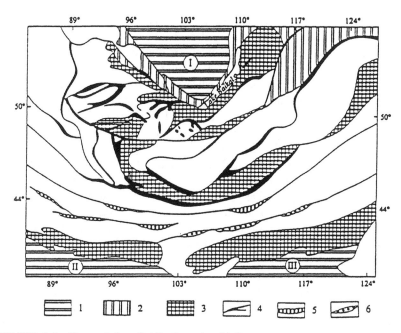

FIGURE 3.2 Present-day distribution of ophiolite zones within the Central Asian Fold Belt. Key: 1 — platforms: (I) Siberian, (II) Tarim, (III) Northern Chines; 2 — marginal salients of the Siberian Platform Basement; 3 — pre-Riphean consolidated blocks in the Central Asian fold belt; 4–6 — ophiolite zones: 4 — late Riphean–early Cambrian, 5 — early-middle Palaeozoic, 6 — late Palaeozoic.

a) Early Caledonian Structures

These are correlated with the initial phase of the Central Asian Fold Belt. Starting from the second half of the Riphean, the continental massif in the region was involved in large-scale extension and subsequent fragmentation into a number of continental blocks. Shear zones were initiated at the same time as the rifts which developed and became widespread on the southern margin of the Siberian Platform (Shpunt, 1987). These have been preserved locally in the continental block structure of the fold belt (Yarmolyuk *et al.*, 1989). Further opening of the rifts resulted in the formation of a system of marine basins underlain by oceanic crust. In the present-day structure, they can be correlated with ophiolite belts in Tuva, in the Sayan Mountains, in Mongolia and in the Transbaikal area, where they separate continental blocks from the pre-Riphean basement. The late Riphean–early Cambrian age of the ophiolites was inferred from stratigraphic relationships. For the ophiolites from the Bayan–Khongor Zone of Mongolia the Sm/Nd isochron age 569 ± 21 Ma on gabbros (a layered complex) was determined (Kepezinskas *et al.*, 1990).

Ophiolite belts are composed of several different rock types. A typical ophio-lite section contains (in ascending order): harzburgites, and, less commonly, depleted lherzolites of the ultrabasic complex; dunites, pyroxenites and gabbros of the layered complex; a sheeted dyke complex; basaltic pillow lavas, andesitic basalts; jaspilites and siliceous and calcareous oozes (deep-water sediments) (Zonenshain and Kuzmin, 1978). In general, the nature of the section and the composition of the ultrabasic and layered complex enable us to correlate them with similar rocks of other ophiolite assemblages around the world (Kepezinskas et al., 1987). Volcanic rocks from ophiolite zones vary in their composition (Table 3.1). They include basic rocks of the tholeiitic and calc-alkaline series, and marianites and boninites are also fairly common.

Studies by K.B. Kepezinskas et al. (1987) have shown that the rocks that can be correlated with basalts of present-day mid-oceanic ridges are actually quite rare in the ophiolite section. In addition, most of the basalts from ophiolite zones are similar to those of marginal seas. These are highly (1.9–2.5 wt.%) or moder-ately titanium-rich (0.9–1.8%) rocks, with relatively low K_2O and Al_2O_3 content (0.1–0.5% and about 15 wt.% respectively).

Volcanic rocks from ophiolite zones contain a substantial volume of basalts and andesitic basalts similar to island-arc tholeiites. As compared with other tholeiites, they have a lower TiO_2 (about 0.9 wt.%) and higher Al_2O_3 (18–21 wt.%) content. In sections they are associated with greywackes and volcanics of the calc-alkaline series.

Calc-alkaline rocks prevail. They in turn are dominated by basalts and andesites, with dacites also being reported. They occur in association with island-arc tholeiites and as andesite pillow lavas. These rocks have low K_2O, TiO_2 and Cr, and relatively low Al_2O_3 content. The MgO content often reaches 10–17 wt.% at an SiO_2 content of 55–64 wt.%. Based on real characteristics (major and trace elements, REE), these rocks are close to the high-Mg and low-Ti andesites of Troodos and to rocks of the marianite–boninite association of the western Pacific (Kepezinskas and Kepezinskas, 1985).

Generally, the study of the composition of volcanic and dyke rocks from the late Riphean–early Cambrian ophiolite zone of Central Asia indicates at least two stages of magmatism in the igneous rock succession of ophiolite assemblages. The earlier stage corresponds with the oceanic phase of crust formation. Its relics are poorly preserved in the present-day structure. They are represented by rocks of ultrabasic and layered complexes and by MORB-type basalts. During the later phases of its evolution, the palaeo-ocean appeared as the melanocratic basement of incipient island arcs. Based on the hypothesis that ophiolites proper (or rocks from the earlier complex) are the basement complexes of ocean basins, it is also assumed that basins on oceanic crust, and small microcontinents with pre-late Riphean continental crust, existed during the late Riphean–early Cambrian in Central Asia. The late Riphean–early Cambrian ocean basins are assumed to have been fairly deep and to have contained siliceous sediments. Also, there were no coarse-grained clastic sediments to suggest the presence of nearby eroding areas.

Table 3.1 Representative compositions of volcanics of the Late Riphean–Early Cambrian ophiolite zone of Mongolia. After Kepezinskas *et al.* (1987).

Components	1	2	3	4	5	6	7	8	9	10	11	12	13	14	15
SiO_2	48.65	52.58	52.82	51.52	52.21	51.04	46.66	52.38	52.63	51.46	56.08	58.27	49.41	52.21	55.81
TiO_2	1.09	2.14	2.09	1.19	1.35	1.59	0.85	0.74	0.70	0.97	0.64	0.51	0.16	0.16	0.23
Al_2O_3	16.05	15.10	14.50	15.90	14.16	13.41	18.68	19.43	21.36	17.19	16.30	15.48	12.88	11.20	13.32
Fe_2O_3	5.10	1.86	2.80	2.71	2.33	6.63	2.01	2.24	3.91	3.42	1.50	2.82	9.39*	11.31*	9.29*
FeO	7.71	9.76	9.12	6.83	7.89	6.01	8.79	7.05	4.83	5.44	6.12	4.75	—	—	—
MnO	0.27	0.17	0.21	0.17	0.17	0.19	0.22	0.13	0.14	0.15	0.15	0.10	0.14	0.20	0.14
MgO	7.30	5.99	6.70	6.98	7.34	5.75	8.40	4.49	4.96	7.20	6.20	6.69	12.49	15.98	13.22
CaO	10.46	8.51	7.46	10.77	9.97	10.56	11.79	9.29	7.01	10.01	6.70	5.46	14.85	7.01	4.64
Na_2O	2.30	3.40	3.77	2.95	3.78	3.73	2.43	3.63	3.80	3.04	4.92	4.34	0.58	1.86	3.12
K_2O	0.26	0.49	0.53	0.79	0.58	0.70	0.26	0.38	0.52	0.82	1.12	1.24	0.06	0.02	0.14
P_2O_5	n.d.	n.d.	n.d.	0.20	0.23	0.30	0.11	0.24	0.13	0.20	0.17	0.32	0.04	0.04	0.09
LOI	3.58	2.66	2.74	1.92	1.96	1.95	2.68	3.64	2.87	3.44	2.21	2.94	4.26	5.59	4.34

*Total iron in the form Fe_2O_3. Note: in the column headings 1–3 — refers to titanium tholeiites, 4–6 — moderately Ti-rich tholeiites, 7–9 — low-titanium tholeiites, 10–12 — calc-alkaline volcanics, 13–15 — volcanics of the boninite series.

Table 3.2 Chemical composition of Early Palaeozoic granodiorite formation rocks (Tocktogenshil Complex, North Mongolia) (from Marinov, 1973).

Components	1	2	3	4	5	6	7	8	9
SiO_2	64.95	63.72	63.30	55.50	68.64	63.30	73.31	74.04	72.00
TiO_2	0.80	0.56	0.55	1.52	0.29	0.38	0.35	0.14	0.35
Al_2O_3	14.82	15.02	14.44	17.28	16.28	15.44	12.76	12.61	14.85
Fe_2O_3	1.98	1.15	1.38	5.21	1.80	2.85	1.19	1.27	0.95
MnO	3.52	4.61	4.04	3.24	0.94	3.12	1.44	1.94	1.41
MgO	0.08	0.12	0.14	0.18	0.06	0.09	0.09	0.16	0.13
CaO	3.96	3.62	4.38	3.02	1.14	2.30	0.51	0.43	0.99
Na_2O	3.18	4.60	3.91	6.40	2.95	5.50	1.94	0.30	2.03
K_2O	2.92	3.13	2.15	4.03	5.67	3.75	3.72	4.28	3.61
P_2O_5	0.06	0.20	0.11	0.88	0.16	0.15	—	—	0.10
LOI	2.01	0.61	0.98	1.08	0.54	1.55	0.47	0.44	0.39
H_2O	0.07	0.21	—	0.32	—	0.12	0.09	0.11	—
Total	100.19	100.35	99.52	99.99	99.85	100.10	99.61	99.44	99.99

Columns 1–4 — rocks of phase I: 1 — tonalite, 2 — quartz diorite, 3 — quartz diorite, 4 — diorite; 5–9 — rocks of phase II: 5 — granite, 6 — tonalite, 7 — granite, 8 — granite, 9 — granite.

Basin closure took place during the mid–late Cambrian and early Ordovician. Apparently the oceanic crust was subducted under island arcs containing calc-alkaline volcanic complexes. At that time granitoid belts were formed, which were related to the margin parts of the continental blocks surrounding the ophiolites. The granitoids could be grouped into the tonalite–plagiogranite, granodiorite and granodiorite–granite assemblages. These assemblages have a lot in common. They are characterized by an early phase of intrusion of small, gabbrodiorite, diorite and tonalite bodies. The massifs are dominated by granodiorites, plagiogranites, adamellites, tonalites and granites of the second (late) phase. The assemblage they are assigned to depends on the relative proportions of different rock types. Based on their chemical composition, they correspond with the calc-alkaline series (Table 3.2).

Ophiolites of Ozernaya zone (western Mongolia) have been characterized in Kovalenko et al. (1996b).

b) Hercynian Structures

These occur mainly in the area of the South Mongolian Hercynides bounding the Caledonian structural terrane to the south. The emplacement of the Hercynian

Palaeotethys took place in the Ordovician and can be correlated in time with the closure of the Caledonian marine basins, and the accretion of the entire complex of Caledonian structures to the southern margin of the Siberian Platform. The formation of the Hercynian Palaeotethys is shown by the presence of Ordovician ophiolites, which are the earliest rocks occurring at the base of the Hercynian section (Zonenshain and Tomurtogoo, 1979). The Silurian ophiolites are better known (Ruzhentsev et al., 1990). They are represented by a mélange, consisting of serpentinized peridotites and, to a lesser degree, of rocks of the layered complex, as well as gabbros. The mélange is transgressively overlain by a tuff–jaspilite horizon and overlying dyke swarms of tholeiitic basalts (upper Silurian–upper Devonian).

A change in the structural framework of the Hercynides then occurred, due to a new extensional phase at the Silurian–Devonian boundary, resulting in the formation of a system of areally scattered dyke swarms that served as conduits for basaltic eruptions. The extensional zones were formed synchronously with the island-arc chains (Ruzhentsev et al., 1990), and contained the following rock types: greywackes, flysch, siliceous and carbonate deep-water sediments, basalt, andesitic basalts and andesites, all occurring in the same sections.

Volcanic sections in island-arc complexes vary in thickness from 500 to 2,500 m and consist mainly of lavas, locally interbedded with sediments containing Lower–middle Devonian fossils (Gordienko, 1987). The volcanics (basalts, andesites, dacites) can be correlated with the calc-alkaline series (Table 3.3) and are similar to island-arc volcanic assemblages, which agrees well with the environment in which they were formed.

c) Late Hercynian Complexes

These are of limited distribution and were reported from the south-easternmost Central Asian Belt within the Solonker Zone (Ruzhentsev et al., 1990). The zone now possesses a complex nappe structure. The autochthon is made up of Late Carboniferous–Permian jaspilites and volcanics. The volcanics are predominantly tholeiitic basalts, usually in the form of pillow lavas and hyaloclastites, which are generally strongly albitized, chloritized and epidotized. Despite the absence of melanocratic basement rocks in the autochthon, this complex is identified as palaeo-oceanic (Ruzhentsev et al., 1990).

The allochthon consists of three thrust nappes. The lower one is composed of biogenic limestones of the middle Carboniferous–upper Permian. The middle nappe comprises limestones and sandstones (middle–upper Carboniferous), transgressively overlain by andesite basalts and their corresponding tuff, interbedded with limestones yielding Kungurian fauna (lower Permian). The upper nappe consists of peridotites. Rock complexes of the two lower nappes can be correlated with island-arc complexes (Ruzhentsev et al., 1990); the palaeotectonic environment of the ultrabasic rocks is as yet unknown.

Magmatism and Geodynamics

Table 3.3 Chemical composition of Devonian volcanic rocks of the Hercynian Transaltai zone of South Mongolia. After Gordienko (1987).

	SiO$_2$	TiO$_2$	Al$_2$O$_3$	Fe$_2$O$_3$	FeO	MnO	MgO	CaO	Na$_2$O	K$_2$O	LOI	P$_2$O$_5$	Total
1.	47.84	1.26	18.60	3.40	4.57	0.15	4.58	9.35	2.71	2.58	2.83	0.49	100.36
2.	47.14	1.25	15.58	2.04	8.71	0.17	7.36	12.30	1.70	0.96	2.21	0.11	99.53
3.	49.70	1.24	14.79	3.57	7.45	0.16	7.37	10.76	1.87	0.67	2.19	0.32	100.09
4.	50.28	1.02	14.47	4.02	6.83	0.16	7.14	10.62	1.67	0.83	2.52	0.09	99.65
5.	53.41	0.78	17.00	3.58	5.96	0.18	5.00	6.84	3.28	1.92	2.18	0.29	100.20
6.	53.55	0.77	16.40	3.58	5.69	0.17	4.81	8.64	3.59	1.20	1.59	0.28	100.27
7.	54.90	0.61	16.17	1.63	8.40	0.16	2.78	6.24	4.52	2.02	2.01	0.44	99.55
8.	57.18	0.70	19.25	1.04	3.88	0.08	1.75	3.52	4.70	4.40	3.79	0.42	99.38
9.	59.48	0.93	16.74	3.40	3.26	0.12	2.81	3.77	3.97	3.47	2.74	0.32	100.11
10.	67.37	0.23	15.80	1.00	1.89	0.08	0.70	1.56	5.50	3.80	2.45	0.11	99.59

Rocks 1– 4 — basalts; 5– 9 — andesites; 10 — dacites.

3.1.2 Continental-Stage Magmatism

The Central Asian Fold Belt is unparalleled in the scope and duration of continental-stage magmatism. It began in the second half of the Early Palaeozoic and continued up to the Holocene, i.e. for more than 450 Ma. Igneous activity passed through several major stages: Early–Middle Palaeozoic, Late Palaeozoic, Early Mesozoic and Late Mesozoic–Cenozoic. Episodes of continental magmatism correspond with those processes that affected the adjacent continental blocks of the marine basins.

a) Early and Middle Palaeozoic Magmatism

The earliest continental magmatism of this time dates back to the Ordovician and Silurian. It is represented mainly by granitoids, whose formation was related to the continuing collision of continental plates at the site of the closed late Riphean–early Cambrian ocean basins.

The granitic magmatism of the collision zone was characterized by granodiorites and granites. The most important igneous event was the formation of the huge ($> 1,300,000 \, km^2$) Angara–Vitim Batholith (Litvinovsky et al., 1992). The early stages of its development were marked by the formation of granodiorites, tonalites and quartz diorites. Two-feldspar biotite granites and gneissoid granitoids (usually with taxitic textures) and subalkaline granites, granodiorites and syenitediorites were formed during the main phase. Rocks with taxitic texture form autochthonous or para-autochthonous masses, mostly in the southern part of the batholith. Allochthonous granitoids of igneous appearance occur mainly in the northern part of the batholith. At the final stage of batholith formation, dyke- and stock-like bodies of leucocratic subalkaline granites were intruded.

In the second half of the Silurian, continental igneous assemblages began to be formed. They were chiefly grouped along the boundary of a continent that had been created by then, together with marine basins of the Palaeotethys, i.e. the Irtysh–Zaisan, South Mongolian and Khangai–Khentei branches, respectively, to the west and south (Fig. 3.3). On the whole, igneous rocks form a marginal belt stretching for more than 3,000 km and 200–300 km in width (Gordienko, 1987; Yarmolyuk and Kovalenko, 1991).

The igneous rocks of this belt can be assigned to the calc-alkaline, subalkaline and alkaline series. The distribution of assemblages of varying alkalinity correlated with the zonal structure of the igneous terrane (Fig. 3.3).

The calc-alkaline zone occurs closest to the boundary of the palaeocontinent with the ancient marine basins. It includes igneous rocks grouped into granodiorite–granite and granite assemblages and andesite–basalt and dacite–rhyolite assemblages.

Massifs composed of rocks of the granodiorite–granite and granite assemblages are dominated by gneissoid plagioclase and microcline–plagioclase granodiorites

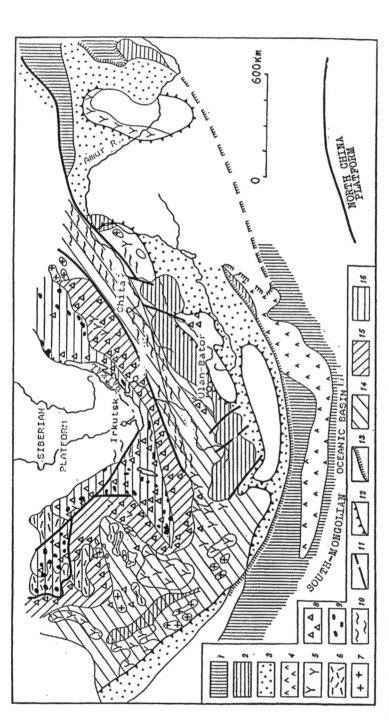

FIGURE 3.3 Palaeotectonomagmatic map for the early Palaeozoic of Central Asia. Modified from Gordienko (1987). Key: 1 — essentially oceanic complexes; 2 — deep-water flysch deposits; 3 — shallow-water terrigenous and carbonate–terrigenous rocks; 4 — arc volcanic complexes; 5–10 — continental formation of marginal belt: 5 — calc-alkaline intermediate and acid volcanics, 6 — calcalkaline, intermediate and acid rocks, 7 — calc-alkaline granitoids, 8 — subalkaline granitoids, 9 — alkali-gabbroid and alkali-granitoid complexes, 10 — ultrametamorphic granitoids, 11 — faults; 12 — boundaries of epicontinental shelf seas; 13 — inferred boundaries between ocean and continent; 14–16 — domains of: 14 — normal series, 15 — moderate-alkali and normal series, 16 — moderate-alkali and alkaline series.

Table 3.4 Chemical composition of representative rocks of Devonian granodiorite–granite and granite formations of West Mongolia (from Luchitsky et al., 1975).

	SiO_2	TiO_2	Al_2O_3	Fe_2O_3	FeO	MnO	MgO	CaO	Na_2O	K_2O	H_2O	P_2O_5	Total
1.	61.85	0.73	17.74	1.41	3.65	0.05	2.46	5.12	4.18	1.48	0.96	0.27	99.80
2.	67.00	0.54	16.06	2.25	1.60	0.03	1.57	3.33	4.94	2.07	0.72	0.08	100.29
3.	68.43	0.68	14.80	1.57	1.47	0.06	0.60	5.20	3.57	3.05	0.53	0.19	99.93
4.	69.47	0.06	12.86	3.71	1.81	0.06	1.83	2.14	3.85	3.28	0.41	0.21	99.68
5.	70.80	0.28	13.65	1.07	1.80	0.05	1.20	2.12	4.00	4.68	0.76	0.07	100.47
6.	73.78	0.02	13.09	1.38	1.15	0.04	0.40	1.44	3.98	3.78	0.49	0.23	99.76
7.	74.80	0.21	12.74	0.47	1.58	0.06	0.41	1.00	3.40	4.90	0.34	0.06	99.95
8.	74.64	0.22	12.75	0.40	1.80	0.03	0.40	1.16	3.60	4.05	0.43	0.07	99.54
9.	75.20	0.10	12.50	0.50	0.80	0.03	0.06	1.90	4.10	4.38	0.51	0.03	100.25

Row 1 — diorite, 2 — granodiorite, 3–6 — plagiogranite, 7–9 — two-mica granites.

and granites; two-mica- and muscovite-granites are less common. The latter are either linked by a gradual transition with granitoids of higher basicity or actually intrude them. In chemical composition of the two-mica and muscovite–granites can be correlated with the calc-alkaline series. The alkali content is 6.0–8.4 wt.%, and the Na_2O/K_2O varies from >1 in the granitoids of high basicity to <1 in the two-mica granites (Table 3.4).

The volcanic rocks of this zone are represented by two assemblages (Table 3.5). The earlier assemblage consists of two-pyroxene and pyroxene–plagioclase basalts, andesites and agglomerates of the same composition. The later assemblage contains lavas, agglomerates and tuffs of dacitic, rhyodacitic and rhyolitic compositions. The age of the rocks was determined as Early–Middle Devonian (Luchitsky, 1983).

The zone of potassic subalkaline rocks encloses that of the calc-alkaline rocks from the edge of the continent. The zone is dominated by granitoids; volcanics are less abundant. The granitoids are represented by assemblages of biotite granites, leucogranites and granosyenites (Table 3.6). The total area of their exposures, in only the western part of the belt, is over 50,000 km^2.

The rocks have the following unusual features: they belong to the subalkaline series ($Na_2O + K_2O > 8\%$, $K_2O > Na_2O$), and are oversaturated with silica and Al_2O_3 (Luchitsky, 1975). The granitoids are rich in trace elements (W, Mo, Be and Sn) and REE.

Within the zone dominated by subalkaline rocks, volcanics occur only in strongly deformed volcanic areas (Gordienko, 1987). The latter are, in turn, dominated by acid volcanics: subalkaline rhyolites, dacites, trachyrhyolites and trachytes (Table 3.7). Usually they consist of fused and usual types of tuff; small lava flows and stocks are less common. In sections, they are associated locally with porphyritic trachyandesite flows. The volcanic strata reach up to 2,000 m thick.

Alkaline Series Zone

This zone correlates with the rear (intracontinental) part of the marginal volcanic–plutonic belt (Yarmolyuk and Kovalenko, 1991). It includes masses of nepheline syenites, alkaline gabbroids (ijolite–urtites, juvites and foyaites), alkali granites and syenites (Yashina, 1982). Volcanics associated with basanites, phonolites, trachytes, trachyrhyolites, pantellerites and comendites, as well as teschenite, mariupolite and pantellerite sills, are also very common (Table 3.8). Apart from the alkaline series, assemblages also contain rocks of the K–Na moderately alkaline series. They are dominated by moderately alkaline olivine basalts and alkali-feldspar trachyrhyolites. The $^{87}Sr/^{86}Sr(0)$ in basic rocks of the zone varies from 0.7033 to 0.706 and is independent of the volcanics (Vorontsov, 1994).

On the whole, the volcanic–plutonic belt follows the boundary of the Palaeo-continent with the ancient marine basins. An active continental margin is assumed to have formed above a subduction zone that gently plunged under the continent

Table 3.5 Chemical compositions of Devonian volcanic rocks from the calc-alkaline of the Mongolian Altai. From Luchitsky (1983).

Components	Andesite–basalt assemblage						dacite–rhyolite assemblage				
SiO_2	48.58	48.90	49.40	55.80	57.50	58.20	73.80	74.88	76.30	74.50	75.50
TiO_2	0.77	0.60	0.67	1.50	1.40	1.90	0.22	0.19	0.16	0.19	0.30
Al_2O_3	11.18	11.20	16.58	13.66	14.70	14.00	11.60	11.54	10.31	11.31	11.60
Fe_2O_3	4.23	2.60	2.32	3.00	2.42	4.50	2.72	1.13	2.47	1.52	1.30
FeO	7.70	8.90	7.28	6.81	5.60	4.00	1.00	1.58	0.95	1.08	1.30
MnO	0.18	0.16	0.15	0.16	0.10	0.11	0.02	0.01	0.07	0.03	0.03
MgO	9.20	10.10	7.10	4.84	3.95	2.90	0.54	0.65	0.38	1.05	0.25
CaO	10.80	11.20	7.50	1.46	4.72	3.60	0.65	0.34	0.34	0.22	0.22
Na_2O	1.30	1.30	2.58	3.00	5.60	5.45	2.78	2.46	0.82	0.14	4.20
K_2O	1.80	0.76	1.04	1.20	0.76	1.28	5.50	6.30	6.50	8.00	5.28
P_2O_5	0.14	0.18	0.18	0.40	0.16	0.30	0.02	0.01	0.01	0.01	0.01
F	0.03	0.03	0.02	0.08	0.04	0.05	0.09	0.02	0.05	0.05	0.03
LOI	3.72	4.38	4.68	4.72	2.72	4.20	0.98	0.51	1.22	1.44	0.45
Total	99.63	100.30	99.50	99.63	99.67	100.49	99.92	99.61	99.57	99.55	100.49

Table 3.6 Chemical composition of representative Devonian rocks of the granite–leucogranite and granosyenite formations of the Tess intrusive complex of Northern Mongolia. From Marinov (1973).

Components	Granosyenites			Leucogranites				
SiO_2	65.10	67.00	67.93	71.18	71.24	71.80	72.42	75.70
TiO_2	0.9	0.56	0.55	0.35	0.38	0.36	0.32	0.26
Al_2O_3	15.50	15.78	16.12	14.93	13.42	12.93	15.40	11.89
Fe_2O_3	3.21	2.50	1.50	2.32	1.94	2.04	0.99	2.16
FeO	0.36	2.06	1.21	0.21	1.61	1.11	0.72	0.58
MnO	0.11	0.08	0.10	0.05	0.10	0.13	0.09	0.13
MgO	0.68	1.07	0.49	0.39	0.44	0.42	0.09	0.27
CaO	1.90	1.77	1.91	0.90	1.22	0.96	0.70	0.30
Na_2O	4.13	7.15	4.20	3.44	4.75	4.24	4.24	4.19
K_2O	6.74	3.20	5.46	5.45	3.75	4.64	4.94	4.09
P_2O_5	0.02	0.03	0.07	0.05	0.02	0.04	0.01	0.04
LOI	0.57	1.10	0.52	0.60	0.74	0.74	0.81	0.26
Total	99.22	100.10	100.06	99.87	99.61	99.41	100.73	99.87

Table 3.7 Chemical composition of rocks of the Devonian trachyrhyolite formation, from the Talmennur Depression of North Mongolia. After Gordienko (1987).

Components	Trachydacites			Trachyrhyolites			
SiO_2	65.00	65.81	66.38	69.18	72.01	72.66	64.96
TiO_2	0.20	0.60	0.76	0.49	0.21	0.24	0.12
Al_2O_3	15.98	16.72	15.94	16.11	11.35	13.87	12.73
Fe_2O_3	2.00	2.22	3.77	1.18	0.91	1.04	0.92
FeO	1.28	1.08	0.70	0.99	1.04	1.17	0.97
MnO	0.09	0.06	0.05	0.03	0.03	0.01	0.08
MgO	0.99	0.63	0.44	0.09	0.07	0.08	0.09
CaO	1.85	0.85	0.65	0.37	0.32	0.21	0.21
Na_2O	4.57	5.16	5.09	5.98	4.73	4.41	3.01
K_2O	5.50	5.19	3.95	4.07	5.19	5.12	5.62
P_2O_5	0.22	0.13	0.26	0.13	0.07	0.04	0.05
LOI	2.02	0.98	1.41	0.91	0.49	0.63	0.58
Total	99.70	99.43	99.40	99.53	99.41	99.51	99.34

Table 3.8 Chemical compositions of Devonian alkaline volcanic assemblages from the Haan–Hohniy Ridge Zone, Northern Mongolia (after Yarmolyuk and Kovalenko, 1991).

Components	1	2	3	4	5	6	7	8	9	10	11	12
SiO_2	47.79	51.55	45.88	44.92	45.54	48.82	54.32	34.03	70.41	70.39	60.74	61.38
TiO_2	2.75	3.03	2.06	2.71	2.48	2.27	1.23	4.55	0.25	0.24	0.05	0.05
Al_2O_3	16.05	13.79	15.49	16.07	16.29	17.58	18.03	8.58	12.72	12.69	18.59	18.43
Fe_2O_3	12.77	13.37*	4.47	4.21	12.39*	10.87*	8.06*	27.85	5.11*	4.92*	4.25*	3.82*
MnO	0.31	0.23	0.23	0.20	0.17	0.19	0.15	0.24	0.99	0.13	0.31	0.23
MgO	5.27	4.40	6.91	6.27	5.82	4.13	1.80	14.23	0.06	0.05	0.23	0.14
CaO	3.23	6.40	8.50	8.58	9.73	5.21	4.11	2.73	0.32	0.46	0.57	0.87
Na_2O	4.95	3.86	2.75	2.36	3.28	4.77	5.89	1.85	6.21	6.00	8.38	8.68
K_2O	1.60	1.47	2.44	3.29	1.52	2.55	3.88	0.93	4.54	4.34	4.61	4.62
P_2O_5	0.76	0.55	0.34	0.69	0.47	0.94	0.37	0.29	0.02	0.02	0.02	0.02
LOI	4.40	1.67	3.70	3.54	2.54	2.82	2.17	3.78	0.32	0.66	2.11	1.73
Total	99.91	100.32	99.77	99.57	100.23	100.15	100.01	99.79	100.05	99.9	99.86	99.97
Li	50	16	52	54	20	44	32	14	30	60	64	66
Rb	22	20	46	40	16	28	46	11	172	178	163	167
Nb	27	—	10	29	13	30	50	12	169	173	300	169

Table 3.8 *Contd.*

Components	1	2	3	4	5	6	7	8	9	10	11	12
Zr	254	206	133	215	148	230	338	127	1,689	1,917	2,300	1,578
Cs	—	—	—	—	—	2	2	4	—	—	—	—
La	56	17	15	45	30	64	73	17	150	240	190	170
Ce	130	26	35	90	72	120	110	250	380	300	270	—
Nd	25	10	13	13	12	30	52	20	130	170	150	90
Yb	1.4	4.7	2.2	2.2	2.1	2.5	3.1	2	9.6	16	16	11
Y	21	30	14	26	21	27	35	7	119	129	130	110
Sr	580	340	780	680	900	880	440	350	10	14	320	510
Ba	600	480	1,100	530	190	920	1,400	100	36	35	110	90
Pb	12	12	12	3.4	3.5	12	9	1.3	20	23	43	90
Zn	270	150	220	170	83	160	95	190	160	190	190	26
Sn	3.9	6.1	2.0	3.6	4.0	3.2	5.6	2.4	34	18	34	180
Ni	16	18	120	83	62	13	4	380	7	3	—	25
Cr	3	14	140	140	110	10	8	330	17	6	—	7
												5

Columns 1–2 — subalkaline basalts; 3–4 — pseudoleucitic trachybasalts; 5–8 — differentiates of a teschenite sill; 9–10 — pantellerites; 11–12 — mariupolites (a sill).

* — FeO + Fe$_2$O$_3$ as Fe$_2$O$_3$.

FIGURE 3.4 Palaeotectonomagmatic map for the late Palaeozoic of Central Asia. 1–11 — igneous complexes of marginal belt: 1 — basalts and andesites, 2 — andesites and dacites, 3 — rhyolites, 4 — subalkaline andesites, 6 — moderately alkaline olivine basalts, 7 — basalt–pantellerite assemblage, 8 — diorite–granodi- orite assemblages, 9 — granodiorite–granite assemblages, 10 — granite–leucogranite assemblages, 11 — alkaline granitoids; 12 — faults; 13 — platform bordering on the fold belt; 14 — late Hercynian rocks of the Palaeotethys; 15 — Dasibalbar marine basins; 16 — southern Mongolian Hercynian rocks. Circled figures: I–III — volcanic belts: I — southern Mongolian, II — central Mongolian, III — northern Mongolian–Transbaikal; IV — Khangai Batholith.

from the edge of the Palaeotethys (Yarmolyuk and Kovalenko, 1991; Kovalenko and Yarmolyuk, 1995; Kovalenko *et al.*, 1995a). The formation of this zone is attributed to the gradual closure of the Palaeotethys which had ceased to exist in the Irtysh–Zaisan Zone and in the South Mongolian Hercynides by the middle early Carboniferous.

b) Late Palaeozoic Magmatism

During the Late Palaeozoic, marine environments, as mentioned above, persisted mainly in the Southernmost Central Asian fold belt, namely, in the Solonker Zone. The majority of the area, including the South Mongolian Hercynides, developed during that time, under continental conditions. Numerous, diverse igneous assemblages were formed, widely ranging in composition. They are concentrated within an extensive volcanic–plutonic belt (Fig. 3.4) located along the continental margin within the Late Palaeozoic Eurasian Palaeocontinent and followed the boundary of the late Hercynian Palaeotethys (Mossakovsky, 1975; Yarmolyuk and Kovalenko, 1991).

Within the belt, two complexes of igneous rocks, identified as differentiated and bimodal, can be recognized. The former consists of differentiated assemblages, varying in composition from basalts through andesites and dacites to rhyolites. The latter is composed mainly of basic and acid rocks; intermediate rocks are actually absent.

Differentiated Complex

The formation of this complex occurred during two stages: i.e. the Carboniferous and Permian. During the Carboniferous igneous activity was concentrated exclusively within the Hercynian terrane, where the South Mongolian volcanic belt formed. During the Permian igneous activity shifted northwards into the Caledonian terrane. This resulted in the formation of a structurally complex igneous terrane, including two volcanic belts, i.e. the Central Mongolian and North Mongolian–Transbaikalian belts, separated by the Khangai Batholith (Fig. 3.5).

Irrespective of their age and geographic position, the Late Palaeozoic volcanic belts of rocks of the differentiated complex of Central Asia are very similar in their structure (Yarmolyuk, 1983). Volcanic areas are mainly composed of two volcanic assemblages: the earlier, basalt–andesite, and the later, dacite–rhyolite. The former consists of basalts, andesitic basalts, andesites and to a lesser extent tuffs of the same composition. The section thickness reaches 2,000 m. The dacite–rhyolite assemblage rests conformably on these rocks. Usually it is composed of tuffs and ignimbrites of dacitic, rhyodacitic and rhyolitic composition, and to a lesser degree, by lavas of this composition. They form sequences up to 2,000–3,000 m thick. Sections in both assemblages show homodromous

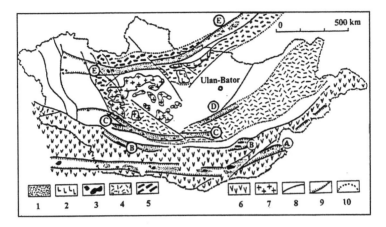

FIGURE 3.5 Distribution of rift zones within the late Palaeozoic marginal belt in Mongolia. After Yarmolyuk and Kovalenko (1991). Key: Rift zones A — Gobi–Tien Shan; B — Main Mongolian lineament; C — Gobi Altai; D — North Gobi; E — North Mongolian–Transbaikal. 1 — bimodal volcanic assemblages; 2 — moderately alkaline olivine basalts; 3 — alkaline granitoids; 4–6 — domains of differentiated volcanic complex: 4 — early Permian Central Mongolian volcanic areas, 5 — early Permian North Mongolian volcanic areas, 6 — early–middle Carboniferous South Mongolian volcanic areas, 7 — Khangai batholith; 8 — faults; 9 — rift zone boundaries; 10 — boundaries of volcanic areas.

differentiation leading to the formation of intermediate rocks in the transitional zones between assemblages.

The chemical composition of volcanics from these assemblages correlated with that of calc-alkaline and K–Na subalkaline series (Table 3.9). Calc-alkaline rocks in both the Carboniferous and Permian terranes are concentrated in the south, close to the boundary of the palaeocontinent with the Palaeotethys.

Rocks of higher alkalinity (Table 3.10) occupy the rear (intracontinental) zone. This distribution of volcanic rock compositions has been interpreted in terms of the relationship of magmatism to subduction zones underthrusting the edge of the late Hercynian Palaeotethys under the continent. During the Carboniferous and Permian the subduction zone was tilted at an angle of 45° and 15°, respectively (Yarmolyuk and Kovalenko, 1991).

The volcanics associated with diverse granitoids grouped into granodiorite–granite, leucogranite and granite–leucogranite, with granosyenite assemblages (Table 3.11).

The Khangai Batholith, located between volcanic belts, occupies a special place in the structure of the Permian igneous terrane. It has an area of 100,000 km². The batholith is composed of diverse granitoids that can be grouped into two multiphase intrusive complexes (Table 3.12): The early

Table 3.9 Chemical composition of representative Carboniferous volcanic assemblages of Southern Mongolia.

Components	1	2	3	4	5	6	7	8	9	10	11
SiO_2	54.96	53.33	51.06	48.64	55.06	60.32	72.40	71.20	73.90	75.50	77.05
TiO_2	0.83	0.77	0.73	0.92	0.76	0.61	0.43	0.41	0.36	0.26	0.18
Al_2O_3	16.10	16.45	17.17	17.56	18.86	16.80	13.60	14.80	13.80	12.60	12.80
Fe_2O_3	4.80	1.65	1.79	2.56	4.94	1.23	1.39	1.69	0.87	0.74	0.91
FeO	2.69	7.48	8.49	8.70	2.69	4.63	1.09	1.09	0.84	0.73	0.58
MnO	0.12	0.28	0.22	0.19	0.13	0.11	0.06	0.09	0.07	0.07	0.04
MgO	4.66	4.58	3.58	4.58	2.89	3.85	0.96	0.64	0.24	0.36	0.36
CaO	6.99	9.00	8.83	9.12	6.82	4.04	0.90	1.70	1.10	0.70	0.65
Na_2O	3.16	3.32	2.84	2.88	3.38	3.83	5.27	3.82	4.11	2.77	2.77
K_2O	2.32	1.36	2.46	1.90	2.68	2.95	2.19	2.47	2.82	4.86	3.38
P_2O_5	0.99	0.34	0.53	0.40	0.38	0.31	0.16	0.02	0.08	0.09	0.10
LOI	3.72	1.59	2.50	2.92	1.50	2.26	1.16	1.02	1.38	1.16	1.00
Total	100.34	100.15	100.20	100.37	100.09	100.94	99.61	98.95	99.57	99.84	99.82

Columns 1–6 — andesite–basaltic assemblage, Edrengiyn–nuuru Ridge. The rocks were sampled going up the section (1983). Columns 7–11 — rhyolitic assemblage, Noyon Ridge. The rocks were sampled going up the section. After Yarmolyuk (1978).

Table 3.10 Chemical composition of representative rocks from the differentiated complex of the volcanic–plutonic belt of the Orkhon–Selenga Depression, North Mongolia. After Gordienko (1987).

Components	Basic and intermediate volcanic suites					Acid volcanic suite				
SiO$_2$	51.50	51.10	58.50	58.54	60.12	66.53	69.70	70.92	71.24	76.40
TiO$_2$	1.17	1.15	1.66	1.63	1.66	0.71	0.64	0.35	0.42	0.17
Al$_2$O$_3$	15.38	14.94	16.61	16.70	17.92	15.85	14.10	13.57	13.36	12.60
Fe$_2$O$_3$	3.66	4.57	3.55	3.54	4.78	2.67	2.67	1.76	1.36	—
FeO	5.64	5.27	2.65	2.09	0.84	1.11	1.65	1.68	1.43	1.43
MnO	0.15	0.18	0.08	0.08	0.07	0.12	0.12	0.22	0.21	0.09
MgO	6.00	6.23	1.77	1.88	0.44	0.12	0.26	0.08	0.21	0.08
CaO	9.61	9.50	3.79	3.88	2.55	1.16	0.69	0.64	0.64	0.41
Na$_2$O	2.22	2.76	3.91	3.84	3.84	5.72	4.36	3.98	4.90	3.60
K$_2$O	1.84	1.31	3.74	3.77	4.55	4.41	4.63	5.66	4.92	4.70
P$_2$O$_5$	0.23	0.22	0.61	0.67	0.65	0.12	0.13	0.06	0.09	0.04
LOI	2.41	2.09	3.43	3.04	2.79	2.14	0.62	0.78	0.64	0.51
Total	99.44	99.08	100.17	99.66	100.25	99.24	99.57	99.70	99.64	100.03

Table 3.11 Chemical composition of representative samples of Permian granite–leucogranite and granosyenite formation, Central Mongolia. From Marinov (1973).

Components	1	2	3	4	5	6	7	8	9	10	11
SiO_2	62.80	68.32	69.45	71.52	73.90	66.86	68.98	68.04	74.08	74.45	76.14
TiO_2	0.80	0.27	0.06	0.26	0.25	0.65	0.37	0.29	0.09	0.30	0.12
Al_2O_3	15.70	14.65	16.68	14.01	13.70	14.45	14.96	15.98	14.46	12.70	12.99
Fe_2O_3	3.00	4.79	1.01	3.06	1.68	2.04	1.50	2.01	1.27	0.81	0.38
FeO	3.20	0.36	0.79	0.22	0.80	2.56	1.97	1.02	0.73	0.72	1.01
MnO	0.07	0.02	0.02	0.05	—	0.08	0.12	0.06	0.06	0.04	0.06
MgO	0.39	0.91	0.20	0.15	0.27	1.60	1.30	0.66	0.02	0.17	0.51
CaO	2.91	0.70	1.30	1.12	1.00	2.82	3.0	1.85	0.85	1.30	0.56
Na_2O	5.31	4.62	4.82	4.56	5.47	3.41	4.22	4.86	3.30	3.36	4.15
K_2O	4.22	3.95	4.22	3.84	3.44	4.47	3.20	4.26	5.10	6.00	3.88
P_2O_5	0.27	0.15	0.02	0.07	—	0.17	0.11	0.20	0.03	0.05	0.03
LOI	0.58	1.38	0.89	0.57	—	1.22	0.90	0.69	0.18	0.33	0.22
Total	99.25	100.11	99.46	99.40	100.51	100.33	100.63	99.92	99.47	100.63	100.05

Column 1 — syenite, 2, 6–8 — granosyenite, 3–5 — biotite leucogranite, 9–11 — granites.

Table 3.12 Chemical composition of representative rocks from the Khangai Batholith. From Marinov (1973).

Components	Khangai Complex						Sharausgol Complex					
SiO_2	65.63	67.47	68.14	69.15	70.42	74.62	67.28	69.84	70.28	72.04	74.48	75.93
TiO_2	0.63	0.50	0.45	0.30	0.35	0.18	0.43	0.39	0.24	0.16	0.28	0.15
Al_2O_3	16.36	15.00	14.92	14.50	15.19	13.90	16.85	15.19	15.42	13.76	13.13	13.71
Fe_2O_3	0.88	0.97	0.68	3.05	0.91	0.36	0.77	1.45	1.70	1.70	0.90	0.20
FeO	4.18	2.84	3.03	2.27	1.85	1.82	1.81	1.63	1.22	1.44	0.72	0.78
MnO	0.12	0.09	0.11	0.09	0.09	0.08	0.05	0.11	0.09	0.04	0.06	0.01
MgO	1.21	1.86	1.71	0.98	1.67	0.33	1.05	0.65	0.69	0.22	0.61	0.08
CaO	3.23	3.06	2.82	2.56	2.00	1.22	1.67	1.53	1.55	0.90	1.00	0.34
Na_2O	3.71	3.08	3.31	3.20	3.54	3.81	5.56	4.28	3.81	3.88	3.26	4.04
K_2O	3.19	3.95	3.44	3.18	4.26	3.72	3.06	4.74	4.83	4.96	4.90	4.28
P_2O_5	0.14	0.23	0.17	0.14	0.05	0.10	0.51	0.09	—	—	0.05	0.02
LOI	0.60	1.14	1.10	1.04	0.58	—	1.34	0.11	0.65	0.66	0.28	0.38
Total	99.78	100.19	99.94	100.46	100.91	100.14	100.38	99.92	100.48	99.76	99.67	99.92

(Khangai) complex, consisting of hornblende–biotite and biotite calc-alkaline granodiorites and granites, and the late (Sharausgol) complex (subalkaline leucocratic and biotite granites) (Fedorova, 1978). Age data from the batholith indicate that it was formed during the Permian.

Igneous Assemblages of the Bimodal Complex

The bimodal complex is made up of volcanic bimodal assemblages (basalt–trachyrhyolite and basalt–comendite) after which the complex is named, and alkaline granites and syenites.

These assemblages are spatially related to linear submeridional graben systems superimposed on Late Palaeozoic volcanics of the differentiated complex. The bimodal volcanic assemblages are separated from the differentiated complex by transitional units of tuffaceous sedimentary rocks or by erosion surfaces.

The bimodal volcanic assemblages are of similar types within different zones (Yarmolyuk and Kovalenko, 1991). They are composed of alkaline basalts, trachyrhyolites, comendites, pantellerites and to a lesser degree trachyandesites, trachydacites and trachytes. The age of the rocks of the bimodal complex in Central Asia was found to change regularly. The earliest (late Carboniferous–early Permian) are assemblages occurring in the Gobi–Tien Shan Zone of South Mongolia. In the late early Permian the bimodal assemblage was formed in the Gobi–Altai and North Gobi zones, and in the late Permian, in the North Mongolian–Transbaikal zone. The isochronous Rb/Sr dates of the volcanic rocks are available for the latter; the age was estimated as 249 Ma (Yarmolyuk *et al.*, 1990). This age distribution for the bimodal complex suggests a successive northward shift of bimodal magma generation zones.

The recognized zones of bimodal assemblages are interpreted as rift zones. These zones are grouped into the Late Palaeozoic rift system of Central Asia (Yarmolyuk and Kovalenko, 1991). Rocks of bimodal volcanic assemblages (Table 3.13) belong to the moderately alkaline and alkaline potassic–sodium series (Yarmolyuk and Kovalenko, 1991). The basic rocks show a high alkalinity ($K_2O + Na_2O = 4$–8 wt.%) and TiO_2 (1.8–3.3 wt.%), ($^{87}Sr/^{86}Sr)_0$ amounts to 0.7047–0.705 (Yarmolyuk and Kovalenko, 1991). Rocks of the acid group have high silica content (66–77% SiO_2), a low CaO content and low Al_2O_3 (9–13.5 wt.%). High Na_2O and K_2O content (up to 13 wt.%) were found in almost equal ratios. The agpaitic factor is close to 1, and even in subalkaline rocks and alkaline rocks it is about 1; ($^{87}Sr/^{86}Sr)_0$ varies from 0.7055 in trachyrhyolites to 0.712 in pantellerites (Yarmolyuk and Kovalenko, 1991).

Intrusive rocks of the bimodal complex are represented almost exclusively by alkaline granites and syenites. These rocks always contain only alkali-feldspar and are leucocratic. They are distinquished by their low content of alkali-earth elements. An elevated content (up to ore concentrations) (Tables 3.14 and 3.15) of REE, Y, Zr, Nb, Hf, Ta, Zn, Sn and a low Sr and Ba content are the main

Table 3.13 Chemical composition of rocks of the bimodal volcanic assemblage of the Egiyn–Gol River area of North Mongolia (after Yarmolyuk and Kovalenko, 1991).

Components	1	2	3	4	5	6	7	8	9	10	11	12	13
SiO_2	49.20	60.69	46.99	47.77	67.08	67.43	68.39	47.50	65.36	69.30	48.62	65.63	70.00
TiO_2	1.81	0.89	1.85	2.13	0.76	0.74	0.73	2.02	0.56	0.57	2.59	1.47	0.64
Al_2O_3	16.76	17.84	16.48	15.85	14.47	14.67	14.42	16.22	15.98	13.17	15.04	15.41	14.70
Fe_2O_3	3.80	3.15	5.14	5.08	2.84	3.13	4.33	5.60	2.16	4.50	6.59	4.57	3.49
FeO	6.37	1.73	5.85	6.81	2.11	1.56	0.90	5.90	2.22	2.15	5.28	—	0.10
MnO	0.17	0.11	0.16	0.21	0.16	0.14	0.12	0.17	0.11	0.25	0.17	0.19	0.15
MgO	5.06	0.87	5.11	5.17	0.57	0.40	0.05	4.57	0.29	0.24	6.24	0.11	0.09
CaO	7.19	1.98	7.78	8.20	0.31	0.34	0.18	7.73	0.34	0.26	8.51	0.56	0.10
Na_2O	3.65	6.30	3.70	3.65	5.76	5.76	5.18	3.54	6.72	5.18	3.20	5.95	4.94
K_2O	2.60	5.13	1.63	1.00	5.20	4.91	4.96	2.14	5.41	4.56	1.43	5.05	5.25
H_2O	0.88	0.36	0.74	1.26	0.10	0.17	0.23	1.38	0.24	0.14	0.76	0.30	0.01
LOI	2.39	0.70	4.57	2.69	0.96	0.73	0.49	3.10	0.88	0.76	1.64	0.94	0.47
La	37.3	—	35.2	25.4	89.5	49.3	65.6	37.7	30.5	79.6	21.8	101.2	123.7
Ce	80	—	79	57	191	100	146	77	64	180	48	216	320
Nd	44	—	46	36	92	59	65	43	35	79	30	95	135
Sm	10.3	—	11.7	8.5	18.9	11.7	12.8	9.5	7.9	16.2	6.7	19.2	29.2

Table 3.13 *Contd.*

Components	1	2	3	4	5	6	7	8	9	10	11	12	13
Eu	2.9	—	3.4	2.6	1.8	2.7	1.9	2.5	2.3	1.3	2.2	1.58	2.2
Gd	5.5	—	5.9	4.8	15.1	9.2	11.7	7	5.7	14.6	5.8	15.4	24.6
Tb	0.95	—	1.1	0.81	2.46	1.6	1.83	1.16	0.9	2.23	0.8	2.5	4.04
Yb	2.8	—	2.9	2.7	10.3	5.4	7.0	3.0	2.4	9.5	1.9	11.5	17.5
Lu	0.35	—	0.4	0.4	1.4	0.8	1.1	0.43	0.3	1.2	0.34	1.4	2.2
Y	34	—	34	25	87	49	54	33	24	80	21	116	162
Zr	294	—	264	246	894	456	563	325	253	761	241	985	1364
Nb	14	—	11	7	50	27	34	12	8	49	6	57	87
Rb	63	—	39	16	96	80	106	27	36	121	12	140	118
Hf	4.9	—	4.6	4.2	24.2	11.0	14.9	5.6	4.1	20.7	3.5	27.0	41.0
Ta	0.7	—	0.5	0.5	2.5	1.2	1.8	0.8	0.4	2.4	0.4	2.9	4.3
Sr	800	—	590	789	32	88	125	943	850	59	1,004	40	31
Sc	28.1	—	28.8	28.3	8.6	8.9	6.5	23.9	24.1	6.7	24.5	6.7	5.8
Co	29.2	—	29.4	40.1	1.25	1.5	2.2	36.3	38.7	2.1	39.0	1.9	0.7
Cr	124	—	112	123	157	39	68	163	300	42	151	51	23

Rocks were sampled up the section of the volcanic suite: 1 — basalt; 2 — trachyte; 3–4 — basalts; 5–7 — pantellerite; 8 — basalt; 9 — alkaline trachydacite; 10 — pantellerite; 11 — basalt; 12 — alkaline trachydacite; 13 — pantellerite.

Table 3.14 Compositions of alkaline granites from the Haan–Bogdo Massif. After Kovalenko (1977).

Components	1	2	3	4	5	6	7	8	9
SiO_2	74.45	73.84	76.90	73.76	71.90	71.28	67.40	71.96	66.20
TiO_2	0.16	0.18	0.18	0.25	0.42	0.44	0.66	0.26	0.87
Al_2O_3	11.62	11.35	12.10	10.77	8.58	8.55	5.36	9.04	4.17
Fe_2O_3	1.90	2.10	0.65	2.43	4.20	3.67	5.28	5.94	12.00
FeO	2.43	2.25	1.35	1.98	4.10	4.00	4.76	0.27	0.89
MnO	0.15	0.18	0.08	0.20	0.27	0.28	0.47	0.21	0.31
MgO	0.03	0.02	0.01	0.40	0.16	0.28	0.01	0.03	0.02
CaO	0.05	0.06	0.05	0.58	0.45	0.58	0.16	0.07	0.16
Na_2O	4.61	4.61	4.31	4.93	5.14	5.04	4.17	5.32	5.54
K_2O	4.59	4.80	4.69	4.77	3.53	3.95	4.69	3.99	3.90
F	0.13	0.08	0.04	0.13	—	—	0.34	0.18	0.03
H_2O	0.10	0.22	0.22	0.52	0.98	1.32	1.90	0.90	1.95
ZrO	—	—	—	—	—	—	5.40	1.06	5.14
Total	100.22	100.40	100.57	100.67	99.73	99.34	100.47	99.26	101.11
a.c.	1.08	1.11	1.00	1.22	1.65	1.73	2.22	1.44	3.17

Columns 1–6 — main-phase granites; 7–9 — rare–metal granites.

Table 3.15 REE content in the rocks of the Haan–Bogdo Massif. After Kovalenko (1978).

Components	1	2	3	4	5	6
La	64	657	2,400	9,400	487	2,900
Ce	138	1,280	4,450	1,300	577	3,700
Pr	12	203	953	1,900	53	560
Nd	58	687	1,933	5,700	303	1,700
Sm	19	203	308	1,000	80	370
Eu	—	—	—	—	—	—
Gd	15	177	230	5,200	109	240
Tb	—	—	—	—	—	—
Dy	17	237	148	450	131	210
Ho	2	47	26	79	33	33
Er	10	53	54	170	64	77
Yb	21	267	73	195	150	93
Y	63	733	918	2,500	593	1,200
(TR+Y)%	0.04	0.45	1.10	2.80	0.26	1.10
La/Yb	3.0	2.5	32.9	48.2	3.2	31.2

Column 1 — main-phase granites; 2–5 — elpeditic ekerites and pegmatites; 6 — elpedite–armstrongite ekerite.

geochemical characteristics of the rocks (Kovalenko, 1977). The concentrations of Li, Rb, Be and F are often quite high.

c) Early Mesozoic Magmatism

Figure 3.6 shows the distribution pattern of Early Mesozoic (Triassic–middle Jurassic) igneous formations. The outbreak of igneous activity occurred in the second half of the Triassic–early Jurassic. Igneous rocks underlie an extensive igneous terrane that occurs towards the Mongol–Okhotsk part of the Central Asian Belt.

The early Mesozoic igneous rocks form a zonally symmetrical area. The core of the area (the Khentei batholith) is made up of large masses of granitoids of the granodiorite–granite assemblage and by smaller intrusive bodies of granitoids of the leucocratic assemblage. This core is bordered to the north, west and south by a zone of smaller intrusions dominated by masses of the leucocratic assemblage.

The outer zone of the igneous terrane consists of volcanic fields and areally scattered, relatively small masses of granitoids. It is also known as a zone of dispersed magmatism, the name reflecting a discrete dispersed distribution pattern of igneous rocks (Kovalenko and Yarmolyuk, 1990). Granitoids consist of alkaline granitoids, granosyenites, amphibole–biotite leucocratic alkali-feldspar

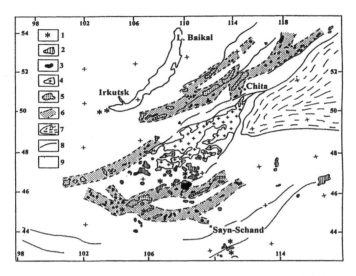

FIGURE 3.6 Distribution of igneous rocks within the Mongol–Okhotsk Belt in the early Mesozoic. Key: 1 — lithium–fluoric granites; 2 — leucogranites; 3 — alkaline granitoids and other alkaline rocks; 4 — granodiorite–granite and gabbro–diorite rock assemblages; 5 — volcanics with high alkalinity (trachybasalts, trachyandesites, etc.); 6 — zones of alkali rocks distribution; 7 — eugeosynclinale zone with spilite–dolerite volcanism and ultrabasic rocks; 8 — main lineaments; 9 — amagmatic terranes.

granites along with Li–F granites. The latter form small intrusions and are the youngest of the structurally complex masses of leucocratic granite assemblage. Therefore, the area is characterized by a concentrically zoned distribution of granitoids of different compositions. The core contains mainly granitoids of the granodiorite–granite assemblage (Table 3.16), the intermediate zone contains of the leucogranite assemblage rocks (Table 3.17) and the outermost zone (zone of dispersed magmatism) contains various granitoids, of mostly subalkaline and alkaline composition, as well as Li–F granites. Most of the Mesozoic granitiod assemblages are similar to those of the late Palaeozoic. An unusual feature of early Mesozoic magmatism is the presence of Li–F granites, with an elevated content of lithophilic elements, particularly lithium and fluorine (Kovalenko, 1977). There are also microcline–albite and amazonite–albite granites with protolithionite, zinnwaldite and lepidolite. The Li–F granites are subalkaline rocks strongly depleted in calcium and oversaturated in Al_2O_3 (Table 3.18). These rocks have elevated lithium and fluorine content (above average for granites: 2–8 and 4–5 times the average, respectively) as well as Rb (300–750 ppm), Sn (4–12.5 ppm), Be (4–11 ppm), Ta and Hf. Barium and Sr content are anomalously low (6–84 and 4–21 ppm, respectively). The REE

Magmatism and Geodynamics

Table 3.16 Chemical composition of Early Mesozoic granite–granodiorites. From Kovalenko (1982).

Components	1	2	3	4	5	6
SiO_2	64.77	66.27	68.29	70.97	74.59	74.27
TiO_2	0.78	0.54	0.40	0.25	0.12	0.22
Al_2O_3	16.95	16.22	17.41	14.99	13.84	12.76
Fe_2O_3	1.77	0.91	0.46	1.40	0.84	0.72
FeO	2.05	2.64	1.98	1.28	0.99	1.25
MnO	0.12	0.05	0.09	0.05	0.03	0.08
MgO	1.39	1.21	0.84	0.42	0.28	0.27
CaO	3.02	3.09	1.63	1.85	0.82	0.91
Na_2O	4.94	4.61	4.33	3.54	3.76	3.56
K_2O	3.59	2.92	3.88	2.95	4.28	4.83
P_2O_5	0.18	0.12	0.14	0.01	0.06	0.02
LOI	0.59	0.47	0.54	0.36	0.51	0.48
Total	100.20	99.06	100.07	99.07	100.19	99.35

Columns 1–3 — granodiorites, 4–6 — granites.

Table 3.17 Chemical composition of representative rocks of Mesozoic leucogranite formation. From Zaitsev and Tayson (1971).

Components	1	2	3	4	5	6	7	8
SiO_2	73.50	73.86	70.94	71.27	73.59	73.25	71.84	74.86
TiO_2	0.30	0.20	0.35	0.25	0.15	0.10	0.15	0.10
Al_2O_3	14.16	13.97	15.01	14.91	16.64	12.46	12.46	12.80
Fe_2O_3	0.57	0.96	1.56	1.59	0.82	0.66	1.63	0.68
FeO	1.54	0.88	1.49	1.07	0.85	1.10	1.28	0.80
MnO	0.03	0.04	0.09	0.10	0.05	0.21	0.17	0.17
MgO	0.40	0.16	0.28	0.08	0.12	0.31	0.07	0.15
CaO	1.14	1.14	1.77	1.60	1.48	2.75	1.21	0.88
Na_2O	3.37	4.22	4.11	4.43	3.84	3.54	5.40	3.80
K_2O	5.04	4.00	4.16	4.31	3.75	3.91	4.17	4.32
P_2O_5	0.08	0.06	0.05	0.02	0.02	0.02	0.18	0.09
LOI	0.30	0.28	0.19	0.21	0.20	0.83	0.80	0.85
Total	100.79	99.77	100.00	99.84	99.51	99.30	99.38	99.50

Columns 1–2 — Ikh-Khaizkhan massif, 3–5 — Onton-Khaizkhan massif, 6–8 — Unzhul massif.

Table 3.18 Chemical composition of lithium–fluorine granites of Mongolia. From Zaitsev and Tayson (1971).

Components	I		II		III		IV	V
SiO$_2$	74.1	75.95	75.89	76.77	74.95	76.15	71.97	70.816
TiO$_2$	0.1	—	0.01	0.10	—	0.10	—	—
Al$_2$O$_3$	12.6	12.74	12.91	11.89	13.24	12.91	15.82	16.149
Fe$_2$O$_3$	0.3	0.20	0.02	0.07	0.15	0.51	0.06	0.143
FeO	1.7	0.86	1.03	1.22	1.05	1.58	0.32	0.507
MnO	0.0	0.02	0.02	0.06	0.02	0.02	—	0.203
MgO	—	0.33	—	—	—	0.01	0.03	0.12
CaO	0.7	0.46	0.29	0.53	0.26	0.28	0.12	0.246
Na$_2$O	4.0	4.73	4.59	4.39	4.32	4.54	6.89	6.005
K$_2$O	5.1	3.64	4.45	3.96	5.18	3.73	3.73	3.248
P$_2$O$_5$	0.0	0.02	—	—	0.01	0.01	—	0.103
F	0.3	0.31	0.28	0.24	0.34	0.25	0.25	2.605
LOI	0.6	0.45	0.26	0.54	0.30	0.16	0.38	1.106
Li	195	338	2.89	88	4.28	396	878	2136
Rb	511	730	795	422	881	753	1.500	2161
Pb	21	84	38	46	39	111	24	43
Sn	12	12.3	50	6.9	102	15.3	102	47
Zn	37	137	64	57.5	63	280	35	44
Tl	1.4	4.2	2.5	1.9	4.4	4.0	3.7	—
Mo	2.0	0.8	—	1.5	1.4	1.0	1.8	—
Be	9.9	5.4	7.0	6.3	6.4	7.9	4.9	22.7
Ba	—	17.9	—	10.3	—	13.6	—	8.7
Sr	—	3.0	—	4.7	—	2.5	—	6.3
B	14	13	10	10	11	10	10	11
K/Rb	86	44	46	81	40	44	23	11.7
(TR + Y)	420	340	220	—	240	340	40	110

I — alaskites with biotite; II — microcline-albite alaskites; III — amazonite–albite granites; IV — albite–lepidolite granites; V — topaz-bearing rhyolite (ongonite).

content is below the average for granites. The initial $^{87}Sr/^{86}Sr$ ratio varies from 0.7107 to 0.7177 (Kovalenko, 1977).

Among the plutonic rocks of the outer zone, other subalkaline (monzonites, granosyenites and syenites) rocks and alkaline granitoids are also very common. The granitoids, unlike the Late Palaeozoic alkaline granites, are depleted in rare alkalies, fluorine and lithophilic trace elements like Sn, Be and Mo. The Sr content is anomalously low (less than 25 ppm).

By contrast the Early Mesozoic volcanic rocks are less common than their plutonic equivalent. Their exposures occur only in the peripheral zone of the igneous terrane (zone of dispersed magmatism). Volcanic areas, as a rule, are small in size and scattered. The volcanics show a wide range of compositions, from basalts to rhyolites, and occur in different assemblages. The basalt–trachyandesite–trachybasalt, trachyandesite–dacite–trachyrhyolite and bimodal basalt–trachyrhyolite–comendite assemblages are the most abundant. On the whole, these rocks correlate with the moderately alkaline potassic–sodium series.

d) The Late Mesozoic–Cenozoic Magmatism

The main igneous events from this period are related to intra-plate activity that affected the whole fold belt east of longitude 94°E. They resulted in the formation of numerous volcanic centres, grouped into the Asian intra-plate volcanic province (Yazmolyuk et al., 1996). The igneous products are irregularly distributed over the whole area. They are grouped in a number of areas differing in their magmatic history (Fig. 3.7). Generally, the province was formed by multiple volcanic events that occurred without major breaks from the late Jurassic to the Holocene.

The Late Jurassic

Igneous events from this period were mainly caused by formation of the marginal volcanic–plutonic belt of the Great Khingan Mountains and widespread intra-plate igneous activity to the west of it (Fig. 3.7). The latter occurred in three isolated areas — the South Khangai, the West Transbaikal and the Central Aldan. Despite the fact that they are spatially and structurally separated, the igneous assemblages formed there have much in common. They are distinguished by their high alkalinity, which led to the formation of silica-depleted rocks such as melanephelinites and melaleucitites, leucitites and nephelinites, phonolites, trachytes, trachyte–latites and their subalkaline equivalents because of the predominance of intermediate rocks. These igneous assemblages vary regionally, for example the rocks of the Central Aldan province are richer in potassium than those of the other two regions.

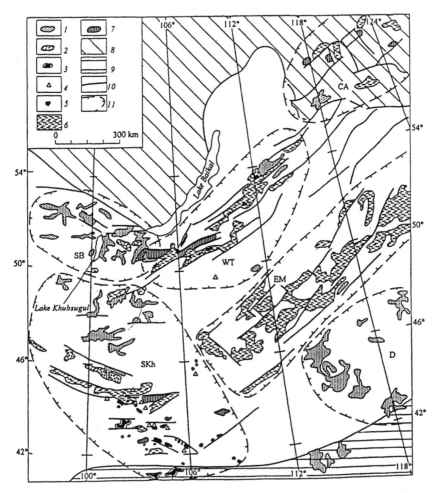

FIGURE 3.7 Scheme of distribution of late Mesozoic-Cenozoic intraplate volcanic rocks in the Central Asia. Key: 1–7 volcanic assemblages: 1 — late Cenozoic, 2 — late Oligocene, 3 — early Oligocene, 4 — Eocene, 5 — late Cretaceous, 6 — early Cretaceous, 7 — late Jurassic; 8–9 — ancient platforms: 8 — Siberian, 9 — north Chinese-Korean; 10 — faults; 11 — boundaries of volcanic areas. Volcanic areas: CA — Central Aldan; WT — Transbaikalian; SB — South Baikalian; SKh — South Khangai; EM — East Mongolian; D — Dariganga.

Early Cretaceous

The intra-plate magmatism of this time was extensive and occurred against a backround of rifting in different parts of the fold belt. There were three conformable phases of magmatic evolution marked by a similar change in compositions and concurrent changes in the scale of volcanic eruptions. The most intense igneous activity occurred at the beginning of the Cretaceous and coincided with the formation of rift zones in the South Khangai and West Transbaikal region, as well as with the simultaneous formation of the East Mongolian region. Eruption products are represented by thick (several hundreds of metres) flows of alkaline basalts which accumulated in the grabens.

The next phase of magmatism (middle early Cretaceous) involved trachyrhyolites, trachydacites and ongorhyolites and their intrusive equivalents: leucogranites, including Li–F granites. Acid rocks occur mainly in areas where the maximum eruption of basalts occurred during the earliest Cretaceous.

The latest early Cretaceous igneous activity in Central Asia was manifested as relatively weak eruptions of moderately alkaline basalts in rift zones of the South Khangai, East Mongolian and West Transbaikal areas. The West Transbaikal area shows unusual nephelinite and phonolite lava flows, which preceded the eruption of basalts in the late early Cretaceous.

Unlike the above-mentioned areas, early Cretaceous magmatism in the Central Aldan area was smaller in scale and differed in composition. During this time, high-alkaline low-Ti potassic rocks: lamproites, leucitites, phonolites and trachytes, etc., were formed (Kanukov, 1992; Makhotkin, 1991). Their formation took place against a background of regional extension accompanied by the emplacement of grabens, which correlates with the tectonic regime of early Cretaceous magmatism in other areas.

Late Cretaceous–Early Cenozoic

This time interval includes no fewer than four phases of volcanism, with a duration of 10 to 15 Ma each. They are all marked by subdued igneous activity and are represented by isolated lava domes, stocks, laccoliths and the remains of small lava fields. These phases of volcanic activity were recorded in the South Khangai and West Transbaikal regions NE China (Fig. 3.7). The volcanic products, apart from moderately alkaline basalts, include alkaline rocks: basanites, alkaline trachybasalts and teschenites.

The Latest Oligocene

The volcanic activity of that time was huge in scale, for example in the Lake Depression of the South Khangai area, a lava field with an area of about 9,000 km^2 was formed (Fig. 3.7). In the West Transbaikal area, the scale of the eruption was relatively small; however, the volcanic activity covered a large area. The late Oligocene witnessed the initiation of the South Baikal area (Yarmolyuk et al., 1990), and probably of the Dariganga area, whose south-eastern extremity on the Geological Map of China (Geological Atlas of China,

1981) is dated as Palaeogene. The volcanic products include Fe–Ti moderately alkaline and alkaline basalts.

Late Cenozoic

This period is characterized by enhanced volcanic activity (Fig. 3.7). The middle Miocene outburst of volcanic activity was probably the most voluminous. It correlated with extensive lava fields in the South Khangai, South Baikal and West Transbaikal areas, and also renewed volcanic activity in the Central Aldan area (the Udokan Plateau) and the first occurrences of volcanism on the Dariganga Plateau of Mongolia. In NW China and in the Far East of Russia extensive basalt plateaux, totalling 25,000 km^2 in area, formed during the Miocene (Esin *et al.*, 1992; Fau and Hooper, 1991). The Miocene volcanics are dominated by K–Na moderately alkaline basalts rich in Fe–Ti; alkaline basanites also occur.

The next phase of large-scale eruptions occurred in the Pliocene. The most intense volcanic activity occurred on the Udokan Plateau and also on the Dariganga Plateau of Mongolia, where the largest lava fields, composed of alkaline basalts, were formed at that time. The Pleistocene–Holocene phase ranks below the middle Miocene and Pliocene phases in the volume of volcanic products. However, there are large volcanic fields of this age in Central Khangai (South Khangai area) and important lava fields in East Tuva (South Baikal area) as well as on the Dariganga Plateau.

To conclude, it should be noted that igneous activity within the province was almost continuous from the late Jurassic to the Holocene, i.e. for the last 160 Ma. It took place in spatially and structurally separated volcanic terranes. In some of them (South Khangai and West Transbaikal) igneous activity of varying intensity could be traced through the entire age interval, from late Jurassic to Holocene inclusive. In the other provinces, either long hiatuses (Central Aldan area) or short episodes of magmatism (mainly due to the late formation of the volcanic terrane) have been reported. Based on the analysis of the relationship between volcanic products from different phases in separated areas, it was assumed that these terranes were related to the activity of mantle hot-spots (Yarmolyuk *et al.*, 1992, 1991, 1996).

Elevated alkalinity is typical of igneous rocks from this province (Table 3.19). Alkaline rocks proper were widespread during the late Jurassic, with the formation of assemblages dominated by nephelinites, phonolites, trachytes and trachyte–latites with the involvement of carbonatites. Moreover, they predominate in late Cenozoic assemblages, being represented by tephrites, basanites and limburgites, etc. They were formed at different times in different areas and were less widespread in their distribution. Only in the Central Aldan region do the alkaline rocks dominate all phases of igneous activity.

Moderately alkaline basalts are the most voluminous igneous rocks of the province. They show high total alkalinity (Na$_2$O + K$_2$O above 6 wt.% at

Table 3.19 Average compositions of rocks of diachronous intra-plate volcanic assemblages of Central Asia (normalized to 100%).

Age groups	Main rocks	Number of samples	Composition (mass%)										
			SiO_2	TiO_2	Al_2O_3	Fe_2O_3	FeO	MnO	MgO	CaO	Na_2O	K_2O	P_2O_5
Late Jurassic	$A_{K-NaHTi}$	87	52.13	2.16	17.17	7.75	2.51	0.12	3.39	6.30	4.35	2.82	1.3
	T	61	59.47	1.02	17.73	4.52	1.05	0.07	1.63	3.45	5.40	4.95	0.71
	T	34	64.67	0.79	17.45	3.42	0.62	0.08	0.08	0.96	5.54	6.11	0.26
Early Early Cretaceous	B	150	51.60	2.28	17.89	3.44	5.87	0.18	3.68	7.40	3.82	2.53	1.26
Middle Early Cretaceous	TP	90	73.38	0.18	14.41	2.45	0.97	0.08	0.16	0.22	3.45	4.65	0.04
Late Early Cretaceous	$A_{K-NaHTi}$	86	52.24	2.06	16.55	5.26	4.84	0.15	4.25	7.01	4.24	2.34	1.07
Early Late Cretaceous	$A_{K-NaHTi}$	21	51.43	1.91	17.11	3.16	5.82	0.16	5.87	7.82	4.27	1.91	0.55
Late Late Cretaceous	$A_{K-NaHTi}$	27	50.89	2.18	15.55	5.91	4.56	0.14	5.69	8.38	4.10	2.07	0.51
	$HA_{K-NaHTi}$	6	44.74	2.15	14.25	6.00	6.74	0.24	7.55	10.61	5.03	1.72	0.99
Eocene	$A_{K-NaHTi}$	20	53.29	1.85	17.26	7.98	2.49	0.15	3.75	7.46	3.87	1.40	0.50
	$HA_{K-NaHTi}$	11	45.72	2.82	13.03	3.90	8.99	0.17	9.90	9.68	3.37	1.60	0.81

Age	Rock type	n											
Early Oligocene	$A_{K-NaHTi}$	21	51.39	2.57	14.87	3.48	6.41	0.12	6.23	6.87	4.29	3.04	0.71
	HA_{K-HTi}	16	46.26	2.81	13.70	4.39	8.14	0.17	8.46	9.75	3.77	1.60	0.94
Late Oligocene	$A_{K-NaHTi}$	37	47.85	2.61	14.16	4.48	6.60	0.15	7.69	9.31	3.82	2.47	0.86
	$A_{K-NaHTi}$	15	50.58	2.70	14.41	4.16	7.05	0.13	6.75	6.94	3.43	3.11	0.74
Early Miocene	$HA_{K-NaHTi}$	52	49.60	2.56	14.45	4.03	6.32	0.14	7.11	8.18	4.18	2.54	0.87
	$A_{K-NaHTi}$	118	48.60	2.35	15.12	3.32	8.11	0.16	8.15	8.30	3.47	1.78	0.64
Middle Late Miocene	$A_{K-NaHTi}$	146	49.04	2.38	14.96	3.59	7.91	0.15	7.78	8.01	3.63	1.93	0.63
Pliocene	$A_{K-NaHTi}$	100	48.21	2.48	14.96	3.07	8.24	0.16	8.00	8.65	3.76	1.85	0.61
	$HA_{K-NaHTi}$	15	46.60	2.51	15.10	5.50	6.80	0.18	8.58	8.64	3.77	1.88	0.53
Pleistocene–Holocene	$A_{K-NaHTi}$	119	51.21	2.44	15.31	2.51	6.95	0.13	6.23	7.20	4.47	2.56	0.69

B — basanites and melaleucitites, HA$_{K–NaHTi}$ — alkali basalts, A$_{K–NaHTi}$ — subalkaline basalts, TL — trachyte–latites, T — trachytes, TP — trachyrhyolites and ongonites.

$Na_2O > K_2O$) and a high TiO_2 content (generally > 2 wt.%). These rocks are typical of large-scale volcanic eruptions (hundreds of cubic kilometres volume) in the early Cretaceous and middle Miocene, but also occurred during other phases of volcanism. Acid igneous rocks (trachyrhyolites, ongonites, leucogranites and Li–F granites) amount to no more than 5% of the total volume of igneous products. They tend to be confined to the early Cretaceous, and were formed during the early Cretaceous riftogenic magmatism. On the whole, the magmatism of the province is comparable in composition with that of intra-plate areas. This is consistent with the overall position of the province within the Asian continent, and with geological data on the relationship of this separate regions to mantle hot-spots.

e) Geodynamic Environments of Continental Magmatism in Central Asia

The geochemical diversity and age differences in igneous activity during the continental developmental stage of the Central Asian Fold Belt was caused by the multiple involvement of this region in tectonomagmatic reworking, mainly due to interaction of oceanic and continental plates (which actually persisted in this area throughout the whole of the Phanerozoic), but also because of intra-plate activity.

As has been shown above, during the Phanerozoic, the continental massif, including blocks of reworked Precambrian crust, and accreted zones of a newly formed crust, were first formed within the Caledonian block of Central Asia. This took place in the Devonian, when a marginal volcanic belt, with zonation typical of an active Andean-type continental margin, was formed along the boundary of the palaeocontinent with ocean basins (Irtysh–Zaisan and Palaeotethys). Data available on the distribution — within this belt — of rocks of similar composition, but differing in K_2O content, allowed Gordienko (1987) to reconstruct a palaeo-Benioff zone, which was subducted beneath the continent of the oceanward side to a depth of 250 km in the rear part of the active continental margin. Subduction magmatism terminated the formation of Caledonian continental crust and lithosphere. This crust was susceptible to brittle deformation, which led to the break-up in its rear part and the formation of continental rift structures, characterized by alkaline magmatism (Fig. 3.8). Further development of the active continental margin of the Palaeotethys took place during the late Palaeozoic. Tectonic events in the late middle Palaeozoic, and at the very beginning of the late Palaeozoic, resulted in the closure of the Irtysh–Zaisan marine basin and the accretion of the South Mongolian Hercinides to the Caledonian paleocontinent. The boundary with the Palaeotethys shifted southwards and followed the northern edge of the Solonker Zone of the late Palaeozoic Palaeotethys.

Geodynamically, the late Palaeozoic period of development of this region was very complex and comprised several phases. The pattern of changes in

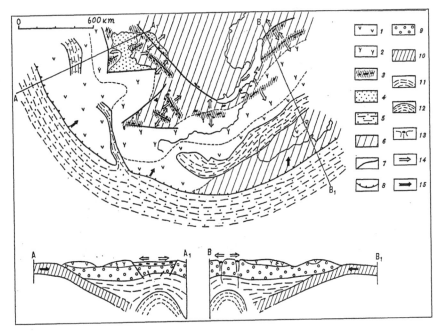

FIGURE 3.8 Palaeogeodynamic map of the development of the Central Asian Belt in the Devonian. Key: 1–4 — igneous rock terranes: 1 — calc-alkaline series, 2 — moderately alkaline series, 3 — alkaline series, 4 — undifferentiated moderate-alkaline and alkaline series; 5 — areas of marine sedimentation; 6 — continental areas; 7 — boundaries of igneous terranes; 8 — boundary of an active continental margin; 9 — continental lithosphere; 10 — Oceanic lithosphere; 11 — asthenosphere; 12 — mantle diapir; 13 — continental rifts; 14 — extensional vector; 15 — compressional vector.

structure of igneous areas, and the geodynamic environments that led to their formation, have been interpreted using a model that accounted for the overriding of the active continental margin of the North Asian Plate by the spreading centre of the late Palaeozoic Palaeotethys (Kovalenko and Yarmolyuk, 1990).

At the early phase of interaction between continental and oceanic plates during the early–middle Carboniferous, crustal stabilization took place within the South Mongolian Hercinides, accompanied by the formation of an active continental margin (ACM) (Fig. 3.9). The distribution of rocks within an asymmetrically zonal belt correlates with the development of the latter above a subduction zone that plunged at an angle of 45° under the continent on the edge of the late Palaeozoic Palaeotethys.

During the late Carboniferous the environment on the ACM changed in the same way as in the Cenozoic of western America, when it was overridden by the

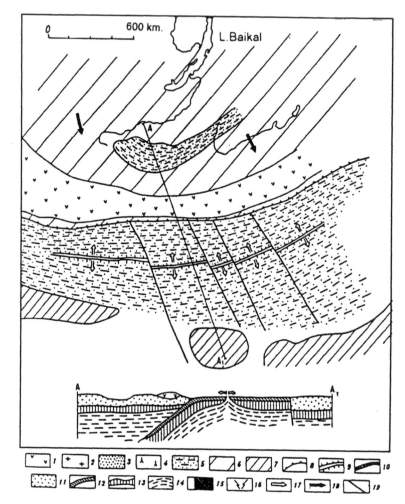

FIGURE 3.9 Palaeogeodynamic map of the development of the Central Asian Belt in the early–middle Carboniferous. Key: 1–5 — distribution of: 1 — volcanic fields of marginal volcanic belts, 2 — batholith magmatism, 3 — alkaline rocks (riftogenic areas), 4 — dispersed magmatism of the calc-alkaline series, subalkaline and alkaline series, 5 — marine sedimentation; 6 — areas developing in the continental regime; 7 — stable regions (microcontinents); 8 — boundary of the active continental margin; 9 — spreading zones; 10 — collision suture; 11 — continental crust; 12 — oceanic crust; 13 — mantle; 14 — asthenosphere; 15 — stacking zones; 16 — rift zones; 17, 18 — vectors: 17 — extension, 18 — compression (relative magnitude of vectors); 19 — transform faults.

plate from the spreading zone of the East Pacific rise (Christiansen and Lipman, 1972). The calc-alkaline and subalkaline igneous zone shifted from the edge of the continent northwards into the Caledonian block of the fold belt, where, during the early Permian, basalt–andesite and dacite–rhyolite volcanic assemblages and granitoids of the calc-alkaline and subalkaline series were very common. At the same time, within the Hercynian Block, i.e. at the edge of it, rift zones with bimodal magmatism were formed (Fig. 3.10). The rift zones were apparently the projections of the impact of the overridden spreading centre: they moved inland with the counter-movement of the North Asian and Chinese continental lithospheric plates, which then led to the cessation of differentiated magmatism.

The overridden spreading zone or "asthenospheric window" seems to have paused beneath Khangai during the late Permian. It may have been related to the collision of the North Asian and Sino-Korean late Palaeozoic palaeocontinents, which resulted in the closure of the late Palaeozoic Palaeotethys, at least in its

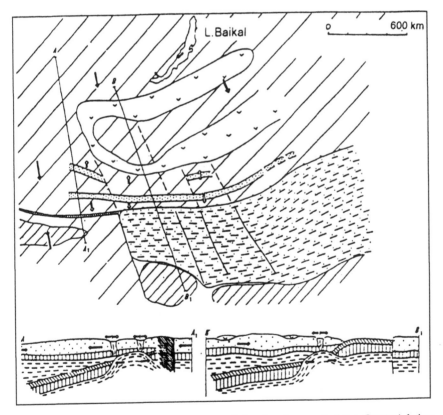

FIGURE 3.10 Palaeogeodynamic map of the development of the Central Asian Belt in the late Carboniferous–early Permian. Symbols as in Fig. 3.9.

western part. This process resulted in the formation of a new zonally symmetrical type late Permian magmatic zone (Fig. 3.11). As mentioned above, its core consisted of granitoids of the Khangai Batholith formed at the site of the Khangai Trough. The rocks of the trough are assumed to have been molten, forming large volumes of granitoid melts due to deep-seated asthenospheric heat sources, including mantle magmas reaching the crust (Kovalenko and Yarmolyuk, 1990). Rift zones with alkaline bimodal magmatism are characteristic of the peripheral parts of the area. Apparently, the extensive midland marine basin was also initially a rift system, marked by poorly developed, and, as yet, poorly known magmatism

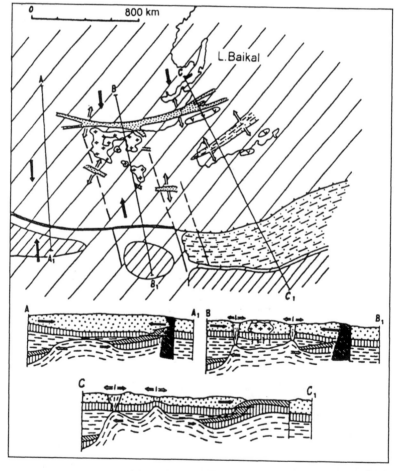

FIGURE 3.11 Palaeogeodynamic map of the development of the Central Asian Belt in the late Permian. Symbols as in Fig. 3.9.

with island-arc characteristics, which can be traced in the eastern part of the axial zone of the area (Djargalantuyn and North Gobi marine basins).

The symmetrical structure of the early Mesozoic tectonomagmatic area is similar to the above-mentioned late Permian area (Fig. 3.12). This similarity can be explained, proceeding from the overall dynamics of formation of continental crust during the late Palaeozoic–early Mesozoic. There is an assumption that the activity of the spreading Palaeotethys zone, which was overridden by the North Asian continent, continued during the Mesozoic (Fig. 3.12) but was attenuated gradually eastwards. Therefore, the specific late Permian collision environment of Mongol–Okhotsk type existed during the early Mesozoic. Under these conditions, a zoned tectonomagmatic region formed there due to the overriding of the spreading zone. Its typical features were intense, diverse granitoid magmatism

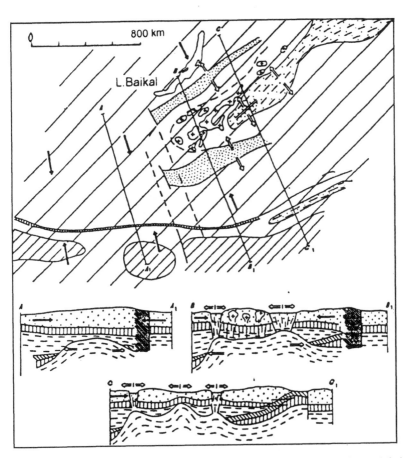

FIGURE 3.12 Palaeogeodynamic map of the development of the Central Asian Belt in the early Mesozoic. Symbols as in Fig. 3.9.

and scattered continental rifting, which were confined to the periphery of the zonally symmetrical tectonomagmatic areas.

During the late Mesozoic–Cenozoic, magmatism within the fold belt was mainly intra-plate in nature. Its occurrence is localized in separately developing areas, differing in their time of inception, continuity or discontinuity of igneous events and in the composition of their coeval volcanics. These events are related to the activity of hot-spots in the mantle (Yarmolyuk *et al.*, 1991, 1995, 1996). A high concentration of hot-spots within an intra-plate province apparently can be explained in terms of the hypothesis that a hot-field existed beneath Central and East Asia (Zonenshain *et al.*, 1990). The existence of such a hot-field is taken to be a regulator of intra-plate activity; it also explains the correlation between the intensity of volcanic events occurring in remote parts of the province. The time of formation of the mantle hot-field has not been determined as yet. Probably the abundant specifically intra-plate products (basalt–comendite assemblages, alkaline and Li–F granites) occurring during the late Palaeozoic and early Mesozoic within the same areas as the late Mesozoic–Cenozoic intra-plate associations suggest the existence of a mantle hot-field at this time.

This assumption agrees with the above-discussed model, accounting for the involvement of a sublithospheric igneous source in the formation of the late Palaeozoic and early Mesozoic igneous regions.

The intra-plate magmatism of hot-spots in Eurasia, at least during the late Cenozoic (the last 40–20 Ma), took place against the background of the collision of Eurasia and India (Molnar and Tapponier, 1975; Yarmolyuk *et al.*, 1995, 1996). In Mongolia and southern Siberia, this collision resulted in the formation of a collection of microplates, accompanied by compression, extension and rotation of their components (Zonenshain *et al.*, 1994). It is also possible to partially relate the intra-plate magmatism of Central Asia to the interaction of microplates (e.g. the Shanxi Rift graben in China). However, it is most probable that hot-spots themselves initiated the formation of microplates in this part of Asia and created triple junctions, manifested as local rift zones, for example, in the South Baikal and South Khangai area (Yarmolyuk *et al.*, 1992).

f) Magmatism Relevant to the Formation of Other Fold Belts in North Eurasia

Urals Fold Belt

This belt provides an example of linear collision fold structures. It extends meridionally (with a bend in the Pai–Khoi and Novaya Zemlya area) for some 3,500 km and has a width of 150 to 450 km. In the west, it borders the Russian Platform; to the east, it borders the Siberian block (Zonenshain *et al.*, 1990). During the late Permian collision of these continents, the intervening Uralian

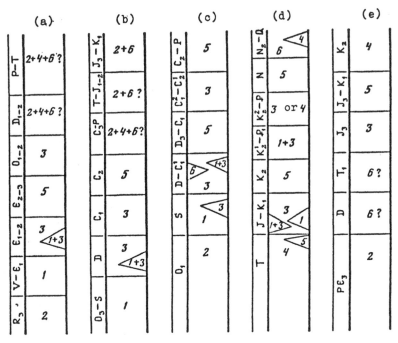

FIGURE 3.13 Evolution of magmatism and geodynamic environments in some fold belts. After Bogatikov *et al.* (1987). Key: a — Caledonides of the Central Asian Belt, b — Hercynides of the same belt, c — South Urals, d — Mediterranean, e — Verkhoyansk area, Chersky Range; environments: 1 — oceanic rifts of marginal basins, 2 — continental rifts, 3 — young island arcs, 4 — developed, mature arcs and Andean type active continental margins, 5 — continental collision, 6 — intra-plate.

ocean was consumed and imbricate thrust sheets of pelagic sedimentary rocks, oceanic lithosphere and island-arc rocks were transported westward over sediments of the ancient passive continental margin of the Russian Craton.

Magmatism played an important part in the formation of the fold belt. This has been demonstrated in detail for the South Urals part of the belt (Zonenshain, 1984; Bogatikov *et al.*, 1984). The following changes in magmatism and geodynamic environments were established there (Fig. 3.13).

1. Early Ordovician — alkaline-basaltic and bimodal (basalt–rhyolite) volcanism (Sakmarian zone, Trans-Ural Uplift), reflecting continental rifting and break-up of the margin of the Russian Craton.
2. The late Silurian–early Devonian — the oceanic complex, making up an allochthonous nappe that consists of fragments of oceanic lithosphere

(Voykar, Kempirsai, Khabarny and other massifs) and island arcs (Polyakov, Surgal and other formations), crops up almost continuously along the eastern slope of the Urals, comprising a single ophiolite zone with a length greater than 2,000 km and a width up to 250 km. The time of crystallization of the ophiolite of the Kempersai massif was shown to be 397 ± 20 Ma and the Voykar massif is 387 ± 34 Ma (Sharma *et al.*, 1995). It correlates with the time of opening of the Ural palaeo-ocean.

3. The Devonian and early Carboniferous (up to early Visean) calc-alkaline andesite–basalt and andesite–dacite–rhyolite series of the internal zones of the Urals and migration of volcanic activity from west to east, from the early–middle Devonian (Irendyk Formation) to the middle–late Devonian (Ulutau and Koltuban formations) and early Carboniferous (Beryezov Formation). They are characteristic of the long-term existence of island arcs and their shift eastwards. In the back-arc zone, close to the Russian Craton, tholeiite basalts and bimodal basalt–rhyolite strata were formed (lower Devonian Baimal–Buribaev Formation and middle Devonian Karamalytas Formation), indicative of extension and spreading in the back-arc zone and the formation of marginal basins. In the southern Urals, in Mugodzary, middle Devonian (Givetian) basalts: swarms of parallel dykes and pillow lavas showing mixed oceanic and island-arc characteristics, are very common. They are relics of the sea-floor of the back-arc marginal basin. Alkali basalts occurring in the Sakmarian zone, and in other areas, are probably related to hot-spots.

4. Late Devonian–early Carboniferous — granite and granite–gneiss domes of the Mugodzar anticlinorium with an age of 370–310 Ma. The growth of these domes could be caused by the collision of the Mugodzar microcontinent with an island arc.

5. The upper Visean, Namurian and partly Bashkirian calc-alkaline andesite–basalt volcanic series, making up the extensive Valeriyanov volcanic belt at the border of the Urals with Kazakhstan. It correlates with an island arc formed at the margin of the Kazakhstan continent.

6. The late Carboniferous and Permian granites and granite–gneiss domes, whose formation was caused by the collision of the Kazakhstan continent and Russian Craton, which began in the mid-Carboniferous and continued into the early Triassic (90 Ma). This long-term collision resulted in the formation of the present-day linear nappe-fold structure of the Urals; in the considerable thickening of the crust; and in the lithospheric layering consisting of a series of overthrust slabs. Some researchers assign the formation of miaskites of the Ilmen and Vishnevye Mountain to this time. Therefore, the Urals are characterized by, firstly, the development of island-arc systems and back-arc seas that can explain the presence of large volumes of basic and intermediate rocks and, secondly, by the very long duration of continental collision.

Mediterranean Fold Belt

This belt is marked by a long-term history of evolution subdivided into the Variscan and Alpine tectonomagmatic cycles (Peive *et al.*, 1978; Zonenshain *et al.*, 1990). We will now discuss the Mesozoic–Cenozoic stage of development of the belt, which reflects the history of closure of the Tethys. The sequence of events, as reflected in magmatic activity, can be represented in the following way (see Fig. 3.13) (Bogatikov *et al.*, 1984).

1. The late Triassic (locally early Jurassic) calc-alkaline volcanism of the northern Caucasus (Nogaisk Formation) and Crimea indicate that the North Eurasian continental margin was active at that time. The composition of the volcanics suggests that crustal subduction was taking place beneath the late Palaeozoic Tethys. During the late Triassic, the Iranian continent and the Eurasian margin experienced a collision that resulted in the Cimmerian orogeny.

2. The first half of the Jurassic was characterized by the formation of ophiolites in the Sevan–Akera Zone, which suggests that spreading continued in the Tethys Ocean. From the middle Jurassic (locally from the early Jurassic) to the Neocomian, the Lesser Caucasus–East Pontian volcanic belt was formed, composed of andesites and andesitic basalts of the calc-alkaline series. The belt correlates with an island arc of corresponding age. In the back-arc zone, within the Greater Caucasus, alkaline and tholeiitic basalts were erupted during the early Jurassic; this could have been caused by the ocean-floor spreading that was responsible for the formation, by the middle Jurassic, of a deep marginal basin, stretching from the eastern Black Sea to the Kopet Dagh.

3. During the middle Jurassic, the first important collisions of the southern passive margin and continental fragments of Gondwanaland with an island arc due to the approach of Africa and Arabia to Eurasia, which led to the closure of most of the Tethys.

4. The late Cretaceous (since the Turonian, or locally since the Campanian) and the Palaeogene were characterized by the formation of a new volcanic belt, stretching from SW Iran through the Transcaucasus (Talysh, Adzhar–Trialet Zone) into Bulgaria (the Burgas Synclinorium). The belt is composed of shoshonite and calc-alkaline lavas in the north and south, respectively, marking thus indicating the presence of an ACM subject to underthrusting of fragments of the Tethys from the south. In the late Cretaceous, a back-arc basin opened in the western Black Sea as evidenced by riftofenic alkaline basalts in the axial part of the Adzhar–Trialet Zone and also those of the Burgas Synclinorium. Granite–granodiorite intrusions in the Transcaucasus and Iran are related to this volcanic belt. By the close of the Palaeogene the continuing covergence of Africa, Arabia and Eurasia resulted in the complete subduction of the Tethyan oceanic crust beneath the Eurasian margin. The area of the Black Sea Basin was reduced considerably and a new phase of continental collision began.

5. The ongoing stage of Caucasian volcanism (10 Ma to Recent) includes, on the one hand, the calc-alkaline series of Armenia and the Greater Caucasus and, on the other hand, alkaline basalts; the former can be easily correlated with the collision of Arabia and Eurasia, and the latter may be assigned to intra-plate volcanism (see Chapter 2).

Hence, several periods of collision of southern continental landmasses with the northern active margin can be recognized in the development of the Mediterranean belt within the Caucasus. The collision is still continuing at the present day.

The Verkhoyansk–Kolyma Fold Belt

This belt is an example of a different type of magmatic evolution (see Fig. 3.13). Over a period of 500 Ma, i.e. from the Riphean to the late Jurassic, this site was occupied by a virtually non-magmatic passive margin, where only two short-lived episodes of intra-plate volcanism took place, namely in the Devonian and early Triassic (trap volcanism). In the late Jurassic, the approach of the Kolyma–Omolon Massif to the passive margin resulted in the formation of a volcanic island arc (Uyanda–Yasach or Ilinotas volcanic belt: the Chersky Range) and then short-term volcanic activity was followed by collision of the massif with the Siberian continent. When continental collision had ceased, the fold belts of the continent were overridden by the Okhotsk–Chuckchee volcanic–plutonic belt that indicates an active continental margin environment of Andean type. The Verkhoyansk–Kolyma Fold Belt is characterized by an extremely short (30–40 Ma) period when island arcs were present and continental collision occurred.

Consequently, similar igneous assemblages typical of major geodynamic environments recur in the history of fold belts and reactivation zones. Their order of occurrence and the way in which they are combined vary greatly from place to place. This is quite evident from the above-discussed examples of magmatic evolution in the different fold belts of Eurasia (Fig. 3.13). All of the fold belts were formed at the site of different types of ocean basin due to the approach and subsequent collision of continents. This explains the presence, in fold belts, of relic oceanic crust in the form of ophiolites and complexes that correlate with conditions of plate collision. For example, the Ural Mountains were formed mainly at the site of island arcs and marginal basins, during a long period of continental collision. The Mediterranean belt developed in a rather similar manner, with several periods of collision. For the Verkhoyansk–Kolyma belt there was a very short period of continental approach and related magmatism. At first the fold structures of the Central Asian Belt developed similarly to those of the Urals, at the site of island arcs and marginal basins. However, the late stages saw the extensive occurrence of mixed magmatism, indicating the preserve of active Andean-type continental margins and continental rifting, probably involving intra-plate environments. This complex situation (seen for example in the

Mongol–Okhotsk) takes place primarily in the case of thrusting of a continental margin over an oceanic spreading ridge. For the late Cenozoic, the margin of North America provides an example of such an environment (see Chapter 2).

The above discussion shows that both the Wilson cycle and other cycles, allowing for a strictly fixed sequence of geodynamic environments and related magmatism, cannot really be applied to mobile belts. This can be useful when discussing regular features in the development of a typical fold belt in general. However, particular fold areas differ primarily in their sequence of geodynamic environments and their characteristic magmatism, as well as in its intensity or absence, which goes hand in hand with the specific magmatism of each area and its metallogeny.

3.2 MAIN TRENDS IN THE MAGMATIC EVOLUTION OF FOLD BELTS

The discussion of changes in the types of igneous assemblages formed during successive stages in the development of fold belt, and their geochemical features, make it possible to recognize general trends in magmatic evolution, during the formation and development of the Earth's continental crust. As mentioned above, the earliest assemblages were formed in ocean basins, but more often in back-arc basin environments and in the early stages of development of ensimatic island arcs. They originated from basic mantle-derived magmas. An unusual feature is the weak differentiation of the magmatic melts, giving a persistent homogeneity to rocks of this stage. Acid rocks (represented by soda rhyolites) are considerably outnumbered by basic rock types, and were associated with the evolution of the basaltic melt. All of the rocks show low alkalinity, a marked dominance of Na_2O over K_2O, a low Sr and Ba content and also a low initial strontium isotope ratio (0.703–0.705) and low LREE content. Assemblages forming part of the ophiolitic assemblages temporally and spatially are closely related to it (ultrabasic rocks or MORB-type basalts, etc). In fact, this stage marks the beginning of the generation of the continental crust within fold belts, resulting mainly from the transformation of the oceanic crust.

During the transitional stage, marked by the presence of arc complexes, the formation of igneous assemblages continued. However, most of the igneous rocks belong to the calc-alkaline series. The number of rock types and the relative volume of intermediate and acid types increase dramatically. Differentiated series of all compositions appear. The K, Sr and Ba content increases considerably, as well as the K_2O/Na_2O ratio in rocks similar to those formed at an early stage. The assemblages that formed as the results of a complex interaction between mantle and crustal material included: gabbro–granite, gabbro–diorite–granite (diorite–granite). Finally, a typical crustal granite–migmatite assemblage was formed, related to the regional metamorphism and anatexis of crustal sialic material. Processes of K-metasomatism that affected the earlier-formed granitoids were widespread. The transitional stage was a period during which most of the continental crust of fold belts was formed.

The continental stage of evolution involves the accretion of continental crust, thus increasing the complexity of its structure, and the final stabilization of the fold belt within an environment of active continental margins and collision zones. This stage is characterized by the wide variety of its igneous rocks. In the early phases basalt–andesite–rhyolite, andesite–dacite–rhyolite and dacite–rhyolite assemblages dominate the volcanic rocks and diorite–granodiorite dominates the plutonic rocks that form large intrusions that are often controlled by fractures. At later stages the volume of dacites and rhyolites in the volcanic assemblages increases and the plutonic formations begin to be dominated by granites, and the importance of leucocratic and alaskite varieties increases with time. This homodromous pattern of igneous evolution holds true for many Phanerozoic fold belts, irrespective of their relationship to any particular tectonomagmatic cycle. Obviously, such changes in composition are due to the involvement of sialic crust in the process of magma generation. For example, the initial Sr-isotope ratio in rocks of the early Andean volcanic belt equals 0.7041–0.7059 and increases in the rocks of the neo-Andean belt up to 0.7050–0.7130 (Klerx et al., 1977). The above data suggest a possible contamination of the deep-seated magma by material from the sialic crust.

The end of sialic magmatism witnessed an important stage in the stabilization of fold belts during continental collision. After that time, conditions were unsuitable for melt generation at the boundaries of interacting plates. This affected the nature of magmatism and mechanisms of formation and transformation of crust during the oceanic, transitional and initial continental phases of the evolution of fold belts. During the later stages of evolution of these areas, melt sources beneath the lithosphere became of primary importance. Igneous products that could have been correlated with these sources showed a distinct intra-plate specificity and were represented by rocks of mantle origin (Fe–Ti subalkaline and alkaline basalts, phonolites and alkali trachytes). Products from crustal sources are subordinate (Li–F granites and leucogranites) and sialic rocks such as comendites, pantellerites, alkaline granites and syenites more often correlate with contamination of mantle magmas to varying degrees by crustal material.

3.3 SUMMARY

1. Tectonomagmatic activity and magmatic assemblages of the Phanerozoic fold belt are rather similar to the present-day situation. However, their order of

occurrence and the way in which they are combined vary greatly from place to place. All of the fold belts were formed at the site of different types of ocean basin due to the approach and subsequent collision of continents. For example, the Urals Mountains were formed mainly at the site of island arcs and marginal basins, during a long period of continental collision. The Mediterranean belt developed in a rather similar manner, with several periods of collision. For the Verkhoyansk–Kolyma belt there was a very short period of continental approach and related magmatism. At first the fold structures of the Central Asian Belt developed similarly to those of the Urals, at the site of island arcs and marginal basins. However, the late stages saw the extensive occurrence of mixed magmatism, indicating the preserve of active Andean-type continental margins and continental rifting, probably involving intra-plate environments.

2. The above discussion shows that both the Wilson cycle and other cycles, allowing for a strictly fixed sequence of geodynamic environments and related magmatism, cannot really be applied to mobile belts. This can be useful when discussing regular features in the development of a typical fold belt in general. However, particular fold areas differ primarily in their sequence of geodynamic environments and their characteristic magmatism, as well as in its intensity or absence, which goes hand in hand with the specific magmatism of each area and its metallogeny.

CHAPTER 4
PHANEROZOIC ANOROGENIC MAGMATISM

E.V. SHARKOV

Phanerozoic orogenic igneous activity was analogous to that occurring at active plate margins at the present day. However, that of anorogenic areas is believed to have been similar to intra-plate magmatism.

On the continents, anorogenic igneous activity is fairly well understood, much better than that of the oceans, with studies covering a wide time span. The whole spectrum of rock types that evolved throughout the whole of the Earth's history can be studied here directly to obtain better understanding of the evolution of igneous activity of this type. As yet some of the rock types have been found only on the continents.

Typical evidence for continental magmatic activity in anorogenic areas has been found in areas associated with the development of palaeorift structures associated with ancient platforms. They can be readily identified, presenting no problems in their interpretation. The situation is different with regard to the vast fields of continental basalts, or traps, which have no direct equivalents in the late Cenozoic (plateau-like basalt fields in Colombia and the Snake River Plain area, of the Western USA, are believed to result from back-arc spreading). Vast submarine plateaux of the Ontong–Java type and the Shatsky Rise are the oceanic equivalents of the traps. The anorogenic units also include large alkaline intrusions of the Khibiny and Lovozero type (Kola Peninsula), massifs of ultrabasic–alkaline rocks containing carbonatites, as well as kimberlites and high-Ti lamproites. The last two types are relatively small in scale, but are very important in the analysis of processes in the upper mantle, because they bring fragments of mantle rocks to the surface.

4.1 PALAEORIFTS MAGMATISM

The Devonian Dnieper–Donets Depression (Dolenko *et al.*, 1991) and the Permian Oslo Graben in the Baltic Shield (Neumann and Ramberg, 1978) are the best studied examples of palaeorift valleys.

The Dnieper–Donets Palaeorift (DDP) forms a long depression (up to 900 km in length but only 20–100 km wide), extending NW–SE (Figs. 4.1–4.4). It forms the boundary between the Ukrainian Shield and the Voronezh Uplift of

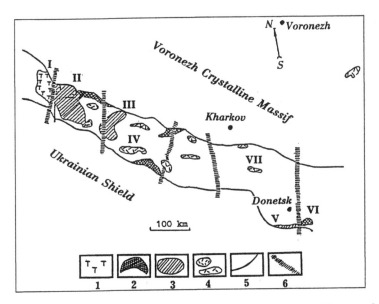

FIGURE 4.1 Distribution chart of Devonian igneous assemblages of the Pripyat–Dnieper–Donets Palaeorift. After Dolenko *et al.* (1991).
Assemblages: 1 — alkali basaltoids and phonolites; 2 — melanonephelines, alkali ultramafic rocks, feldspathoid gabbroids and carbonatites; 3 — trachy-basalt–trachyandesite–trachyrhyolite; 4 — basalt–dolerite; 5–6 — fractures: on the boundaries of the depression, deep; transverse fractures.
Igneous units: I — Pripyat, II — Chernigob, III — Belaya Tserkov, IV — Dyke, V — Volonovakh-Elanchic, VI — Pokrov-Kirey, VII — tholeiite basalt, VIII — Voronezh.

crystalline basement of the Russian Platform (the Voronezh Crystalline Massif, VCM). The origin of this structure has been assigned to the late Precambrian. The formations of that time are represented by diabase dykes, 566 Ma old, which are believed to be coeval with the trap magmatism of the SW Russian Platform. Although the onset of the main stage of palaeorift development is referred to the Devonian, the onset of its development as an independent rift structure probably occurred in the Givetian. Most of the volcanics occur in the upper parts of the Devonian sections, where they form two successions associated with two major stages of volcanic activity, i.e. late Frasnian (c. 344 Ma) and late Famennian (c. 324 Ma). Volcanics occur over an area of about 16,000 km², with a volume of 6,000 km³, constituting 20% of that of the Devonian formations in the depression.

Lyashkevich (Dolenko *et al.*, 1989) classified the igneous rocks of the DDP into a number of assemblages, such as:

1. basalt–dolerite,
2. trachybasalt–trachyandesite–trachyrhyolite,

3. melanephelinites, alkali–ultramafics, alkali gabbroids and carbonatites,
4. tephrites, basanites and phonolites.

The basalt–dolerite assemblage is represented by the Pripyat–Donets dyke swarm which is well developed over the entire area of the DDP; it is composed of tholeiitic non-differentiated basalts. The trachybasalt–trachyandesite–trachyrhyolite

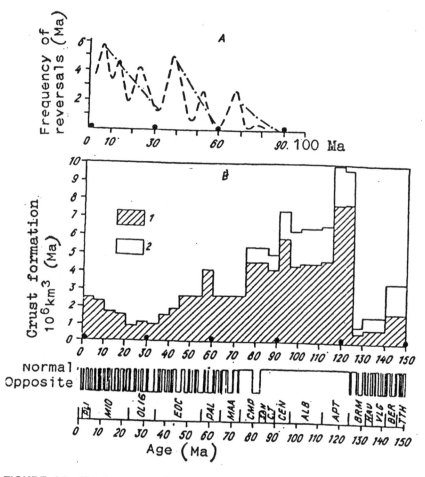

FIGURE 4.2 The frequency of occurrence of magnetic inversions (A) and their time scale (B), their comparison with the intensity of the mantle igneous activity (oceanic crust, submarine plateaux, hot-spots, continental flood basalts (CFB); each step of the histogram (B) corresponds with the volume of oceanic crust generated over a period of 5 Ma (after Dobretsov, 1994). 1 — oceanic crust; 2 — the volume of already subducted crust.

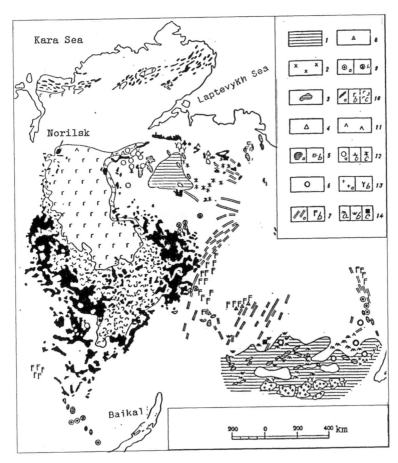

FIGURE 4.3 The distribution of the Permian–Triassic rocks of the Siberian Platform and adjacent areas. After Kovalenko (1987). 1 — protrusion of crystalline basement; 2 — basalt–rhyolite assemblages of the middle and late Proterozoic; 3, 4 — late Proterozoic: 3 — traps, 4 — agpaite nepheline syenites; 5, 6 — late Proterozoic–early Cambrian: 5 — traps (a — intrusions, b — volcanics), 6 — assemblages of ultrabasic–alkaline rocks (a — intrusions, b — volcanic rocks); 7–9 — middle Palaeozoic: 7 — traps (a — intrusions, b — volcanic rocks), 8 — trachybasalt assemblage (mostly volcanic rocks), 9 — assemblages of ultrabasic–alkaline rocks (a–central type intrusions, b–kimberlites); 10–12 — late Palaeozoic–early Triassic: 10 — traps (a — intrusions, b — lavas, c — tuffs), 11 — trachybasalt assemblage (predominantly volcanics), 12 — assemblages of ultrabasic-alkaline rocks (a — central-type intrusions, b — volcanic alkali basaltoids and meimechites, c — kimberlites); 13–14 — late Mesozoic: 13 — intrusions of granitoids and syenites (a), volcanic rocks of acid and medium composition (b), 14 — alkali gabbroids and alkali syenites (a), alkali basaltoids (b), and kimberlites (c).

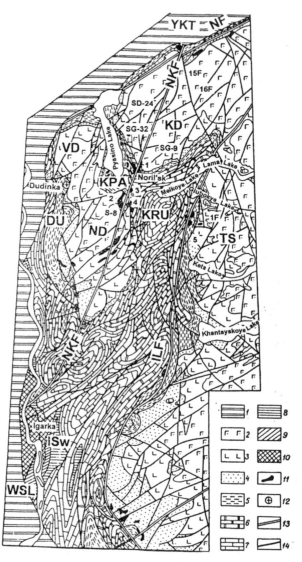

FIGURE 4.4 Simplified geological map of the Noril'sk region (after Distler and Kunilov, 1994). 1 — Jurassic and Cretaceous sediments; 2–3 — upper Permian to lower Triassic flood basalts: 2 — Morongovsky to Samoedsky suites; 3 — Ivakinsky to Nadezdinsky suites; 4 — middle Carboniferous to upper Permian (Tungusskaya series) formations (lagoonal and continental sediments); 5 — Devonian to lower Carboniferous sedimentary formations (calcareous and dolomite marbles, sulfate-rich evaporites, limestones); 6 — Ordovican and Silurian sedimentary formations (marine dolomites, limestones and argillites); 7 — Cambrian marine sedimentary formations (limestones); 8 — Vendian marine formations (carbonate sediments); 9 — upper Rifean to lower Vendian marine formations; 10 — Riphean (Baikalian) volcanosedimentary metamorphic basement; 11 — differentiated trap intrusions

assemblage comprises high-Ti moderately alkaline differentiated series with basalts predominant, together with highly acid variaties. An assemblage of melanephelinite, alkaline ultramafic and gabbroids is developed at the base of the section. It comprises a differentiated alkaline series, including high-Ti picrites, limburgites, augitites, basanites, analcime basalts, trachytes, alkali trachytes, meimechites, phonolites, nephelinites, leucitites, tephrites and their tuffs, and carbonatites. The last assemblage (of alkali basites and phonolites) is very similar in all of its parameters to that of melanephelinite, differing only in its more leucocratic composition, with nephelinites and trachytes being predominant. The trachybasalt–trachyrhyolite assemblage is found up section, being composed of a high-Ti, moderately alkaline differentiated series with predominant basalts and acid varieties.

The assemblage of melanephelinites, alkali-ultramafic rocks and carbonatites makes up the lower volcanic sequence. The upper sequence comprises an alkali basalt assemblage. Igneous rocks from other assemblages are found during the entire rifting stage. It took all of these assemblages 10–20 Ma to form.

The picrites and basalts are characterized by the very high titanium content of the rocks ($TiO_2 = 4.89–5.42$ wt.%). Similar properties are known to be typical of the volcanics of the Maimecha–Kotui Province (see below) and of the western branch of the East African Rift. The DDP volcanites are similar to those of the East African Rift in their low magnesium content ($Fe_2O_3 + FeO > MgO$), and also in many cases their K_2O/Na_2O of more than 1, i.e. the alkalinity is mostly of potassic type. However, more commonly, the K_2O/Na_2O is less than 1, i.e. more typical of continental rifts. As in many other rift zones, high-alkaline rock types (nephelinites, limburgites and phonolites, etc.) are characteristic of the earlier stages of rift-zone development.

The Permian Oslo Graben is another good example of a palaeorift (Neumann and Ramberg, 1978). Its geological history dates back to the latest Carboniferous–early Permian. Major eruptions of alkali picrites, olivine nephelinites, melilitites, tephrites and leucitites, forming the Shien lava plateau, which was 1–2 km, and some places, up to 3 km high, occurred during the next stage in the

FIGURE 4.4 (*continued*) (numbers mark the Noril'sk-type intrusions with massive sulfide ores or significant disseminated ores: 1 — Kharaelakh and Talnakh, 2 — Noril'sk I, 3 — Noril'sk II, 4 — Chernogorka, 5 — Imangda); 12 — Bolgokhtokh intrusion; 13 — principal faults: NFK — Noril'sk–Kharaelakh, NF — North-Kharaelakh, ILF — Imangda–Letnaya; 14 — other faults. Principal structural elements: WSL — West Siberian plate: YKT — Yenisei-Khatanga trough; TS — Tunguska sineclise; KD — Kharaelakh depression; KPA — Kauerkan–Pyasino anticline; DU — Dudinka uplift; KRU — Khantayka–Rybnaya uplift; ISw — Igarka swell. Black dots — location of typical flood basalt sections: 1F, 15F, 16F, S-8, SD-24, SG-9, SG-32, shown on the Fig. 4.5.

area to the NE, with moderately alkaline olivine basalts, hawaiites and trachy-basalts prevailing. Finally tholeiitic basalts were erupted in the Oslo area. The type of igneous activity there appears to have been very similar to that of the DDP. However, unlike the latter, rocks of medium and acid composition, like those already erupted, continued to predominate during the later stages of the Oslo Rift development, represented by trachyandesites (rhomben-porphyries), erupted over the whole rift terrane during the next (third) stage of its development, as well as by the larger Lavrika monzonite–mariupolite batholith area and by the syenites and the alkali granites of the Norling–Hurdal and Drammen areas.

It is interesting to note that in the Oslo Graben area the present-day section displays a predominance of intrusive rocks over volcanics, while trachyandesites prevail among the latter, constituting 82% of the total volcanics, with basalts (15%) and acid lavas contributing to the remainder. The intrusive rocks show different relationships. Monzonites and alkali granites are represented in almost equal quantites, comprising 99% of all the intrusive rocks, the proportion of basic rocks being no higher than 1%.

Thus, the palaeorifts can be divided into two different types:

1. a basic type similar to that from present-day continental rift areas, and
2. a mainly sialic type. A similar type was formed during specific stages of Ethiopian Rift development, although the widespread occurrence of trachyandesite has not been observed.

By and large the palaeorift magmatism is believed to be inherently similar to Cenozoic rift magmatism.

4.2   CONTINENTAL AND OCEANIC FLOOD BASALTS

Traps are very extensive areas of intrusive and volcanic rocks, mainly of basic composition, formed on stable cratons. Tholeiitic basalts are the predominant rock type, the composition of their major components being similar to those of MORB. As with other examples of intra-plate magmatism, they are associated with alkaline or even high-alkaline units. The boundary between continental flood basalts (CFB) and plateau basalts associated with the development of hot-spots is fairly arbitrary, being mainly determined by their scale of occurrence, which is relatively small in this case.

The age of the trap fields ranges from early Cenozoic (Palaeogene: 58 ± 2 Ma: Baffin Bay traps, North American Platform, Francis, 1985) to early Proterozoic (eastern Baltic Shield: 1670–1650 Ma: Kratz et al., 1978). Detailed geo-chronological studies have shown that some of the largest trap fields were formed extremely rapidly, taking only 1–2 Ma (Renne and Basu, 1991), coinciding in time with fauna extinction events at the boundaries of the Cretaceious and Palaeogene (Deccan traps) and the Permian and Triassic (Siberian traps). Moreover, some authors suggest that the formation of the traps should be attributed to the rifting that caused ocean opening, e.g. Upper Cretaceous Deccan traps and their subsequent development resulted in the opening of the Indian Ocean (Krishnamurthy and Cox, 1977), and the late Jurassic–early Cretaceous traps of the Parana River (South America) and the Etendeka (South Africa) are regarded as a single trap field that eventually led to the opening of the South Atlantic (Harry and Sawyer, 1992).

It is worth mention that the development of the Mesozoic traps coincides in time with the so-called Cretaceous Superchron, i.e. the time when the Earth's magnetic field appeared to be "frozen" at its normal polarity for 40 Ma (Larson et al., 1991). That period is known to be characterized in general by abnormally high mantle igneous activity: the rate of the young oceanic crust formation was twice as high in comparison to that at present (Fig. 4.2); large submarine plateaux were formed, the latter being regarded as the analogues of trap assemblages, such as the Ontong–Java in the SW of the Pacific and Kerguelen in the south of South Indian Ocean, as well as others (Dobretsov, 1994). In terms of area, all of these traps, both continental and submarine, tended to form in the vicinity of the largest upwelling of the lower mantle, i.e. African–South Atlantic and Pacific, which were active during Mesozoic–Cenozoic times (see Chapter 9).

4.2.1 The Siberian Traps

The late Palaeozoic–early Mesozoic (Permian–Triassic) Siberian flood basalt province (SFBP) represents the world's largest province of trap magmatism (Fig. 4.3). After its formation, SFBP covered an area of $1.5 \times 10^6 \, km^2$, with an average thickness of about 1 km. Judging by the geological and isotope-geochronological evidence, these traps were formed from the end of the late Permian to the beginning of the early Triassic, in the interval of 254–220 Ma ago (the K–Ar method: Dyuzhikov et al., 1988). Later Renne and Basu (1991) showed that the main, tholeiitic upper parts of these traps were formed at the boundary of the Permian and the Triassic in the time interval from 248.3 ± 0.3 to 247.5 ± 0.7 Ma ago ($^{40}Ar/^{39}Ar$ method), that is, over 0.9 ± 0.8 Ma, with an average eruption rate of about $1.3 \, km^3$ per year. The lower part of the SFBP section, mainly alkaline and subalkaline in composition yielded a $^{40}Ar/^{39}Ar$ plateau age of 253.3 ± 2.6 Ma (Basu et al., 1995), distinctly older than the main tholeiitic pulse of the SFBP at c. 250.0 Ma.

The Siberian traps actually cover the entire western area of the Siberian Platform, which was stabilized by the end of the Precambrian in Baikalian time. The platform cover comprises early Palaeozoic dolomites, limestones and shales of oceanic origin, overlapped by the Devonian calcareous and dolomitic marls, sulphate evaporites and lower Carboniferous shallow-water limestones. They are overlain, with angular non-conformity, by middle Carboniferous–lower Permian coastal oceanic and continental sediments, including carbonaceous formations. The entire succession is overlain by the SFBP of the late Permian–early Triassic, composed of alkali basalts, as well as tholeiitic basalts and picrobasalts. In other words, the volcanic eruptions were preceded by an upwelling of the whole area.

The most complete section of Siberian traps is found in the NW part of the trap field, in the Noril'sk area (Fig. 4.4). The volcanic formations here represent a layered succession of flood basalts and tuff beds, their thickness about 3.7 km (Fig. 4.5). A large number of sill-like intrusions are believed to have formed at the same time as the lavas. Two of them, the Noril'sk and Lower Talnakh intrusions, are associated with the largest deposits of sulphide (Cu–Ni–PGE) ores.

According to Fedorenko and Dyuzhikov (1981) and Dyuzhikov et al. (1988), the trap succession was formed in three stages: in the late Permian and during two stages in the early Triassic. Volcanics from the first stage (Ivakinsky, Syverminsky and Gudchikhinsky formations) correspond in their composition mainly with high-Ti, moderately alkaline (less frequently tholeiitic) basalts; up section (the Gudchikhinsky Formation) picrite lavas are widespread. This part of the formation coincides with that of the sulphide-bearing intrusions that seem to be their intrusive analogues (Dyuzhikov et al., 1988).

Picrites are also found in the base of the second-stage volcanic series (Khakanchansky, Tuklonsky and Nadezhdinsky formations). The volcanics of these two stages, excluding the abnormal alkali basic rocks and high-Mg rocks, differ from the middle Siberian traps in their higher alkalinity and silica content and their lower iron content. Sporadic alkali-ultramafic–mafic assemblages are found there, their number increasing eastward and becoming most numerous in the Maimecha–Kotui Province (see below).

During the third stage the volcanics of the Morongovsky and Mokulayevsky formations were formed, with thickness exceeding 2,000 m; their composition is fairly homogeneous. They extend beyond the Noril'sk region and form part of the single basalt plateau of the Siberian Platform. These are typical traps, being composed mainly of tholeiitic basalts. Slight variations in the mineral composition of the rocks include a regular increase in titanium, and from the middle of the section the barium and phosphorus also regularly increase upwards.

Intrusive Formations

Dolerite and picrodolerite sills, which are known to be the intrusive analogues of the lava series, are widespread among the volcanic cover and its underlying sedimentary formations (sulphate evaporites, terrigenous-carbonate and carbonaceous sediments). Sills of picrodolerites (Fig. 4.6) are of particular interest,

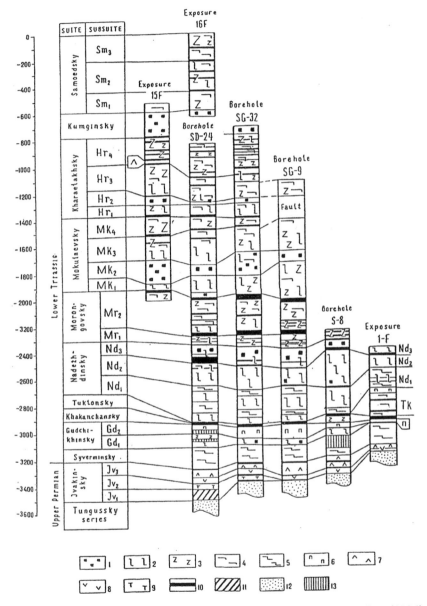

FIGURE 4.5 Typical flood basalt section. After Distler and Kunilov (1994). Locations shown on Fig. 4.4. 1 — glomeroporphyritic basalts; 2 — porphyritic basalts; 3 — aphyric basalts; 4 — poikilophytic basalts; 5 — tholeiitic basalts; 6 — picritic and oliviniophyric basalts; 7 — subalkaline andesitobasalts, labradore, two plagioclase and andesine basalts; 8 — subalkaline titanium–augite basalts; 9 — alkaline trachybasalts; 10 — tuffs; 11 — interlayering of tuff breccia and trachydolerites; 12 — terrigenous sedimentary rocks of the Tungusskaya series; 13 — intrusions.

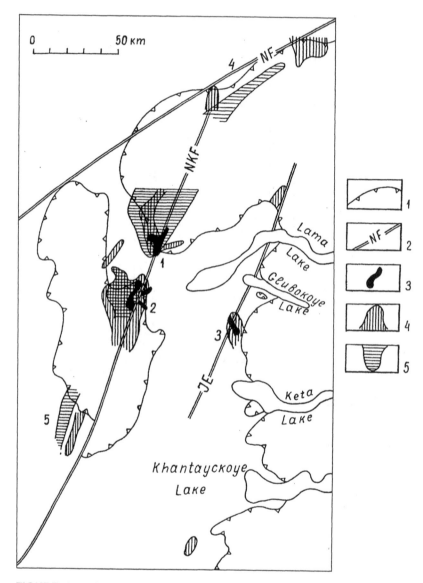

FIGURE 4.6 Schematic location map of the Noril'sk and Lower Talnakh type intrusions (after Distler and Kunilov, 1994). 1 — modern boundaries of the volcanic rocks; 2 — magma-ore controlling faults: NFK — Noril'sk–Kharaelakh; NF — North-Kharaelakh; JF — Imangda; 3–4 — Noril'sk type intrusions: 3 — with massive ores or significant disseminated mineralization; 4 — weakly mineralized or barren; 5 — lower Talnakh type intrusions. The numbers on the map mark ore junctions (1–3) and potential ore junctions (4–5): 1 — Talnakh, 2 — Noril'sk, 3 — Imangda. 4 — North-Kharaelakh, 5 — South-Noril'sk.

as they are frequently associated with sulphide Cu–Ni–PGE mineralization, including the unique deposits of Noril'sk and Talnakh (Dyuzhikov *et al.*, 1988).

These ore-bearing intrusions are frequently believed to represent the intrusive analogues of picrites of the Gudchikha Formation, being differentiated intrusions forming structurally complex branching bodies. These intrusions are lens-shaped (or U-shaped) in cross-section, with steep edges, and have a maximum thickness (up to 200–300 m) in the deeper areas (Fig. 4.7). The latter typically have an autonomous inner structure with the development of ultrabasic differentiates at the base. Normally, intrusions are subconcordant with their enclosing rocks. Dyke facies, igneous breccias, thick (up to 300–400 m) exocontact aureoles, including Ca–Mg and Ca-scarns, various metasomatites and magnesian hornfels are all widespread there.

Differentiated intrusions are characterized by distinct and also cryptic layering, including (downward): non-olivine dolerites and quartz dolerites, olivine dolerites, troctolite and picrite gabbro–dolerites, forming lower cumulate beds. Endocontact zone formations are well developed, including, down the section, endocontact ore-bearing taxitic gabbro–dolerites and, up section, Crt-bearing eutaxitic gabbros with locally increased Mg content, accompanied by sulphide and platinum mineralization.

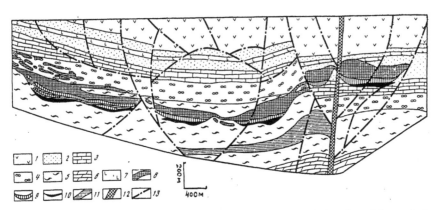

FIGURE 4.7 East–west section of the Talnakh ore body (after Dyuzhikov *et al.*, 1988). 1 — volcanic formations of the Upper Permian–Lower Triassic; 2 — terrigenous coaliferous sediments of the Upper Carboniferous–Upper Permian; 3 — carbonate Middle/Upper Devonian sediments; 4 — Middle Devonian sulphate–carbonate sediments; 5 — Lower/Middle Devonian terrigenous carbonate sediments; 6 — Silurian carbonate sediments; 7 — titanium–augite dolerites; 8–9 — Talnakh Intrusion: 8 — contact gabbro–dolerites, upper taxite, gabbro–dolerites, olivine– and olivine–biotite–gabbro–dolerites with intrusions of Cu–Ni sulphide ores; 10 — heavy Cu–Ni ores; 11 — Lower Talnakh Intrusion; 12 — main junction of the Noril'sk–Kharaelakh Fracture; 13 — other fracture dislocation.

Commercial Cu–Ni–PGE mineralization is typical of the picrite and taxite layers as well as of the lower exocontact zone. The main ore bodies include both massive and disseminated types. The disseminated ores are mainly developed in the rocks of intrusions in the lower section, while the massive ores are more characteristic of the exocontact zones, where they form large lenses in metamorphic and metasomatic rocks underlying the intrusion (Fig. 4.8).

Geochemical analysis of the Noril'sk traps (Lightfoot et al., 1990, 1993) (Table 4.1) has shown that occurrence of incompatible elements and the ratio of LILE to HFSE (large-ion lithophilic elements to high field-strength elements) is higher here than for most of the tholeiite and picrite basalts.

Picrite lavas are widespread within the Gudchikhinsky Formation. This is 700–800 metres thick, about 200 m of which are lavas. The incompatible element content in the picrites is fairly low, although they are significantly depleted in HREE (Yb = 1.0–1.1 ppm); the HREE profile is more gentle that for most other rocks. Both picrites and basalts are characterized by high ratios of Gd/Yb (2.3–3.1), slightly enriched by TiO_2 (1.2–2.3 wt.%), and have small Nb and Ta anomalies (Nb/La = 0.8–1.1), as well as by radiogenic Nd (ε_{Nd} = 3.7–7.3). Unlike these, the picrites and basalts of the higher Tuklon Formation of the second stage are characterized by a moderate Gd/Yb ratio (1.6–1.8), lower TiO_2 (0.45–0.95 wt.%), a significantly negative Ta and Nb anomaly (Nb/La = 0.42–0.57) and a low radiogenic Nd content (ε_{Nd} from 0 to 4.6). The Nadezhdinsky basalts have higher concentrations of Ce than those of the Morongovsky Formation of the third stage with similar Yb, although, while the concentration of SiO_2 is higher and more variable, the Ce/La ratio systematically decreases up section.

The basalts of the upper Morongovsky and Mokulayevsky formations have a similar composition, although they are not identical. Basalts in the lower sections of the Morongovsky Formation are similar to those of the Nadezhda Formation, while the upper section has more in common with the Mokulayevsky Formation, with a mg# ranging from 0.63–0.68, and the ε_{Nd} is more radiogenic, ranging from +0.64 to +1.61. In the lavas of these formations a progressive decrease of SiO_2 content has been observed, with the La/Sm being in the range of 4.6 to 2.0 and ε_{Sr} in the range +67 to +13.

Thus the above analysis has shown that the two lower formations (Ivakinsky and Syverminsky) have a similar composition and differ from the Guchikhinsky Formation, while all three formations of the first stage are significantly different from the tholeiite basalts of the second and third stages, which constitute about 50% of the entire section. According to Lightfoot et al. (1990, 1993), the variations in the rocks of the series are not related to fractional crystallization of melt in intermediate magma chambers, or to magma mixing. The isotope-geochemical characteristics of the Guchikhinsky Formation appear to indicate that source rocks should be attributed to asthenospheric plume material, while the rocks of the Tuklonsky Formation are contaminated by continental lithospheric material.

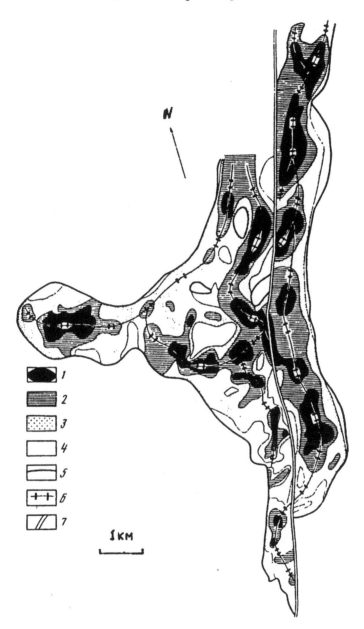

FIGURE 4.8 A schematic representation of the structure of the Talnakh Intrusion (after O.A. Dyuzhikov *et al.*, 1988). 1–4 — the thickness distribution: 1 — more than 150 m, 2 — 100–150 m, 3 — 50–100 m, 4 — less than 50 m; 5 — the outline of the intrusion; 6 — axial intrusion zones; 7 — main junction of the Noril'sk–Kharaelakh Fault.

Table 4.1 Average composition of trap rocks from the Noril'sk district (after Lightfood *et al.*, 1990).

Components	iv(5)	sv(2)	gd(5)	gd(p)	nd(11)	mr(11)	mk(9)
SiO_2	51.83	52.97	49.48	47.3	51.75	49.20	48.68
TiO_2	2.16	1.61	1.70	1.21	0.98	1.11	1.21
Al_2O_3	15.36	16.50	13.13	9.67	16.74	16.47	16.43
Fe_2O_3	13.40	11.14	12.88	15.30	10.70	12.36	12.87
MgO	3.84	6.41	10.86	18.29	6.11	7.02	6.83
MnO	0.20	0.15	0.18	0.20	0.16	0.18	0.19
CaO	7.85	6.44	7.76	6.91	10.32	11.25	11.37
Na_2O	3.09	3.35	2.02	0.64	2.31	1.94	2.02
K_2O	1.62	1.20	0.85	0.41	0.83	0.36	0.30
P_2O_5	0.65	0.24	0.12	0.09	0.09	0.11	0.10
LOI	2.64	3.20	4.16	6.60	2.33	1.54	1.78
Mg*	—	—	—	—	—	—	—
Ni	24	55	391	775	38	95	105
Cu	24	36	89	72	66	117	141
Cr	91	219	543	850	115	162	181
Co	26	34	54	79	36	42	44
Sc	25	25	28	26	32	37	37
V	157	188	248	211	218	262	265
Zn	145	99	110	145	93	97	103
Rb	38	25	15	7.1	20	5	4
Sr	428	420	302	218	240	210	206
Y	41.0	25.0	17.0	12.4	22.0	23.0	23.0
Zr	327	186	114	66	125	95	90
Nb	31.4	16.1	9.5	5.0	9.1	5.5	4.5
Th	4.7	2.8	1.1	0.87	3.1	1.4	1.0
U	1.3	0.7	0.4	0.32	0.9	0.5	0.4
Ta	1.56	0.85	0.53	0.30	0.52	0.32	0.26
Hf	7.77	4.87	3.10	1.91	3.33	2.55	2.43
La	41.4	20.7	8.4	4.97	16.6	8.8	6.9
Ce	0.8	48.6	19.8	13.02	35.7	20.1	16.5
Pr	11.4	5.5	3.4	1.81	3.7	2.3	2.0
Nd	47.4	26.1	13.2	9.07	17.4	11.9	11.2
Sm	10.0	5.89	3.50	2.44	3.91	3.18	3.21
Eu	2.87	1.84	1.26	0.87	1.12	1.06	1.11
Gd	9.41	5.27	3.93	2.88	4.06	3.76	3.79
Tb	1.44	0.82	0.62	0.45	0.68	0.65	0.65
Dy	7.97	4.82	3.51	2.63	4.08	4.09	4.13
Ho	1.60	0.97	0.69	0.50	0.88	0.88	0.90
Er	4.21	2.62	1.83	1.37	2.45	2.56	2.57
Tm	0.88	0.35	0.23	0.17	0.36	0.37	0.38
Yb	3.60	2.29	1.48	1.04	2.24	2.35	2.36
Lu	0.51	0.32	0.20	0.12	0.33	0.37	0.36

iv — Ivakin Formation, sv — Syverma Formation, gd — Gudchikha Formation, gd(p) — picrite from the Gudchikha Formation, nd — Nadezhda Formation, mr — Morongov Formation, mk — Mokulaev Formation.

The evolution of melts from the upper formations occurred in environments where crustal contamination by primary magma (which originated from a source of T-MORB type and similar in composition to the basalts of Mokulayevsky Formation) became less important. However, in their publications, Lightfood *et al.* (1993) suggest that the crustal component of this contamination can be attributed to older areas of mantle lithosphere, which contain relic remnants, rather than to the continental crust that overlies it.

The primary $^{87}Sr/^{86}Sr$ ratio of the traps range widely, from 0.70470 to 0.71027, with a maximum of 0.705–0.707 (Almukhamedov *et al.*, 1992). Higher values of this ratio are characteristic of the lower part of the lava section, where basalts of higher alkality are widespread. Undepleted mantle and areas of mantle enriched due to the recycling of oceanic crust are believed to be the most probable source rocks for the Siberian traps. Almukhamedov *et al.* (1992) believe that the direct contribution of continental crust to changing the composition of the melt is insignificant.

On the other hand, Sharma *et al.* (1992), who studied the variations in Sr-, Nd- and Pb isotope contents in these traps, concluded that the contaminating lithospheric basement contributed greatly to the formation of the Siberian traps. According to their data, these magmas were generated from two sources: firstly, the almost undepleted mantle plume material and secondly, continental lithospheric material, particularly the Archaean and Palaeoproterozoic lower crust that was assimilated in the course of the ascent of the magmas to the surface.

According to Basu *et al.* (1995), olivine phenocrysts of nephelinite lavas of the NE part of SFBP showed He-3/He-4 ratios up to 12.7 times the atmospheric ratio; these values suggest a lower mantle plume origin. The neodymium and strontium isotopes, rare earth element concentration patterns, and Ce/Pb ratios of the associated rocks were also consistent with their derivation from a near-chondritic, primitive plume. Geochemical data from the c. 250 Ma-old volcanic rocks higher up in the sequence indicate interaction of this high-He-3 SFBP plume with a suboceanic-type upper mantle beneath Siberia.

The observed sequence of alterations of high-Fe and high-Ti basalts and picrobasalts of moderate alkalinity with tholeiitic basalts is an ordinary phenomenon typical of the development of large-scale intra-plate volcanic activity, for example in the Red Sea Rift (see Chapter 2). This model can be accepted in this case, although, unlike the Red Sea region, there was no ocean opening. According to isotopic analyses and incompatible-element ratios, continental-crustal contamination contributed greatly to melt petrogenesis, particularly at the early and middle stages of Siberian trap development.

4.2.2 The Karoo Province (African Platform)

There is a similar assemblage of volcanic and intrusive picrites and basalts within the Karoo Province of South Africa. Volcanics in the Nuanetsi area are the best studied; they are known to have a number of unusual features, e.g. in a number of cases they are rich in potassium, which makes them different from similar rocks in other regions. The best-studied intrusions belong to the massifs of the Insizwa Unit, which, like those of the Noril'sk Region, are associated with large deposits of Cu–Ni sulphide ores.

The Karoo Trap Province of South Africa covers an area of 140,000 km^2, with a trap thickness of up to 1 km. The main peak of activity is dated at about 190 Ma ago (Ellam and Cox, 1989). It is believed that the origin of the trap province is closely related to the opening of the Indian Ocean, at the time when Gondwana broke up into Africa and Antarctica. In the Nuanetsi area, at the base of the trap section, nephelinites occur in small quantities, overlapped by Letaba high-magnesian basalts (9% MgO). Tholeiite basalts of the Sabie River Formation are found up section, interbedded with rhyolites in the upper part of the series.

It has been established by numerous authors (Cox and Jamieson, 1984; Bristow, 1984; Ellam and Cox, 1989) that large volumes of magma with an MgO content of about 15 wt.% were erupted in the area. Rocks with MgO 18 wt.% contain large amounts of olivine phenocrysts; they seem to be associated with the cumulus processes. Large volumes of lava, with an MgO content ranging from 15 to 18 wt.%, probably represent the primary liquid originated from the partially melted mantle; this is confirmed by the fact that the Fe/Mg ratio in picrites is in equilibrium with the olivine of the mantle. Typical rock compositions are shown in Table 4.2.

The trends in composition of the picrobasalt series seem to depend significantly on the fractionation of olivine and orthopyroxene, present as xenocrysts in the picrites (Ellam and Cox, 1989). Using chemical analyses, the picrites are subdivided into two types, which tend to intergrade so-called low SiO$_2$ and high SiO$_2$ types. For a magnesium content of 14–16 wt.%, these on average correspond with 47 and 50% SiO$_2$. They differ significantly in their geochemical characteristics, with the K, Sr, Ba, Zr and REE content being the most important; they are much higher in the high-SiO$_2$ varieties, while the low-SiO$_2$ varieties of picrobasalts are enriched with basaltic components such as Fe, Ca and Al, which is believed to be associated with the different sources of the melts in the mantle.

Ellam and Cox (1989) studied the Sr, Nd and Pb isotope picrite systems of the Nuanetsi area, in order to determine the mantle source for the magmas. It was assumed that since the series evolution was associated with the fractionation of olivine and orthopyroxene the latter should not greatly affect the isotope systems. The ratios of high $^{87}Sr/^{86}Sr$ (0.70523–0.70793, $\varepsilon_{Sr} = 13.6$–52.0) and low $^{143}Nd/^{144}Nd$ (0.51182–0.51217) were revealed. This suggests the important contribution of sources with high Rb/Sr and low Sm/Nd ratio as compared with those of MORB. These sources should have remained isolated from the rapidly transforming and homogenizing asthenosphere, from which the MORB originated. Hence they should have been located within the subcontinental lithosphere, which was not subject to reworking.

Table 4.2 Chemical composition of picritic lavas and dykes of the Nuanetsi area (after Cox and Jamieson, 1984; Ellan and Cox, 1989).

Components	Lavas						Intrusions	
	N76	N34	N124	N356	N77	N296	N22	N95
SiO_2	47.6	48.70	48.37	49.6	46.87	48.00	47.6	47.5
TiO_2	2.60	3.34	3.10	2.54	1.64	0.71	2.62	2.53
Al_2O_3	6.52	7.12	6.65	7.99	9.34	13.86	5.83	7.40
Fe_2O_3	4.7	11.09*	11.56*	2.6	13.52	11.12	3.4	4.4
FeO	6.48	—	—	7.39	—	—	8.40	7.74
MnO	0.14	0.14	0.15	0.13	0.17	0.16	0.15	0.16
MgO	9.0	16.47	18.75	16.9	15.91	14.81	20.3	17.0
CaO	5.94	6.28	5.70	5.62	9.06	8.73	6.05	8.13
Na_2O	0.98	1.16	0.52	1.44	1.32	1.53	1.09	1.52
K_2O	2.08	2.89	3.70	2.78	0.56	0.21	1.99	1.06
P_2O_5	0.46	0.48	0.34	0.36	0.20	0.09	0.42	0.37
H_2O	2.86	—	—	1.94	—	—	1.48	1.45
Total	99.4	97.67	98.96	99.3	98.56	99.22	99.3	99.3

Table 4.2 *Contd.*

Components	Lavas						Intrusions	
	N76	N34	N124	N356	N77	N296	N22	N95
Ba	960	1067	1148	1540	272	133	760	380
Cr	1130	859	878	870	848	875	1,265	960
Ni	1108	738	880	962	675	478	1,145	830
Rb	36	85	80	53	23	17	30	18
Sr	884	1206	1115	1146	445	137	945	690
Zr	346	460	454	363	129	63	370	220
$^{87}Sr/^{86}Sr$	—	—	—	—	0.70523	0.7079	—	—
$\varepsilon^{190}Sr$	—	—	—	—	13.6	52.0	—	—
$^{143}Nd/^{144}Nd$	—	0.51201	—	—	0.51217	0.5118	—	—
$\varepsilon^{190}Nd$	—	6.1	—	—	—	11.2	—	—
$^{206}Pb/^{204}Pb$	—	17.569	—	—	17.531	17.719	—	—
$^{207}Pb/^{204}Pb$	—	15.536	—	—	15.526	15.818	—	—
$^{208}Pb/^{204}Pb$	—	38.404	—	—	38.253	40.714	—	—

Samples N76, N34, N124, N356 — high SiO_2; samples N77, N296 — low SiO_2.

The study has shown that, like the Siberian traps, these lavas display large variations in their isotopic ratios and incompatible elements, indicating that the magma sources were heterogeneous. The Sm–Nd isotopic ratio corresponds with the late Precambrian isochron (1.1 Ga). This determines the age of the source of magma material and also shows evidence of the important role of old continental mantle in the source. This is the time of subcontinental lithospheric stabilization under the Nuanetsi, when major processes of upper crust formation occurred to the south, in the Namaqua–Natal mobile belt, and in Lesotho the lower crust was formed during the combined evolution of the continental crust and lithospheric mantle.

Similar evidence for the strong involvement of crustal and lithospheric material in trap petrogenesis, particularly at the early stages of their evolution, was found for other trap areas such as Deccan and Parana, etc. (Wilson, 1989). As in the cases discussed, trap igneous activity is found to evolve from K to Na moderately alkaline and sometimes high-alkaline series, or from alkaline series to tholeiitic ones. At the early stages of evolution the parental melts were derived from slightly depleted or more often enriched substrates, but later, as the process progressed, these melts became more and more depleted, becoming more like similar sources of T-MORB, and even N-MORB.

This evidence seems to imply the onset of spreading of asthenospheric plumes, with their ascent being related to the appearance of trap fields. Spreading started in the continental lithosphere beneath the crust. It is the interaction with this lithosphere that resulted in the formation of Ni-, Cu- and PGE-enriched picrites of the type that occur in the Gudchikha and Nuanetsi formations. The spreading plumes continued to ascend, partially assimilating the lower-crustal rocks and giving rise to basalt floods typical of the upper parts of the sections. Their chemical properties are primarily determined by plume material which by that time was depleted and contaminated with upper mantle and crustal material.

Judging by the fact that not all of the areas of trap igneous activity had developed into oceanic spreading zones, we can assume that the tops of plumes, in some cases, such as the Siberian traps, did not reach the base of the crust, and that their spreading did not result in its rupture.

It is of interest, however, that the peripheries of the fields display igneous activity significantly different from that in the central part of the fields. As indicated by observations of the Siberian Platform, the formation of alkali-ultrabasic assemblages and kimberlites was concurrent with the eruption of traps.

4.2.3 The Oceanic Equivalents of Continental Traps

Oceanic basaltic plateaux are found in all oceans, in those areas where crustal thickness ranged from about 18 to 40 km. They are made up of tholeiitic basalts, similar to those that are being formed in Iceland at the present time. The early Cretaceous Ontong–Java Plateau in the Pacific Ocean is the best studied now. It is located north of the Solomon islands, at depths of about 1.7–2 km. They are much higher than the adjacent ocean floor which occurs at depths of 4.5 km (Mahoney et al., 1993). The plateau is 1.5×10^6 km^2 in area, which is even larger than the Siberian trap field. In its central part the plateau basalts are overlapped by carbonate sediments of Mesozoic–Cenozoic age, their thickness reaching about 1 km. The basalts were penetrated during the Deep Sea Drilling Project, and are exposed at the surface in the region of Malaite Island. The thickness of the crust within the plateau, as shown by seismic data, reaches 40 km, which is much greater than that typical of the oceans.

The plateau is made up of homogeneous tholeiite basalts (Table 4.3). Variations of major- and rare-elements in the lavas are close to those of E-MORB and to transitional tholeiites (Mahoney et al., 1993). Isotopic analysis has shown that $\varepsilon_{Nd(T)}$ is $+4.0$ to $+6.3$; $(^{87}Sr/^{86}Sr)T = 0.70423$, and $^{206}Pb/^{204}Pb = 18.245–18.709$. The sources of hot-spot type are found to be important in their formation. The combination of such a source with high degrees of melting, and the absence of an obvious age progression, would indicate, as believed by some researchers, rapid plateau formation above the head of the spreading super-plume of the Louisville hot-spot. The results of isotope-geochronological and palaeomagnetic analyses, as well as of geological studies, agree with the above conclusion that it took only 3 Ma for the main part of the plateau to form, in the interval from 120 to 124 Ma, with a peak of activity occurring at 122 Ma. During that period, about 51×10^6 km^3 of basaltic magma were erupted, at the rate ranging from 8 km^3/year to 22 km^3/year, which is comparable to or higher than the average rates of basalt melt flow during the formation of continental trap fields (Mahoney et al., 1993). The superplume material as judged from xenoliths in the alneite dykes from Malaite Island, re-presented by garnet and spinel lherzolites, is similar to the deepest substrate occurring in areas of intra-plate magmatic activity, e.g. the Baikal Rift (Dobretsov et al., 1989).

In the north of the Ontong–Java Plateau, basalts of about 90 Ma in age are found, thus indicating that in some areas magma eruption continued for an additional 30 Ma or that there was a second independent stage of plateau formation related to the Louisville hot-spot (Mahoney et al., 1993).

Thus trap assemblages are by no means a unique feature of continental crustal segments. They are also found to develop within stable areas of the ocean floor. As for continental formations, they appear to develop over large astheno-spheric uplifts (superplumes), the process taking place over a relatively short time period of a few first million years.

Table 4.3 Chemical composition of rocks of submarine plateau of Ontong–Java (after J. Mahoney et al., 1992).

Components	74R-2	75R-4	79R-3	90R-1	69R-2
SiO_2	48.55	48.75	48.31	49.87	49.86
TiO_2	1.61	1.62	1.611	1.13	1.34
Al_2O_3	14.23	14.16	14.42	13.70	15.01
Fe_2O_3	13.01	13.43	13.53	12.97	11.01
MnO	0.22	0.21	0.23	0.18	0.17
MgO	6.82	6.74	6.90	7.71	7.36
CaO	12.12	12.13	12.12	12.11	12.72
Na_2O	2.46	2.38	2.40	2.07	2.25
K_2O	0.14	0.13	0.30	0.05	0.31
P_2O_5	0.14	0.14	0.14	0.10	0.11
LOI	0.59	0.45	0.070	0.07	0.36
Total	99.99	100.1	100.00	100.00	100.5
Cr	164	162	162	145	241
Ni	99	99	86	92	119
Rb	<4	<4	12	5	6
Sr	203	174	173	108	157
V	327	313	358	339	301
Y	30	30	28	24	26
Zr	98	98	98	64	78
Sc	42		46	47	53
Y	31		29	25	26

Components	74R-2	75R-4	79R-3	90R-1	69R-2
Zr	101	—	97	63	79
Nb	6.2	—	6.1	3.4	4.1
Cs	0.04	—	0.21	0.01	0.09
Rb	5	—	10	2	8
Ba	24	—	25	17	13
La	6.1	—	6.0	3.4	3.8
Ce	15.0	—	15.4	8.7	10.1
Pr	2.3	—	2.4	1.4	1.7
Nd	11.2	—	11.5	7.0	8.3
Sm	—	—	3.5	2.5	2.7
Eu	1.3	—	1.3	0.9	1.1
Gd	4.5	—	4.3	3.3	3.8
Tb	—	—	0.8	0.6	0.7
Dy	4.9	—	5.0	3.9	4.7
Ho	1.1	—	1.0	0.9	1.0
Er	2.9	—	2.8	2.4	2.8
Yb	2.8	—	2.8	2.3	2.7
Lu	0.42	—	0.43	0.35	0.37
Hf	2.64	—	2.54	1.66	2.04
Ta	0.44	—	0.36	0.20	0.31
Th	0.45	—	0.51	0.23	0.33
U	0.13	—	0.12	0.09	0.35

4.3 ALKALINE PROVINCES

4.3.1 Alkali-Ultrabasic Units

These represent complex mineralogically varied assemblages, with predominantly plutonic rocks and subordinate volcanic rocks; their formation is believed to be gradual, from ultrabasic rocks to carbonatites (Vasil'ev, 1988). These units are centred within stable areas of the Earth's crust, i.e. ancient cratons.

The main volume of alkali-ultrabasic rocks occurs on shield and ancient platform. Their maximum occurrences were in the late Precambrian, Devonian–Carboniferous, Permian–Triassic and Eocene–Miocene. These intervals coincide with major regional events, e.g. the onset of the middle palaeozoic intra-plate igneous activity throughout the Baltic Shield, the Permian–Triassic Siberian traps and the Jurassic–Carboniferous trap formations in the Karoo Province, Etendeka and Parana in Africa and South America.

Most alkali-ultrabasic intrusions are cylindrical in shape, but sometimes they are funnel-shaped, forming vertical and steeply dipping bodies, in cross-section rounded or elongated and oval, ranging from a few hundreds of square metres to tens of square kilometres in size. Drilling and geophysical studies have shown that these intrusive units can occur as deep as 10 km. It is established that the contacts between the intrusions and enclosing rock are always discordant and, as a rule, are very sharp (Vasil'ev, 1988).

The physical energy of the intruding magma is confirmed by dome-like swelling and radial fractures of the country rocks, the country rocks being subject not only to deformation but also to strong thermal effects as well as metasomatism (intense contact metamorphism to hornfels grade, and fenitization).

The massifs are composed of various petrographic series (from older to younger):

1. ultramafic series, including dunites, olivinites, peridotites, pyroxenites and jakupirangites;
2. alkali-mafic feldspathoid series, including melteigite–jotunites, urtites and the melilite series;
3. nepheline and alkali syenite series;
4. ore-bearing series, including calcite and magnetite–apatite rocks containing forsterite;
5. carbonatite series, including dolomite and ankerite carbonatites.

The scale of the series, their quantitative ratios, and the completeness of their series vary according to province, as well as within massifs of the same unit.

A strongly pronounced concentric zoning is observed in the structure of some of the bodies, which seems to be due to alternation of rocks of different composition. Dunites or olivinites occupy a central position in such massifs, forming the core of the body, which is surrounded by circular or arc-like zones of peridotites (wehrlites), pyroxenites (clinopyroxenites and their olivine-bearing varieties), melilite rocks, jakupirangites, melteigite–jotunites, alkali-gabbroids, syenite-porphiries and other. Carbonatites and their associated apatite–magnetite–forsterite rocks can be located in both the ultramafic cores and at the periphery of the massif, within the endo- and exocontact zones. The origin of the carbonatites in these environments could be either metasomatic or magmatic. The occurrence of carbonatite lavas in the Oldonyo Lengai Volcano in the Kenya Rift and similar magmatism of other rifts (Le Bas, 1987) make magmatic models for their origin rather more probable.

Units with a pronounced concentric zoning include such massifs as Kovdor and Pesochny (Kola Peninsula), Inagly and Konder (Yakutia), Shawa (East Africa) and Jakuperanga (Brasil) and many others (Fig. 4.9). Other intrusive bodies do not exhibit a strong zoning in their rock distribution, this zoning being smoothed and obscured by metasomatic processes; however, there are some cases where zoning is completely absent.

Intrusions of complex composition are normally accompanied by a series of dyke rocks that form swarms of linear bodies, arranged radially with respect to the intrusion centres; the morphology of the dykes themselves is rather variable. The dykes of the endocontact zone form thin, steep bodies (several metres in thickness), extending for tens and hundreds of metres. They are composed of ultrabasic, alkali-ultrabasic and alkaline rocks, such as meimechites, picrites, kimberlite-like rocks, monchikites, alnoites, various syenites, etc. The dykes that are located between the intrusions are represented both by steep linear bodies, intruding the inner zone, and by arc-shaped bodies, with the dyke walls concordant to the zoning and intrusion contact.

It should be noted here that rocks similar to those observed within the alkali-ultrabasic units, i.e. dunites, wehrlites, clinopyroxenites, melteigites, urtites and nepheline syenites, are widespread in the form of xenoliths in the alkaline lavas and tuffs of the Kenyan Rift. This, as shown by Polyakov (Belousov et al., 1974), can be regarded as evidence to indicate their close relationship with continental alkaline volcanic activity.

Most of the ultramafic rocks of the units in question have an undersaturated SiO_2 content (Table 4.4), while the contents of MgO and CaO vary strongly from dunites to pyroxenites. The same is true of the clinopyroxenites, where higher contents of Na_2O and K_2O are due to the addition of alkaline minerals. Also, it is only the higher alkalinity of the jakupirangites that seems to be original feature of this variety. Another important feature of ultramafic rocks is their commonly (except for dunites from massifs of the Aldan Province) higher iron and titanium content.

There are significant differences in the distribution and content of a number of trace elements in ultramafic rocks (Table 4.4). Thus the olivinites are characterized by a higher titanium content and lower Cr and Ni content. The same

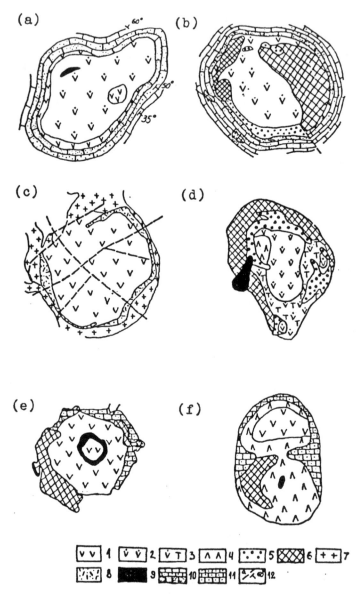

FIGURE 4.9 The structure of the alkaline–ultrabasic intrusions of the (a) Bor–Uruakh, (b) Kugda, (c) Inagli, (d) Kovdor, (e) Shawa, (f) Yakupiranga. By Vasiljev (1988). 1 — dunites and peridotites; 2 — olivinites; 3 — phlogopitization zones in the olivinites; 4 — clinopyroxenites, jakupirangites; 5 — melilites; 6 — urtites, ijolites and melteigites; 7 — alkaline syenites (pulaskites, etc.); 8 — alkali gabbroids; 9 — carbonatites; 10 — country rocks; 11 — thermally metamorphized rocks and fenites; 12 — country rock elements.

Table 4.4 Average composition of ultramafic rocks of alkaline–ultrabasic complexes (Laz'ko, 1988b).

Components	1	2	3	4	5	6	7	8	9	10	11
SiO_2	39.26	40.52	37.00	35.92	38.74	40.35	43.90	40.86	41.09	40.79	46.09
TiO_2	0.46	0.02	0.32	1.31	0.27	1.74	0.98	2.45	4.30	3.96	3.48
Al_2O_3	0.52	0.50	0.39	0.28	0.63	2.10	1.12	3.50	4.06	6.14	2.58
Cr_2O_3	0.60	0.62	0.54	0.17	0.24	0.32	0.42	0.21	—	0.44	—
Fe_2O_3	6.52	2.42	3.76	5.47	2.74	6.40	2.29	8.55	8.79	8.51	5.50
FeO	8.06	5.14	12.78	12.58	10.71	8.59	10.0	10.28	5.77	7.86	6.02
MnO	0.19	0.16	0.25	0.30	0.40	0.20	0.20	0.12	0.17	0.22	0.22
NiO	0.30	0.21	0.34	0.13	0.16	0.20	0.20	0.10	—	—	—
MgO	42.35	50.14	43.48	42.67	44.79	34.03	30.55	14.67	13.76	12.34	11.28
CaO	0.83	0.34	0.74	0.82	0.93	5.46	9.15	18.62	20.69	15.98	23.01
Na_2O	0.15	0.10	0.14	0.15	0.12	0.22	0.30	0.49	0.54	2.49	1.06
K_2O	0.09	0.13	0.21	0.14	0.12	0.21	0.20	0.02	0.70	1.18	0.33
P_2O_5	0.13	0.08	0.03	0.06	0.15	0.18	0.01	0.13	0.12	0.09	0.44
Number of analyses	36	37	6	37	35	20	12	14	12	8	5

Columns 1–3 — dunites: 1 — Gulin Massif, 2 — massifs of the Aldan province, 3 — massifs of the Kola–Karelia Province; 4, 5 — olivinites from the massifs of Maimacha–Kotui (4) and the Kola–Karelia (5) provinces; 6, 7 — wehrlites of the Gulin Massif (6) and massifs of the Kola–Karelia Province (7); 8, 9 — clinopyroxenites from massifs of the Maimecha–Kotui (8) and Karelian–Kola (9) Provinces; 10–11 — jakupirangites from massifs of the Maimecha–Kotui and Karelian–Kola Provinces.

feature is typical of pyroxenites and jakupirangites. There is a general trend of a higher content of REE in the series, from dunites to clinopyroxenites.

The massifs of the Maimecha–Kotui Province are a typical example of such formations (Kogarko *et al.*, 1995). This is a large area with the widespread occurrence of alkaline rocks in the north of the Siberian Platform, to the west of the Anabar Shield. More than twenty massifs of ultrabasic-alkaline rocks are recognized there; in fact, all of the large bodies are centred in areas of cross-crossing of large fractures (Fig. 4.10). According to Dyuzhikov *et al.* (1988) and Basu *et al.* (1995) the alkaline rocks in the province were formed at the same

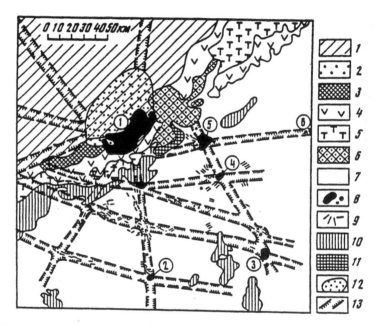

FIGURE 4.10 The geological structure of the Maimecha–Kotui Igneous Province by Vasiljev (1988). 1 — Mesozoic sediments of the Khatanga Depression; 2–6 — Triassic volcanic sediments: 2 — meimechites, the Maimecha Formation, 3 — alkaline basaltoids, trachytes, andesites, subalkali and alkali picrites and pthers of the Delkana Formation, 4 — basalts of the Kogotoka Formation, 5 — basalt tuffs and tuffaceous Sandstones of the Pravy Boyar Formation, 6 — alkaline basaltoids, augitites, subalkali and alkali picrites, as well as other rocks of the Arydzhang Formation; 7 — Palaeozoic carbonate and terrigenous sediments; 8 — the largest alkali-ultrabasic intrusions (numbers are in circles): 1 — Gulin, 2 — Bor–Uryakh, 3 — Magan, 4 — Kugda, 5 — Odikhincha, 6 — Nemakit; 9 — dykes of alkali and ultrabasic rocks; 10 — sills of dolerites; 11 — sills of alkali and ultrabasic rocks; 12 — the boundary of the Gulin Massif as indicated by geophysical data; 13 — zones of deep faulting.

time as the traps of the Noril'sk area, where alkaline rocks are insignificant, and increased eastward, reaching a maximum within the province itself.

The units of the Maimecha–Kotui Province that have been studied possess a number of common structural and morphological features, i.e. they are located in an area of gently dipping Palaeozoic terrigenous–carbonate formations, which were deformed so that the dip was almost vertical, close to the contact zones, with stock-like bodies of oval and circular cross-section (up to a few square kilometres in area, Fig. 4.9). The only exception is the Gulin Pluton with an area of $600\,km^2$ (according to geophysical data, its size could actually be up to $2,000\,km^2$.

They all have quite a variable, although similar, suite of rocks, which are present in the massifs in various ratios. Ultramafic rocks (dunites, olivinites, wehrlites, clinopyroxenites) predominate or are present in significant amounts in the intrusions of Bor–Uryakh, Kugda and the Gulin massifs. In other massifs (Nemakit, Odikhincha, Magan, Yessey, etc.), ultramafic rocks (mostly olivinites) play an insignificant role. They occur among other rocks as relics in the endo-contact zones of older ultrabasic intrusions or as disintegrated blocks of earlier bodies that had been subjected to metasomatism.

4.3.2 Kola–Karelia Alkaline Province

This province is located in the eastern Baltic Shield (Fig. 4.11). It was formed in the middle–late Devonian, i.e. about 380–360 Ma ago (Kogarko et al., 1995). Alkaline rocks are widespread over the entire province, including large intrusions of nepheline syenites in the Khibiny and Lovozero Tundras, dykes and diatremes of varied alkaline picrites, nephelinites, lamprophyres and alkaline lavas and diamondiferous kimberlites in the Arkhangel'sk area, as well as concentrically zoned confocal complexes of ultrabasic–alkaline rocks containing carbonatites of the type common in the Maimecha–Kotui Province. There are twenty similar units in the Kola–Karelia region (Fig. 4.9), such as Kovdor (with an area of $40\,km^2$), Lesnaya Varaka ($12\,km^2$), Africanda (about $6.5\,km^2$), etc. These units are centred in basement fractures and where they cross, form a number of sublatitudinal belts (Vasil'ev, 1988).

A zonal distribution of alkaline rocks within the province has been revealed. Large intrusions of nepheline syenites of the Khibiny and Lovozero Tundras occupy the centre of the province; dyke swarms of various alkaline picrites and other alkaline rocks radiate from this. The centre is surrounded by the above-described numerous confocal massifs of ultrabasic–alkaline rocks, which may represent the deep feeder zones of volcanoes, while kimberlite diatremes are found at the SE margin of the Arkhangel'sk District. The area where these rocks occur appears to coincide in size with their parental asthenospheric plume (diapir), being about 600 km long and about 300 km wide (Fig. 4.11).

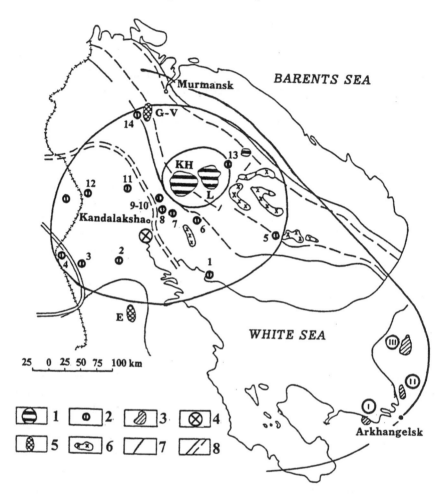

FIGURE 4.11 Kola–Karelia Alkaline Province. 1–4 — the Middle Devonian Formations: Key: 1 — nepheline syenite intrusions (KH — Khibiny Tundra, L — Lovozero Tundra); 2 — confocal alkaline–ultrabasic massifs with carbonatites; (numbers on the map: 1 — Tur Cape, 2 — Kovdozero, 3 — Vuoyarvy, 4 — Sallantlavy, 5 — Pesochny, 6 — Ingozero, 7 — Salmagor, 8–10 — Khabozero Group; 11 — Mavr–Guba, 12 — Kovdor, 13 — Kurgin, 14 — Seblyavo); 3 — kimberlite fields; 4 — breccia pipe, Elovy Island; 5 — Early Proterozoic (about 1950 Ma old) layered basic and ultrabasic, alkaline intrusives (Gr–V — Gremyakha–Vyrmes, E — Eletozero); 6 — Early Proterozoic (about 2.4 Ga) alkali granites; 7 — suggested contours of the Devonian mantle plume; 8 — constraints of major Precambrian structures (cf. Fig. 5.1).

Unique intrusions of nepheline syenites from the Khibiny and Lovozero Tundras represent neighbouring massifs, 1,330 and 650 km² in size, respectively (Kogarko *et al.*, 1995). They are funnel-shaped in cross-section (Fig. 4.12) and the results of geophysical studies show them to extend downwards as far as 7–8 km.

The Lovozero Massif is the best studied. It occupies a graben-like structure that overlaps the early Precambrian formations of the Baltic Shield. The graben is infilled by a late Palaeozoic volcanic–terrigenous succession, and its remains can be found at the top of the massif. Various slates and sandstones, tuffs and lavas of alkaline picrites and basalts are found here; there are also alkaline trachyte and phonolite porphiries and trachyandesites. The age of the massif is 365 ± 15 Ma (Kramm *et al.*, 1993).

The Lovozero Massif appears to be a typical layered intrusion with a well-pronounced Marginal and central Layered series, which are frequently recognized as independent stages of intrusion. The Lateral marginal series is basically made up of fine- and medium-grained poikilitic nepheline–sodalite syenites. This series extends along the periphery of the massif, covering about 10% of its area. These rocks are not strongly layered, resting concordantly on the contacts. The main volume of the Lovozero Intrusion is composed of rocks of the Layered series, with their layering being unconformable with the Marginal series, as in the rocks of the Skaergaard Intrusion.

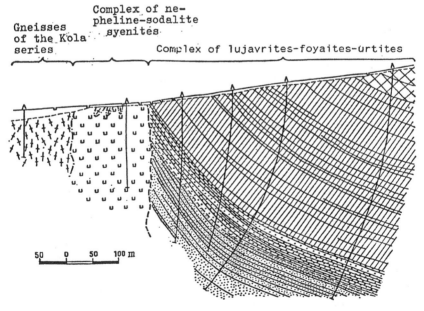

FIGURE 4.12 The structure of the endocontact zone of the Lovozero Intrusion (from Bussen A. and Sakharov S., 1972).

This group is characterized by a well-developed rhythmic layering, as shown in the alternation of urtites (cumulus nepheline), foyaites (nepheline + alkali feldspar) and lujavrites (nepheline + alkali feldspar + aegirine) which corresponds with the order of evolution of the paragenesis of these minerals, from the phonolitic melt recognized as the pluton's parental magma (Sharkov, 1980). This unit is characterized by a very continuous section; even thin layers (20–40 cm) can be traced for a long distance without any change in thickness of texture. An apparent decrease in layer thickness occurs near the endocontact zone, where the layers are observed to pinch out (cf. Fig. 4.12).

Rocks of the Upper marginal group consist mainly of eudialyte lujavrites, forming a poorly layered sheet-like body in the upper part of the intrusion, overlapping the formations of the Layered series. Table 4.5 displays the average chemical composition of the Lovozero Intrusion rocks.

The Khibiny and Lovozero massifs are associated with a number of different dykes of shonkinite, theralite, trachyte, melanephelinite, alkali picrite, carbonatite, etc. Xenoliths of spinel peridotites have been found in one of the picrite dykes; they consist of olivine (75%), enstatite (12%), Cr-bearing pargasite and Cr-spinel (3%), as well as small amounts of Cr-diopside, phlogopite, apatite and calcite (Arzamastsev et al., 1988). This suggests that the peridotites are fragments of mantle represented by strongly depleted rocks that were subjected to strong metasomatism prior to magma formation. The temperature and pressure parameters of these rocks were 1000°C and 15 ± 1 kbar.

Detailed geochemical evidence suggests that the formation of alkaline rocks of the Khibiny–Lovozero Unit was associated with differentiation of parental melts whose composition was similar to that of olivine nephelinites (Kogarko, 1977).

However, it should be noted that the composition of deep-seated xenoliths in igneous rocks of the province varies between the plume centre (the Khibiny Intrusion) and its periphery. Fragments of metasomatized asthenosphere are found at the centre, while in the picrite and kimberlite-like dykes at a distance of 100–150 km, there are xenoliths of eclogites and garnet granulites from the lower crust (Sharkov and Pukhtel, 1986; Kempton et al., 1995). At the plume margin, 400–500 km from the centre, the Arkhangel'sk kimberlites contain diatoms, from the deep ancient lithosphere. Judging by inclusions in the diamonds (pyrope and chromian spinels) their host rocks were depleted garnet peridotites and eclogites, the oldest representatives of ancient lithospheric mantle, formed at pressures of approximately 50 kbar (Sobolev et al., 1992).

We can conclude from these data that the modern upper mantle of the northern Baltic Shield is made up of both ancient lithospheric basement and material from the Devonian asthenospheric plume. The pattern of deep xenolith indicates that the top of this diapir dips in a south-easterly direction. It comes closer to the surface (30–40 km) in the Lovozero–Khibiny area, where seismic data indicate a protrusion of extremely dense rocks at a depth of 30 km (Solodilov, pers. comm., 1994). This can be recognized as the remains of the Devonian plume

Table 4.5 Average chemical composition (wt.%) and trace element content (ppm) in the rocks of the Lovozero massif (after Kononova, 1984).

Components	1	2	3
SiO_2	56.79	54.14	53.21
TiO_2	1.03	0.81	1.26
ZrO_2+HfO_2	0.170	0.290	1.36
Al_2O_3	19.94	17.19	15.51
Fe_2O_3	1.74	6.50	6.90
FeO	1.28	1.50	1.73
MnO	0.20	0.27	0.27
MgO	0.98	0.98	0.13
CaO	1.45	1.47	1.89
Na_2O	8.13	9.58	9.58
K_2O	5.94	5.23	4.81
P	0.13	0.096	0.044
F	0.21	0.12	0.18
Cl	0.10	0.13	0.29
S	0.52	0.078	0.087
Sr	0.140	0.055	0.13
Ba	0.11	0.006	0.07
Cr	24	28	29
Ni	5	7	34
Co	<5	n.d.	0–5
V	56	109	120
Nb	320	645	1,016
Tr_2O_3	1,200	2,350	2,640

1 — fine-grained porphyritic nepheline syenites,
2 — layered foyaites, urtites and lujavrites,
3 — layered eudialyte lujavrites.

material. The xenoliths do not include any lower-crustal rocks; these rocks are identified only at the periphery of the asthenopheric protrusion. The P–T conditions during the formation of lower-crustal rocks (Bindeman *et al.*, 1990) suggest that the top of the diapir occurred at a depth of 60–70 km here, reaching 150 km in the vicinity of Arkhangel'sk. From geophysical measurements the thickness of subcrustal lithosphere could have been as great as 250 km in the area (Egorov *et al.*, 1994).

The mechanism for the formation of the concentrically zoned confocal alkali-ultrabasic units is less obvious than that for the Lovozero Intrusion. Although they are undoubtedly related to alkaline volcanism, the petrogenetic processes that led to the formation of such an usual inner structure are still unclear.

Ultrabasic volcanics, spatially related to alkali-ultrabasic complexes, are likely to be similar in composition to the parental mantle-derived melts that gave rise to varied alkaline rocks. These volcanics include meimechites, subalkaline and alkaline picrites, which are most widely developed in the Maimecha–Kotui Province, although less extensive occurrences of similar volcanics are present in a number of other alkali-ultrabasic provinces, especially in the Kola–Karelia Province. Meimechites and picrites, together with volcanics of different composition, infill linear zones of the subsidence caldera type.

Not all the rocks in the assemblage were studied thoroughly. Data available, however, show that these volcanics are quite unusual. Geochemical data imply that they were enriched by typical components of ultrabasic rocks (Cr, Ni) and by incompatible elements (Rb, Sr and Zr), in amounts similar to those in alkaline basites (Table 4.6). The concentrations of Ti, Cr, Co, Sr and Zr in the

Table 4.6 Average composition of ultramafic volcanics of alkaline–ultrabasic complexes (after Laz'ko, 1988).

Components	1	2	3	4
SiO_2	40.44	40.52	40.94	40.44
TiO_2	1.60	1.75	3.69	3.27
Al_2O_3	2.36	2.56	4.53	4.58
Cr_2O_3	0.40	0.29	0.14	0.20
Fe_2O_3	6.71	5.23	9.00	5.77
FeO	6.52	7.81	7.08	9.58
MnO	0.18	0.17	0.25	0.21
NiO	0.22	0.21	0.13	0.12
MgO	36.79	36.56	22.71	25.14
CaO	4.25	4.14	9.53	9.36
Na_2O	0.20	0.23	0.72	0.63
K_2O	0.15	0.29	0.83	1.25
P_2O_5	0.18	0.24	0.45	0.45
Cr	2,230	1,970	1,800	640
Ni	1,770	1,490	1,470	620
Co	130	120	120	94
V	180	160	240	120
Sc	96	15	15	24
Rb	15	31	28	48
Sr	370	690	680	1,200
Zr	185	260	260	360
Number of analyses	38	14	19	32

Columns 1–2 — lava and dyke meimechites; 3–4 — lava and dyke picrites.

Table 4.7 REE abundances in meimechites and picrites of the northern Siberian Platform (Laz'ko, 1988).

Components	1	2	3	4	5	6
La	41.6	29.3	24.8	31.5	46.0	120
Ce	97.4	68.3	50.3	70.0	105	220
Nd	51.3	37.4	—	—	—	—
Sm	9.15	7.33	5.0	5.6	9.2	14.0
Eu	2.21	2.02	1.7	1.8	3.2	3.3
Gd	6.07	5.20	—	—	—	—
Tb	0.82	0.76	0.66	0.61	1.0	0.82
Yb	0.92	0.88	—	—	—	—
Lu	0.121	0.113	0.088	0.193	0.15	0.33

Columns 1–4 — meimechites; 5, 6 — picrites.

meimechites are noticeably different from those in the picrites. All the volcanics are rich in light REE — their concentration in the picrites is higher than that in the meimechites (normally 2–5 times higher), but there is a readily identifiable gradual decrease in their concentration toward the heavy REE (Table 4.7).

4.4 KIMBERLITES

It is difficult to name another igneous rock with equally variable appearance, texture, mineral and chemical composition. The range of composition, together with frequent similarities with subalkaline and alkaline picrites, melilite-bearing volcanics, basanites, ultrabasic lamprophyres and subvolcanic members of alkali-ultrabasic units as well as the constant presence of xenolith material and extensive rock alteration, tend to blur definition of the term "kimberlite". Nowadays kimberlite is defined as an ultrabasic rock, tending to be alkaline, frequently diamondiferous, with inclusions of olivine, magnesian ilmenite, phlogopite, pyrope garnet, chromian diopside and other compositions, embedded in a fine-grained phlogopite–serpentinite–carbonate matrix (Sobolev, 1974; Laz'ko, 1988c).

Kimberlite magmatism is characteristic of stable continental areas. As mentioned above, these rocks are frequently located at the periphery of large provinces of CFB (Fig. 4.3). The distribution of rocks is extremely patchy with the kimberlite bodies being concentrated in small areas, widely separated from each other. This allows us to distinguish the independent kimberlite field and the

areas where they occur (Fig. 4.13). Kimberlites found in the central areas of
a number of cratons (Siberia, South Africa, Tanzania, etc.) are diamondiferous
(e.g. Clifford's Rule), while those located on the periphery of the craton are
much less productive.

Most of the kimberlites can be referred to relatively young Phanerozoic
formations, although Precambrian magmatism of this type has also been recognized. The oldest pipes seem to be those of the Kuruman Field (South Africa)
which originated about 1.63–1.67 Ga ago (Extended..., 1986). The middle
Palaeozoic, the late Permian and Triassic epochs turned out to be the most
productive; at this time most of the kimberlite bodies of the Central Yakutia
Province were formed. In late Jurassic and Cretaceous times, the kimberlites
of South and West Africa, Angola, Brazil, China and the northern Siberian
and Russian platforms were formed. The youngest Cenozoic rocks of this
type are dated as middle Oligocene (the Colorado Plateau) and Eocene
(Montana, USA; and Igvicy, Tanzania). However, they differ in a number of
properties from typical diamondiferous kimberlites (Dawson, 1980; etc.).

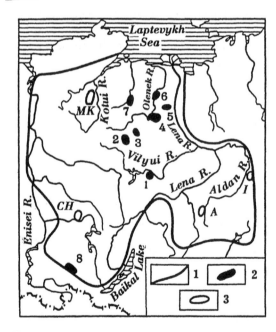

FIGURE 4.13 The distribution of kimberlites on the Siberian Platform (from
E.E. Laz'ko, 1988). 1 — the platform border; 2 — areas where kimberlites were
formed (figures on the scheme): 1 — Malo–Botuob, 2 — Dadlyno Alakitsky, 3 —
Verkhy Mun, 4 — Sredney–Olenok, 5 — Prilensky, 6 — Nizhne–Oleneksky, 7 —
Kuoman, 8 — Near Sayany; 3 — the areas where kimberlite-like rocks are devel-
oped (letters on the scheme): MK — Maimecha–Kotui, A — Aldan, I — Ingil, CH —
Chadobets.

Although they occur as intrusions, extrusions and pyroclastics, they most frequently occur as tubular, subvertical bodies, i.e. kimberlite pipes or diatremes (Fig. 4.14). In plan their contours are highly variable, changing from round and oval to extended and frequently with irregular boundaries (Fig. 4.15).

According to their mica content, kimberlites are subdivided into lamprophyric (micaceous) and "basaltoid" (poor in phlogopite), although the second term, as indicated by Dawson (1983), is not very appropriate, since there is no plagioclase in these rocks. In the extreme varieties of micaceous kimberlites the volume of phlogopite in the main body sometimes exceeds 50%, while in the basaltoid kimberlites it occur, as disseminated scales; there are also transitional forms between the two extremes. The third variety is calcitic kimberlite which is rich in igneous calcite (more than 30%), with primary melt inclusions.

In terms of their chemistry, the kimberlites can be classified with the alkali picrites. Their main properties are: high magnesium content, low SiO_2 and Al_2O_3 content and higher concentrations of Cr and Ni, which indicate the ultrabasic nature of the rocks (Tables 4.8 and 4.9). This is combined with very high TiO_2 and P_2O_5 content and particularly K_2O (up to 6 wt.% in some micaceous varieties), which suggests that they resemble alkali picrites. However, their concentration of volatile components (CO_2 and H_2O) is unusually high. The high content of CO_2 is due to the saturation of the rocks with $CaCO_3$. The chemical

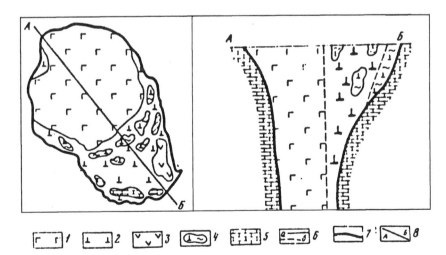

FIGURE 4.14 The structure of the Mir Kimberlite Pipe. According to Laz'ko (1988): (a) plan view, (b) cross-section; 1–3 — kimberlite breccia emplaced at different stages: 1 — first stage, 2 — second stage and 3 — third stage; 4 — areas enriched by xenogenic material; 5 — enclosing rocks and their xenolith; 6 — geological boundaries (a — observed, b — assumed); 7 — the pipe outline; 8 — the cross-section along the line A–B.

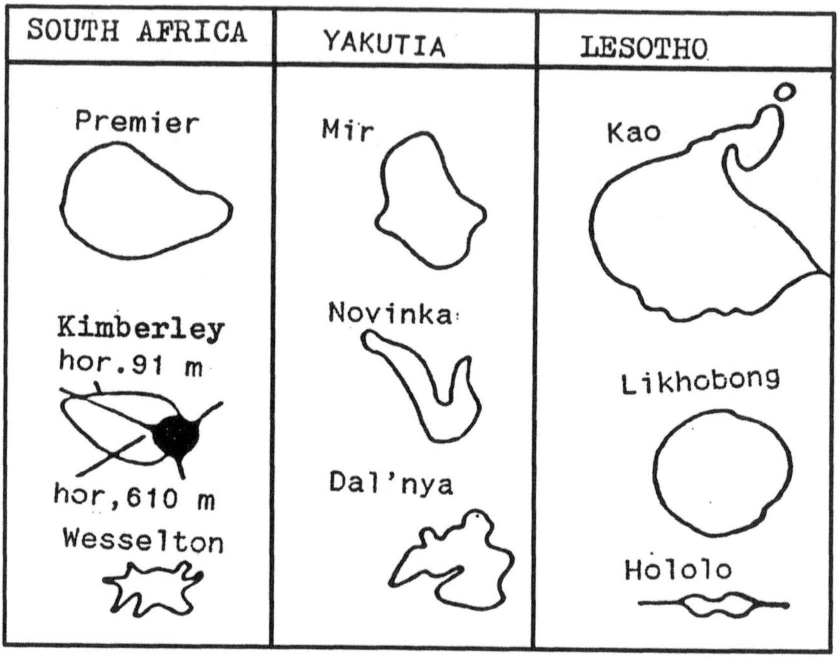

FIGURE 4.15 The surface contours of some of the kimberlite bodies (after Dawson, 1983).

composition of kimberlites is extremely variable, not only within different bodies, but also from different stages of the same pipe. This can be attributed mainly to variations in the mineral content, to contamination by xenogenic material and to intensive alteration.

Kimberlites are rich in the following elements: K, Ti, P, Rb, Sr, Ba, Zr, Y, LREE, Nb, Na, Th and U (Table 4.9); this is characteristic of other intra-plate ultrabasites, but kimberlites show a unique degree of trace-element concentration. This is best demonstrated by their REE content (Fig. 4.16); their concentration in some kimberlites reaches 1,000 ppm, accompanied by a strong enrichment in light lantanoids (Dawson, 1980; Laz'ko, 1988). It is of interest that the African and Indian kimberlites have a higher average REE content than those of Yakutia and North America. This has allowed researchers to distinguish Gondwana and Laurasia megaprovinces (Ilupin et al., 1978).

Isotope ratios in kimberlites are variable (Smith, 1983; Fraser et al., 1985; Mahotkin et al., 1995; etc.). This is especially true of the $^{87}Sr/^{86}Sr$ ratio (about 0.703–0.721); low ("mantle") values of the $^{87}Sr/^{86}Sr$ ratio (up to about 0.7050) are found only in the slightly altered basaltoid kimberlites, while the higher ratios are typical of micaceous kimberlites. Correlations between the Rb and Sr

Table 4.8 Average chemical composition of kimberlites (Laz'ko, 1988).

Components	1	2	3	4	5	6	7	8	9	10	11	12	13	14
SiO_2	33.68	25.72	27.33	28.63	36.59	48.62	44.93	39.78	28.26	33.35	31.60	17.04	4.51	24.63
TiO_2	1.29	1.27	1.44	2.32	0.12	1.76	1.71	2.11	3.75	2.42	3.87	0.34	2.02	0.56
Al_2O_3	2.95	2.12	1.66	2.87	1.55	4.13	4.48	2.89	3.42	3.98	3.06	2.68	1.75	1.65
Cr_2O_3	0.11	0.11	0.24	0.16	0.07	0.15	0.13	0.18	0.16	0.42	0.22	0.09	0.28	0.18
Fe_2O_3	4.71	3.96	3.09	5.83	2.93	8.54	8.69	9.35	5.85	9.39	3.98	2.68	4.96	5.06
FeO	2.98	3.19	6.54	2.32	3.94	—	—	—	4.98	—	9.05	1.24	0.69	—
MnO	0.10	0.12	0.18	0.14	0.15	0.15	0.14	0.15	0.14	0.16	0.19	0.07	0.24	0.24
MgO	25.50	27.76	32.71	26.87	39.62	23.11	22.54	27.73	22.86	23.65	26.60	17.02	3.81	11.98
CaO	9.21	13.87	13.03	11.35	2.13	4.47	6.66	6.87	12.67	9.86	9.25	24.44	41.43	25.46
Na_2O	0.13	0.32	0.38	0.09	0.08	0.57	0.85	0.55	0.20	0.48	0.25	0.23	0.08	0.06
K_2O	0.47	0.82	1.07	1.12	0.12	0.74	1.32	1.65	1.78	1.88	2.10	1.16	0.90	0.32
P_2O_5	0.29	0.34	0.49	0.33	0.23	0.18	0.28	0.23	0.65	1.69	0.48	0.43	0.74	0.83
H_2O	9.71	7.06	1.52	9.12	9.23	6.75	7.94	8.37	6.41	7.54	3.40	5.07	1.82	11.13
CO_2	8.53	11.22	10.35	8.84	3.02	0.83	0.33	0.36	8.79	5.18	5.95	27.45	32.39	17.94
LOI*	0.34	2.12	0.07	0.01	0.22	—	—	—	0.08	—	—	0.06	4.38	—
Number of analyses	198	141	16	7	14	4	7	4	6	14	7	9	12	16

*S, F, Cl — nd. Columns 1–7 — basaltoid kimberlites: 1 — Mir Pipe; 2 — Udachnaya–Vostochnaya Pipe, as a whole; 3 — Udachnaya–Vostochnaya Pipe' slightly altered kimberlite of deep horizons; 4 — porphyry kimberlite Dalnyaya Pipe; 5 — Ruslovaya Pipe; 6, 7 — "grey" and "brown" kimberlite, Premier Pipe; 8–11 — micaceous kimberlite: 8 — "black" kimberlite, Premier Pipe; 9 — Flogopitovaya Pipe, 10 — autoliths in the basaltoid breccia of Wesselton Pipe; 11 — kimberlite dykes of Greenland; 12–14 — calcite kimberlites: 12 — Aikhal Pipe, north-eastern block, 13 — vein 4 in the Udachnaya Pipe, 14 — vein rocks from the Bellsbank Quarry.

Table 4.9 Abundance of trace elements in kimberlites (Laz'ko, 1988).

Components	1	2	3	4	5	6	7	8
Cr	835	910	—	1,060	1,210	—	1,190	1,270
Ni	675	1,100	—	950	1,170	—	1,010	1,050
Co	87	81	76	73	86	76	53	60
V	79	120	—	—	—	—	—	—
Sc	14	11	19	8.8	9.5	18	8.8	19
Rb	21	16	—	51	150	120	4.0	53
Sr	400	620	—	180	490	770	430	1,230
Ba	330	590	—	210	1,020	940	130	4,720
Ta	1.8	5.3	12	6.5	8.0	2.7	1.9	8.9
Nb	48	86	—	54	71	—	41	170
Zr	67	57	12	120	96	—	140	280
Hf	1.6	1.9	—	2.9	2.0	4.1	3.0	6.4
Y	6.3	5.7	—	8.0	5.0	—	9.0	19
Ta	6.0	7.5	17	5.5	5.3	7.4	14	46
U	0.75	2.3	3.2	0.8	0.8	2.6	2.5	5.3
La	33	48	150	28	32	51	110	315
Ce	77	120	360	49	55	110	200	520
Nd	26	43	—	—	—	48	70	150
Sm	3.8	6.4	21	3.7	3.7	6.9	8.1	17

Table 4.9 *contd.*

Components	1	2	3	4	5	6	7	8
Eu	0.70	0.80	5.4	0.95	1.0	1.5	2.0	4.1
Gd	3.6	3.8	—	—	—	—	—	—
Tb	0.45	—	2.0	0.39	0.49	0.44	1.0	2.0
Yb	0.55	0.2	1.5	0.90	0.70	0.85	1.0	2.1
Lu	—	—	0.18	0.14	0.15	—	0.11	0.20
Number of analyses	48	70	6	4	4	7	10	6

Columns 1–4 — basaltoid kimberlites: 1 — Mir Pipe; 2 — Udachnaya–Vostochnaya Pipe; 3 — Wesselton Pipe; 4 — kimberlite Premier Pipe; 5–6 — micaceous kimberlites: 5 — "black" kimberlite from the Premier Pipe; 6 — Finsch Pipe; 7–8 — calcitic kimberlites from the Bellsbank Quarry: 7 — Water vein; 8 — Bobbijan Dyke.

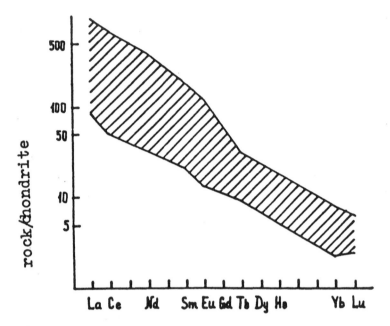

FIGURE 4.16 The distribution of rare-earth elements in kimberlites (shaded area) (after Laz'ko, 1988).

content, Rb/Sr ratios, and values of the $^{87}Sr/^{86}Sr$ ratio observed in fresh micaceous kimberlites provide evidence that high isotope ratios characterized the geochemistry of proto-kimberlite magmas.

Study of the Nd isotope composition of basaltoid kimberlites ($\varepsilon_{Nd(T)}$ from -0.6 to $+4.1$) indicates that they are not differentiated or slightly depleted in particular deep zones in comparison with the primary chondritic mantle (Basu *et al.*, 1984). Study of micaceous kimberlites (Smith, 1983; Fraser *et al.*, 1985) has shown that they originated from a source rich in REE (it was REE enriched as far back as the Precambrian) ($\varepsilon_{Nd(T)}$ -5.4 to -8.6), which remained in the subcontinental lithosphere for a long time. Two similar separate rock groups have been distinguished using Pb isotope analysis (in basaltoid and micaceous kimberlites $^{206}Pb/^{204}Pb = 18.62–20.01$ and $17.21–18.42$; $^{207}Pb/^{204}Pb = 15.52–15.76$ and $15.47–15.62$; and $^{208}Pb/^{204}Pb = 38.511–39.58$ and $37.64–38.23$).

The REE spectra of the kimberlites indicate that they originated by partial melting in the presence of garnet. According to experiments by Ringwood and Kesson (1992) this garnet phase was represented by majorite that contained only 10–12 wt.% of Al_2O_3, testifying to kimberlite formation at depths of 400–650 km.

The deep-seated xenoliths from the kimberlites, which represent primarily fragments of ancient continental lithosphere, have highly variable compositions.

Together with lower-crustal formations, represented mainly by garnet granulites and eclogites, there are various mantle rocks, originating from a depth of 260 km and more. The xenoliths include all possible peridotites and pyroxenites (containing both garnet and spinel and transitional spinel-garnet), chrominum-bearing dunites and olivinites, all types of eclogites (both bimineralic type and those containing corundum, kyanite, coesite and sanidine as essential rock-forming minerals), specific high-Ti ultramafic rocks with ilmenite, phlogopite, amphibole, and glimmerite, as well as marides (phlogopite–garnet–pyroxene rocks), alkremites and eutectoid (graphic) ilmenite–pyroxene rocks (Sobolev, 1974; Laz'ko, 1988; etc.).

As indicated by the above data, mantle ultramafic rocks can be divided, like the xenoliths in the intra-plate basalts, into two assemblages: one with high magnesium–chromium content and the another with a high titanium content. This makes it possible to suggest that the lithospheric upper mantle was formed by two types of material, most likely of different origin. The third type of the material is represented by eclogites.

Among deep-seated xenoliths in kimberlites, garnet peridotites and other ultramafic rocks generally dominate, although there are pipes with mantle rocks represented mostly by eclogites (the Zagadochnaya Pipe in Yakutia, the Roberts–Victor and Orapa pipes in South Africa, etc.). The diamondiferous rocks, i.e. some of the eclogites and the garnet peridotites, are found to be the oldest; their model age is about 3.4 Ga (Pokhilenko, 1989). In this case, the peridotites are represented by extremely depleted varieties of the dunite–harzburgite assemblage. However, Shimizu and Sobolev (1995) suggest that some of these diamonds could have a younger age. They show that there was a heterogeneous distribution of tiny trace elements both within and among discrete garnets from inclusions in single diamonds with Sm–Nd model age 3.2 Ga from the Mir Pipe (Yakutia). The conditions for survival of these heterogeneities within individual garnet crystals are tightly constrained by diffusion kinetics, which would rapidly homogenize the trace-element distributions in the sub-continental mantle. Thus, while it seems clear that peridotitic diamonds and their inclusions are derived from ancient lithospheric material, these data require that they crystallized shortly before the eruption of the kimberlite 360 Myr ago.

Yakutia Kimberlite Province (Siberian Platform)

According to Sobolev (1974) and Laz'ko (1988), most of the kimberlite locations are centred on West Yakutia, in the basins of rivers such as the Olenek, Anabar, and the western tributaries of the Lena River (Fig. 4.12). A few hundred intrusions are known from the area, mostly pipes; veins and dykes are less frequent. Kimberlite volcanic activity occurred mostly within the Anabar Shield.

The Yakutian kimberlites are of different ages. It has been suggested that there were three well-developed stages of kimberlitic igneous activity in the province (the poorly studied late Riphean veins in the Pri-Sayanian area are not included): they are middle Palaeozoic and early and late Mesozoic. However, precise U–Pb dates on zircons have demonstrated that there were at least five such stages in the Phanerozoic alone: late Ordovician, late Silurian, late Devonian, late Permian–early Triassic and late Jurassic (Davis *et al.*, 1980). The duration of each stage did not exceed 17.5 Ma, with the age of the kimberlites becoming younger northwards, as confirmed by the available geological evidence.

The upper Devonian Mir Pipe, located in the Malo–Botuobin Region, is a typical example of a kimberlite body from the Central Siberian Platform. Together with the neighbouring Sputnik Pipe and three kimberlite veins, they form a single subvolcanic system. The Mir Pipe intrudes carbonate platform sediments. Around the pipe the contact zone is locally uplifted with respect to the country rocks, although more often their subhorizontal bedding is preserved. The diatreme itself is oval in plan, representing a tubular, bell-mouthed body in cross-section. The internal structure of the Mir Pipe can be interpreted in several different ways. The diatreme is composed of kimberlite breccia, among which different researchers have distinguished two to six varieties (intrusion stages). One of the most typical interpretations of the pipe structure is shown in Fig. 4.13. The varieties distinguished do not only differ in their occurrence in the pipe; the heavy-mineral concentrations, the amount of xenolithic material, and the amount of autoliths also differ. In the breccia of the south-eastern part of the body, there are fresh deep-seated xenoliths, with an extremely diverse petrographic composition (Sobolev, 1974; Laz'ko, 1988; etc.).

The composition of kimberlites of the Central Siberian Platform is dominated by varieties of basaltoid type, although micaceous types also occur (Zagadochnaya, Udachnaya–Vostochnaya etc.), while veins associated with diatremes are often infilled by calcite kimberlites. The basalt breccias of the pipes (Udachnaya–Vostochnaya and Mir) differ from conventional picrites in their somewhat higher carbonate content, though in certain parameters (TiO_2, P_2O_5, K_2O) they resemble medium-alkaline rocks. Micaceous kimberlites, which are widespread in the north of the province, are enriched in TiO_2, Al_2O_3 and K_2O; also some of them are rich in iron. High CaO and CO_2 content (together sometimes comprising up to 90% of the rock) is specific to calcite kimberlites.

Table 4.9 shows the geochemical parameters of the Yakutian kimberlites. Isotopic evidence suggests that proto-kimberlite magmas were partial melts from a somewhat depleted (compared with chondrites) metasomatized mantle source, which was strongly affected by the addition of incompatible elements before both of the pipes (Mir and Udachnaya–Vostochnaya) were formed (Laz'ko, 1988).

The Arkhangel'sk Kimberlite Province is located at the north-eastern periphery of the Baltic Shield (Fig. 4.11). The kimberlite pipes are usually overlapped by platform cover sediments, as indicated mainly by drilling (Sinitsyn *et al.*, 1992). At present three kimberlite fields can be distinguished: Nenok on Onezhsky Peninsula of the White Sea, west of Arkhangel'sk, and Imzhozero and Zolotitskoye north of Arkhangel'sk. Only the Zolotitskoye Field is known to be diamondiferous. Here all the pipes intrude late Proterozoic (Vendian) sediments, and are overlapped by Carboniferous, lower Permian and Cretaceous Russian platform sediments. About 50 diatremes are recognized there; 15 of them are diamondiferous. The area of the pipes varies from 0.5 to 170 hectares. Porphyritic and autolithic high-K micaceous kimberlites are found to dominate the rock composition of the pipes. The age of the kimberlites is estimated to be upper Devonian (Makhotkin and Zherdev, 1993).

The diamondiferous and non-diamondiferous kimberlite fields in this province occur at different structural stages, e.g. the Zolotitskaya Field occurs within the Archaean Kola craton, while the two other fields are located within the Belomorian Belt, which was active in the early Proterozoic (see Chapter 5). Thus, on the basis of available data, we can claim that the Clifford Rule, concerning the diamondiferous pipes of Archaean cratons, is valid in this case.

According to Mahotkin *et al.* (1995), Sr and Nd isotope composition in the Arkhangel'sk kimberlites, and sinchronous pipes of melilitites and basalts of all rock types, have similar $\varepsilon_{Sr(T)}$ varying from 0 to $+30$ whereas they have variable $\varepsilon_{Nd(T)}$ ranging from ~0 and $+3$ for basaltoid kimberlites and -3–4 for micaceous ones. Zero figures of ε_{Nd}, combined with elevated La/Yb ratios, indicate that the source of basaltoid kimberlites and melilite picrites was either the isotopically homogenous thermal layer of the lithospheric mantle or asthenospheric mantle, both of which were enriched in LREE and probably HFSE shortly before magmatism. Nd model ages of this rock source region are equal to 380–360 Ma.

4.5 LAMPROITES

High-K high-Mg rocks, known as lamproites, are relatively rare. Like kimberlites, they contain inclusions of mantle xenoliths and diamonds. The lamproites vary in age from Proterozoic to Cenozoic; they occur mostly as dykes and diatremes. Unlike the kimberlites, which are predominantly located in the central parts of the cratons, lamproites usually occur at their peripheries.

It should be noted that at present the lamproite series is taken to include all rocks that are rich in K and Mg. They are divided into two types: low-titanium miaskites and high-titanium agpaites (Wilson, 1989; Bogatikov *et al.*, 1991). The first are typical of igneous activity at active plate boundaries, where they are associated with calc-alkaline and shoshonite-latite rocks, for example in Spain

(Wilson, 1989). High-titanium lamproites are typical of intra-plate environments. Diamond shows have been recognized in this type of environment (Wilson, 1989; Bogatikov *et al.*, 1991; etc.) and will be discussed in greater detail.

Lamproites represent a fairly varied assemblage of volcanic and hypabyssal rocks, from three groups (ultrabasic, basic and predominantly of medium alkaline and high-alkali composition). Lamproites are characterized by the following petrochemical properties; they contain much more potassium than sodium, mostly $K_2O/Na_2O > 3$; they have a high Mg and Cr content and they are saturated or undersaturated with SiO_2. In the normal composition of lamproite-series rocks there is often hypersthene present and less frequently small amounts of leucite. This property distinguishes them from the kimberlites; they also have low Al_2O_3 and CaO content.

Lamproites form breccia pipes and subvolcanic bodies, lava flows and piro-clastics (Arkanzas, USA; Aldan, Russia; Western Australia). In some cases they form autonomous units of sills and dykes (e.g. in Kanzas, Greenland and Zambia). In the Kuruman Province in South Africa (Bristow *et al.*, 1984) they sometimes form small lava flows (e.g. the magupites of Wyoming, USA).

The composition of the lamproites is illustrated in Table 4.9, common chemical properties of interest being; high MgO content (up to 30 wt.%), K_2O (up to 7 wt.%), K_2O/Na_2O (4.4–35), and a somewhat higher SiO_2 content, which might be up to 45 wt.%, as compared with alkali picrites and kimberlites. Unlike the kimberlites, the lamproites, particularly the juvenile formations of the Ellendale Field, Western Australia (Table 4.10), have a much higher H_2O than CO_2 content ($H_2O/CO_2 = 10–15$) testifying to differences in composition between the lamproites and kimberlites.

Ultrabasic rocks of the lamproite series also have a high content of incompatible elements (Ni, Cr and Sc), whose concentrations often reach more than 1%.

The incompatible elements of the lamproites can readily be classified into two groups. One group, K, Rb, Ba and Sr, is present in all known lamproites at high concentrations, this being the major distinguishing feature of these rocks. Another group of incompatible elements, Zr, Nb, Ta, Hf and Th, i.e. the LREE, are extremely variable in their concentrations. Thus, for instance, the $(La/Yb)_N$ ratio ranges from 14 to 209, overlapping compositions typical of kimberlites (Bogatikov *et al.*, 1991). The ratio of Sr isotopes ($^{87}Sr/^{86}Sr$) varies strongly from 0.7038 to 0.714 with small $\varepsilon_{Nd(T)}$ changes from -8 to -16, thus on the curve of ε_{Nd}–$^{87}Sr/^{86}Sr$ it lies within the enriched group.

The ages of the ultrabasic rocks of the lamproite series also differ. According to Bergman (1987) and Bogatikov *et al.* (1991), the oldest (1,818–1,870 Ma) are found in the Aldan Shield and in the Kuruman Province (1,600 Ma) where they are associated with coeval kimberlites. Late Proterozoic lamproites have been recognized in Western Australia (the diamondiferous Argaile Pipe with ages of 1,253 and 1,048 Ma) and in West Greenland (1,227 ± 12 Ma). Jurassic lam-proites are found in the Aldan Basin and late Cretaceous lamproites in Kanzas,

Table 4.10 Average chemical composition of lamproites and phlogopite melaleucitites (Bogatikov *et al.*, 1988).

Components	1	2	3	4	5	6	7	8
SiO_2	42.25	43.43	45.60	41.47	40.34	45.09	43.39	39.26
TiO_2	0.50	0.70	2.32	3.62	2.23	2.76	2.30	2.72
Al_2O_3	4.60	4.08	4.32	3.64	5.03	7.26	8.36	3.71
Fe_2O_3	6.50	1.64	n.d.	n.d.	1.39	1.40	5.19	6.59
FeO	4.83	4.48	7.45	8.10	6.42	5.91	1.08	2.00
MnO	0.17	0.11	0.14	0.13	0.10	0.02	0.13	0.13
MgO	23.23	30.23	22.56	24.98	23.15	16.06	11.10	26.83
CaO	7.32	7.92	4.84	4.99	4.86	5.37	11.62	4.99
Na_2O	0.67	0.35	0.11	0.46	1.02	1.35	0.71	0.49
K_2O	2.95	3.07	3.42	4.12	6.32	7.28	6.95	2.64
P_2O_5	0.47	1.13	0.91	1.68	0.96	1.19	1.93	0.51
CO_2	3.10	n.d.	3.42	0.45	5.28	3.03	0.48	0.22
H_2O	3.45	2.86	4.57	6.36	2.90	3.28	5.14	9.91
Cr	1,642	—	1,430	1,000	760	530	520	270
NI	630	—	990	1,000	860	520	170	1,020
Co	96	—	—	70	32	—	32	100
V	8	—	110	85	20	22	23	13
Sc	22	—	12	21	—	—	20	15
Rb	96	—	260	480	170	135	190	180
Sr	860	—	770	1,310	2,560	2,190	3,820	1,210
Ba	1,820	—	1,080	10,300	3,230	6,200	6,700	2,320
Zr	97	—	630	1,130	490	790	1,490	720
Nb	6	—	200	180	—	—	130	110
Th	2	—	17	60	—	—	—	14
La	10	—	120	420	260	370	400	—
Ce	19	—	240	730	345	450	850	—
Nd	7	—	—	—	—	—	290	—
Cu	73	—	39	56	28	36	—	50
Zn	63	—	54	71	88	84	—	73
Number of analyses	3	2	8	89	1	3	6	8

Note: Additional elements analysed in 1: Hf 0.7; Ta 0.06; Sm 2.3; Eu 0.76; Gd 1.3; Tb 0.21; Dy 1.3; Yb 0.50; Lu 0.08 $(La/Yb)_N$ 14; in 8 — Hf 14.2; Ta 3.9; Sm 11.9; Eu 2.8; Yb 0.72; Lu 0.15; $(La/Yb)_N$ 134.5. Columns 1–5 — lamproites: 1 — Central Aldan; 2 — Lomamsky Massif, Aldan Province; 3 — average for Argail Pipe tuff, Australia; 4 — average for the Ellendall Pipe of Australia; 5 — Western Greenland; 6–7 — phlogopite melaleucitities: 6 — Leucite Hills, Wyoming, USA; 7 — Silver City, Kansas, USA; 8 — Prider Greek, Arkansas, USA.

USA, while most of the lamproites, particularly basic and intermediate varieties of the rocks of lamproite series, formed only during the Tertiary (West Kimberley in Australia, Gaussberg in Antarctica, Wyoming, USA, etc.).

As has been mentioned above, lamproites occur in two different geodynamic environments: intra-plate environments and active plate boundaries. The first type includes the lamproites of Australia, West Greenland and South Africa. These are usually agpaitic-type rocks with high-Ti content, also with higher Zr and Nb content. They are centred mostly within old fold belts, consolidated in the Proterozoic, which then were joined to the Archaean cores of the cratons (West Kimberley in Australia, the North America Platform, etc.). They form in an intra-plate environment and are often associated with kimberlite fields. It is these varieties that turn out to be diamondiferous. However, miaskite-type lamproites (low-Ti lamproites) are generally related to collision tectonics, where they are associated with the rocks of the calc-alkaline and shoshonite-latite series as is the case for the lamproites of Spain, Italy, East Kimberley, Australia, and the Mesozoic lamproites of the Central Aldan Shield.

4.6   ACID IGNEOUS ROCKS

Anorogenic igneous assemblages sometimes contain numerous acid rocks. Their typical representatives were described in Chapter 3, since it is common to ascribe them to orogenic environments. In Russian literature they are often speculatively connected with tectonically activated structures. A few examples of these assemblages recognized in Siberia, Kazakhstan and Mongolia.

Anorogenic igneous activity can be expressed by the formula:

$$T_{HTi} + MORB + A_{K-NaHTi} + A_{KHTi} + HA_{K-NaHTi} + HA_{KHTi}$$

4.7   SUMMARY

1. Phanerozoic anorogenic continental magmatism is a feature of the intra-plate environment: magmatism of this type is essentially determined by the emplacement of intrusions, which, in turn, are caused by deeper erosion of igneous systems than occurred in the late Cenozoic.
2. Palaeorifts are known to be typical of anorogenic magmatism. They have a graben-like structure and are characterized by the same igneous rock assemblages as modern rift areas, with the widespread occurrence of rocks

of high and medium alkalinity. Two types of mafic profiles can be distinguished among them, represented by the typical Dnieper–Donets Palaeorift and another, salic, type, the Oslo Graben. The latter type is not characteristic of the Cenozoic. So far it has been found only within the East African Rift Zone.

3. Continental flood basalts (trap assemblages) are the most cogent examples of Phanerozoic anorogenic igneous activity. They cover huge areas (tens and hundreds of thousands of square kilometres), their thickness varying from 1 to 4 km. In a number of cases (in the Siberia and Deccan traps) isotopic evidence indicates that they were generated in a (geologically) very short time (about 1 Ma).

4. The trap section displays a transition from medium-alkaline (less frequently highly alkaline) picrites and basalts to MORB-type tholeiite basalts. Isotope-geochemical data indicate the contamination of both continental crust and lithospheric mantle by these melts. We have already dealt with changes in melt composition, typical of the large-scale intra-plate magmatism (cf. the Red Sea Rift, see Chapter 2), which could be attributed to gradual depletion of the underlying mantle during progressive melting. In some cases the process of trap magmatism has led to oceanic spreading (the Parana–Etendeka traps and the Deccan traps, etc.), while in other cases similar evolution has net resulted in the crust rupture (Siberian traps). In the latter cases the top of the asthenospheric diapir probably did not reach the crustal floor.

5. Vast submarine plateaux of the Ontong–Java type, the Shatsky Rise, etc. can be regarded as oceanic analogues of traps. They are typically dominated by E-MORB-type tholeiite basalts and "transitional" tholeiites. Similarly to the continental traps, these plateaux evolved very rapidly, taking only 2–3 Ma.

6. At the edges of the trap areas there are medium-alkali K–Na series picrite basalts as well as highly alkaline varieties, and related concentrically zoned massifs of alkali-ultrabasic rocks including carbonatites. Fragments of similar rocks from these massifs are found as xenoliths in the lavas of some alkaline volcanoes, associated with present-day continental rifting (e.g. the Oldonyo Lengai Volcano in the Kenya Rift). These massifs are believed to be the deep eroded parts of alkaline volcanoes.

7. Similar massifs and alkaline K–Na series intrusions may be associated with ancient hot-spots, i.e. asthenospheric plumes, or located at the periphery of the plumes, or may be totally unrelated to the plumes. The Devonian Kola–Karelia Alkaline Province is one example of this type of igneous activity, with the distribution pattern of the igneous formations apparently affected by the relief of the top of the plume; another example of this is the Permian–Triassic Maimecha-Kotui Province.

8. Kimberlites and titanium lamproites are typical of intra-plate environments. This type of igneous activity is characterized by a highly volatile content

(e.g. CO_2 for the kimberlites and H_2O for the lamproites). They occur mainly in the form of breccia pipes (diatremes). Kimberlites normally contain a high proportion of xenoliths of ancient lithospheric material (at depths down to 260 km and more) and of diamond. In titanium lamproites the xenoliths are less frequent, while the diamonds may occur in significant amounts. Typically, the parental diamondiferous rocks (garnet dunites and eclogites) are extremely depleted, as they are the oldest representatives of the continental lithospheric mantle material ($> 3.4-3.2$ Ga).

9. Anorogenic igneous assemblages often contain numerous "extra-geosynclinal" granitoids, with alkalinity varying from normal to alkaline.

CHAPTER 5
EARLY PROTEROZOIC MAGMATISM AND GEODYNAMICS — EVIDENCE OF A FUNDAMENTAL CHANGE IN THE EARTH'S EVOLUTION

E.V. SHARKOV and O.A. BOGATIKOV

Proterozoic magmatism (2.5 ± 0.1 to 1.0 ± 0.1 Ga ago) represents a very important transition from the magmatism of the early stages of the Earth's evolution to the Phanerozoic magmatism that continues to the present day. Data currently available suggest that by 2.6–2.5 Ga ago, the Earth's crust had stabilized and this enabled the formation of the huge dyke swarms and large layered intrusions that are very typical of that time. The type of tectonomagmatic activity in all the Precambrian shields allows the Proterozoic to be clearly divided into two radically different stages: from 2.5 to 2.2–2.0 Ga ago, and from 2.0 to 1.0 Ga ago, differing both in their type of magmatic processes and in their geodynamics.

Distinctive igneous siliceous high-magnesian series (SHMS) are highly typical of early Palaeoproterozoic (2.5–2.2 Ga ago) magmatism. When comparing them with present-day rock types, they are close to boninite-series rocks (a high-magnesian variety of the calc-alkaline series, see Chapters 1 and 2). However, in contrast with the Phanerozoic, when similar rocks occur only in island-arc environments, in the early Palaeoproterozoic they were widespread, although there is no geological evidence of plate tectonics.

The first reliable geological evidence of plate tectonics roughly coincides with 2.2–2.0 Ga stage. This period witnessed the first large-scale emplacement of geochemically enriched melts of Fe–Ti picritic composition, moderate- and high-alkaline basites and associated Ti-rich alkaline rocks of the K–Na and K series, in contrast to the variably depleted magmas typical of earlier epochs. Some 1.7–1.5 Ga ago, after most cratonization had been completed, gigantic magmatic belts of acid potassic composition, including anorthosite–rapakivi-granite complexes, were emplaced.

The sequence of Proterozoic geological events is now best understood in the Baltic Shield, where the whole record of igneous activity can be observed (Kratz et al., 1978; Gorbaschev and Bogdanova, 1993; Mitrofanov, 1995). Therefore, this chapter concentrates on data from the Baltic Shield, which is a classic area for studying the Early Precambrian rocks of Europe.

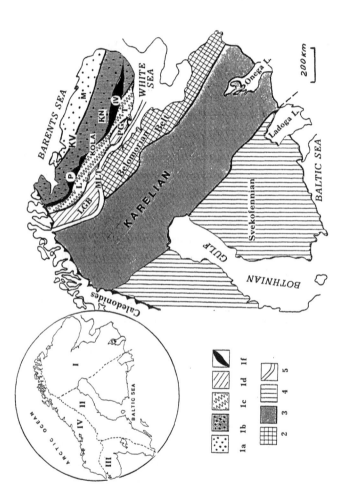

FIGURE 5.1 Tectonic sketch of the eastern Baltic Shield. After Sharkov and Smolkin (1997), with corrections. Key: 1–3 — Kola-Karelian geoblock: 1 — Kola Domain: (a) Murmansk Block (M), (b) Central Kola Block (CK), (c) Lotta (L) and Tersk (T) blocks (domains), (d) Lapland-Umba Granulite Belt, consisting of Lapland Belt (LGB) and Por'egubsky Block (PG), (e) Pechenga-Varzuga Belt with Pechenga (P) and Imandra-Varzuga (I-V) structures, and Kolmozero-Voron'ya Belt (KV); 2 — Belomorian (White Sea) Mobile Belt; 3 — Karelian Domain; 4 — Svecofennian Geoblock; 5 — Main Lapland Thrust. For details see text.
In the circle — scheme of geoblocks of the Baltic Shield: I — Kola-Karelian, II — Svecofennian, III — Dahlsland, IV — Caledonian. After Kratz et al. (1978), with corrections.

5.1 PALAEOPROTEROZOIC IGNEOUS ASSEMBLAGES (2.5–2.3 Ga OLD)

The Baltic Shield is the largest Precambrian basement inlier of the Russian Platform. According to Kratz *et al.* (1978), it is divided into four geoblocks of differing ages (Fig. 5.1): the Caledonian, Dahlsland, Svecofennian and Kola–Karelian. The nappes of the Caledonides are mainly developed within the Caledonian Geoblock; the Dahlsland Geoblock is dominated by Late Proterozoic metamorphic rocks; most of the shield is made up Early to Middle Proterozoic rocks of the Svecofennian Geoblock; the oldest, Paleoproterozoic and Archaean, rocks occur in the eastern Baltic Shield, in the Kola–Karelian Geoblock.

During the early Palaeoproterozoic (2.5–2.3 Ga ago) three types of coeval structural provinces developed in the eastern Baltic Shield: (i) Karelian and Kola granite–greenstone cratons, (ii) Lapland-Umba granulite belt (LUGB) of the moderate pressures among them, and (iii) specific transitional nappe-folded zones of tectonic flowage between these high- and low-grade terranes, which consist mainly of reworked granite–greenstone lithologies of adjacent cratons (Fig. 5.2). Among the latter, the Belomorian Mobile Belt (BMB) between the Karelian craton and the LUGB is the best preserved. The cratons were vast areas with mantle-derived magmatism of siliceous high-Mg (boninite-like) series (SHMS) in the form of volcanosedimentary sequences in graben-like structures, large layered intrusions and gabbronorite dyke swarms. A very specific variety of such magmatism, synkinematic with deformations, was characteristic for the BMB; it was represented by thousands of small intrusions of drusite complex, disseminated through all the territory of the belt. Instead of these domains, within the LUGB crustal-derived magmatism of enderbite–charnockite series predominated.

The Kola craton consists now of five domains separated by shear zones and thrusts (Fig. 5.2). They are from NE to SW: (1) the Murmansk domain — a fragment of the Archaean granite–greenstone terrane; (2) the Keivsko–Porosozerskay zone — a fragment of late Archean greenstone belt; (3) the Central Kola domain in which is a fragment of late Archaean granulitic–gneiss terrane partly reworked during the Early Proterozoic; (4) the Pechenga–Varzuga Belt — composed of the Pechenga and Imandra–Varzuga Early Proterozoic volcanosedimentary rocks; and (5) the Tersky and Lotta domains (in the west possibly correlative with the Inari terrane, cf. Gaal *et al.*, 1989) basically comprising retrograde metamorphosed Archaean tonalitic gneisses.

The LUGB consists of two fragments: the Lapland Granulite Belt (LGB) in the north and the Por'egubsko-Umbinsky block in the south (Fig. 5.2). This belt, formed of medium pressure granulites upon mainly metapelitic sedimentary rocks, was emplaced and metamorphosed during the early Palaeoproterozoic about 2.5–2.2 Ga ago (Huhma and Merilainen, 1991; Sharkov *et al.*, 1997).

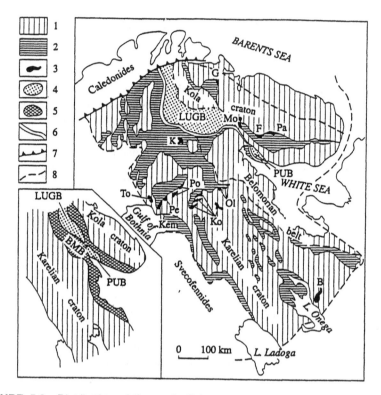

FIGURE 5.2 Distribution of the early Palaeoproterozoic magmatic rocks in the eastern Baltic Shield; below: generalized structural scheme for this time. After Sharkov *et al.* (1997). Key: 1 — Archaean metamorphic basement; 2 — the Palaeoproterozoic volcanosedimentary structures with SHMS rocks at the base; 3 — the early Palaeoproterozoic layered intrusions: Bur — Burakovka, Gm — Generalskaya Mt., Koil — Koilismaa Complex, Mo — Monchetundra, Pa — Pana Tundras, Fed — Fedorova Tundras, Por — Portimo Complex, Kem — Kemi, Koit — Koitilainen, Oul — Oulanka Group, Pen — Penikat, Tor — Tornio; 4 — Lapland-Umba Granulite Belt with abundant enderbite–charnockite magmatism; 5 — Belomorian Mobile Belt (on the scheme) with small intrusions of the drusite complex; 6 — the late Palaeoproterozoic Main Lapland Thrust; 7 — Caledonidian Thrust; 8 — boundary of the Baltic Shield.

From the south-west the LUGB is bounded by the 1.9 Ga old Main Lapland Thrust (MLT); a narrow (10–15 km wide) zone of compression, thrusting, folding and high-pressure granulite metamorphism. This thrust separates the Kola craton from the Karelian–Belomorian one (Kratz *et al.*, 1978; Priyatkina and Sharkov, 1979; Barbey and Raith, 1990).

The volcanic and intrusive rocks of the eastern Baltic Shield, described below, are related to the so-called Karelides, emplaced during the Karelian, or Sumian–Sariolian, tectonomagmatic cycle. The Karelides include major extensive graben-like zones, containing volcanosedimentary rocks; these zones separate major sialic domains of Archaean basement where there was Karelian activation of basement rocks with the formation of large layered intrusions and dyke swarms as well as intrusions of hyperstene diorites and charnokites.

5.1.1 Volcanic Series

Palaeoproterozoic volcanic series are widely developed at the base of sections of major graben-type structures in both the Kola and Karelian cratons. They include the Pechenga and Imandra–Varzuga structures, making up the Pechenga–Varzuga Belt, the Pana–Kuolajarvi, East Karelian, and Vetreny Poyas (East Karelian Belt) structures (Smolkin, 1992), as well as the Laplandian greenstone belt in northern Finland (Lehtonen et al., 1992). These volcanosedimentary complexes formed during at least two stages, namely, the Sumian (2.5 to 2.4 Ga) and the Sariolian (2.35 to 2.2 Ga).

During the Sumian, the volcanics of all the structures were rather closely related and represented by SHMS rocks; they formed a continuous sequence, of normal alkalinity, from low-Ti picrite and picrobasalt to high- and moderate-magnesian basalts, andesite–basalt, andesite, dacite and rhyolite. They are characterized by considerable variations of MgO content, high SiO_2 and Cr, moderate Ni, Co, Cu and V, low TiO_2. The andesitic-basalts and andesites generally have Mg* = 53–65.

Of the above-mentioned rocks, the picrites, as parental melts of the series, are the most interesting. In the Vetreny Poyas Formation the picrite and basalts make up a 2.3 km-thick volcanic sequence (Smolkin, 1992) their Sm–Nd age is 2448 ± 42 Ma (Pukhtel et al., 1991). Similar volcanics are typical of the Seidorechka (Imandra–Varzuga structure) and some other Sumian formations, where they alternate with andesites, as well as acid lavas (Tables 5.1 and 5.2). Volcanics alternate with conglomerates, subaerial crossbedded sandstones and gritstones, showing the subaerial conditions of their formation. The oldest dates for them were obtained from dacites of the Madetkoski Formation in central Finland: 2526 ± 46 Ma (Lehtonen et al., 1992). The age of volcanites from the Seidorechka Formation is 2424 ± 5 Ma (Balashov et al., 1993).

The high-Mg volcanics bear certain similarities to Archaean komatiites (see Chapter 6) due to the presence of flow banding, spinifex textures and the relatively low proportion of pyroclastics; therefore, in some cases they are described as "Proterozoic komatiites" (Bogatikov, 1988).

The Sariolian stage appears to be similar to the Sumian. The complete volcanic sequence is: low-Ti picrite, basalt, andesite–basalt and dacite of age 2324–2314 Ma (the Akhmalakhti Formation in the base of the Pechenga

Table 5.1 Typical chemistry of Sumian volcanic rocks of the Baltic Shield.

Components	1	2	3	4	5	6	7	8	9
SiO_2	49.14	51.38	52.60	54.58	56.84	64.04	69.97	68.31	71.48
TiO_2	0.49	0.72	0.69	0.80	0.59	0.94	0.72	0.86	0.58
Al_2O_3	15.73	14.11	13.31	13.87	11.59	13.16	11.61	11.66	11.06
Cr_2O_3	0.015	0.13	0.10	0.018	—	0.003	0.002	0.002	0.003
Fe_2O_3	1.51	1.66	2.02	2.76	1.24	2.63	2.07	2.99	1.43
FeO	8.65	8.11	8.62	7.22	8.15	6.04	4.33	4.15	4.07
MnO	0.22	0.10	0.13	0.16	0.15	0.11	0.07	0.09	0.07
MgO	10.20	9.54	9.12	5.60	9.60	1.67	0.78	0.63	0.79
CaO	10.92	11.20	10.26	8.08	5.89	3.19	1.60	3.41	1.03
Na_2O	2.11	1.90	1.82	3.15	3.27	3.18	4.66	2.85	2.88
K_2O	0.30	0.30	0.47	1.33	0.80	3.03	2.67	2.34	4.42
P_2O_5	0.11	0.02	0.04	0.13	0.04	0.21	0.08	0.13	0.09
LOI	0.80	0.94	0.60	2.04	2.01	1.61	1.23	1.98	1.84
Total	100.19	100.11	99.78	99.74	100.17	99.81	99.79	99.40	99.74
Cu	94	—	47	—	—	—	—	—	—
Co	—	47	—	—	—	—	—	—	—
Ni	157	235	235	70	—	22	12	9	11
V	327	119	119	200	—	85	8	7	5
Mg*	66	65	63	53	66	29	27	17	23

Columns 1–3 — metabasalts of the Vetreny Poyas Formation: 1 — from Bogatikov (1988), 2–3 — from Smolkin (1992); 4–9 — rocks of the Strelna Group, from Smolkin (1992): 4 — andesite-basalt, 5 — high magnesium andesite, 6 — andesite, 7 — granophire (imandrite), 8 — rhyodacite, 9 — rhyolite.

Table 5.2 Chemistry of Sumian and Sarolian high-magnesium volcanites (from Smolkin, 1992).

Components				2090				3014			
	1	2	3	1	2	3	4	3	8	15	18
SiO_2	49.19	50.37	51.03	45.76	47.35	46.88	48.06	46.35	48.48	47.95	52.38
TiO_2	0.48	0.48	0.61	0.56	0.53	0.62	0.66	0.35	0.43	0.46	0.61
Al_2O_3	9.44	9.43	10.43	7.53	9.01	9.78	10.94	7.26	9.19	8.96	12.49
Fe_2O_3	2.10	2.42	1.47	3.71	1.35	3.99	2.20	0.97	2.27	3.22	1.04
FeO	7.74	6.72	8.79	7.12	8.73	7.34	8.74	10.54	8.85	8.08	9.75
MnO	0.18	0.17	0.18	0.19	0.17	0.17	0.18	0.18	0.16	0.18	0.18
MgO	16.17	15.53	13.52	20.24	18.76	14.07	13.56	21.50	16.49	16.17	8.09
CaO	7.76	8.17	6.98	7.53	7.27	8.34	7.79	6.94	8.03	8.45	10.92
Na_2O	1.37	2.03	2.44	0.25	0.66	1.48	1.50	0.87	0.91	0.89	1.80
K_2O	0.83	0.23	0.53	0.17	1.13	0.77	1.33	0.28	0.34	0.21	0.29
H_2O	4.33	4.04	3.76	5.81	5.20	4.67	3.79	3.39	3.53	3.99	1.43
P_2O_5	0.07	0.06	0.05	0.07	0.06	0.09	0.09	0.05	0.07	0.07	0.08
CO_2	0.07	0.12	0.09	0.17	—	0.73	0.57	0.14	0.22	0.19	0.15
Cr	1900	2900	1200	2100	1900	1400	1200	2580	1730	1710	541
V	170	170	180	130	110	170	190	140	170	170	230
Ni	560	630	370	800	720	500	470	744	484	502	111
Co	80	61	60	90	80	70	70	110	90	100	70
Cu	50	100	90	250	40	70	70	70	90	90	120

In addition, the rocks contain 0.01–0.03% S. 1–3 are massive and globular lavas of the Polisar Formation; 2090 is from a layered flow 30 m thick (2090, columns 1–4: 5 m, 10 m, 15 m and 20 m from the bottom of the flow, respectively). Khosijarvi Formation of Mt. Khaukkatunturi; 3014 is a layered flow, 37 m thick (3014, columns 3, 8, 15, 18: distances in metres from the bottom of the flour), from the Vetreny Poyas Formation of Mt. Golets.

structure: Mitrofanov and Smolkin, 1995). Some volcanics in northern Finland have yielded a 2330 Ma Sm–Nd age (Huhma *et al.*, 1990). Presumably simultaneously, the basement of the Pechenga zone was penetrated by granites and affected by retrograde metamorphism (Duk *et al.*, 1989).

In the Sariolian volcanics, picrite can account for as much as 80% of the section (e.g. Polisar Formation of the Imandra–Varzuga structure; Smolkin, 1992). Here picrite occurs as massive and pillowed lava flows; tuffs and explosian breccias are less common. In places, the massive lava flows contain scoria and lower and upper amygdaloidal zones. The thickest flows may be layered. A combination of massive and pillow lavas and laminated tuffs and tuffites locally with signs of slumping suggest predominantly submarine volcanism here. Rocks of differing basicity generally alternate within each geological section, but, on the whole, high-magnesian volcanics make up the lower and the middle parts, and andesite–basalts make up the upper part.

Small sill- and dyke-like subvolcanic bodies of various pyroxenites and gabbronorites are related to the high-magnesian volcanics both spatially and genetically.

In layered picrite flows, the content of refractory elements (Mg, Cr, Ni and Co) decreases from the bottom upwards, but there is an increase in the content of less refractory elements (Ca, Al, V, Cu), alkalies (Na), and radioactive (Th and U) elements, as well as SiO_2, Zr, Sc and Ba; this is consistent with the fractional crystallization model (Table 5.2).

When plotted on a CaO–MgO–Al_2O_3 diagram (Fig. 5.3), rocks of layered lava flows show a typical differentiation trend, showing a CaO/Al_2O_3 ratio close to 0.8; the value is lower than that for Archaean komatiites (1.5–1.0) but higher than for Phanerozoic boninites (0.6). The average Al_2O_3/TiO_2 ratio in the high-Mg volcanics (18–20) is close to chondritic (20) (Smolkin, 1992).

Table 5.3 presents data on the REE content, and Fig. 5.3 displays chondrite-normalized REE spectra. Data suggest that early Palaeoproterozoic high-Mg volcanics are noted for their moderate LREE enrichment; this distinguishes them from Archaean komatiites, but indicates that they are closer to some varieties of present-day boninite (Crawford, 1989; Fryer *et al.*, 1992).

5.1.2 Layered Basic and Ultrabasic Intrusions

Large layered intrusions of peridotite–pyroxenite–norite–anorthosite, representative of intrusive analogues of SHMS volcanics, are common in the eastern Baltic Shield, within the Kola and Karelian domains, which had stabilized by the beginning of the Palaeoproterozoic. They are exemplified by intrusions of:

1. the Monchetundra, the Fedorovo–Pansky Tundras, Generalskaya and the Imandra lopolith (in vicinity of the monchetundra massif) on the Kola Peninsula;

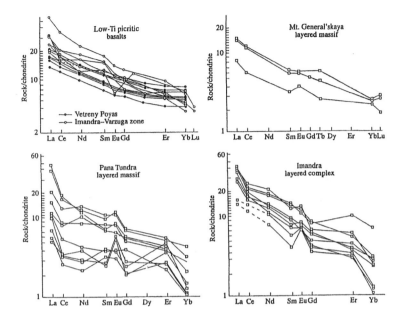

FIGURE 5.3 REE spectra of chondrite-normalized high-magnesian Early Proterozoic volcanics and rocks of layered intrusions of the Baltic Shield. After Sharkov *et al.*, 1997.

2. the Oulanka Group (Lukkulaisvaara, Tsipringa, Kivakka) in northern Karelia;
3. Burakovsky in southern Karelia; and
4. Kemi–Suhanko, Penikat and Suote–Narankavaara (Koilismaa complex) in central Finland (Fig. 5.2).

The intrusions usually measure tens to hundreds of square kilometres in area, with their thicknesses varying from 3 to 5 km. As a rule, they have undergone variable degrees of reworking during the successive phases of Svecofennian Orogeny; at present they often represent tectonic blocks that are sheared and metamorphosed along the margins. Primary igneous rocks have been preserved mainly in the central parts of these blocks. Isotopic dates on the layered intrusions suggest that there were two main stages of formation, namely 2.5–2.47 Ga (the Monchetundra, Generalskaya and Fedorovo–Pansky massifs, the Kola Peninsula) and 2.45–2.43 Ga for the Oulanka group and Burakovsky in Karelia and for the Finnish intrusions (Alapieti *et al.*, 1990; Balashov *et al.*, 1993; Amelin *et al.*, 1995). Only the Imandra intrusion yielded an age of about 2 395 Ma (Balashov *et al.*, 1993). Thus, most layered intrusions are of Sumian age.

Most of the intrusions have an asymmetrical cross-section with the lower part being dominated mainly by ultrabasic cumulates (Ol + CrSp, Ol + Opx + Opx ± Cpx ± CrSp), the middle part by basic cumulates (Opx + Pl ± Cpx ± Ol, Pl,

Table 5.3 REE content in high-magnesian volcanic rocks (ppm) (from Smolkin, 1992).

Elements	Polisar Formation				Kirech Formation			Vetreny Poyas Formation			
	P1-1	P1-2	3	4	345-4	345-18	3516	3014-3	3014-8	3014-15	3014-18
La	6.1	7.6	5.4	6.3	2.1	11.4	9.3	4.74	5.41	5.93	7.55
Ce	13	14	12	14	5.5	22.3	17.9	7.91	6.43	10.5	12.6
Nd	4	6.8	—	2.5	11.87	9.6	3.63	3.63	3.02	4.29	4.47
Sm	2.1	2.3	1.4	2	0.6	2.77	1.99	1.38	1.63	1.65	2.2
Eu	0.42	0.75	0.5	0.52	0.06	0.9	0.73	0.37	0.39	0.46	0.801
Gd	—	—	—	—	0.42	2.9	2	—	—	—	—
Tb	0.49	0.34	—	0.35	—	—	—	0.261	0.295	0.329	0.402
Yb	1.3	0.95	0.5	1.1	0.65	1.71	1.22	0.845	0.937	1.07	1.36
Lu	0.13	0.12	—	0.16	0.1	0.27	0.19	0.101	0.143	0.147	0.221

1–4, 3014 — from neutron activation analysis data; 345, 3516 — from high-precision induction flame spectrometry data.

Pl + Pig + Pig — Aug) and the upper part by Mgt gabbronorites and gabbro-diorites, Py diorites and, sometimes, granophyres (Sharkov, 1980; Alapieti *et al.*, 1990; Mitrofanov, 1995).

It is precisely this rock sequence which is characteristic of the Monchetundra Intrusion (Fig. 5.4), located in the centre of the Kola Peninsula. It cuts high-grade metamorphosed Archaean rocks and is involved in deformation related to the Svecofennian Lapland Thrust. Detail geological-petrological and isotope-geochemical studies (Sharkov, 1980; Tolstikhin *et al.*, 1992) indicate that the Monchetundra Intrusion consists of two major tectonic blocks:

1. the Monchegorsky Pluton of ultramafic-mafic rocks (dunute, harzburgite, brozitite and norite) forming the lower part of the intrusion, and
2. the westerly gabbronorite–gabbronorite-anortosite intrusion of the Monche–Chuna–Volchya Tundras (of the Glavny Ridge), representing its upper part.

The remainder of the section has been removed by erosion.

All the rocks of the Monchetundra complex are related via zones of rhythmic layering; they represent a single rock sequence about 5–6 km in thickness. The rock composition of the chilled zone is relatively high in magnesium at the level

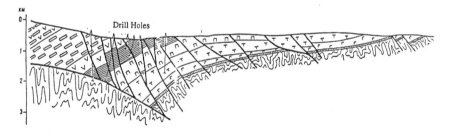

MONCHETUNDRA LAYERED INTRUSION

1 Plagioclase cumulates (gabbro-norite-anorthosites)
2 Plagioclase cumulates (anorthosites)
3 Cpx + Opx + Plag ± Ol (gabbro-norites)
4 Rhythmic alternation of Plag, Cpx+Opx+Plag, Opx+Plag, Opx and Opx+Ol cumulates
5 Opx cumulates (bronzitites)
6 Opx+Ol±CrSp cumulates (harzburgites) mainly, which rhythmically alternate with Opx and Ol+CrSp cumulates
7 Ol+CrSp cumulates (dunites)
8 Chilled Zone (mainly fine-grained melanocratic norite)
9 Archaean metamorphic wall rocks

FIGURE 5.4   EW section through the Monchetundrovsky Intrusion in the area of the Loipeshnyun–Nittis–Kumuzhya–Travyanaya Mountains. From Sharkov (Laz'ko and Sharkov, 1988).

of Monchegorsky pluton and high in alumium at the level of the main Ridge massif (Table 5.4). In its mineral composition and rock texture, the Monchetundra Intrusion as a whole is similar to the classic Bushveld and Stillwater intrusions (see below).

Numerous discordant bodies of micronorites and microgabbronorites (the "Critical Horizon" of Mt. Nyud) are typical of the eastern Monchegorsky

Table 5.4 Typical chemistry of endomorphic zones of the Sumian layered intrusions and microgabbro from their satellite intrusions.

Components	1	2	3	4	5	6	7
SiO_2	52.31	50.71	49.51	49.26	48.65	44.82	50.97
TiO_2	0.45	0.30	0.24	0.08	0.02	0.29	0.19
Al_2O_3	13.30	16.95	14.75	19.86	28.60	10.99	16.51
Cr_2O_3	—	—	0.13	0.076	0.0056	0.20	0.037
Fe_2O_3	0.50	0.51	1.08	0.81	1.55	9.78	5.82
FeO	9.80	4.97	7.24	6.63	1.02	—	—
MnO	0.19	0.14	0.12	0.09	0.035	0.18	0.14
MgO	13.42	9.69	15.99	9.41	1.86	16.94	10.46
CaO	5.76	13.36	6.12	8.03	11.77	11.78	12.76
Na_2O	2.13	2.34	1.30	2.17	3.15	1.14	1.03
K_2O	0.33	0.18	0.15	0.20	1.50	0.24	0.10
P_2O_5	—	—	0.03	0.07	0.09	0.03	0.03
LOI	1.38	0.70	2.74	2.78	1.18	2.38	1.40
Total	99.56	99.85	99.58	99.56	99.52	99.50	99.41
Zr	—	—	29	44	52	22	32
Nb	—	—	2	2	3	4	2
Y	—	—	12	12	13	12	13
Sr	—	—	201	364	527	150	233
Ba	—	—	100	100	302	107	100
Cu	—	—	—	—	—	4	10
Co	—	—	71	61	29	98	36
Ni	—	—	299	424	667	352	167
V	—	—	196	144	26	—	—
Mg*	70	77	79	71	66	76	76

Column 1 — chilled zone of the Monchegorsk Pluton; 2 — endomorphic gabbro-norites of the Monche–Chuna–Volch'i Tundras Massif; 3 — melanocratic Ol-bearing micronorites (microbronzitites) from the "Critical horizon" of Mount Nyud; 4 — micronorite from the same location; 5 — microanorthosite from the same location; 6 — melanocratic fine-grained gabbronorite of the Koilismaa Complex; 7 — micro-gabbronorite, Lukkulaisvaara Intrusion. Sources: 1 — Laz'ko and Sharkov, 1988; 2 — Sharkov and Ledneva, 1993; 3, 4, 5, 7 — Sharkov et al., 1994; 6 — Alapieti, 1982.

Pluton. The earliest and the later bodies are respectively magnesian and aluminiferous in composition (Table 5.4). In places the bodies contain xenoliths of the country rocks (hornfelsed cordierite gneiss) and were probably the result of repeated injections of fresh melt into a crystallizing magma chamber. Similar bodies of microgabbronorite are common in other layered intrusions of the Baltic Shield, particularly the Lukkulaisvaara Intrusion in northern Karelia, where they also contain xenoliths of underlying lower-crustal rocks (garnet granulites) (Sharkov and Ledneva, 1993).

The appearance of the peculiar Sopcha "peridotite" or "ore" bed in the central part of the Monchegorsky Pluton is probably associated with one of these pulses of fresh melt. It is a 2–3 m layer of peridotite cumulates with rich dissiminated Cu–Ni sulphides and PGE mineralization of the middle part of a thick (about 700 m) macrolayer of very uniform bronzite cumulates, in the Mt. Sopcha area. Slumping structures and autobreccias are widespread here. It is evident that the occurrence of this horizon of high-temperature cumulates among lower-temperature rocks cannot be explained by differentiation in a magma chamber. However, it could have been induced by pulses of a new batch of higher-density high-magnesian melt into the crystallizing magma chamber. In this case, fresh melt of the composition of parental magma of early high-magnesian micronorites of Mt. Nyud could have moved across the chamber floor, displacing the old melt upwards (Fig. 5.5), and become immediately involved in solidification; as the melt crystallized, the density of magma would have decreased and the melt would have been involved

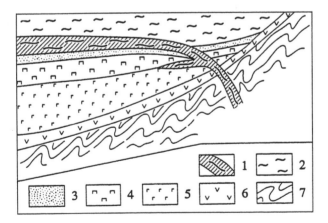

FIGURE 5.5 Scheme illustrating the character of pulses of fresh melt during the solidification process of the Monchegorsk Pluton. 1 — fresh batch of melt; 2 — main body of melt in the magma chamber; 3 — zone of crystallization; 4–5 — consolidated parts of the intrusion: 4 — Opx-cumulates, 5 — Opx+Ol+CrSp-cumulates; 6 — rocks of the chilled zone; 7 — country rocks of the Kola Group.

in a common convection system, mixing with the old melt, and the crystallization trend would have returned more or less to its original position. It is clear that such a situation could take place only if the volume of the batch was relatively small and if the fresh magma melt differed greatly in density from the old melt in the chamber.

The cases of non-systematic rock alternation, as a whole, are typical of sections of practically all the layered intrusions of the Baltic Shield; they are probably related to recurrent pulses of fresh magmas.

The Fedorovo–Pansky Intrusion, also faulted into two fragments, namely the intrusions of the Fedorov and Pana Tundras, are similar in structure to the Monchetundra Intrusion (Mitrofanov and Torokhov, 1994). The Mt. Generalskaya intrusion represents an elongated body of thickness about 1.0–1.5 km. It is formed mainly of gabbronorites with zones of alternation of $Ol + Pl$, $Ol + Opx + Pl$, Opx and $Opx + Pl \pm Cpx$ cumulates in the central part. It is the rare type of the massifs in the Baltic Shield with Ol–Pl cumulates.

The Koilismaa Complex in north-eastern Finland underwent stronger reworking. According to Alapieti (1982), it is separated into large blocks, namely the Narankavaara, Syote, Porttivaara, etc. Within the layered series, the lower zone is composed of ultrabasic and basic cumulates, similar to those of the Monchegorsk pluton. In the Narankavaara block, no well-defined anorthosite macrolayer can be observed, and late differentiates of pyroxene dioritic, granodioritic, quartz-monzonitic and tonalitic composition occur above the gabbronorite. In the layered series of western intrusions of the complex, namely, the Syote, Porttivaara and others, the lower parts of the sections resemble Narankavaara, whereas the upper parts are dominated by anorthosite with thin interlayers of magnetite gabbronorite. Their uppermost parts are formed by thick (up to 1 km) layers of granophyre. The granophyre shows an REE distribution pattern similar to those of the underlying layered basic rocks. Alapieti (1982) suggested that the granophyre was comagmatic to the basic rocks, although contamination by enclosing gneisses could have played an important role in their genesis.

To summarize this brief description of the early Palaeoproterozoic layered intrusions of the eastern Baltic Shield, it should be emphasized that, in their rock and mineral composition, in the character of the layered series, etc., they are fairly similar, differing only in the proportions of particular rock types. All of the intrusions have a similar internal structure, and their rocks form a single continuous sequences, via intermediate varieties and zones of rhythmic alternation. This suggests that the intrusions crystallized in a normal fashion, within a single magma chamber by fraction crystallization. However, the presence of unpredicted layers of high-temperature olivin-bearing cumulus mineral assemblages within lower-temperature parts of sections, different disturbances of layering, etc., suggest that pulses of new magma batches of fresh melt were injected into the magma chambers of the crystallizing plutons.

It can be deduced from the succession of micronorites bodies (feeders of such pulses) in the Monchegorsky pluton that new batches of melt would have evolved over time, from high-magnesian to lower-magnesian and aluminiferous varieties, with the latter being characteristic of the later stages of magma chamber development. The final stages of evolution could have witnessed the appearance of intermediate and acid rocks, of dioritic and granophyric type, also related to the general sequence of intermediate varieties. It is evident that these combined factors would have changed the melt composition in the magma chambers and would have led to composition variations in the chilled zones of the intrusions up section (see above).

Figure 5.6 is an isochron diagram, based on the published results of studies of Sm and Nd isotopes in rock-forming minerals from the volcanics and layered intrusions under consideration. The graph shows the isotopic ratios of various minerals, which plot on the isochron straight line, giving an age of 2440 ± 41 Ma. Along with the above-mentioned evidence of REE distribution (see Fig. 5.3), the

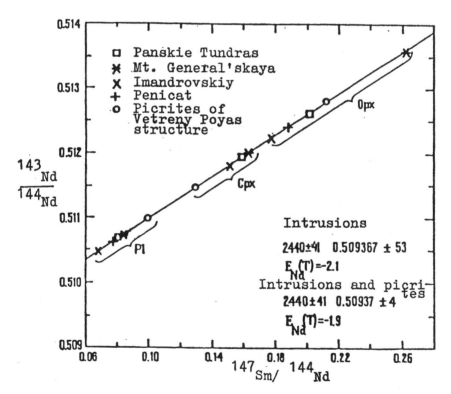

FIGURE 5.6 Isochron for rock-forming minerals and rocks of layered intrusions and high-magnesium volcanic rocks of the Vetreny Poyas Fm (after Mitzofanov and Torokhov, 1994).

data imply that the intrusive and volcanic rocks have a genetic affinity and a common mantle source, which differs drastically in its ε_{Nd} (-1.9) from the then-depleted mantle, according to the DePaolo–Wasserburg model ($+4$ for $t = 2.44\,Ga$).

Economic deposits of Cu–Ni ore (the Monchegorsk Pluton), chromite (the intrusions of the Kemi–Suhanko, Burakovsky and Imandra group) and platinum-group elements (Fedorovo–Pansky, Penikat and Lukkulaisvaara) are related to the above-mentioned intrusions.

5.1.3 Minor Intrusions (Dyke Swarms, etc.)

Small (usually tens or hundreds of metres, rarely a few kilometres across) intrusions of peridotite, orthopyroxenite, norite, gabbronorite, anorthosite, diorite and granophyre are common in rocks of the Archaean basement around the periphery of large layered intrusions, particularly the Monchetundra and Burakovsky bodies within the BMB, and within the Sumian–Sariolian volcanosedimentary complexes. Unlike large layered intrusions, where all the rock varieties occur within a single pluton, here the rocks generally form individual bodies. For typical chemical analyses see Table 5.5.

a) The Drusite Complex of the Belomorian

The Drusite Complex of small intrusions is one of the most unusual cases of the Early Precambrian magmatic activity. The complex is composed of small (a few hundred metres to 1 km in length), rootless intrusions, basic, ultrabasic and, less commonly, intermediate in composition, which impregnate the high-grade metamorphic rocks of the Belomorian Belt (Shurkin et al., 1962; Sharkov et al., 1994; etc.). The bodies number in the thousands. They are dominated by intrusions of norite and gabbronorite; bodies of ultramafic rocks (plagioclase harzburgite, bronzitite, websterite, and for the most part, plagioclase lherzolite) and anorthosites, marnetite gabbronorite-diorite and pyroxene diorites, are subordinate. As a whole, in their petrographical and geochemical features these rocks are rather close to the cumulates of layered intrusions described above (Monchetundra, Koilismaa, Burakovka, etc). According to estimates made in northern Karelia over an area of about 6,000 km², ultramafic rocks account for 17%; olivine gabbronorite for 43%; norite and gabbronorite for 30%; gabbronorite-anorthosite for 6%; Mgt gabbro-diorites and diorites for 4% of the intrusions (Malov and Sharkov, 1978). The largest bodies display primary igneous layering, caused by alternation of leucocratic and melanocratic rocks. In some cases, peridotite and pyroxenite occur at the bases of the bodies, and gabbroids with segregations of gabbronorite–pegmatite occur close to the contact with

Table 5.5 Typical chemistry of chilled zones of the intrusions of the White Sea drusite complex and the dyke complexes (after Sharkov et al., 1994).

Components	1	2	3	4	5	6	7	8	9	10	11	12
SiO_2	49.94	50.42	50.20	51.30	51.72	53.32	51.23	51.85	49.27	45.76	52.92	48.88
TiO_2	0.66	0.43	0.51	0.69	0.25	0.53	0.97	0.45	0.51	0.40	0.85	0.99
Al_2O_3	11.27	11.29	12.57	12.20	16.57	16.34	15.99	14.87	10.49	12.28	15.78	17.35
Cr_2O_3	—	0.21	—	—	0.027	0.015	0.024	0.061	0.18	0.38	0.028	0.049
Fe_2O_3	0.73	1.48	1.00	1.91	0.96	2.30	1.61	1.78	1.68	2.30	1.56	2.49
FeO	9.68	7.98	9.74	9.00	6.32	6.74	10.03	8.28	8.33	10.97	8.54	7.43
MnO	0.17	0.14	0.16	0.16	0.13	0.16	0.14	0.14	0.17	0.20	0.15	0.10
MgO	15.98	16.23	14.31	13.83	9.04	7.87	8.66	8.63	15.81	14.63	6.24	8.66
CaO	7.97	7.11	8.64	8.53	12.40	8.73	7.44	7.70	7.60	10.06	8.78	8.80
Na_2O	1.62	2.11	1.67	1.81	2.07	2.79	2.77	3.78	1.16	1.21	2.89	2.46
K_2O	0.56	0.47	0.53	0.57	H.O.	0.94	0.49	0.80	0.31	0.15	0.70	0.86
P_2O_5	0.09	0.13	—	—	0.05	0.11	0.12	0.12	0.09	—	0.10	0.11
LOI	1.25	1.30	0.96	—	0.84	0.68	1.00	1.23	4.98	1.41	1.92	1.94
Total	99.92	99.56	100.29	100.00	100.18	100.54	100.49	99.69	99.88	99.75	100.46	100.12
Zr	—	65	—	—	—	82	59	65	87	—	64	58
Nb	—	3	—	—	—	6	3	3	—	—	—	—
Y	—	15	—	—	—	16	16	15	—	—	—	—

Table 5.5 *contd.*

Components	1	2	3	4	5	6	7	8	9	10	11	12
Sr	—	141	—	—	—	165	226	141	—	—	—	—
Ba	—	233	—	—	—	241	288	233	—	—	—	—
Cu	—	—	—	—	—	—	—	—	—	110	—	—
Co	—	72	—	—	55	48	60	48	—	100	—	—
Ni	—	565	—	—	134	314	165	189	560	320	110	200
V	—	—	—	—	131	154	178	118	140	300	230	120
Mg*	74	77	72	71	70	64	59	63	76	68	55	64

Columns 1–3 — chilled zones of lherzolite–gabbronorite intrusions: 1 — Kopatozero, 2 — Southern Pezhozero, 3 — Malye Sal'nye Tundras; 4 — average composition of lherzolite–gabbronorite massifs; 5–6 — endomorphic zones of gabbronorite intrusions: 5 — Boyarsky, 6 — Deda; 7 — endomorphic leucogabbronorite of the Pezhostrov anorthosite massif; 8 — xenolith of "hypersthene porphyrite" from this massif; 9–12 — rocks from dyke swarms (after Mitrofanov, 1989), 9 — melanocratic gabbronorite from the Pechenga District, 10 — Ol-gabbronorite from the Monchegorsk District, 11 — gabbronorite from the Central Kola District, 12 — gabbronorite from the Pechenga District.

country rocks. However, intrusions composed of only one of the main rock varieties are the most common.

Individual bodies of the drusite complex are irregular-oval and boudin-like in shape; essentially, they represent large boudins, metamorphosed and migmatized along their margins. Fig. 5.7 shows typical examples of the structure of drusite bodies. They were named after their drusitic (coronal) textures, formed as a result of metamorphic alternation of primary igneous mafic minerals.

Zones of eruptive breccia, with xenoliths of the plagiomigmatite wall rocks, are locally preserved within the internal zones of the intrusions. The 0.15–0.5 m wide exocontact zones show hornfelsing and anatexis of the wall rocks. Anatectic veinlets of kyanite-bearing granophyres cross-cut the xenoliths and chilled facies of the intrusions.

Judging by the best-preserved bodies, originally they may have been sills or dykes; sometimes, they may have been horseshoe-shaped in plan view, infilling exfoliation caves in the hinge parts of relatively large folds. This suggests that the intrusions were syntectonic. As a result, the intrusions suffered intense amphibolite-facies metamorphism, and relics of the original igneous rocks were preserved only locally, usually in the centres of the intrusions. Besides, both drusites and country rocks underwent repeated reworking 2.0–1.9 Ga ago as part of the development of the Main Lapland Thrust (Priyatkina and Sharkov, 1979; Bogdanova and Kaulina, 1995).

Statistical studies showed that, against the general pattern of NW–SE-, E–W- and NE–SW-trending bodies, one cluster of intrusions has a markedly different areal distribution. The NW–SE-trending zone is the main zone, i.e. most satured with drusites (5–30% of the area). It is also characterized by ultrabasic igneous

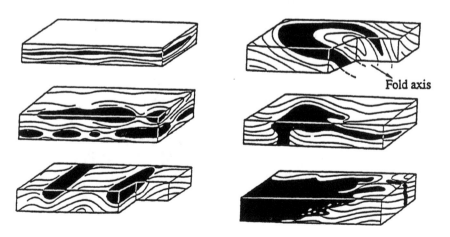

FIGURE 5.7 Structure of drusite bodies of the White Sea area. After Shurkin *et al.* (1962).

rocks, dominated by peridotite and pyroxenite (Fig. 5.8). Transverse E–W-
and NW–SE-trending zones are connected with the main NW–SE-trending zone
to the east and, to a lesser degree, to the west. Unlike the main zone, the transverse zones are dominated by gabbroids, with ultrabasic varieties subordinate.
However, we should note that all types of intrusions are observed inside all
zones.

The orientation of these zones coincides with the dominant directions of the
Sumian–Sariolian deformations in the Belomorian Belt (Volodichev, 1990).
According to these data, the earliest F_1 deformations have N–W to submeridional orientation, the later F_2 folds have a E–W direction and the latest F_3 open
folds are oriented in NE–SW to submeridional direction. From this position
three groups of drusites could be noted; however, there are no precise data about
their ages. From these observations it may be assumed that a large magma-
generated area, close to the areas beneath adjacent cratons, existed during the
formation of the drusite complex; the melts supplied from the area tended to be

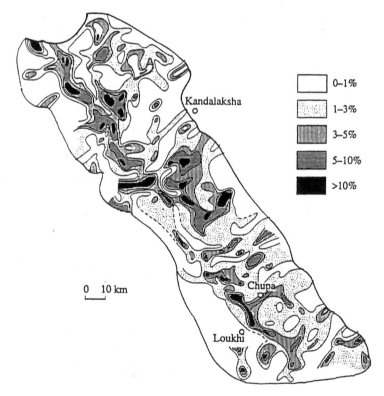

FIGURE 5.8 Density of distribution of bodies of the drusite complex (percent of
the total area) in the Belomorian Belt. After Malov and Sharkov (1978).

localized in a direction coinciding with the orientation of the main tension field at each stage of its development.

There are only two available age data on drusite massifs in the northern Karelia (Mgt gabbronorite-diorites of the Tolstik Peninsula and lherzolites of Tupaya Guba) from the sublatitudal zones. According to Sm–Nd and U–Pb data, their age is about 2.45 Ga (Bibikova et al., 1993a; Arestova et al., 1995), which coincides with the time when most of the Karelian layered intrusions were forming.

Typical examples of drusite intrusions of lherzolite and anorthosite were encountered on Pezhostrov Island in the White Sea, (Sharkov et al., 1994). Plagioclase lherzolites, with subordinate olivine gabbronorite, form an oval boudin-shaped body, 50×100 m in area; the body, together with the enclosing rocks, is sheared along its margins. The rocks are $Ol + Opx + Cpx$ and $Ol + Opx + Cpx + Pl$ cumulates with minor chrome spinels; intercumulus phases in the lherzolite are mainly presented by plagioclase and, less commonly, by phlogopite. Fine-grained rims of metamorphic minerals, such as garnet, clino-pyroxene and green hornblende, are developed along the boundaries of plagio-clase and ferromagnesian minerals. Where it comes into contact with olivine and plagioclase the assemblage always follows an earlier concentric orthopyroxene-clinopyroxene–spinel rim (corona), suggesting a subsolidus reaction between these minerals. Experimental data indicate that this reaction takes place at 8 ± 2 kbar (Green and Hibberson, 1970). On the other hand, the presence of an equilibrium assemblage of $Ol + Pl$ suggests that during magma crystallization the pressure did not exceed 7 kbar (Presnall et al., 1978). From this it follows that the intrusions were formed at pressures between 6 and 7 kbar; this corre-sponds with lithostatic pressure at depths of about 20–24 km.

The drusite body of gabbronorite–anorthosite on Pezhostrov Island is repre-sented by a tectonic block, 0.5×1 km in size, deformed and migmatized along its margins (Fig. 5.9). The anorthosites represent plagioclase cumulates contain-ing interstitial pigeonite (partly inverted), pigeonite–augite, and minor biotite, quartz and ilmenite. Fine- to medium-grained leucogabbronorites of the chilled zones, with eruptive breccia, where angular xenoliths of fine-grained porphyry gabbronorite are cemented by leucogabbronorite, are preserved locally.

Prismatic crystals of hyperstene and inverted pigeonite occur as micropheno-crysts in the porphyry microgabbronorite of the xenoliths, and the groundmass is made up of a fine-grained aggregate of basic plagioclase and pyroxenes, mainly Opx, Pig and Pig–Aug with minor chromite and phlogopite. The rocks resemble some microgabbroid varieties of Monchegorsk Pluton and Lukkulaisvaara, but they do not resemble any of the country rocks of the Belomorian Belt. They are prob-ably rock fragments from the feeder of the intrusion, as evidenced by the fact that they are chemically similar to rocks of the chilled zone of the intrusion (Table 5.5).

The rocks of the Pezhostrov Intrusion resemble those of the intrusions of the Fedorovo–Pansky and Mt. Generalskaya in their REE composition (Table 5.6, Fig. 5.8), thus suggesting the genetic affinity of their primary melts. On the other

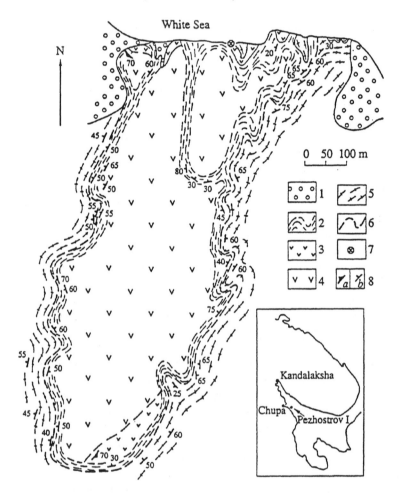

FIGURE 5.9 Structure of the drusite anorthosite body of the Pezhostrov Island, White Sea. From Sharkov *et al.* (1994). Key: 1 — Quaternary deposits, 2 — plagischists; 3 — chilled-margin gabbronorite; 4 — anorthosite; 5 — tonalitic gneisses; 6 — contacts; 7 — breccia of hyperstene porphyry; 8 — primary layering (a) and shearing (b).

hand, they resemble the REE spectra of the boninites and island-arc tholeiites of the Izu-Bonin Arc (Fryer *et al.*, 1992) (Fig. 5.10).

b) Dyke Complexes

Sumian–Sariolian SHMS dyke swarms are common within the Kola and Karelian craton (Ein, 1984; Aro and Laitakari, 1987; Mitrofanov, 1989). Their

Table 5.6 Trace and rare-earth element content in microgabbroids of the Monchegorsk Pluton and in rocks of the White Sea drusite complex (ppm). From Sharkov (1994).

Elements	1	2	3	4	5	6	7	8
Sc	22.40	1.65	28.20	24.00	23.80	24.30	38.60	26.50
Cr	522	24.2	915	1970	427	188	213	135
Co	57.30	28.10	69.00	69.90	45.80	54.50	53.70	40.00
Ni	408	656	252	387	182	299	—	—
Ga	18.1	16	11.7	10.2	13.7	16.8	13.7	14.6
Rb	5.5	34.4	8.5	14.7	16.7	13.7	—	26.4
Sr	637	842	334	213	313	116	65	182
Zr	—	—	—	149	61	136	85	123
Cs	0.310	0.250	0.230	0.375	0.698	0.326	—	0.544
Ba	200	299	—	193	277	200	108	202
La	1.40	1.01	0.54	8.78	7.97	7.82	0.90	12.80
Ce	2.50	1.82	1.53	21.02	17.30	17.39	2.31	26.13
Nd	—	—	—	8.31	8.18	7.82	1.42	10.34
Sm	0.134	0.125	0.447	1.976	1.841	2.059	1.560	2.486
Eu	0.550	0.420	0.240	0.555	0.760	0.960	0.620	0.714
Tb	—	—	0.37	0.69	0.27	0.30	0.54	0.38
Dy	—	—	—	—	—	1.7	2.9	2.4
Yb	0.188	0.103	0.072	1.156	1.188	1.555	2.322	1.339
Lu	0.038	0.006	0.006	0.172	0.169	0.262	0.334	0.210
Hf	0.109	0.158	0.093	1.449	1.030	1.554	1,229	1,848
Ta	0.32	0.04	—	0.20	0.11	0.18	0.05	0.19
Th	0.19	0.29	0.23	1.82	0.58	0.79	0.26	2.92
U	—	—	—	0.48	—	1.18	—	0.87

Columns 1–3 — microgabbroids of the Monchegorsk Pluton; 4–8 — rocks of the White Sea drusite complex. For chemistry see Tables 5.6.

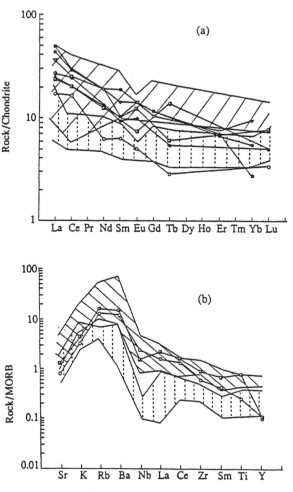

FIGURE 5.10 (a) Chondrite-normalised (after Sun, 1982) REE spectra for the Pezhostrov drusite bodies. Hatched field — plot of the Lower Zone of the Bushveld intrusion (Sharpe and Hulbert, 1985); speckled field — boninites from the Izu-Bonin island arc (Murton *et al.*, 1992). (b) MORB-normalized (after Taylor and McLennan, 1985) incompatible element plots for the same rocks. Hatched field — plots for the Mariana volcanic arc; speckled field — various boninites from eastern Pacific (Murton *et al.*, 1992). After Sharkov *et al.* (1994).

thickness varies from 0.5 to 35 m, locally reaching 60 m; their dip is subvertical. Their composition varies from plagioclase lherzolite and olivine gabbronorite to leucogabbronorite, dominated by fine-grained gabbronorite; each dyke is usually made up of one rock type. The rocks generally have a porphyritic texture. The mineral assemblages that make up the cumulus minerals in all of the above-mentioned intrusions usually tend to be phenocrysts.

Like drusites, these swarms, both in Karelia and the Kola Peninsula, trend NW–SE, E–W and NE–SW (Ein, 1984; Mitrofanov, 1989). Similar dyke swarms in all three directions occur in north-eastern and northern Finland (Aro, Laitakari, 1987). Dykes of NE–SW direction in northern Finland are about 2.44 Ga old (Vuollo *et al.*, 1994). The oldest swarms in the Kola Peninsula and northern Karelia are about 2.55 Ga old (Amelin *et al.*, 1995).

Dyke swarms are often associated with large layered intrusions and oriented in the same way. So, in the vicinity of the Burakovsky massif (southern Karelia), dyke swarms are oriented in the same NE–SW direction as the massif, while dykes in the Oulanka area have E–W orientation; the same situation occurs in the Kola Peninsula (Mitrofanov, 1989). Obviously, gabbronoritic dyke swarms of all three directions and layered intrusions with the same orientation associated with them could be synchronous in time. The small amount of chronological data available supports these ideas. It suggests that drusite "swarms" could be synchronous with them as well, and unit isopic age data are not contrary to it. From this it follows that about 2.5–2.47 Ga ago, magmatic activity was controlled by NW–SE and N–S deformations, and 2.45–2.43 Ga ago the W–E and NE–SW directions were dominant. All this time, the eastern part of the Baltic Shield was an area of boninite-like magmatic activity and could be called the early Palaeoprotherozoic Baltic province of siliceous high-magnesian series magmatism.

This province is about 1,000 km in length and c. 700 km in width (Sharkov *et al.*, 1997). It is located within three principal domains of the eastern part of the Baltic Shield: the Kola and Karelian Archaean cratons and the Belomorian Belt between them. The south-western border of the province occurs approximately along the line from Ladoga Lake to the northern part of the Bothnia Gulf of the Baltic Sea and coincides with the margin of the younger Svecofennian domain where the western continuation of the Karelian craton was strongly reworked by Svecofennian orogeny 1.9–1.8 Ga ago. Its north-eastern border is a tectonic one — the Kolmozero–Voronia Thrust here. Its north-western continuation was overlapped by the Norwegian Kaledonides nappes and its south-eastern one by the sedimentary cover of the Russian platform. Thus the original size of the province must have been very substantial.

5.1.4 Geodynamics of the Baltic Shield in the Early Palaeoproterozoic

The geodynamic situation of the Baltic Shield at the early Palaeoproterozoic was as follows. Two principal types of structures occurred here: (1) the large, relatively stable Kola and Karelian cratons with belts of essential mafic magmatism of siliceous high-magnesian series (the Pechenga–Varzuga, East–Karelian belts, Vetreny Poyas, etc.) and large layered intrusions and dyke swarms in their basement and flanks, and (2) the Lapland–Umba Granulite Belt (LUGB) between them. The Belt is formed of medium-pressure pyroxene granulites, which evolved upon metasedimentary rocks (mainly metapelites). The accumulation of the rocks forming the LUGB and their subsequent metamorphism probably occurred in the Early Proterozoic about 2.5–2.2 Ga ago (Huhma and Merilainen, 1991; Timmerman and Balagansky, 1994; Mitrofanov, 1995).

At this time, the Belomorian Belt was a kind of transition zone between the Karelian craton and the LUGB, which were undergoing intense deformation and amphibolite-facies metamorphism. It was characterized by a specific disseminated type of sinkinematic ultramafic–mafic magmatism (drusite complex) and the existence of local zones of high-pressure metamorphism (Bogdanova, 1995; Miller et al., 1995).

So, in the early Palaeoproterozoic on the Baltic Shield two extensional areas (Kola and Karelian cratons) existed, and a compensated compression zone (Lapland-Umba granulite belt) occurred between them (Fig. 5.2). The latter was surrounded by specific transitional zones (belts) of tectonic current; the Belomorian Belt, between the Karelian craton and the LUGB, was the best preserved. Most geologists believed that the BMB was a collision zone between the Kola and Karelian cratons (Mitrofanov, 1995; Glebovitsky, 1997). However, in contrast to such zones, it evolved in the condition of an extension detachment zone (Balagansky et al., 1996), and represented a structure of another type (Sharkov et al., 1997).

There is no geological evidence of plate tectonic activity. The geodynamic situation was characterized by the two relatively stable Kola and Karelian cratons with visible extensional areas resembling Phanerozoic intra-plate (continental rift and continental flood-basalt areas) activity, and strongly deformed granulite belt among them, surrounded by zones of tectonic flowage. It was rather like the Archaean situation (see Chapter 6). The differences were: the presence of the rigid stable crust within the cratons, which was marked by large layered intrusions, dyke swarms and volcanosedimentary piles in graben-like structures; and a different type of magmatism.

5.2 SILICEOUS HIGH-MAGNESIAN SERIES OF OTHER SHIELDS

This type of igneous activity is also widely known in other early Palaeoproterozoic terranes, where it was mainly studied in layered intrusions and dyke swarms.

Intrusive bodies of the boninite-like series are very widely developed in the Voronezh Crystalline Massif (VCM), an uplift of the Precambrian basement in the Central Russian Platform. Here they are represented by numerous layered intrusions of harzburgite–bronzitite–norite and norite–diorite composition, locally containing Cu–Ni sulphide and PGE deposits (Chernyshov et al., 1991).

The oldest body of this type in North America is the classic 2.7 Ga old (DePaolo and Wasserburg, 1979) Stillwater layered intrusion, which occurs within one of the uplifts of the Archaean Wyoming Craton, in the SW Rocky Mountains (Lambert et al., 1989). In character, it resembles the Monchetundra intrusion. The basal endocontact zone, 50 m thick, is overlain here by the Ultramafic zone, about 1,100 m thick, containing the Peridotite and Bronzitite members. The overlying Banded Zone, about 800 m thick, comprises rhythmically intercalated layers of bronzitite, norite, anorthosite and, in place, troctolite and plagioclase-bearing dunites. The latter occur in the middle part of the Banded Zone; a 1–3 m thick horizon of economic PGE mineralization, known as the J-M Reef, is associated with these rocks (Lambert et al., 1989). Slumping structures and autobrecciation are observed here; as a whole, these features resemble the Sopcha "ore bed" in the Monchegorsk Pluton and the Merensky Reef of the Bushveld Intrusion. The appearance of cumulus clinopyroxene characterizes the Lower Gabbro Zone, about 660 m thick. Above this comes the Anorthosite zone, about 1,900 m thick, with two interlayers of gabbronorite, and the Upper Gabbro Zone, formed by alternating thick layers of leucogabbro and anorthosite. The remainder of the intrusion section is overlain by platform sediments.

Many dykes of high-magnesian and high-Al rocks, which are considered as representative of the magmas forming the intrusion, were observed in the Archaean country rocks (Longhi et al., 1983).

Sm–Nd and Re–Os isotopic studies of mafic and ultramafic cumulates of the intrusions and associated dykes showed that it was formed by at least two magma types, referred to as U-type (ultramafic magma) and A-type (anorthositic magma). They differ in their major element, trace element, and precious metal abundance and isotopic composition and were derived from at least two distinct sources (Lambert et al., 1989, 1995). For U-type magma, the typical combination of low initial Os isotopic values with low initial Nd isotopic values ($\varepsilon_{Nd} < -1$), high $^{207}Pb/^{204}Pb$ for given $^{206}Pb/^{204}Pb$ (Wooden et al., 1990), and high $(Ce/Yb)_n$ ratios in the U-type cumulates and dykes is more consistent with the involvement of Re-poor but trace-element-enriched portions of the subcontinental lithospheric mantle in the petrogenesis of these magmas (Lambert et al., 1995). Following these investigators, the radiogenic initial Os isotopic composition of the J-M Reef and other portions of the elevated PGE concentrations suggest that A-type parental magmas incorporated Os from radiogenic early Archaean crust. Heterogeneous Nd and Os isotope data from the rocks below the J-M Reef suggest that the A-type magmas injected into the Stillwater magma chamber during crystallization of the U-type melts.

In the Canadian Shield, the boninite-like rocks are represented by the huge Matachewan dyke swarm, 2.45 Ga old, stretching for 800 km, with a width of about 450 km (Halls and Fahrig, 1987). For the most part, the dykes are composed of norite and gabbronorite, which are locally very leucocratic or, occasionally, extremely melanocratic. Their feeder chambers were located at moderate depths in the crust, at depths of no less than 20 km. These feeder chambers were differentiated and permanently supplied from the mantle. The composition of the dykes is very homogeneous over a distance of 100 km along a particular fracture, but adjacent individual dykes may differ greatly from each other. It should be noted that the formation of the dyke swarm, as suggested by isotopic dating, took place no more than 5 Ma ago. The dyke swarm's formation was related to the rise of a large mantle plume.

The earliest documented geomagnetic reversal was established for this dyke swarm by Halls (1992). Only one reversed polarity event could be recognized; this, together with the results of U–Pb precision dating, also points to the relatively rapid formation of the dyke swarms. Similar reversals have been known to be very common in the Phanerozoic, as indicated by well-known zones of magnetic anomalies in ocean-floor basalts.

The large Sudbury nickel-bearing norite–diorite pluton, 1.85 Ga old, has been regarded as an astrobleme by many investigators. It also contains rocks similar to those derived from boninite-like melts (Krogh *et al.*, 1982).

In Antarctica, dykes of boninite-like rocks make up a swarm in the Vestfold area of East Antarctica (Kuehner, 1989). Their age is estimated to be 2.4 Ga. They were intruded immediately after an intense episode of amphibolite to granulite metamorphism at 2.5 Ga.

Dyke swarms described as SHMS porphyritic orthopyroxene–clinopyroxene basalts, with an age of 2.4 Ga, occur in the eastern Gold Field, West Australia. Layered intrusions of harzburgite–bronzitite–norite–anorthosite were emplaced simultaneously with the dykes; the Jimberline Intrusion is the best known of these intrusions (Fletcher *et al.*, 1987).

In West Greenland, similar dykes of boninite-like composition, with an age of 2.1 Ga, occur in Archaean cratonic rocks (Hall and Hugnes, 1987). The dyke swarms extend into Scotland (Scoury Swarm, 2.45 Ga: Haerman and Tarney, 1989) and into North America, where they are known to occur in Wyoming, USA.

In South Africa, the series studied is represented by famous layered intrusions, such as the Great Dyke in Zimbabwe with an age of 2.461 ± 16 Ma (Wilson, 1982) and the classic Bushveld Intrusion (Fig. 5.11). According to von Gruenewaldt and Harmer (1992), the latter comprises two compositional suites: the Rustenburg Layered Suite of ultramafic to mafic cumulates, and the Lebowa Granite Suite, a younger sequence of sheeted intrusive granites which occupies an area of about 30,000 km^2 and about 5 km thick. It was intruded into a thick, up to 12 km, sequence of Late Archaean–Palaeoproterozoic sedimentary and volcanic rocks of the Transvaal Supergroup, which unconformably overlies the Early Archaean basement of the Kaapvaal Craton. Volcanic activity occurred intermittently through the deposition of sediments in an intracratonic basin and culminated in the voluminous extrusion of andesitic and felsic lavas of the

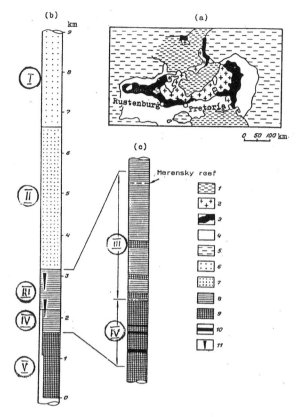

FIGURE 5.11 Structure of the Bushveld Intrusion. From Sharpe (1982). (a) plan view, (b) generalized section, (c) Critical Zone section. Key: 1 — platform cover rocks; 2 — red granite; 3 — Bushveld Intrusion; 4 — Transvaal Group rocks; 5 — Archaean rocks; 6–11 — rocks of the intrusion : 6 — gabbro, olivine–diorite and titanomagnetite beds; 7 — gabbronorite, gabbro, leucogabbro, 8 — essentially bronzitite with thin layers of norite, anorthosite and chromitite, 9 — orthopyroxenite and harzburgite, 10 — chromite layers, 11 — dunite pipes; I–V — zones: I — Upper, II — Main, III–IV — Critical: III — Upper part, IV — lower part; V — Lower. 1 — essentially norite and anorthosite with bronzitite and chromitite; 2 — essentially bronzitite; 3 — ultramafic rocks.

Dullstroom Formation and Rooiberg Group, respectively. The ages of the various volcanic components are 2224 ± 21 Ma, and Dullstroom volcanics have yielded ages of 2089 ± 26 Ma. The extensive Dullstroom–Rooiberg volcanicity was followed by the emplacement of first the Rustenburg Layered Suite and then the Lebowa Granite Suite of the Bushveld Complex. The age of the Rustenburg Suite is well constrained at 2061 ± 27 Ma with granites of 2052 ± 48 Ma.

In the character of the section and the composition of the rocks, the Rustenburg Layered Suite of the Bushveld Complex is similar to the Stillwater and Monchetundra intrusions, although in contrast to these, it has preserved its upper zone. Its thickness varies between 7 and 9 km. The suite can be divided into four zones. The Lower Zone consists of ultramafic rocks, including dunite, harzburgite and bronzitite. Above comes the Critical Zone, composed of norite, anorthosite and bronzitite, containing about 20 chromite seams, each about 2 m thick. The well-known Merensky Reef, the world's largest platinum deposit, occurs in the upper part of the Critical Zone.

Another specific feature of the Critical Zone is the presence of hortonolitic dunite pipes, which cross-cut the layering; the origin of the pipes is related to sub-solidus metasomatism of the rocks by Cl-bearing fluids (Boudreau et al., 1986).

The Main Zone is dominated by gabbronorite; the anorthosite is subordinate. They rest concordantly on the rocks of the Critical Zone although not univer sally, as "transgression", i.e. cutting across the layering of the underlying rocks, is rather common. The Upper Zone is composed of anorthosite, magnetitic gabbro and ferrodiorite; it contains about 25 thin layers of vanadium-bearing magnetitite.

Like the Stillwater Intrusion, the Bushveld Complex is believed to have been formed by recurrent magma injection (Sharpe and Hubert, 1985). The first basic intrusion of U-type high-magnesian magma is thought to have formed the Lower Zone and the lower part of the Critical Zone. The injection of A-type magma took place during crystallization of the upper part of the Critical Zone. The Main Zone was formed as a result of the injection of a major pulse of tholeiitic magma, and the formation of the Merensky Reef was associated with this pulse. Some petrologists believe that the Upper Zone was also formed due to an injection of an individual batch of melt (von Gruenewaldt et al., 1985).

The Lebowa Granite Suite comprises the second main component of the Bushveld Complex. According to Gruenewaldt and Harmer (1992), two main granite types predominate; the Nebo Granite, a major unit of coarse-grained, hypersolvus, mildly alkaline granite; and a more evolved, sometimes aplitic variety, the Klikloof Granite. The Nebo granites exhibit a well-developed and fairly systematic chemical and mineralogical zonation, characterized (from base to top) by a decreasing modal plagioclase concomitant with increasing albite component in the plagioclase; decreasing hornblende and increasing biotite; and increasing quartz. These variations are also reflected in the geochemical trends, i.e. Si, K, Rb increase, and Re, Ti, Ca, P, Ba, Sr and Zr decrease upwards through the sheet. The entire granite mass was probably emplaced as an unusually fluid, very hot (perhaps > 900°C), relatively anhydrous (initial content ~2.2%) restite-free magma. The granites exhibit all the classical features of mildly alkalic A-type magmas.

The Dominion Group represents the upper part of the Witwatersrand System. It was formed in continental and offshore conditions; the system is famous for its metalliferous conglomerates with high gold and uranium content

(Tankard *et al.*, 1982). Similar sedimentary basins, containing metalliferous sediments, were widely developed in the Early Precambrian, especially in the early Palaeoproterozoic; they have been found in Canada, Brazil, Finland, Siberia and elsewhere (Kazansky, 1988).

5.3 GENERAL FEATURES OF PALAEOPROTEROZOIC SILICEOUS HIGH-MAGNESIAN SERIES

From the aforesaid it follows that the SHMS of igneous rocks are very similar to each other in their time of formation and in their mineral and chemical compositions. It is important that all the rocks, from ultrabasic to acid types, are related by transitional varieties and form a single series. The character of the volcanics suggests that they are dominated by magnesian basalts and andesite–basalts. The mineral composition of the intrusive rocks is characterized by abundant orthopyroxene, inverted pigeonite and pigeonite–augite. Only these rocks, which crystallized at moderate and low pressures, contain coexisting magnesian olivine and orthopyroxene, which are "forbidden" in other series of igneous rocks. This distinguishes them from the products of crystallization of tholeiitic magmas, dominated by the dunite–troctolite–gabbro assemblages of cumulates (Laz'ko and Sharkov, 1988).

It is also important to note that the early Palaeoproterozoic boninite-like melts, unlike their Phanerozoic equivalents, were practically anhydrous in many cases. This is suggested by an almost complete absence of primary hydrous minerals in many of the intrusive rocks, where minor hornblende and biotite are locally observed in the intercumulus material. Even intermediate and acid rocks (hyperstene diorite and charnokite) are usually dominated by hyperstene. It is true that primary hornblende was relatively common in some lavas and intrusions, in particular in the Voronezh Crystalline Massif, which suggests that the water content did not play an essential role in the petrogenesis of these rocks.

In their major, rare and rare-earth element contents, the rocks of the SHMS are close to the Phanerozoic boninite series. The only difference is in isotopic characteristics: $\varepsilon_{Nd(T)}=$ from -1 to -2.5 (Balashov *et al.*, 1993; Amelin *et al.*, 1995) here instead of $\varepsilon_{Nd}=$ from $+6$ to $+8$ in the Phanerozoic boninites. In the latter it links with a present of subduction components (sediments and fluids in subduction zone: Pearce *et al.*, 1992); and in the former, with assimilation of Archean crustal material (Amelin and Semenov, 1996; Puchtel *et al.*, 1996, 1997). Contamination of crustal rocks is evidenced in an enlarged initial ratio $^{87}Sr/^{86}Sr = 0.704$ (Smolkin, 1992). It suggests a rather different origin for the SHMS from the magmatic rocks linked with subduction zones. Alternatively, the SHMS rocks rather differ from Archaean komatiite-basaltic series; for example,

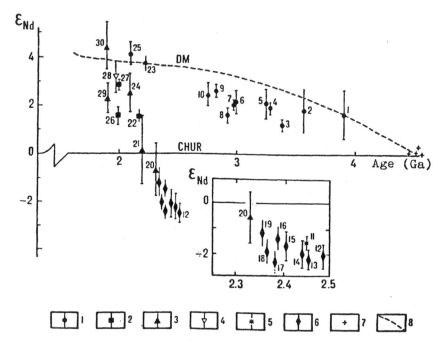

FIGURE 5.12 Variation in $\varepsilon_{Nd(T)}$ value in mantle-derived volcanic and intrusive rocks (after Smolkin, 1992). 1 — komatiite and komatiitic basalt (1 — Barberton, 2 — Vodlozero Block, 3 — Olonka structure, 4 and 6 — Kambalda, 5 — Kenozero, 7 — Shilos, 8 — structures of Karelia and Abitibi, 9 — Ontario Alexo, 10 — Karasjok) 2 — picrite and ferropicrite (22 — Olondo, 26 — Pechenga, 27 — Onega Trough); 3 — basalt and gabbro of Central Finland (20 — Runkaus, 21 — Koli, 23 — Oravaara, 29 — Suokkio, 30 — Erijarvi) and North America (23 — Cape Smith); 4 — ophiolites of Central Finland (28 — Jormua); 5 — boninite-type picrobasalt of Vetreny Poyas (11); 6 — layered intrusions (12 — Panskie Tundras, 13 — Mt. Generalskaya, 14 — Imandra, 15 — Penicat, 16 — Kivakka, 17 — Lukkulaisvaara, 18 — Burakovsky, 19 — Tsipringa); 7 — chondrites; 8 — mantle evolution, rated curve from De Paolo and Wasserburg (1979). DM — depleted mantle, CHUR — chondrite uniform reservoir.

on the Baltic Shield they differ by their enriching by LREE, positive Zr anomaly, $(Ce/Sm)n = 2$ ratio and negative value of ε_{Nd} (in komatiites it varies from $+2$ to $+3$: Samsonov *et al.*, 1997).

Thus, the early Palaeoproterozoic siliceous high-Mg series magmatic melts had a mixed origin and derived from highly depleted mantle material with the addition in different proportions of crustal (mainly lower-crustal) material, which was assimilated during the ascent of high-temperature mantle-derived melts to the surface.

5.3.1 Geology of the High-Magnesian Series

As stated above, Phanerozoic boninite series are only related to island-arc magmatism and do not occur in other tectonic settings. In contrast to these, the evolution of the Palaeoproterozoic boninite-like series most resembles Phanerozoic intra-plate magmatism. It occurred predominantly in the rigid stabilized continental crust as graben-like volcano-tectonic depressions, large dyke swarms and layered intrusions. The character of magmatism also did not vary in the relatively mobile parts of the Earth's crust, where the melts were injected simultaneously with crustal deformation; this is exemplified by the Belomorian Drusite Complex in the Baltic Shield. Of particular importance is the fact that the palaeo-equivalents of the boninite series represent the main type of the Early Proterozoic magmatism, both in the Baltic Shield and in other Precambrian shields, such as the Canadian, Greenland, Australian and Antarctic ones; it was probably the prevailing type of igneous process. However, no reliable geological evidence, either of subduction zones or oceanic-type crust (ophiolites, etc.), has been found for that time, either in the Baltic or in other shields.

Another noteworthy phenomenon is the multistage structure of large layered intrusions, formed as the result of successive injections of fresh magma batches into a solidifying magma chamber. In all cases, the composition of the melts evolved from high-Mg to high-Al; this led to a single type of variation in the intrusion cross-sections, although, in each particular intrusion, variation was irregular and unpredictable in character. The origin of the evolving melt series, which was probably supplied from one and the same magma chamber, is of great importance, but is not yet completely understood. The series must have evolved at a relatively rapid rate, since new batches of fresh melt were fed into the intrusive chambers, where the main magma body was still present (see Chapter 1). The new batches mixed with this residual melt, significantly changing its fractional crystallization trend.

In contrast to large layered intrusions, a completely different type of situation occurred in the small intrusions and dyke swarms: here the same rock varieties were distributed in individual bodies. The time taken for the formation of the large dyke swarms was also very short, being no longer than 5 Ma and possibly even shorter (Halls, 1992). This also suggests very fast rates of evolution for the large, deep-seated magma chambers, accompanied by the opening of new magma conduits. However, in these cases and in synkinematic drusite intrusions, magma pulses were not stored in larger chambers and did not mix with residual melts, and as a result formed solo bodies.

The same assemblage of ultrabasic, basic, intermediate and acid lava flows is also characteristic of volcanic grabens. This suggests a similar type of magmatic evolution.

As shown above, on the Baltic Shield the igneous process in question occurred over vast areas, covering almost all of its eastern part and forming the

Baltic SHMS province. We do not know the primary size of the province because all its boundaries are secondary ones; now its preserved area is about $1.0 \times 10^6 \, \text{km}^2$. This points to the presence, in early Palaeoproterozoic times, of a giant mantle diapir (superplume). The fact that the volcanic–plutonic complexes within the Kola and Karelian cratons of that time are often represented by sub-aerial formations (Kratz et al., 1978; Smolkin, 1992) implies that these areas represented large uplifts, a few hundreds (thousands?) of kilometres across; the Lapland–Umba Granulite Belt was a kind of sedimentary basin between them (Timmerman and Balagansky, 1994). This type of tectonomagmatic activity (plume tectonics) was very typical of the Early Precambrian (Khain, 1993).

Similar structures are believed to occur now on Venus. For instance, Aphrodite Land, the largest uplift on the planet, is formed by large oval structures, 2,500–3,500 km across. Their origin is associated with the uprise of mantle diapirs. There are no plate-tectonics processes operating on Venus (Marchenkov and Zharkov, 1989; Phillips et al., 1991) and Mars (Marchenkov et al., 1992), and also there is no geological evidence for the existence of plate tectonics on the Earth in the early Palaeoproterozoic.

This implies that, during the period under consideration, the geodynamics of the Earth's crust may have evolved on the same principle, giving rise to the unusual magmatic activity whose effects we can now observe. As shown above for the Baltic Shield, we suggest two spreading areas above mantle diapirs, which coincides with the Kola and Karelian domain; LUGB was a compression zone formed above the descending flows in the mantle between two diapirs; the Belomorian Belt was a transitional zone between them. This is a very unusual structure in its dynamics and disseminated magmatism which have no analogues in the Phanerozoic environments.

5.4   HIGH-K ROCKS OF THE EARLY PALAEOPROTEROZOIC

In the early Palaeoproterozoic, the more alkaline igneous rocks of were repre-sented mainly by low-Ti, high-potassium rocks such as alkali granites and syen-ites, which are similar to rocks from present-day convergent plate margins. In the early Palaeoproterozoic they may have occurred within different ancient compression zones.

Thus, potassic alkaline granite and charnokite–syenite magmatism occurred simultaneously with the formation of layered intrusions in the Baltic Shield (Batieva, 1976). Essentially, potassic alkali granites and monzonite–granites, with an age of 2.44–2.35 Ga (Balashov, 1996), occurred at the boundary of large Early Proterozoic structures such as the Keivy structure in the Central Kola domain and the Imandra–Varzuga structure, and along the northern boundary of the Keivy structure at its contact with the Murmansk structural domain

Table 5.7 Major- and rare-element content in Sumian charnockites (1–2) and peralcaline granites (3–6) of the Kola Peninsula (after Batieva and Vinogradov, 1991).

Components	1	2	3	4	5	6
SiO_2	64.11	70.72	70.83	73.21	76.16	75.47
TiO_2	0.63	0.39	0.61	0.43	0.18	0.08
Al_2O_3	16.07	14.28	12.40	11.67	11.05	8.33
Fe_2O_3	1.69	0.75	1.56	1.81	1.80	1.05
FeO	3.25	2.47	3.29	2.52	1.63	5.86
MnO	0.07	0.07	0.10	0.07	0.03	0.09
MgO	2.39	0.72	0.24	0.06	0.10	0.02
CaO	4.12	1.91	1.48	0.54	0.17	0.13
Na_2O	3.65	3.05	3.41	4.35	4.38	3.37
K_2O	2.99	4.73	5.05	4.98	4.42	4.07
H_2O^-	0.94	0.80	0.38	0.20	0.06	0.28
H_2O^+	0.10	0.12	0.32	0.15	0.26	0.66
P_2O_5	0.23	—	0.16	0.08	0.02	0.01
Li_2O	—	—	—	—	—	0.091
Rb_2O	—	—	—	—	—	0.128
F	—	—	—	—	0.09	0.24
Cl	—	—	—	—	—	—
Cs_2O	—	—	—	—	—	0.0005
ZrO	—	—	0.20	0.04	0.11	0.19
BaO	—	—	0.11	—	—	—
Total	100.69	100.02	100.14	100.13	100.42	100.27

(Fig. 5.2). They form sills, localized along the thrust between domains (Batieva, 1976). The typical chemistry of these rocks is shown in Table 5.7.

Small bodies of potassic "red" granites and granodiorites of age 2.41–2.44 Ma (Bibikova *et al.*, 1993a; Levchenkov *et al.*, 1995) occurred in the Belomorian Belt; they often cut Sumian drusites (Bibikova *et al.*, 1993a).

Intrusions of potassic granites and syenites 2.4–2.5 Ga old were found in the Aldan Shield in the boundary zone of the Aldan granulite–gneiss terrane and the Olekma granite–greenstone terrane and (Kotov *et al.*, 1993), The two-stage Sm–Nd model ages of the granites vary in the range 3.1–3.6 Ga with negative $\varepsilon_{Nd(T)}$ values varying from −5 to −10. This suggests the presence of melted material, equivalent to the Archaean rocks of both the Aldan granulite–gneiss and the Olekma granulite–greenstone terranes (see Chapter 6), which coincided the formation of thrusts 2.5 Ga ago.

Low-Ti potassic alkali rocks are exemplified by three small intrusions in the Superior Province (Lake Superior). Their age is estimated at 2398 ± 72 Ma (Funerton and Barry, 1984). However, the oldest, essentially potassic rocks ($K_2O/Na_2O = 0.6$) are andesites and dacites of the Oxford Lake Group, with an age of 2.64 ± 12 Ga, and an initial Sr-isotope ratio of 0.7055 ± 34 (Brooks *et al.*, 1982). In their mean content of Rb, Sr, Zn and Y, and in their La/Yb, La/Sm and Sm/Nd ratios, these rocks are similar to those of the shoshonite–latite series and probably also occur in an ancient boundary zone.

Thus, the early Palaeoproterozoic, alkaline, low-Ti potassic magmatism occurred within a relatively well-defined geodynamic environment, being associated with zones of geoblock interaction.

5.5 MAGMATISM OF THE TRANSITIONAL PERIOD OF TECTONOMAGMATIC PROCESSES (2.2–2.0 Ga OLD)

The 2.2–2.0 Ga stage shows early evidence for the transition to a Phanerozoic type of tectonomagmatic activity. Firstly, this stage saw the large-scale appearance of geochemically enriched Fe–Ti picrites and basalts, with normal, moderate and high alkalinity. These rocks are equivalent to those from Phanerozoic intra-plate regions, and previously had been virtually absent. It was also the time when large, locally arcuate belts of zonal high-pressure metamorphism appeared along the different geoblocks' (domains') boundaries. The back-arc zones of some of these belts include synchronous marine or continental sedimentary basins, generally distinguished by their intense basaltic (including MORB) volcanism. As a whole, the geological setting was similar to that of Phanerozoic active lithospheric plate margins. Thus, from this stage onwards, the tectonomagmatic activity grew similar to that of the Phanerozoic, although some features were rather different (for example, the extensive development of anorthosite assemblages and the development of granulite metamorphism in suture zones, etc.). This ancient variant on Phanerozoic tectonomagmatic activity was terminated about 1 Ga ago.

The geological history of this stage is best understood by looking at the Baltic Shield, hence this region will be used as an example in discussing the problems related to this transitional period. Here it is associated with the Svecofennian stage of regional development, occurring 2.2–1.8 Ga ago (Kratz *et al.*, 1978).

5.5.1 Magmatism of Ancient Collision Zones and Back-Arc Basins in the Baltic Shield

In the Baltic Shield the Svecofennian stage (2.2 to 1.8 Ga ago) is characterized by three types of tectonomagmatic activity, namely:

1. the appearance of a large stable region in the Karelian Craton (domain);
2. the formation of two regional complementary structures in the Kola Craton: a zone of compression, associated with the Main Lapland Thrust,

with associated large gabbro–anorthosite intrusions, and a zone of extension: the volcanosedimentary Pechenga–Varzuga Belt (Figs 5.1 and 5.2);

3. in the central Baltic Shield the Svecofennian ocean opened ~2 Ga ago and after it closed an orogenic belt formed in its place 1.9–1.86 Ga ago (Gaal and Gorbatchev, 1987; Gorbatchev and Bogdanova, 1993).

In its geology, the first region was similar to the Phanerozoic zones of intraplate magmatism, the second closely resembles the Phanerozoic plate collision zone, and the third resembles the zone of oceanic spreading and the further orogenic belt after its closure.

a) Magmatism within the Karelian Craton

During the Svecofennian stage, Jatulian subaerial volcanosedimentary complexes, infilling many narrow linear NW-trending grabens, were formed within the Karelian craton (Kratz et al., 1978; Sharkov, 1984a). Iron–titanium moderately alkaline basalts and picrite–basalts occur predominantly there; Fe–Ti tholeiites are subordinate. These rocks have a massive amygdaloidal structure; horizons of tuff and tuff breccia are common. The eruptions were subaerial in character. For typical analyses see Table 5.8.

In addition to lava flows, numerous sills and dykes, with similar compositions, occur in the grabens. They are also extensively developed outside the grabens themselves, both in Karelia (Ein, 1984) and in eastern and Northern Finland (Aro and Laitakari, 1987; Vuollo et al., 1995), suggesting that a basaltic terrane, comparable in scale to the Phanerozoic traps, occurred in the eastern Baltic Shield at the time. This has led some scientists, to believe that the Jatulian magmatism represents one of the oldest trap provinces.

The Jatulian complexes overlie the above-mentioned Sumian–Sariolian rocks, or directly overlie the Archaean crystalline basement with minor unconformity. A weathering crust is present at the point of contact. Isotopic dates suggest an age range of 2.2–2.1 Ga, averaging 2.15 for these rocks (Balashov, 1985; Vuollo and Piirainen, 1992). Isotopic studies of the rocks of the Konchezesk Sill in the Onega Plateau, from the upper part of the section (Suisar Formation), gave an Sm–Nd age of $1\,974 \pm 27\,Ma$, with an initial $\varepsilon_{Nd(T)}$ value of $+2.87 \pm 0.12$ (Pukhtel et al., 1992). The positive $\varepsilon_{Nd(T)}$ value suggests that a mantle reservoir, which had long been poor in LREE (and into which an incompatible-element-enriched component was supplied shortly before melting), served as the source for picrite–basalt magmas, as for the similar rocks of Pechenga (see below). Tholeiitic dykes of the same age (1.97 Ga) are found in North Karelia (Vuollo et al., 1995).

Swarms of Fe–Ti tholeiitic and moderately alkaline basalt dykes, with an age of 2.1 Ga, have also been found in Scotland (Heaman et al., 1989), Greenland

Table 5.8 Average composition of differentiated of layered lava flows of the ferropicritic assemblage and tholeiitic metabasalts (Sharkov and Smolkin, 1989).

Components	1(3)	2(17)	3(21)	4(5)	5(6)	6(4)	7(7)	8(150)
SiO_2	43.04	43.04	43.06	43.37	47.50	47.55	47.70	47.73
TiO_2	2.28	1.77	2.20	4.20	3.44	2.96	2.20	1.55
Al_2O_3	7.08	5.68	6.87	11.35	10.30	11.94	7.50	13.08
Fe_2O_3	1.58	2.92	3.04	4.46	2.23	4.74	3.04	3.48
FeO	12.58	11.46	11.37	12.40	10.88	10.15	11.10	10.87
MnO	0.19	0.17	0.19	0.20	0.17	0.16	0.18	0.20
MgO	16.94	19.87	13.97	6.20	7.80	5.77	11.38	6.68
CaO	9.17	6.99	10.47	10.87	10.02	8.48	10.10	10.62
Na_2O	0.12	0.16	0.16	0.39	2.90	2.96	0.23	2.33
K_2O	0.04	0.05	0.10	1.54	0.30	0.28	0.05	0.22
H_2O	5.80	5.84	5.32	3.60	3.00	3.00	4.00	2.49
P_2O_5	0.21	0.16	0.21	0.45	0.42	0.33	0.21	0.16
F	0.038	0.045	0.060	0.160	0.100	0.070	0.060	—
S	0.20	0.35	0.28	0.08	0.36	0.38	0.64	0.12
CO_2	0.15	0.22	1.80	0.10	0.20	0.84	1.30	0.37
Cr	1370	1980	1300	205	410	140	1440	100
V	220	200	340	450	340	280	220	340
Ni	900	1380	595	85	135	135	970	80
Co	95	115	90	60	55	70	110	—
Cu	185	150	165	115	170	105	220	—
CaO/Al_2O_3	1.3	1.2	1.5	1.0	1.0	0.7	1.4	0.8

Column 1 — ferropicrite–basalt of the lower chilled zone; 2 — ferropicrite and ferropicrite basalt (MgO > 16%) of the lower parts of the lava flows; 3 — clinopyroxene ferropicrite basalt (MgO < 16%) of the middle parts of the lava flows; 4 — picrite–basalt of the matrix of the liquation zone; 5 — ferrobasalt containing globules from the liquation zone; 6 — ferrobasalt with spinifex structure; 7 — low-alkali contact ferrobasalt; 8 — tholeiitic metabasalt. The number of analyses is in parentheses.

(Hall *et al.*, 1990) and the Canadian Shield (Boily *et al.*, 1991), suggesting that these regions made up part of the same Early Proterozoic continent: Laurentia-Baltica (Gorbatchev and Bogdanova, 1993).

b) Magmatism within the Kola Geoblock

As stated above, there are two large complementary structures, differing both in their magmatism and in the trend of their tectonic processes, namely, the Main Lapland Thrust (MLT) and the Pechenga–Varzuga Belt (PVB) of basaltic volcanism.

c) The Main Lapland Thrust

An important tectonic feature is the Main Lapland Thrust (MLT) which defines the southern edge of the Lapland–Umba Granulite Belt (see above, Fig. 5.3). It is marked by the exposure of high-pressure granulites (Lapland type) in the Kolvitsa, Kandalaksha, Sal'niye and Taudash Tundras, etc. (Priyatkina and Sharkov, 1979; Gaal and Gorbatschev, 1987). The structures and mineral assemblages of these granulites overprint the older (2.5–2.3 Ga) moderate pressure granulites of the LUGB.

The MLT represents a major thrust structure: an intense shear zone that plunged relatively gently (30–50°) to the north-east. Supracrustal rocks, similar to Sumian–Sariolian rocks in their petrography and geochemistry, lie directly beneath it (the Tana Belt in LGB and the Kandalaksha Formation in the Por'egubsko-Umba block (PUB)) with an age of 2.46 Ga (Kratz *et al.*, 1978; Barbey *et al.*, 1984; Timmerman and Balagansky, 1994).

Altogether, the MLT has a length of some 700 km, its northern continuation being buried beneath Caledonian nappes, and the southern one beneath the waters of the White Sea. Thus, the original length of the Thrust belt must have been extremely large.

A wide range of isotopic date constrains the age of the high-pressure metamorphism and igneous activity to 2.0–1.9 Ga (Tugarinov and Bibikova, 1980; Bernard-Griffiths *et al.*, 1984; Barbey and Raith, 1990; Petrov, 1995). It is thought that the peak of the high-pressure Lapland type of metamorphism occurred 1.95–1.93 Ga ago (Gorbatschev and Bogdanova, 1993; Bibikova *et al.*, 1993a). However, Bogdanova and Kaulina (1995) showed that postconsolidation plagiopegmatites, cutting high-pressure granulites of the Kolvitsa massif, yield a U–Pb age of 2056 ± 2 Ma. It suggests that the MLT could run at c. 2 Ga ago.

In the Kolvitsa, Sal'niye and Taudush tundra areas (Russia), and in the Lake Inari area of Finland, anorthosite–mangerite intrusions make up the bulk of the rocks associated with the development of the MLT, whereas the later hornblende pyroxenite and peridotite are subordinate. The anorthosite–mangerite massifs are 2–3 km thick lenticular sheet-like bodies with structures consistent with the Thrust (Priyatkina and Sharkov, 1979). They have gneissic textures, while their primary magmatic cumulate textures can be observed as relics only (Fig. 5.13). The cumulus phase is represented by plagioclase An_{60-65}, and the intercumulus phases by inverted pigeonite and pigeoinite–augite. Contacts with the country rocks are via a zone of common shearing, although rare areas of eruptive breccia, where fragments of andelying amphibolites are "cemented" by anorthosites, are preserved there (Fig. 5.14). The middle section is formed by mesocratic norite (Pl + Opx cumulates), replaced by pyroxene–diorite up section. The uppermost section is composed of quartz- and orthoclase-bearing rocks of mangerite, quartz-mangerite and quartz-diorite types.

The intrusions are concordant with the thrust structure, and their internal structures are conformable with the contact zone. They were emplaced in the early stages of thrusting and were involved (together with the enclosing country rocks) in all the processes of deformation and metamorphism that took place during the evolution of the structure (Fig. 5.15). As a result, the intrusive rocks were largerly metamorphosed to form crystalline schists, and the primary igneous textures and structures can survived only locally. The schistosity as a whole is parallel with igneous layering or crosses it at an acute angle.

FIGURE 5.13 Cumulate textures of gabbronorite–anorthosites of the Kolvitsa Intrusion. Photo by E.V. Sharkov.

FIGURE 5.14 Eruptive breccia in the endomorphic zone of the Kolvitsa anorthosite — mangerite Intrusion. Photo by E.V. Sharkov.

FIGURE 5.15 Type of deformation in the rocks of the Kolvitsa Intrusion (dark — deformed lens-shaped xenolith of amphibolite from the base of the intrusion). Photo by L.A. Priyatkina.

An inverted metamorphic zoning is observed with a transition from garnet amphibolites (Hbl + Gr + Cpx + Pl) through garnet granulites (Gr + Cpx + Pl) to Hy + Gr + Pl granulites. PT conditions are estimated to be 10–11 kbar and 850–900°C (Priyatkina and Sharkov, 1979; Petrov, 1995). The metamorphic isograds are also parallel with the stratification. All these features suggest that intrusion and solidification of these anorthosite bodies were concommitent with the movements along the Thrust.

The U–Pb age of magmatic zircon from Sal'nye Tundras mangerites is 1925 ± 1 Ma, and zircon from granulites upon them is 1916 ± 1 Ma (Bibikova et al., 1993a). Recently new U–Pb data of 2.40–2.45 Ga were obtained for some anorthosite bodies of the Kolvitsa and Sal'nye Tundras (Mitrofanov et al., 1993). This means that there are at least two groups of anorthosites — the synkinematic anorthosite–mangerite complexes and more ancient anorthosite bodies (Pyrshin, Vaskojoki, etc.), which probably belonged to the early Palaeo-proterozoic Drusite Complex of the Belomorian Belt (see above).

These granulites were intruded by dykes of titaniferous hornblende peri-dotites and pyroxenites. They are also synkinematic intrusions which derived from titaniferous ferropicrites, probably similar to the ferropicrites from Pechenga–Varzuga. Later, lenticular bodies of potassic granites conformable with schistosity appeared. Thus, during the movements along the MLT there were intrusions of calc-alkaline melt, followed by high-titanium picrites and finally high potassium acid magmas.

As movements along the Lapland Thrust waned, domes of gneissic granites ascended mostly along the southeastern flank of the PVB and beneath the Main Lapland Thrust in the Belomorian block. This resulted in the fragmentation of the Lapland–Umba Belt and the suture zone marked by the Lapland Thrust into a number of separate granulite complexes, including the LGB and PUB. Thrusting, dextral transcurrent shearing, folding and faulting also played an improtant role in forming the new geological framework.

d) The Pechenga–Varzuga Belt

This belt, which may have represented an area of back-arc spreading, was initi-ated on a heterogeneous, mostly c. 2.9–2.7 Ga old, Archaean basement (Freeman et al., 1989; Mitrofanov and Smolkin, 1995; Sharkov and Smolkin, 1997). The two halves of the PVB — the Pechenga and the Imandra–Varzuga zones (Figs 5.1, 5.2 and 5.16) — which are now separated by faulting are infilled by volcanic (picrite–rhyolite) rocks and subordinate sediments. The Kola Superdeep Borehole and seismic data suggest that the overall thickness of the Pechenga and the Imandra–Varzuga rocks may reach 8.5 and 11.5 km, respectively (Mitrofanov and Smolkin, 1995). Outside the Pechenga and

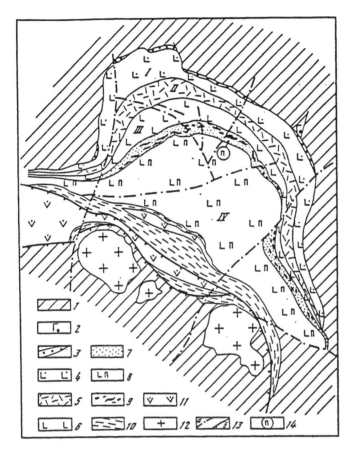

FIGURE 5.16 Geology of the Pechenga area. Simplified from Hanski and Smolkin (1989). 1 — Archaean gneiss and schist, Archaean and early Proterozoic granitoids; 2 — conglomerate; 3 — andesite and basalt; 4 — moderately alkaline basalt, andesite and picrite; 5 — quartzite and dolomite; 6 — basalt and picrite; 7 — "productive" pile (phyllite, sandstone, siltstone, tuff and tuffite); 8 — graphite-bearing phyllite, dolomite and silicite; 9 — basalt; 10 — psammite, phyllite and sericitic schist; 11 — picrite, tuff, tuffite and basalt; 12 — granite-gneiss domes; 13 — faults: (a) slip-strike, (b) thrust; 14 — Pilgujarvi intrusion. I — early Palaeoproterozoic Akhmalakhti Suite, II–VI — late Palaeoproterozoic suites: II — Kuetsjarvi, III —Kolasjoki, IV — Pilgujarvi.

Imandra–Varzuga complexes there are minor occurrences of similar rocks in the Keulik and at Ust'Ponoy. This suggests that these complexes once formed a single, E–W-trending belt that extended across all of the Kola Peninsula and Lapland.

The Pechenga and Imandra–Varzuga complexes are stratigraphically simi-lar (Fig. 5.17). Close to the bottom are the Televi–Majarvi, Polisarka and Seidorechka Formations of Sumian–Sariolian age, see above.

The Jatulian Kuetsjarvi volcanosedimentary Formations were formed after a hiatus which resulted in the development of a weathering crust atop the older rocks. The volcanics show an alkaline chemical affinity (Table 5.8) and have alkali-basaltic, picritic-basaltic, trachyandesitic-basaltic, and, rarely, trachdacitic

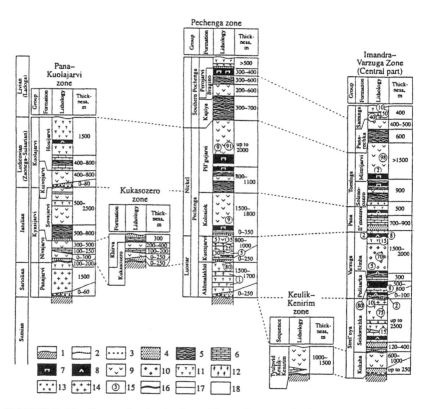

FIGURE 5.17 Generalized stratigraphic columns of the Palaeoproterozoic Pechenga–Varzuga Complex. After Smolkin (1992). 1 — Archaean basement; 2 — weathering crusts; 3 — conglomerate and tuff-conglomerate; 4 — sedimentary, essentially sandstone rocks; 5 — sedimentary, essentially silt-stone rocks; 6 — marbles; 7 — high-Ti ferropicrite, ferropicrobasalt and tuff; 8 — low- to moderate-Ti picrobasalt and basalt; 9 — basalt; 10 — moderately-alkaline basalt; 11 — andesite–basalt; 12 — moderately alkaline andesite; 13 — dacite and andesitic dacite; 14 — rhyolite; 15 — proportion of volcanics in specific for-mations; 16 — stratigraphic breaks; 17 — stratigraphic boundaries; 18 — correla-tion of secions.

compositions. Flows with massive and amygdaloidal textures are prominent, while local slag rinds demonstrate episodic weathering. All these volcanics are characterized by pronounced sodic trends and by the predominance of iron over magnesium. The volcanics yield an Rb-Sr age of 2214 ± 54 Ma (Mitrofanov and Smolkin, 1995).

Red, cross-bedded arkoses, quartzose metapsammites and siltstones with lenses of gravelstones, red dolomites with lenses of dolomitic conglomerates, and jaspers are the predominant sedimentary rocks; terrigenous-carbonate rocks (marly and stromatolitic carbonate rocks, etc.) are minor components. These rocks have been correlated with the lower and upper Jatulian sequences in Karelia and Finland (Sviridenko, 1984).

The upper parts of the Pechenga and Imandra–Varzuga complexes differ from the lower parts. These rocks are represented mainly to the Kolosjoki (Zapolyarny) and Pilgujarvi (Matert) formations in the Pechenga complex, which reach a total thickness of c. 5 km (Fig. 5.17). The volcanic Zapolyarny Formation rests with a minor angular unconformity on the volcanics of the previous formation. The Zapolyarny Formation commences with a tuffite unit followed by pillow-lavas with interbeds of tuffite schists mostly of tholeiitic basaltic and, rarely, picritic-basaltic compositions and normal alkalinity. These volcanics yield a Rb-Sr age of 2114 ± 52 Ma (Mitrofanov and Smolkin, 1995).

The Pilgujarvi Formation, conformable with the underlying Zapolyarny one, begins with a 1 km-thick sedimentary member (called the "Productive pile") which is made up mainly of sandstones, phyllites, siltstones and silicites, along with basaltic and picritic tuffites. These rocks are often enriched in carbonaceous matter and contain syngenetic sulfides. The rock sequences of the "Productive pile" and the Solenoozera Formation, displaying varying bed thickness, alternation of thin flyschoid intercalations with distinct rhythmic bedding, slump structures and prominent graded bedding, are interpreted as turbidites.

The dominantly volcanic upper part of the Pilgujarvi Formation is 1.9–2.0 km thick. It comprises pillowed, massive and variolitic lavas of tholeiitic (MORB-type) composition. Minor lenticular interbeds of black shales pass into tuffaceous silicites with abundant slumping breccias. The latter occur in the middle part of the formation where they form interbeds within globular ferrous picritic basalts. The latter differ from all the other volcanics by virtue of their higher TiO_2 content (1.5% to 3.0%, rarely 5 wt.%), and their high total iron content that reaches 14 wt.% FeO (Table 5.8). These picrites and picrite basalts are therefore of a ferrous type. The upper parts of these volcanic formations feature lava flows that are enriched in potassium and have relatively alkali-rich compositions.

Within the "Productive pile" and the Solenoozera Formation there are abundant dolerite sills and minor layered nickeliferous intrusions of wehrlites, clinopyroxenites, gabbros and orthoclase-gabbros. The chilled zones of these intrusions represent chemical analogues of the ferropicrites (Smolkin, 1992).

Similar intrusions cut across the underlying volcanics. The large Gremyakha–Vyrmes layered intrusion of olivinites, wehrlites, clinopyroxenites, gabbro, alkali gabbro and nepheline syenites with rich Fe–Ti deposits and associated alkaline granites (Sharkov, 1980; Kogarko et al., 1995) is located in the north-eastern foreland of the PVB (Figs 5.2 and 4.9). The U–Pb age of the Gremyakha–Vyrmes is $2\,000 \pm 40$ Ma (Kogarko et al., 1995), and the rocks of this intrusion differ markedly from the Sumian layered intrusions and resemble those of the Pechenga intrusions; the parental magma of the Gremyakha–Vyrmes was obviously, ferropicritic basalt also.

The ferropicritic and tholeiitic basaltic volcanics of the Pilgujarvi Formation have yielded Sm–Nd, Os–Re and Rb–Sr isochron ages of c. 1980 Ma, the nickeliferous intrusions Pilgujarvi, etc. 1960–1980 Ma (Hanski et al., 1991; Smolkin, 1992; Mitrofanov and Smolkin, 1995). All these events are virtually coeval.

Thus, the submarine eruption of the abundant, undifferentiated, mantle-derived magmas of tholeiitic and picritic basalts, the formation of abyssal black shales, turbidites and cherts, and the emplacement of large amounts of mafic–ultramafic melt all took place during a late stage of the Pechenga–Varzuga evolution. This suggests the existence at that time of a relatively deep, non-compensated basin in the PVB.

Intrusions of granitoid domes (e.g. Kaskeljavr, Shuonijavr, etc.) took place simultaneously with the active volcanism in the Pechenga zone in its base and the southern flank 2000–1940 Ma ago (Mitrofanov and Smolkin, 1995).

The topmost units of the sequences are calc-alkaline volcanics represented by andesites and dacites of the Southern Pechenga Group and andesites, dacites and rhyolites of the Panarechka and Saminga Formations (Imandra–Varzuga Complex). There are also ferropicrites (Mennel Formation) and abundant tuffites, tuffaceous sandstones and tuffite-conglomerates. These rocks are considered to be a molasse. Volcanics of the Southern Pechenga Group yield a Rb–Sr age of 1887 ± 58 and 1855 ± 54 Ma (Mitrofanov and Smolkin, 1995).

The consistency of geochemical patterns over a distance of some 250 km is worth noticing. The geological and petrological data provide evidence that the evolution of the supracrustals in the Pechenga zone proceeded from a continental facies (Kuetsjarvi Formation) to transitional (Kolosjoki Formation) and to an oceanic one with sediments of turbidite type (Pilgujarvi Formation). This suggests deposition in a relatively deep-sea sediment-starved basin, induced by spreading behind the Lapland Thrust. On the southern flank of the Pechenga basin calc-alkaline volcanism occurred in nearly continental conditions (the Southern Pechenga Group).

In their geochemistry, the basalts of the Pilgujarvi Formation can be divided into two groups (Smolkin, 1992). The bulk of them are classified as MORB-type basalts (Table 5.9, Fig. 5.18). The rest are grouped together with iron- and titanium-rich picrite, picrobasalts and basalts; they have increased alkalinity and include alkali-basalts. The ferropicrite lava-flows are usually layered (Fig. 5.19).

Table 5.9 Average composition of tholeiitic basalt of the Pilgujarvi Formation (Smolkin, 1992).

Components	1(31)	2(85)	3(28)	4(7)
SiO_2	46.60	47.38	46.30	47.69
TiO_2	1.06	1.33	1.78	1.19
Al_2O_3	12.84	13.73	13.07	13.64
Fe_2O_3	1.69	2.38	2.62	1.96
FeO	11.05	12.17	12.62	11.56
MnO	0.15	0.20	0.23	0.16
MgO	7.65	7.09	6.18	6.61
CaO	10.66	9.70	10.01	11.02
Na_2O	2.49	2.73	2.30	0.57
K_2O	0.26	0.34	0.34	0.45
P_2O_5	0.13	0.16	0.24	0.27
H_2O^+	3.97	3.26	3.71	3.62
CO_2	0.05	0.14	0.12	0.35
Total	100.00	100.00	100.00	100.00
Li	14.0	12.5	9.2	14.1
Rb	11.7	11.1	11.3	15.0
Sr	41.6	68.4	86.4	13.4
Ba	21.6	32.5	27.1	62.9
Sc	8.0	8.9	7.5	8.4
Ga	13.8	17.2	13.5	18.4
Zr	37.1	43.9	45.7	42.9
Nb	6.2	2.7	2.9	2.2
Cu	255.5	271.8	311.8	372.9
Zn	152.2	128.8	131.8	141.4
V	195.2	234.6	134.6	242.9
Cr	205.4	96.6	90.6	141.7
Ni	235.1	76.6	88.1	93.0
Co	60.7	42.0	46.4	47.0

Column 1 — olivine basalt; 2 — differentiated basalt; 3 — basalt; 4 — low-alkali basalt of the fourth sequence. The number of analyses is in parentheses.

As a rule, the following can be recognized: upper and lower chilled zones (A1 and A2); a zone of ferrobasalts, showing microlitic, globular and spinifex textures (B–D); a zone of ferro-picrobasalts (E); and a zone of cumulus olivine-ferropicrites (F).

Spinifex textures are formed by the packing together of tabular and skeletal crystals of titanaugite, plagioclase and ilmenite aggregates; acicular pseudomorphs

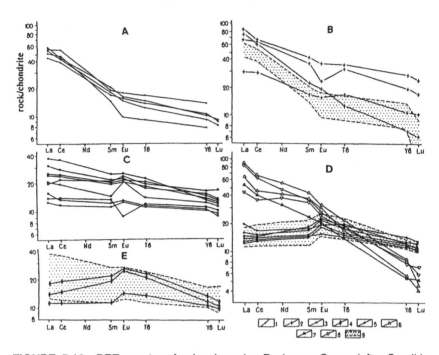

FIGURE 5.18 REE spectra of volcanic rocks, Pechenga Group (after Smolkin, 1992). A — Akhmalakhti Suite; B — Kuetsjarvi Suite; C — Kolasjoki Suite; D — Pilgujarvi Suite (shaded area — basalts of the MORB-type); E — basalts of the MORB-type of the Pilgujarvi Suite. 1 — Ahmalahti andesitic basalts; 2 — Kuetsjarvi basalts and alkali basalts; 3 — Kolasjoki basalts; 4 — basalt interbeds from black shales of the "Productive Pile"; 5 — Pilgujarvi tholeiite basalts; 6 — ferropicrites and their tuffs in the "Productive Pile"; 7–8 — Pilgujarvi ferropicrites: 7 — of the basal unit of the suite, 8 — of the middle part of the suite; 9 — generalized REE-pattern of the basement rocks.

of chlorite and serpentine upon olivine are uncommon. The globular textures are the result of the presence of more leucocratic globules, striae and cross-bedding (Fig. 5.20); they may have been caused by liquid immiscibility (Smolkin, 1992).

The rocks are very rich in LREE and incompatible elements (Zn, Ba, Sr, U, P, F) and resemble Phanerozoic intra-plate igneous rocks (Fig. 5.18) of the oceanic-island and continental-rift types. The chilled zones of Ni-bearing intrusions, representing the intrusive equivalents of these ferropicrites, show similar types of REE spectra (Table 5.10). In addition to differences in their REE spectra, the ferropicrites and ferrobasalts are quite different from the associated tholeiitic basalts in their alkalinity and their higher incompatible element content, thus confirming their independent sources of magma generation. Thus, here we are probably dealing with one of the earliest complex geodynamic settings

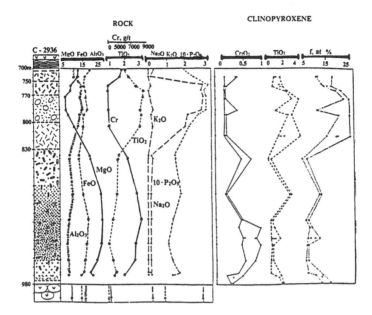

FIGURE 5.19 Cross-section of a layered ferropicrite lava flow from the Pilgujarvi Formation (after Smolkin, 1992).

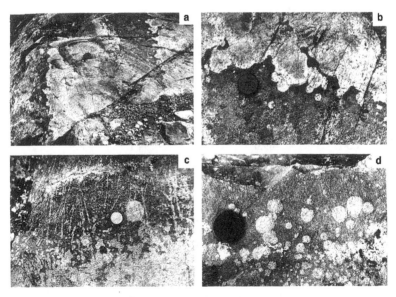

FIGURE 5.20 Liquid immicibility textures in Pechenga ferropicrite. Photo by V.F. Smolkin.

Table 5.10 Composition of chilled ferropicrite and metapyroxenite (Smolkin, 1992).

Components	1748/1	Ca-243	Ca-244	Ca-245
SiO_2	38.90	37.80	38.60	38.10
TiO_2	1.98	2.34	2.12	1.72
Al_2O_3	8.70	9.27	8.14	6.26
FeO	15.18	13.76	14.53	15.43
MnO	0.19	0.23	0.22	0.15
MgO	18.60	17.40	17.50	17.90
CaO	7.21	7.68	8.29	7.90
Na_2O	0.13	0.08	0.04	0.07
K_2O	0.01	0.05	0.03	0.02
P_2O_5	0.24	0.25	0.22	0.18
Fe_2O_3	1.81	3.39	4.78	6.52
FeO	11.60	10.79	10.56	10.28
H_2O^+	5.97	6.08	5.94	5.45
CO_2	0.18	0.10	0.06	0.09
S	170	2580	8890	17300
Cs	0.43	1.21	0.663	1.12
Rb	4.64	7.1	6.8	7.1
Sr	10	30	30	20
Ba	26.2	44.6	40.6	42.3
Th	1.91	2.29	1.78	1.52
Sc	34.4	44.1	38.3	32.9
Sb	0.143	0.062	0.058	0.058
Zr	130	140	120	110
U	0.355	0.935	0.48	0.497
Ta	1.46	1.32	1.12	0.892
Au	—	0.0038	0.0033	0.00435
V	310	420	380	310
Cr	1010	1260	1460	2620
Cu	90	220	270	660
Zn	—	51	49	51
Ni	661	487	815	1700
Co	89.0	75.5	94.1	129
La	20	25	18.5	13.8
Ce	42	51	38	27
Nd	28	32	25	20
Sm	5.4	7.5	6.2	4.6
Eu	3.4	2.4	1.90	1.82
Tb	0.76	0.86	0.69	0.56
Yb	1.50	1.38	1.23	0.90
Lu	0.18	0.15	0.12	0.12

1748/1 — lower chilled zone of layered lava flow, Lammas Lake; Ca-243–245 lower chilled zone of the unmineralized northern Kotselvaara gabbro–wehrlite intrusion (Ca-243 0–5 cm from the contact Ca-245 140–150 cm from the contact). Fe_2O_3, FeO, H_2O^+, CO_2 — from chemical analysis data.

with intra-plate magmatism superimposed on a back-arc area, characterized by MORB-type magmatism.

Many U–Pb, Pb–Pb, Sm–Nd and Re–Os isotopic dates were obtained on ferro-picrites and nickel-bearing intrusions in the Pechenga area (Hanski et al., 1990; Smolkin, 1992; Hanski, 1991). The results of dates obtained at different labora-tories are normally close to each other, falling in the range 1990–1970 Ma, with an initial $^{87}Sr/^{86}Sr$ ratio of 0.70303 and $\varepsilon_{Nd(T)} = +1.6 \pm 0.4$.

The Pb whole-rock isochron passes through the most primitive Pb-isotope composition of sulphide composition of sulphide ($^{206}Pb/^{204}Pb = 15.0$; $^{207}Pb/^{204}Pb = 15$) and yields the isotope composition of primary Pb, showing a rela-tively low $^{207}Pb/^{204}Pb$ ratio as compared with the mean terrestrial ratio for 2 Ga. These data suggest that the ferropicritic melt did not contaminate ancient radi-ogenic crustal Pb, and that the primary ratio $\varepsilon_{Nd(T)} = +1.6$ characterizes the iso-topic composition of the mantle source. The source evolved for a long time to become depleted in LREE and was LREE enriched for only a short time before melting. The model time of enrichment of depleted mantle (DM) is about 2.2 Ga ago when some components were supplied into it (Hanski et al., 1990). Judg-ing from the higher Ti, P and F, the enrichment may have been associated with mantle metasomatism. Thus, the first enrichment of the ancient DM of the Baltic Shield took place at the 2.2 Ga boundary, when the alkali basalts first appeared there.

Magmatism of the PVB culminated in the emplacement of porphyritic potassic granites of the Litsa–Araguba Group and leucogranites of the Strelna Group of age 1840 ± 50 Ma (Mitrofanov, 1995).

e) Geodynamics and Magmatism of the Baltic Shield in the Age Range 2.2–1.9 Ga

All data suggest that the MLT was a suture zone which could have been gener-ated by the collision of the Kola and Karelian–Belomorian cratons, and the sub-duction of the latter beneath the former in a northern to north-east direction (Kratz et al., 1978; Sharkov and Smolkin, 1997). Both the formation of MLT and maximum extension of the PVB appear to have taken place practically simulta-neously c. 2.0–1.9 Ga ago. In both of the structures, some of the igneous rocks are similar. The MLT and PVB were deformed and partly broken into minor structures during the upwelling of granitic gneiss domes, sometimes with granite cores, emplaced at the end of the period of geodynamic activity, when tectonic processes (thrusting, folding and fracturing) also played an important role.

It is important to note that, starting from 2.0 Ga, the Svecofennian ocean began to open in central Finland and Sweden (Marker, 1985; Gaal and Gorbatschev, 1987). The upper section of the Svecofennides here contains tur-bidites and slabs of typical ophiolites 1.97 Ga in age (Kontinen, 1987).

The Svecofennian stage in the Kola–Karelian region started to develop c. 2.2 Ga ago after a break following the end of Sumian–Sariolian activity. In early Jatulian times, moderately alkaline Fe–Ti basalts, typical of continental rifts, were erupted in continental subaerial environments. This type of magmatism was characteristic of the whole of the eastern Baltic Shield, where, in addition to the above-mentioned volcanic grabens, dyke swarms of similar composition were also widely distributed (Ein, 1984; Aro and Laitakari, 1987; Mitrofanov, 1989; Vuollo et al., 1995). This implies that a vast active magmatic region, probably associated with the ascent of a major asthenospheric plume, existed here at the time. Judging from the essentially increased thickness of volcanics in the PVB, the rise of the top of the plume reached its maximum in the region; this probably predetermined the formation of a spreading zone here. Then the lithosphere was thinned, and a marine basin with submarine outflows of MORB and Fe–Ti basalts appeared. These data suggest the extension of continental crust and the emergence of ocean-type crust here, with volcanic islands and seamounts.

Geological and geochronological data indicate that the time was related chronologically to the initiation of subduction along the MLT and the occurrence some time later in the Southern Pechenga Compression Zone of subaerial calc-alkaline magmatism. The further closure of the Pechenga–Varzuga back-arc basin led to a considerable reorganization of the region's geological structure, which evolved during tangential movements, accompanied by uprising of the gneissic–granitic domes, thrusting, folding and fracturing. These processes has the effect of separating the Kola geoblock into structural domains such as the Murmansk, Central Kola, Pechenga-Varzuga, etc., mentioned above (Fig. 5.1), which are characteristic of the present-day structure of this part of the Baltic Shield, and which differ greatly from the previous stage.

The initial dimensions and morphology of the interacting crustal blocks and the back-arc basin, as well as other parameters of collision and back-arc spreading, are still speculative. However, data currently available suggest that the geodynamic processes as a whole resemble those of the Alpine–Himalayan Belt (see Chapter 2). The situation with the Arabian plate, which is subducting beneath the Eurasian Plate along the Periarabian Suture zone, appears to be the closest equivalent. Subduction forms the Anatolia–Caucasus–Elburs andesitic–latitic volcanic arc with the Black and Caspian back-arc seas. At the back of the plate, the Red Sea is opening as a new ocean. A similar situation may have occurred in the eastern part of the Baltic Shield some 2.0–1.9 Ga ago: i.e. Svecofennian ocean–Karelian–Belomorian plate–MLT–LPGB–the southern Pechenga andesitic arc and the Pechenga–Varzuga marine basin (basins?) with submarine volcanism. In the geodynamic context, the succession may be interpreted as follows: oceanic spreading centre–continental subducting plate–suture zone–continental wedge–back-arc basin (basins?). So, the available geological evidence of plate tectonics on the Baltic Shield appeared only from this particular time (Sharkov and Smolkin, 1997).

Within the stabilized Karelian craton, magmatism preserved its intra-plate character (Fig. 5.1), and the geodynamic conditions were consistent with continental rifting. Many narrow graben-like structures were the sites of the outpouring of essentially Fe–Ti picrites and basalts, with normal, moderate and increased alkalinity (Golubev and Svetov, 1983), including occurrences of alkali-ultrabasic magmatism with carbonatites (Akhmedov et al., 1992). Dyke swarms of similar composition were formed simultaneously in basement rocks between the graben-like structures.

5.5.2 Analogous Magmatism in Other Shields

a) The Aldan Shield

In general terms, a similar situation took place at that time in the Aldan Shield, in connection with the Udokan stage of tectonomagmatic activity. The huge Dzhudzhur–Stanovoi Belt, incorporating large anorthosite–mangerite–charnockite intrusions, namely, the Kalar, the Upper Undytkan and the Dzhudzhur, here extends along the boundary of the Early Proterozoic Stanovoi Zone and the ancient Aldan granulite–gneiss terrane (Bogatikov, 1979; Rundquist and Mitrofanov, 1993) (Fig. 5.21). The then Aldan Shield represents a stabilized region where dyke-like intrusions of hornblende titaniferous basic and ultrabasic rocks, with increased alkalinity, developed in the area of southern Yakutia (Buldakov and Kotova, 1991).

The Stanovoi Belt represented a zone of at least two stages of high-pressure metamorphism of age c. 2.2 Ga and 1.95 Ga respectively (Rosen et al., 1994), at the rear of which the large Udokan subaerial sedimentary basin was formed, infilled with red beds of the Udokan Group (2.2 Ga old) containing rich copper deposits (Berezhnaya et al., 1988). According to Rosen et al. (1994), these rocks were deformed and metamorphosed at about 1.95 Ga during the collision of the Olekma and Aldan terranes, and intruded with post-tectonic granites 1.8 Ga ago.

Magmatism here was less intense than in the Baltic Shield and was manifested mainly as intrusions. Dykes of titaniferous picrite, forming thin (0.25 to 25 m) bodies, occur in the Udokan Group and in the Archaean Olekma granite–greenstone terrane (Pukhtel and Zhuravlev, 1992). They are noted for higher TiO_2 concentrations (on average 2 wt.%) and their high enrichment in the most incompatible elements (Zr and LREE). According to Pukhtel and Zhuravlev (1992), the nature of the variations in the petrogenetically important components (Al, Y, V and LREE) points to the leading role of garnet fractionation in the formation of the parental magmas of picrite. An isochron Sm–Nd age of 2202 ± 41 Ma gives the time of dyke emplacement. A positive value of $\varepsilon_{Nd(T)} = +1.6$, along with strong enrichment in the most incompatible elements, provides evidence of the supply of an enriched component, probably induced by mantle metasomatism, into a long-lived depleted mantle only a short time before partial melting.

Almost coeval high-titanium igneous rocks, of increased alkalinity, also occur within the Chiney layered intrusion, cutting through the red sandstone of the Udokan Formation, which was formed 2.2–2.1 Ga ago (Konnikov, 1986). The sandstone proper to some extent was derived from volcanic material, which locally shows some relics of tuffaceous and ignimbritic textures (Sochava, 1986).

Dyke-like bodies of tiniferous hornblende–peridotites and gabbroids, 2.0 Ga in age, occur also in the Sino-Korean Shield (Zimin, 1973).

As in the Baltic Shield, the large anorthosite intrusions are widespread within the high-pressure metamorphic terrane of the Stanovoi Belt (Fig. 5.21). One of them, the Kalar intrusion, represents a coarsely layered body, about 7–8 km thick and 150 km long, concordant with the structure of the country rocks (Sharkov, 1981). Figure 5.22 shows a typical cross-section of the intrusion. The lower two-thirds of the section is made up of anorthosite and gabbronorite–anorthosite; above this comes norite, two-pyroxene diorite (jotunite) and monzonite (mangerite). The upper section is composed of intermediate and acid rocks (quartz-mangerite and charnokite). After its formation the intrusion was sheared and deformed, and therefore its primary textures and minerals have been preserved

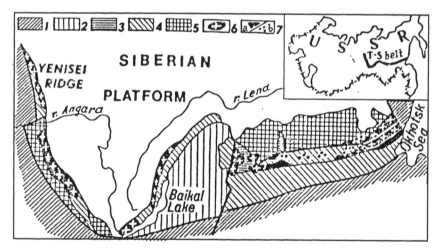

FIGURE 5.21 Index map showing Precambrian basement outcrops along the southern margin of the Siberian Platform (after Larin and Neymark, 1992). 1 — Palaeozoic fold belts; 2 — late Proterozoic (Riphean)–early Palaeozoic fold belts; 3 — early Proterozoic aulacogens; 4 — early Proterozoic fold belts; 5 — Archaean basement; 6 — early to middle Proterozoic acid igneous assemblage; 7 — anorthosite (a — large intrusions: 1 — Kalar, 2 — Upper Undytkan, 3 — Dzhugdzhur; b — small intrusions and dykes). T-S belt — Proterozoic Trans-Siberian belt of acid magmatism.

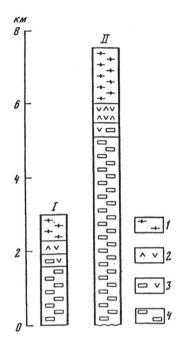

FIGURE 5.22 Typical section of (I) the Kolvitsa (Kola Peninsula) and (II) the Kalar (Eastern Siberia) anorthosite–magnetite intrusions (after Sharkov, 1981). Key: 1 — charnockite; 2 — jotunite; 3 — norite and gabbronorite; 4 — anorthosite and gabbronorite-anorthosite.

only as relics. Like all similar intrusions, during shearing the Kalar intrusion experienced high-pressure zonal metamorphism, from amphibolite facies at the bottom to granulite facies at the top. According to Biryukov *et al.* (1991), the P–T parameters were as high as T = 785–810°C and P = 9–12.5 kbar.

The Sm–Nd whole rock ages of the anorthosite intrusions are in the range 2.22–2.4 Ga, and the mineral ages are 1.7–2.0 Ga (Sukhanov *et al.*, 1991).

b) Canadian Shield

Here the first evidence of ocean opening (Purtuniq ophiolite) is dated to about 2 Ga ago (Helmstaedt and Scott, 1992) and evidence of subduction was found in rocks 1.9–1.7 Ga in age; it was the time of formation of the Penokean, Trans-Hudson, Wormay and Ungava orogens, which extend into Greenland (Hoffman, 1988; Barovich *et al.*, 1989; Hegler *et al.*, 1989; Lucas *et al.*, 1992; etc.).

Dyke-like bodies of tiniferous hornblende–peridotites and gabbroids 2.0 Ga in age occur in Canadian Shield (Pactung, 1987).

However, the events of particular interest took place much later, i.e. 1.3– 1.1 Ga ago; in character, they are almost identical with those in the eastern Baltic Shield, in the Main Lapland Thrust and its back-arc zone 2.2–1.9 Ga ago. Their development started with the formation of the gigantic Mackenzie dyke swarm, 1270 ± 4 Ma ago, and associated large Muscox layered intrusion, 1270 ± 4 Ma ago (Halls, 1992). The dyke swarm stretches for 2,400 km. The Coppermine River flood basalts are probably the effusive equivalents of these dykes (Baragar and Ernst, 1992). The formation of the dykes and volcanics took about 5 Ma. In composition, they vary from moderately alkaline Fe–Ti basalts to continental tholeiites; as a whole, they are fairly similar to the early Jatulian rocks of the Pechenga–Varzuga Belt.

In the period between 1.2 and 1.1 Ga ago, when the large Grenville Belt of zonal high-pressure metamorphism, similar to that within the Lapland Trust, with associated anorthosite–charnockite intrusions (Isachsen, 1969), was formed along the south-eastern boundary of the Lake Superior Province. Unlike the MLT, the Grenville Belt occupied a much larger area because of its subhorizontal dip. As in the MLT and Stanovoi Belts, the metamorphic grade of the rocks increases up section of the shear zone, i.e. towards the boundary with the Archaean Lake Superior Province. Isotopic U–Pb zircon ages of 1.64, 1.36 and 1.16–1.13 Ga have been obtained for the igneous activity (Emslie and Hunt, 1989). The bulk of the anorthosite emplacement is associated with the latest, youngest period, almost coinciding with the maximum of metamorphism (1.1 Ga ago).

The Adirondack intrusion, comprising anorthosite and associated syenite, mangerite (monzonite), monzodiorite (jotunite) and charnockite, is situated in the southern Grenville Belt (Ashwal, 1993). Ample data indicate that the intrusion is a sheet, 3.5–4 km thick, emplaced in the high-grade rocks of the Grenville Group. A transition to norite and gabbronorite is observed in the upper part of the anorthosite body. As in other similar anorthosite intrusions, acid and intermediate rocks occur higher up the section. The intrusion underwent deformation and metamorphism to the granulite facies; as a result, the rocks are now crystalline schists and gneisses. Metamorphism reached a maximum shortly after emplacement of the intrusions, i.e. only 0.05–0.10 Ga later (Morrison and Valley, 1988). The U–Pb ages fall in the ranges 1.125–1.135 Ma for anorthosites and 1.125–1.140 Ma for the mangerites and charnockites, i.e. more or less the same (McLelland and Chiarenselli, 1990).

Study of oxygen, strontium, lead and neodymium isotopic data in the rocks of the Adirondack Complex and the similar Morin Complex showed that the $^{87}Sr/^{86}Sr$ ratio in the basic rocks varies from 0.7029 to 0.7060, $^{207}Pb/^{204}Pb =$ 15.482 to 15.527 and $\varepsilon_{Nd(T)} = +5.5$ to 0: in intermediate rocks (jotunites) the values are 0.7040–0.7077, 15.524–15.578, $\varepsilon_{Nd(T)} = +0.7$ to -2.2, and in charnockites 0.7053–0.7090, 15.463–15.514, $\varepsilon_{Nd(T)} = +1.1$ to -0.8, respectively (Duchesne *et al.*, 1985) whereas the ^{18}O value is about 2.6‰ higher as compared with that of common gabbroids (Morrison and Valley, 1988). The data imply that anorthositic magma underwent essential crustal contamination at great depths, whereas the mangerite–charnockite magma series resulted from the melting of crust at depth, probably due to anorthosite magma injection. As a whole, the composition of rocks of the anorthosite–charnockite complexes is close to the composition of the upper parts of the early Palaeoproterozoic layered intrusions, where anorthosites, pyroxene diorites and granophires play an important rule. The main difference between their compositions is the presence of potassium feldspar in mangerites and charnockites, which is typical for magmatic rocks of compression environments. From this it follows that the parental magmas of the "autonomous" anorthosite complexes could be analogues of the A-type magmas of the layered intrusions, in other words, the leucocratic varieties of the SHMS.

A large spreading province with widely developed basalt magmatism appeared at the same time at the rear of the Grenville Belt (Halls and Fahrig, 1987). A trough developed and Keweenawan volcanosedimentary strata, up to 15 km thick, made up of sedimentary red beds at the bottom and at the top and an extremely thick sequence of basalt volcanics in the middle part, were formed at this time in the Lake Superior area. Dyke swarms of similar composition (Abitibi, Keweenawan and others), with an age of 1.2–1.1 Ga, probably represent the feeders of lava plateaux, and are also common here. In composition, all of basalts, along with the previous ones, vary from moderately alkali-olivine Fe–Ti basalts to typical continental tholeiites. There was a tendency for a transition, with time, from basalts rich in Fe, Ti, P and alkalies to varieties richer in Al, Mg and Ca (Halls and Fahrig, 1987); this suggests the progressive depletion of the source. In composition, the basalts resemble those of the Columbia plateau, thus suggesting a similar origin and weak crustal contamination. This magmatism led to the emplacement of the gigantic layered Duluth Lopolith which covers $>5,000 \, km^2$ with a thickness of 4–5 km, and whose age is $1.099 \pm 0.5 \, Ma$ (Miller, 1995). Instead of the early Palaeoproterozoic layered intrusions, $Pl + Ol$ cumulates play an important role in this intrusion.

The Dahlsland Belt in the south-western Baltic Shield has been recognized as an easterly continuation of the Grenville Belt (Gorbatschev and Bogdanova, 1993). Similar anorthosite–charnockite complexes here occur in the Bergen Arc area: the Rogaland and Bjerkreim–Sohdal complexes (about 1 Ga old) from southern Norway have received the most attention (Ashwal, 1993). Studies of Sr, Pb, and Nd isotopes showed that all the rocks of the series, from anorthosite to charnockite, underwent contamination by lower-crustal material at depth, and the degree of contamination increased from basic to acid rocks (Duchesne *et al.*, 1985). The pattern as a whole is identical to that in the rocks of the Grenville Belt.

c) Ukrainian Shield

Here titaniferous hornblende rocks (peridotite, pyroxenite and gabbro) make up the dyke-like bodies of the Devladov complex, 2.0–1.9 Ga old (Gonshakova,

1973), which are quite close to analogous rocks of the Baltic Shield. The alkaline rocks of the K–Na series that form the dyke-like bodies of the Chernigov Complex (phlogopite–peridotite and pyroxenite, nepheline–syenite and carbonatite), with an age of 1920 ± 80 Ma, also appeared at the same time to the north of the Sea of Azov (Scherbak *et al.*, 1989).

d) South Africa

Evidence of a subduction zone 2 Ga old was recently found in central Tanzania (Moller *et al.*, 1995). The Proterozoic Usagaran belt rims the eastern margin of the Archaean Tanzania craton and encloses a 35 km long belt of amphbolied eclogites intercalated with layers of metapelites, semipelites, and rare marbles. They were interpreted as remnants of the Early Proterozoic subduction zone.

In South Africa, there is the Palabora complex of alkaline–ultrabasic rocks with carbonatites of 1.9 Ga of age, which is similar to the above-mentioned bodies both in age and rock composition.

Thus, high-titanium, moderate- and high-alkaline rocks of the K–Na series appeared on a large scale almost simultaneously world-wide, at about 2.2–2.0 Ga ago. These age estimates also apply to the first appearance of geochemically rich matter in the upper mantle.

Intense processes of lithosphere plate collision, accompanied by the development of an unusual, anorthosite–mangerite (charnockite) type of magmatism and the formation of back-arc basins with the oceanic- and continental-type crust, and the formation of oceans proper, are also very typical of the age range 2.2–1.9 Ga. These phenomena are characteristic not only of the Laurasian but also of the Gondwanana continents. According to Master (1992), a large active magmatic belt, with Andean-type volcanicity and back-arc sea, and oceanic crust, appeared in Equatorial and South Africa, and in the neighbouring regions of South America 2.2-2.0 Ga ago. They recurred in Grenville times, 1.3–1.1 Ga ago, in northern Laurasia, and were also accompanied by the formation of back-arc basins with tholeiitic, alkali-basaltic and alkaline magmatism; gigantic dyke swarms and large intrusions of nepheline–syenites could occur within their basement inliers, similar to the Ilimaussaq Intrusion in the Gardar Province of southern Greenland (Upton and Emeleus, 1987).

5.6 ACID POTASSIC MAGMATISM OF THE PERIOD OF CRATONIZATION OF ANCIENT PLATFORMS

Unlike the Early Proterozoic, when basic and — less commonly — intermediate and acid melts played a leading role, the middle Proterozoic was dominated by acid magmatism, predominantly of the potassic affinity, with an age of 1.8–1.4 Ga. This type of magmatism developed at the sites of closed palaeo-oceans and fold belts, completing their development; it stretched for long distances within the Baltic, Ukrainian, Greenland and Canadian shields, along the southern part of the Siberian Platform, in the Sino-Korean, African, Indian and Brazilian shields, and in western Antarctica (Khain and Bozhko, 1988). Huge intrusions of anorthosite–rapakivi granite (ARG) represent the most characteristic igneous rock assemblage of that period. (Ramo and Haapala, 1996).

On the central part of the Baltic Shield, after closure of the Svecofennian ocean c.1.9 Ga ago and cratonization of the Svecofennian orogen, a huge area of acid magmatism developed, which included the Trans-Scandinavian Belt of acid volcanism and large AGR intrusions (Gorbatschev and Bogdanova, 1993). Combined with the AGR massifs of the Ukrainian Shield, they build up a vast belt, over 2,000 km long, located on the western margin of the Russian Platform (Fig. 5.23). As a rule, the ARG intrusions occur within the Svecofennian fold belt, and only the Salmi massif is located at the point of contact of the Svecofennides and the ancient Karelian Craton. Ukrainian Zapakiviez located within the early Proterozoic formations. The emplacement of dyke swarms of diabase and quartz porphyry, locally combined within a single body, where the acid rock form the central parts of a basic dyke, occurred at the same time.

The plutons of this rock assemblage have well-defined intrusive contacts with the Early Proterozoic and Archaean metamorphic country rocks; however, they did not undergo regional metamorphism. Potassic, essentially acid volcanics are closely associated with them. They infill gently plunging troughs on the surface of rapakivi plutons, generally overlying them with angular unconformity (Velikoslavinsky et al., 1978). The study of Pb–Nd–Sr systems yielded an age range of 1750–1540 Ma for the evolution of most rapakivi and acid volcanic rocks in the Baltic and the Ukrainian shields (Bogatikov et al., 1988; Scherbak et al., 1989; Suominen, 1991; Neymark et al., 1994). This was followed by cratonization in the regions, accompanied by a transition to the platform regime. The sole exception is the western Baltic Shield, where the Ragunda Intrusion is 1320 ± 30 Ma old, and dykes of high-titanium olivine quartz-bearing diabases are 1260 ± 10 Ma old (Suominen, 1991).

Intrusions of anorthosite–rapakivi granite are formed by a series from ultramafic rocks to granites, including all the transitional varieties of the monzonite, quartz monzonite and diorite types. They form large bodies, irregular-oval in plan view, ranging from a few tens to hundreds of thousands of square kilometres in area. Data on the size of the plutons at the present-day erosional surface are given in Table 5.11; it is also evident from Table 5.11 that the plutons are dominated by granite composition. Basic rocks, mostly gabbronorite–anorthosites, are uncommon in the Viborg Intrusion and are entirely absent in places.

Most geologists believe that the rocks in question represent a single series and that they are related by common generation. This is confirmed by the presence of transitional rock varieties and their regular change across the area and

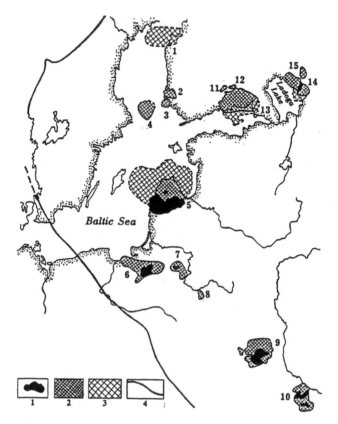

FIGURE 5.23 Distribution of anorthosite–rapakivi–granite assemblage on the Eastern European Platform (after Velikoslavinsky *et al.*, 1978). 1 — gabbro-anorthosite; 2 — rapakivi–granite; 3 — the same, suggested by geophysical data; 4 — western boundary of the Eastern European Platform; plutons (shown by numbers): 1 — Bothnian, 2 — Laigir, 3 — Vokhmaas, 4 — Allan, 5 — Riga, 6 and 7 — Polish, 8 — Belorussian, 9 — Korosten, 10 — Korsun–Novomirgorodsky, 11 — Ahvenisto, 12 — Suomenniemi, 13 — Viborg, 14 — Salmi, 15 — Ulyaleg.

within a section, and the presence of zones of rhythmic layering, including that in granites. According to the current prevailing hypothesis, the intrusions are multiphase bodies, and intermediate rocks (monzonite and quartz monzonite) are usually regarded as hybrid rocks, which originated though the reaction of the granite magma with earlier basic rocks.

a) The Viborg (Wiborg) Pluton

Rapakivi granites ("rotten stone" in Finnish) have received the most study within the Viborg Pluton. According to Velikoslavinsky *et al.* (1978), the pluton is irregular-oval in shape and occupies an area of about $16,000\,km^2$ (Fig. 5.24). However, the total area of the pluton increases to $30,000\,km^2$, taking into consideration its southern part, hidden by the waters of the Gulf of Finland. Two associated intrusive bodies, namely, the Ahvenisto, which has a horseshoe-shaped flank, where rocks of the gabbro-anorthosite complex are present, and the Suomenniemi, wholly composed of rapakivi granite, share a common boundary with the pluton to the north. The age of the Vibrog Pluton is estimated as 1640 ± 40 Ma (Suominen, 1991).

The pluton cuts through the Early Proterozoic gneisses and schists. Geophysical data suggest that the pluton is sheet-like in form, and is composed of several separate intrusions of rapakivi granite, about 30 km in aggregate thickness (Elo and Korja, 1993). A body of high-density rocks, about 6 km thick, interpreted as a gabbro-anorthosite, is located at a depth of 10 km. Beneath the pluton, the Earth's crust is about 41 km thick, i.e. 6–20 km thinner than that to the west, north and south of the pluton.

The Viborg Pluton is believed to have been formed as the result of three intrusive phases. The rocks of the earliest phase are actually quartz-syenites ("lapee-granites"). Ovoid hornblende–biotitic rapakivi is classified as part of the second

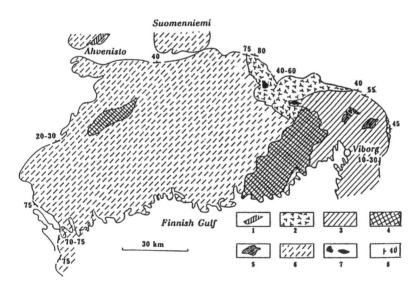

FIGURE 5.24 Sketch map showing the geological structure of the Viborg Pluton and its associated intrusive bodies (after Velikoslavinsky et al., 1978). Key: 1 — gabbronorite–anorthosite of the Ahvenisto massif; 2 — lapee-granite; 3 — ovoid rapakivi; 4 — small-ovoid rapakivi; 5 — biotite equigranular granites; 6 — non-articulated rapakivi (predominantly ovoid); 7 — volcanics, related to anorthosite-rapakivi complex; 8 — dipping of contacts.

Table 5.11 Major and trace element chemistry of the Salmi massif, after Neymark et al. (1994).

Sample	7-122	4-182	4-119	4-159	16710	16711	16709	8-210	8-165	41	2002	403-13
SiO_2	52.59	46.85	47.50	54.45	66.68	67.53	71.54	71.11	71.49	76.45	74.03	75.60
TiO_2	1.79	4.24	4.51	1.67	0.61	0.42	0.48	0.41	0.34	0.15	0.12	0.07
Al_2O_3	21.80	11.60	13.30	16.80	13.02	14.81	12.34	13.70	14.50	12.65	13.55	13.60
Fe_2O_3	1.52	0.38	0.78	1.14	1.34	1.02	1.14	1.49	1.05	0.71	1.40	0.38
FeO	6.14	18.56	14.15	8.78	5.33	3.11	3.22	2.43	2.18	0.85	0.90	1.79
MnO	0.08	0.26	0.23	0.14	0.09	0.06	0.06	0.07	0.06	0.04	0.02	0.03
MgO	0.82	3.40	4.13	1.44	0.33	0.23	<0.20	0.28	0.40	0.26	0.33	0.22
CaO	9.15	8.35	8.39	6.08	2.43	1.69	1.52	1.51	1.56	0.97	0.75	0.97
Na_2O	4.11	2.43	2.75	3.85	2.73	2.67	3.46	3.05	3.15	3.00	3.25	2.23
K_2O	0.89	1.05	1.10	2.77	5.44	6.64	4.39	5.21	5.78	3.50	4.71	4.45
P_2O_5	0.63	1.87	1.72	0.63	0.06	0.06	0.04	0.07	0.07	<0.03	<0.03	<0.03
F	0.05	0.11	0.14	0.05	0.06	0.04	0.09	0.30	0.14	0.21	0.38	1.10
LOI	0.75	1.07	1.15	1.44	<0.50	<0.50	<0.50	1.10	0.78	0.51	0.99	1.19
H_2O^-	0.09	0.18	<0.01	<0.01	—	—	—	0.11	0.13	<0.01	0.15	0.14
Total	100.41	100.35	99.85	99.24	98.12	98.28	98.28	100.84	101.49	99.30	100.58	101.77
$-O+F_2$	0.02	0.05	0.06	0.02	0.03	0.02	0.04	0.13	0.06	0.09	0.16	0.46
Total	100.39	100.30	99.79	99.22	98.09	98.26	98.24	100.71	101.43	99.21	100.42	101.31
Rb	8	35	25	6	174	138	132	171	140	238	377	893
Li	9	—	19	9	—	—	—	42	28	42	80	686
Sr	576	549	339	396	140	168	128	119	124	34	11	42
Y	26	37	44	59	81	47	40	—	51	59	143	—

Zr	41	901	397	363	1044	653	375	621	453	400	255	59
Nb	19	32	36	33	73	37	20	—	33	84	80	—
Pb	<5	12	6	16	25	25	24	—	26	—	38	—
Th	<5	<5	<5	7	8	<5	<5	—	22	16	31	—
Ba	434	4610	452	195	19.73	2530	1057	—	1063	—	302	—
Ni	84	7	22	10	45	27	23	—	15	—	84	—
Co	10	12	29	26	<10	<10	<10	—	30	<10	<10	—
V	46	41	153	172	14	13	11	—	<30	—	<30	—
La	23.3	—	57.7	—	—	56.7	—	—	153	95.6	—	30.0
Ce	57.5	—	132	—	—	120	—	—	318	215	—	45.1
Nd	33.0	—	75.6	—	—	61.3	—	—	111	85.8	—	39.8
Sm	6.68	—	14.7	—	—	11.8	—	—	16.9	17.4	—	8.02
Eu	3.14	—	3.41	—	—	3.04	—	—	1.91	0.62	—	0.126
Gd	6.13	—	12.9	—	—	9.85	—	—	12.4	15.2	—	4.23
Dy	4.67	—	9.66	—	—	8.23	—	—	10.4	16.1	—	5.83
Er	2.34	—	4.95	—	—	4.46	—	—	5.66	9.45	—	4.77
Yb	1.91	—	4.31	—	—	4.12	—	—	5.06	8.91	—	8.13
Lu	—	—	—	—	—	0.635	—	—	—	—	—	1.20
$(Eu/Eu^*)_N$	1.50	—	0.74	—	—	0.86	—	—	0.40	0.12	—	0.07
$(La/Yb)_N$	8.24	—	9.05	—	—	9.28	—	—	20.4	7.25	—	2.49

Samples: 7-122 — anorthosite, 4-182 — olivine gabbronorite, 4-119 — gabbronorite, 4-159 — monzonite, 8-210 — pyterlite, 8-165 — wiborgite, 41 — coarse-grained biotite granite, 2002 — fine-grained biotite porphyry granite, 403-13 — coarse-grained albite–lithian siderophyllite granite, 16709 — coarse-grained amphibole–biotite granite, 16710, 16711 — quartz syenites.

intrusive phase of the pluton, accounting for about 76% of the area studied. These rocks are actually standard rapakivi. They are noted for the presence of large (2–4 cm) oval crystals (ovoids) of Na–K feldspar, generally concentrically zoned. In many cases, the ovoids are mantled with oligoclase (Fig. 5.25). Rapakivi with and mantled and unmantled ovoids are called viborgite and pyterlite, respectively. Schlieren layering, caused by alternating layers rich in or almost devoid of ovoids, is common in these rocks. The trachytoid rare-ovoid rapakivi is classified as the third phase.

The origin of the ovoids of orthoclase presents a particular petrological problem. It is well known that the rounded form of crystals is a disequilibrium form and cannot have appeared during magma crystallization. Moreover, the distinct corrosion and solution by these crystals of surrounding earlier cumulus phases, and the presence of the latter within ovoids, are common. This contradicts the popular hypothesis of the ovoids as phenocrysts (Velikoslavinsky *et al.*, 1978). Rather they could have been formed by the crystallization of intercumulus melt in a crystallization zone that has not yet entirely solidified, whereas the origin of the oligoclase rims (marginal structures) is caused by the elevation of partial water pressure in the residual liquid and by the development of an immiscibility field in the system Or–Ab–An–Q (Kravtsova, 1992). Corrosion of more

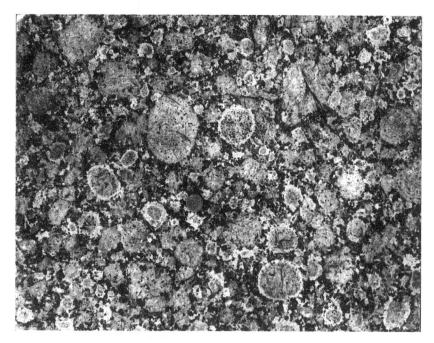

FIGURE 5.25 Typical structure of rapakivi granites, Viborg pluton. Photo by D.A. Velikoslavinsky.

high-temperature cumulus phases by intercumulus material is a relatively common phenomenon in cumulates, as is the rounded form of the oikocrysts. But the presence of similar ovoids is locally observed in the direct contact-metamorphic zone of the rapakivi, in the host rocks and xenoliths. This suggests that the fluid phase played an essential role (partly penetrating into the host rocks) in ovoid formation. Quite other theories have been presented for the origin of mantled alkali feldspar megacrysts, including resorbtion, crystallization from a high-viscosity melt, exsolution, magma mixing, and stability changes related to pressure decrease in the ascending water-undersaturated magma (Ramo and Haapala, 1996). Hybridization or other changes in composition (accumulation of crystals in compositional stratified magma chambers, or in a system of chambers of different depths and rapid decrease of pressure as a result of crystallizing magma transfer from deep-seated chambers to more shallow levels) can also explain the porphyric texture and corrosion of alkali feldspar.

b) The Korosten Pluton

This Ukrainian pluton is an irregular body, almost rectangular in shape and about $12,000 \, km^2$ in area (Velikoslavinsky et al., 1978). It is situated in the centre of a large ring-shaped Proterozoic depression, recognized from Landsat imagery. The rapakivi cuts through the gneiss and migmatite of the early Proterozoic Teterev Formation (2410 ± 20 Ma in age: Scherbak et al., 1989), and the ancient granitoids of the Kirovograd–Zhitomir and Osnitsa complexes. The intrusive contacts are accompanied by hornfelsization of the host rocks, the development of numerous apophyses of rapakivi granites and associated alkali metasomatites. A sketch map of the geological structure (Fig. 5.26) shows that the host rocks are locally encountered as large xenoliths (?) or erosional windows in the pluton. In the northern part, the pluton is overlain by the volcanosedimentary rocks of the Ovruch Formation, 1650 Ma old (Scherbak et al., 1989).

Basic rocks, anorthosites and gabbro–anorthosites, occur mostly in the southern and central parts of the pluton as five "massifs" or "blocks", accounting for about 25% of the area of the complex. Medium-grained gabbronorite, occurring in the marginal parts of the "massifs", is subordinate. Biotite–hornblende fine-ovoidal rapakivi granite is the common granite phase in the Korosten Pluton. Coarse-granited biotite–hornblende ovoidal granites (viborgite and pyterlite), similar to those in Fennoscandia, are less common. Chambered pegmatites with piezoquartz are found among them. Gradational contacts between them can be seen in many places (Lichak, 1983). Geophysical data indicate that the alternation of thick (0.5 to 3 km) sheets of basic rocks and granites can be traced to a depth of about 20 km and over (Orovetsky, 1990). A deep structure of this pluton resembles the Viborg massif's one: it is a huge cone-like transcrustal magmatic system, which can be traced to depths of 35–40 km; it is located above a mantle

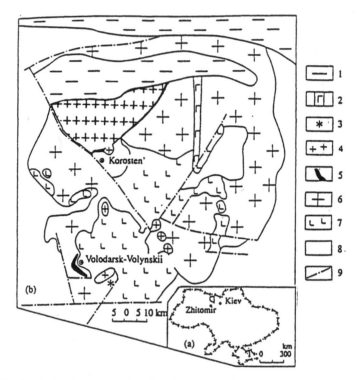

FIGURE 5.26 Sketch map showing the geological structure of the Korosten Pluton. After Kovalenko *et al.* (1996b). Key: 1 — platform cover (in the north part of the pluton — volcanosedimentary rocks of the Ovruch Formation); 2 — gabbro-diabase dyke; 3 — dykes of quartz-microcline-albite granite with zinnwaldite; 4 — small-ovoid grey rapakivi; 5 — field of Volhynia chamber pegmatites; 6 — undifferentiated rapakivi (mostly pink ovoid rapakivi); 7 — gabbronorite-anorthosite and anorthosite; 8 — Palaeoproterozoic gneisse, migmatites and granites.

high (Orovetsky, 1990). Numerous dykes of titaniferous diabase, quartz-porphyry and granite-porphyry are associated with the faults.

The Volodarsk–Volynsky "massif" is the largest basic body, about 1,250 km^2 in area and 2–3 km thick. Its central part is composed of dark, iridescent coarse-grained anorthosite and gabbronorite–anorthosite. They pass into a zone of rhythmically interlayered gabbronorite and monzonite from the centre of periphery in plan view and up section. They also contain layers of pyroxenite (hypersthenite) and ferruginous harzburgite, ranging from a few centimetres to 10 m in thickness, and veins of the same rock types. The layered zone reaches up to 1,000–1,300 m in thickness. Due to the layering, locally it can be observed that the rocks of the intrusion are deformed into gentle folds in places (Fig. 5.27).

SW   NE

FIGURE 5.27

FIGURE 5.27 Structure of the layered complex of the Volodarsk Volynsk "massif" in the Irsha River area. After A.A. Polkanov. 1 — layered gabbronorite; 2 — interbeds and schlieren of anorthosite.

As a rule, rocks of the anorthosite complex plunge gently beneath the enclosing rapakivi granite. It is not unusual to observe a particular type of vertical and horizontal zoning: for example monzonite changing into quartz-monzonite and diorite; above this come fine-ovoidal, green and grey rapakivi; grey, rare-ovoidal rapakivi; pink, large-ovoidal biotite–hornblende rapakivi, and, lastly, equigranular biotite–granite (Lichak, 1983). Large bodies of chambered topaz-bearing pegmatites with piezoquartz are common in the rapakivi granite of the Korosten Complex.

However, the relationships between the basic and acid rocks are ambiguous. On the one hand, the bodies display a conformable layering of granites and basic rocks, accompanied by the widespread occurrence of rhythmically interlayered transitional varieties of monzonite, quartz-monzonite, quartz-diorite, and fine-ovoidal rapakivi types. On the other hand, it is not unusual to observe that the granites cross-cut the rocks of the "anorthosite complex", resulting in the formation of typical intrusive contacts. Separate varieties of granite can also cross-cut each other, often in a fixed order: grey rapakivi is cut by pink rapakivi which, in turn, is cut by equigranular biotite–granite. This order of intrusion has served as a basis for the recognition of the intrusive phases of granitoids, i.e. I, II and III.

Thus, rather contradictory relationships exist between all of the rock types of the Korosten Pluton. On the one hand, they are related to each other via transitional varieties and zones of rhythmic interlayering and, in general, look like a huge layered intrusion. On the other hand, they can have intrusive contacts: in this case they cross-cut each other in the usual set order. This type of intrusive structure is probably caused by: the multiphase formation of the pluton; the supply of fresh magma to the crystallizing magma chamber; and changes in its shape, caused by movements in the country rocks. This is also suggested by the above-mentioned deformation of layering within the gabbroids.

Geophysical data indicate that the Korosten Pluton, like the Viborg Pluton, shows a deep gravity low and represented a large transcrustal anomaly, shaped like an inverted truncated cone, with the base lying at a depth of 40 km (Orovetsky, 1990). The upper part of the cone is composed of alternating sub-horizontal layers of granite and ahorthosite, complicated by large xenoliths of the metamorphic country rocks. The development of norite and gabbronorite is believed to have taken place at higher levels, and the formation of mantle ultrabasic rocks, forming a scarp beneath the pluton, is inferred to have occurred at lower levels. Thus, the deep-seated structure of the Korosten Pluton has a lot in common with that of the Viborg Pluton, suggesting a similar mechanism for the formation of these unusual complexes.

c) The Salmi (Pitkyaranta)

This pluton is situated on the eastern shore of Lake Ladoga (southern Karelia). The relatively small Ulyaleg and the Lodeinoe Pole intrusions of rapakivi, which are both satellite bodies of the Salmi Pluton, are located to the north and south of it. The pluton extends over 125 km north-westward; it is about 5,000 km² in

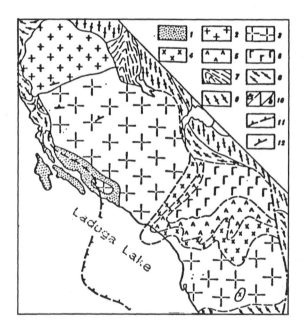

FIGURE 5.28 Sketch map showing the geological structure of the Salmi Pluton. From Sharkov (1983a). 1 — volcanosedimentary rocks of the Salmi Formation; 2 — biotite granite; 3 — pink ovoid biotite–hornblende granite; 4 — grey small ovoid rapakivi, quartz monzonite and diorite; 5 — monzonite; 6 — gabbronorite–anorthosite; 7 — Jatulian volcanosedimentary rocks; 8 — early Proterozoic amphibolites, schists and quartzites; 9 — Archaean granite–gneiss and migmatite; 10 — boundaries between rock varieties (a — proven, b — inferred); 11 — limit of the pluton as suggested by geophysical data; 12 — the orientation of trachytoidic textures.

overall area. Like the Korosten Pluton, the Salmi Intrusion displays a well-defined zoning in the distribution pattern of its main rock varieties (Fig. 5.28). Its northern part is made up of equigranular biotite–granite, giving way further south to pink, ovoidal granite (viborgite), and then to grey, fine-ovoidal granite, replaced by quartz-diorite, quartz-monzonite and monzonite. Dark gabbro-norite–anorthosite and anorthosite, containing thin layers of norite and, rarely, bronzitite, occur in the southern part of the intrusion, as suggested by drilling logs. Albite–lithium leucocratic granites and greisens, making up small dyke-like bodies, are known to occur within the pluton and enclosing metamorphic rocks; the granites and greisens are related to the final phase of rapakivi emplacement (Velikoslavinsky et al., 1978; Sviridenko et al., 1984).

Data on the internal structure of the intrusion are less certain. It is interpreted as being roughly sheet-like in form. By analogy with the Korosten Pluton, horizontal zoning, as observed from the distribution pattern of the major rock types, is thought to be a vertical zoning as well, as shown by the change, in ascending order, of gabbronorite–anorthosite to norite and, further up, to monzonite, quartz-monzonite and quartz-diorite, grey, fine-ovoidal rapakivi, pink rapakivi and biotite–granite. Using the above-mentioned data, the general morphology of the body, which displays a wedge-shaped outcrop of basic rocks in the SW part and a granitoid protrusion (which is conformable with the outcrop and plunges beneath the Lake Ladoge Basin), suggests the deformation of the body into a very gentle anticlinal fold with its hinge plunging south-westwards (Sharkov, 1983).

The Salmi Intrusion is one of the youngest bodies of this type: its U/Pb zircon age is 1540 ± 10 Ma (Suominen, 1991). However, detailed isotopic data (Neymark et al., 1994) show that a U/Pb apatite age is 1563 ± 9 Ma for the gabbro–anorthosite and a Sm–Nd internal isochron is of 1552 ± 69 Ma. Zircons from the rapakivi fall on the U/Pb age concordia of 1543 ± 8 Ma; this is in agreement with the data of Suominen. The Rb/Sr internal eochron in the granite has given an age of 1455 ± 17 Ma, probably showing the time of completion of post-magmatic processes on the emplacement of the batholith. The data suggest a long break, of about 20 Ma, between the emplacement of basic and acid rocks; however, this is not in line with the geological and petrological data, pointing to the presence of a whole range of intermediate rocks and zones of rhythmic inter-layering (Sharkov, 1983).

d) The Riga Pluton

This has been described by Bogatikov and Birkis (Bogatikov, 1979), and is located in western Latvia. The pluton, which is completely overlain by a platform cover, 900–1,800 m thick, has been recognized from drilling records. The pluton is irregular-oval in shape. Its northern part is formed of rapakivi granite. In addition to rapakivi, drilling logs indicate the presence of granosyenite, quartz-syenite, quartz-monzonite and monzonite to the south. Abundant basic rocks (predominantly anorthosite and norite–anorthosite, in places supplemented with norite and gabbronorite containing thin layers of troctolite and plagioclase olivinite), appear in the southern part of the pluton. As in the Korosten pluton, the rocks form separate sublatitunidal bodies (blocks). The bodies are 100–200 km^2 in size, with the largest of them (Priekule) reaching 1,000 km^2 in area.

5.6.1 Geochemistry and Isotopic Geochemistry of Anorthosite–Rapakivi Granite Complexes

In their geochemistry, the rapakivi granites show a high iron content, a predominance of potassium over sodium, and a relatively high alumina content (Table 5.11). The granites are noted for their elevated Zn, Pb, Mo and Zr content; an increase in Be, Sn, Y, Nb, Rb, F and W concentrations is noted in the late differentiates and the albitized varieties, and Li and U in albitized granites only. Decreased concentrations of V and, to a lesser extent, Cr, are typical of these rocks (Velikoslavinsky et al., 1978).

Similar rocks from different plutons vary in composition. For example, the rocks of the Riga Pluton show the highest magnesian and calcium content and the lowest TiO$_2$ and Fe$_2$O$_3$ concentration (Velikoslavinsky et al., 1978). The rocks of the Salmi Pluton show an opposite tendency. The Ukrainian intrusions are intermediate in composition. The Salmi Pluton and, in particular, the Ukrainian plutons are characterized by their predominant concentrations of titanium in the rocks and the high titanium content of the ore oxide phase; a predominance of iron over titanium is noted in the basic rocks of the Riga Pluton. The high TiO$_2$ concentrations in the rocks are always accompanied by increased P$_2$O$_5$ content. The potassium and iron content in the rocks increases concomitantly in all of the anorthosite complexes. In some cases, the basic rocks of the anorthosite–rapakivi granite assemblages show a deficiency in Mn, Ni, V, Cu and Y and some excess Ce, Sn, Mo and La, and, locally, P, Ti and Pb (Velikoslavinsky et al., 1978).

The distribution of rare elements in the major rock varieties of the Salmi Pluton is illustrated in Table 5.11. The elements have been classified into two groups. The first group includes elements (Mn, Cr, Ga, Mo, Zn and La) whose concentrations are similar in both the basic and acid rocks; and the second group includes elements that vary in concentration. For example, P, Ti, Co, Ba and Sr, and Sc concentrations increase and the Pb, Nb, Sn, Y, Yb and Cu decrease from granites to basic rocks.

Figure 5.29 shows the REE distribution of these rocks. Unlike the rapakivi of Finland, the late lithium–fluoric granites of the Salmi Intrusion show a Eu anomaly, a decrease in LREE and an increase in HREE. The general evolutionary trend of the REE was probably determined by the fractionation of feldspar,

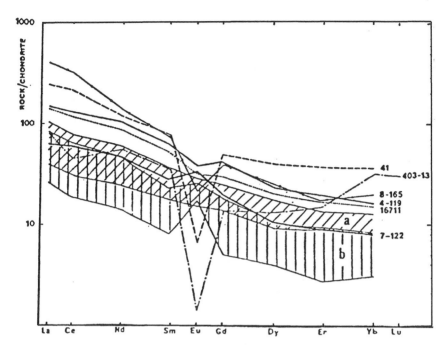

FIGURE 5.29 Character of chondrite-normalized (Taylor and McLennon, 1985) REE spectra in the rocks of the Salmi Pluton (after Neymark *et al.*, 1994). 7-122 anorthosite; 4-119 gabbronorite, 16711 quartz-syenite; 8-165 coarse-grained viborgite; 41 equigranular biotite-granite; 403-13 Li-F granite. Stippled areas show south-western (a) south-eastern (b) gabbro−anorthosites of similar intrusions in Finland (Ramo, 1991).

clinopyroxene and hornblende, as well as accessory zircon and fluorite (HREE concentrations) and allanite (LREE concentrator) (Neymark *et al.*, 1994).

In their geochemical parameters, the rapakivi granites and, in particular, their late fractionates as a whole, are very similar to Phanerozoic lithium–fluoric granites and ongonites (Kovalenko, 1987). This is also suggested by the results of a study of melt inclusions in topazes from the chambered pegmatites of the Korosten Intrusion (Kovalenko *et al.*, 1996a). They are noted for: their fluorine content (5 wt.%), which is high even for ongonite, with concomitant high contents of Al, alkalies and chlorine; extremely low concentrations of ferromagnesian components; and a low agpaitic coefficient. The water content of the melt is estimated to have been 9 wt.%. This suggests, even taking into account the fact that the pegmatite melt is an end-product of differentiation, that the rapakivi belongs with the lithium–fluoric granites.

Ramo (1991) conducted extensive isotope-geochemical studies of the rapakivi and associated rocks of the Baltic Shield. He showed that the granites have $\varepsilon_{Nd(T)}$, varying from -3.1 to -0.2, and the T_{DM} model ages (apart from the topaz-bearing granites) average 2.06 Ga. The Pb isotopic composition of alkali feldspar is very close to those obtained for the Svecofennian crust, thus suggesting that it was a main source for the Finland rapakivi. Modelling of the origin of the Suominiemi granites, performed by Ramo using major and minor elements, suggests a felsic (about 73 wt.% SiO_2) parental magma, which could have been generated with about 20% melting of an granodiorite source. Calculations show that the further evolution of parental magma was probably controlled by the fractionation of alkali feldspar, quartz, mafic minerals and Fe–Ti oxides (in the ratio $68:15:15:2$), with minor apatite, zircon and allanite. The resulting suite of rocks varied from cumulates (hornblende–granites and biotite–hornblende granites) to the crystallization products of the residual liquids (biotite–granite, topaz-bearing granite).

Detailed isotopic studies were also made on the Salmi Pluton, which is located at the boundary of the Svecofennian terrane and the Archaean Karelian craton (Ramo, 1991; Neymark et al., 1994). According to these data, the granites represent a $1:1$ mix of Proterozoic and Archaean crustal material with Archaean lower crust, which was low in radiogenic Pb. Gabbro–anorthosites and granites that occur here have similar primary compositions of Nd, Sr and Pb isotopes in their feldspars ($\varepsilon_{Nd} = -6.5$ to -8.2; $\mu_2 = 8.6$ to 8.9; $k_2 = 3.9$ to 4.0; $(^{87}Sr/^{86}Sr)_0 = 0.7052$ to 0.7057 in basic rocks, and $\varepsilon_{Nd} = -6.2$ to -8.9; $\mu_2 = 8.4$ to 9.2; $k_2 = 4.0$ to 4.4; $(^{87}Sr/^{86}Sr)_0 = 0.7050$ to 0.7072 in the granites). The two-stage neodymium T_{DM} model ages for both of these rock types fall in the range 2.60–2.80 Ga. An ancient LREE-enriched source, with its U/Pb and Rb/Sr ratios poorly integrated in time and an increased Th/U ratio, was involved in the formation of melts that were sources of material for both the basic rocks and the granites. According to the above authors, the data for the basic rocks cannot be explained merely by crustal contamination. Selective assimilation of Pb, Sr and Nd from the Archaean lower crust is required; what is more, the parental melt of the gabbro–anorthosites was derived from an isotopically anomalous subcontinental mantle source. Fusion of the granitic magma mainly resulted from the anatexis of lower-crustal material in the process of mantle diapir rising. This is suggested, in particular, by the finding of inherited zircons with increased Th/U ratios, which were separated from ovoids of orthoclase (Neymark et al., 1994).

Diabase dykes crystallized from evolving Fe–Ti tholeiitic basalts, typical of continental terranes (Ramo, 1991). They are moderately rich in LREE; $\varepsilon_{Nd(T)}$ varies from -1.2 to $+1.6$. The composition of the isotopes coincides with a growth curve for intermediate crustal Pb and yields a secondary isochron with an age of 1.85 Ga, which exceeds the age of actual crystallization by about 200 Ma.

Thus, the geological, petrological and geophysical data indicate that the plutons of the assemblage in question are unique geological bodies, both in their composition, dominated by material from the underlying crust, and in the structure of the magmatic system itself. They probably represent a series of large, multiphase, layered intrusions of basic, intermediate and acid rocks. Data currently available from isotopic dating suggest the possibility of extremely prolonged (up to 20 Ma) existence of the magmatic systems, which evolved uniformly at the same site. In this sense they resemble the Andean-type batholiths, whose formation was rather prolonged (see Chapter 2).

Haapala (1977) described a topaz-bearing quartz-porphyry in south-western Finland, which is similar in composition to the ongonites of Mongolia (Kovalenko, 1987). The rocks are genetically related to the youngest shallow-depth granite complex, associated with the Laitila rapakivi intrusion, with an age of 1570 Ma. As stated above, similar Li–F granites are also associated with other intrusions of rapakivi.

The presence of late Li–F granites and the geochemical specialization of the rapakivi (high K, Rb, Nb, Y, Zr, F, Sn, Be and Li, and the general character of the REE spectra, with the exception of Eu) indicate that the granites are similar to A-type subalkali-granites (Neymark et al., 1994). According to the tectono-magmatic classification, proposed by Pearce et al. (1984), the granites lie in the field of "within-plate" granites.

5.6.2 Similar Processes in Other Shields

The enormous Elsonian volcano–plutonic belt of significantly acid magmatism was formed on the south-eastern and the southern margins of North America in the age range 1.7–1.3 Ga. It extends from Labrador to California for a distance of about 6,000 km and its width is about 1,000 km (Gover and Owen, 1984; Khain and Bozhko, 1988). Large bodies of Kiglapait-type anorthosites, Nain-type anorthosite–diorites and rapakivi granites are its main components. Virtually undeformed rhyolites, felsites, dacites and their tuffs, with an age of 1.4–1.3 Ga, were described over a distance of 1,700 km, from Texas to western Ohio. The formation of most of the plutons lies in the age range 1.5–1.3 Ga (Gover et al., 1990).

Similar magmatism took place 1.9–1.7 Ga ago, also in the southern part of the Siberian Platform; it resulted in the formation of the gigantic Trans-Siberian belt, stretching for 3,000 km from the Sea of Okhotsk to the Yenisei Mountains (Larin and Neymark, 1992) (Fig. 5.21). The belt is made up of rapakivi granites, subalkali granites, anorthosites and, for the most part, acid volcanic rocks. The U/Pb zircon ages increase from east to west: from 1.7 to 1.91 Ga, with 1.7 Ga being the predominant value. The geochemical signature of the acid rocks (high K, Rb, Li, Be, Sn, W, Nb, Ta, Zr, REE and F) indicates that they are typical A-granites and lie in the field of "within-plate" granites on Pearce's diagram. Igneous rocks are localized within various Proterozoic structures (fold belts, remobilized blocks of Archaean basement, etc.). The rocks of the belt show negative $\varepsilon_{Nd(T)}$, varying from -0.2 to -9.2, and the T_{DM} model age falls in the range 2.2–2.6 Ga. These isotopic data vary markedly as compared with similar bodies of the Eastern European and the North American platforms and suggest the variable contribution of ancient crustal protolith to magma generation.

The belt was formed following the collision processes at the 2.2–2.1 Ga stage; this may have led to the rejuvenation of mineral isochrons of the adjacent earlier anorthosite–charnokite intrusions of the Dzhugdzhur–Stanovoi Complex and, hence, they were included in the later Trans-Siberian Belt by Larin and Neymark (1992).

Similar sialic belts, with an age of 1.5–1.8 Ga (mostly 1.7 Ga) for the most part, are extremely characteristic of all the Precambrian shields (Khain and Bozhko, 1988). They are widely developed in the Sino-Korean, the South American (Rio Negro–Guruena), the African shields, both in its northern and southern parts, the Australian craton (Gouler Range), and the Antarctic Shield.

Thus, middle Proterozoic sialic belts occur world-wide; they appear to be peculiar to this period of the Earth's evolution. It is very important that their greatest development took place 1.7 Ga ago, when all of the newly formed oceans disappeared, all the ancient cratons closed up, and Pangaea 1, a single massif of continental crust, may have come into existence; Panthalassa, a prototype of the future Pacific, may have arisen simultaneously (Khain and Bozhko, 1988).

5.7 SUMMARY

1. Three stages, clearly differing in the character of their igneous and geo-dynamic processes, can be traced within the Early Proterozoic.
2. The earliest Palaeoproterozoic stage (2.5–2.2 Ga) is noted for a widespread development of differentiated picrite–basalt–andesite–rhyodacite series, very close to the Phanerozoic boninite series in their geochemistry. Their intrusive equivalents occur as: large layered intrusions of harzburgite–bronzitite–norite–anorthozite–diorites; small intrusions of similar rocks of the drusite complex in the White Sea area; major dyke swarms of predominantly norite–gabbronorite composition of Matachewan type; and volcanic complexes located in large graben-like structures of the Pechenga–Varzuga type.
3. Isotopic data suggest that early Palaeoproterozoic layered intrusions were formed due to very rapid evolution of magma system beneath them. The evolution is responsible for a change from high-magnesian melts, with mantle characteristics (U-type), to high-Al moderate- and low-Mg (A-type) melts with lower-crustal material playing an important part in petrogenesis.
4. The appearance of acid subalkaline, locally high-alkaline and low-titanium potassic-series igneous rocks of the miaskite type is noted locally along the boundaries of major tectonic zones.

5. Mantle sources from this stage are characterized by a different, but usually high degree of depletion. Crustal material is believed to have played a highly significant role in melt formation. As distinct from the Phanerozoic, this type of magmatism displays all the geological features of intra-plate rocks and forms large provinces within the ancient cratons.

6. There is no geological evidence of plate tectonics, although data currently available suggest the development of areas that are undergoing moderate extension, causing the emplacement of graben-like structures, large layered intrusions, and dyke swarms, as well as terranes undergoing compression, similar to the Lapland–Umba moderate-pressure granulite belt of the Baltic Shield. The situation as a whole may have resembled a plume-tectonic regime, similar to that on Venus at the present day.

7. The following stage, 2.2–1.9 Ga, is noted for the first appearance of Fe–Ti picrites and K–Na series basalts, with normal, moderate and high alkalinity; and geochemically enriched mantle sources. This stage witnessed a change from early Palaeoproterozoic plume tectonics to mainly linear structures that are associated with plate tectonics. The development of extensive lithospheric plate-collision zones and subduction phenomena, accompanied by back-arc spreading and the first appearance of typical ophiolites, MORB and intra-plate basalts of the Phanerozoic type, was pervasive. The development of belts of zonal high-pressure metamorphism and the emplacement of large bodies of "autonomous" anorthosite–mangerite–charnockite intrusions took place along the suture zones.

8. Geochemical and isotopic data indicate that the primary anorthositic melts contained a major portion of lower-crustal material. However, mangerites and charnockites, which are common in the complexes, suggest that sialic crust also played an active role in petrogenesis. These rocks are the probable ancient equivalents of calc-alkaline rocks which, in contrast to the early Palaeoproterozoic, had already been associated with subduction zones. This type of magmatism persisted to the end of the middle Proterozoic at 1.2–1.1 Ga ago (Grenville–Dahlsland Belt).

9. Basins floored by oceanic-type crust (for example, the Pechenga–Varzuga Zone in the Baltic Shield) or continental-type crust (Udokan in the Aldan Shield), emerged in the back-arc zones of the structures. It is interesting that the formation of the basins in general began earlier, in the early Palaeoproterozoic, thus probably predetermining the development of arc systems and back-arc basins here.

10. The third stage of development of endogenetic processes occurred from the end of the Early Proterozoic to the beginning of the middle Proterozoic, and reached a peak about 1.7–1.6 Ga ago. It was a time of: the ubiquitous closure of ancient intercratonic oceans; the development of orogenic belts at the sites of their closure; and the formation of huge mainly acid potassic-series volcanic belts, such as the Baltic–Ukrainian, Greenland, North American

and Trans-Siberian belts, etc. Large-scale anorthosite–rapakivi granite assemblages are the typical variety of igneous rock here; geochemically the assemblage resembles Phanerozoic lithium–fluoric granites. The plutons may be unique long-lived igneous systems that developed over mantle highs. According to geophysical data, these magmatic systems often represent huge cone-like transcrustal anomalies which can be traced to depths of 30–40 km.

It is important to note that such a type of sialic magmatism was evolved only at the places of former oceans; at adjacent stable parts of the platforms at this time, common continental rift zones or the CFB were forming.

11. The development of all these igneous rock assemblages could have been at least partly synchronous in time (the earlier assemblages continued to evolve in parallel with later ones, but not vice versa). The Proterozoic as a whole appears to represent the most important stage in the development of the upper layers of the Earth, both in the context of deep-seated petrogenesis and geodynamic activity.

CHAPTER 6
MAGMATISM AND GEODYNAMICS IN THE ARCHAEAN

E.V. SHARKOV and O.A. BOGATIKOV

The Archaean, which spans the time from the formation of the Earth to c. 2600 Ma ago, was a time of crucial importance in the formation of the Earth's crust and mantle. The most ancient dates on Archaean formations were obtained from detrital zircons from Archaean metasediments. For example, zircons with a minimum age of 4.28 Ga (Pidgeon *et al.*, 1990) were discovered in Western Australia in the Yilgarn Block. They are only 0.3 Ga younger than the most ancient zircons in the Solar System which were found in the Vaca Muetra meteorite (Ireland and Wlotzka, 1992). The ancient gneiss remnants from the Acasta Gneisses in Canada yield zircons dated at $3,962 \pm 0.003$ Ma (Bowring *et al.*, 1990); somewhat younger ages of 3.96 Ga have been obtained from tonalitic gneisses of the Slave Province, North-West Territories of Canada: 3.96 Ga (Bowring *et al.*, 1989a), and Amitsoq gneisses in Greenland: 3.82 Ga (Kinny, 1986), etc. This is an indirect argument in favour of the existence of very ancient sialic crust.

The upper boundary of the transition from an Archaean type of geodynamics and magmatism to a Proterozoic type remains the subject of debate. In the literature, the concept of the Archaean–Proterozoic boundary was developed using the example of the Canadian Shield, where a "major unconformity", dated as 2.5 Ga, exists between ancient, highly deformed granite–greenstone complexes and younger, usually less-deformed supracrustal complexes of the Huron supergroup and the Animac Group. A similar situation occurred in the Baltic and Aldan shields where the age of the Archaean–Proterozoic boundary is accepted to be 2.8–2.6 Ga (Semikhatov *et al.*, 1991). However, in the Kaapwaal Province of South Africa a similar unconformity exists between Archaean greenstone basement of the Swaziland and Pongola systems (Witwatersrand triad), with Proterozoic-type sedimentation occurring between 3.0 Ga and 2.8 Ga (Tankard *et al.*, 1982). In Australia the age of the analogous Mount Bruce Supergroup has been estimated as 2.75 Ga (Taylor and McLennan, 1985).

On the other hand, even within one region, for example the North American platform, this stabilization may have occurred for a longer period of time. Thus, in contrast to the Canadian Shield, a large Stillwater layered intrusion of Proterozoic type with an age of 2.7 Ga (DePaolo and Wasserburg, 1979; Lambert *et al.*, 1989, see Chapter 5) has been found in the Precambrian basement in the Rocky Mountains, which indicate that Archaean crustal stabilization occurred very irregularly and over a long time interval of about 300 Ma. In other words, the Late Archaean represents an area of partial overlap of Early Archaean and Early Proterozoic types of geodynamics and deep-seated petrogenesis.

In general, the Archaean is characterized by the high ductility of the upper part of the Earth's crust, which is related to widespread migmatization and, hence, extensive development of various types of ductile deformation, including almost all of the rock complexes except for the central parts of the greenstone belts. This may be explained by the predominance of basic and ultrabasic rocks, which were formed at temperatures significantly exceeding the solidus temperature of the surrounding granitoids. It is obvious that this crust could not have formed the vast topographic highs necessary for the development of a large body of sedimentary rocks, which may account for the relatively minor development of sedimentary rocks (Ronov, 1980), which would have required large areas of erosion for their formation.

Ophiolitic assemblages, which are inherent to the oceanic lithosphere, are unknown in the Archaean and so the existence of oceanic spreading zones at an early stage of the Earth's evolution would be highly problematical. A lack of spreading zones consistent with palaeomagnetic data suggests that, at that time, all Archaean cratons made up a single whole (Piper *et al.*, 1983).

Within the ancient shields (Fig. 6.1) two types of Archaean structural terrane have been distinguished: granite–greenstone and granulite–gneissic (Condie, 1994).

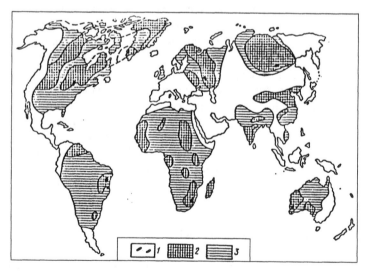

FIGURE 6.1 Archaean cratons and early Proterozoic mobile belts in cratonic basement. After Khain and Bozhko (1988). 1 — exposed areas of earliest (>3.5 Ga) crust (i.e. "grey gneisses"); 2 — Archaean cratons; 3 — early Proterozoic mobile belts.

They differ in their internal structure, their structural and metamorphic history, the ratio of exogenetic to endogenetic factors in the generation of crustal material and the type and form of the igneous activity and metallogeny. From the Early Archaean onwards this type of area can be divided into two coexisting geodynamic systems, of predominantly extensional type (granite–greenstone terranes, GGST) or predominantly compressional type (granulite–gneiss terranes, GGT) (Fig. 6.2). Their sedimentary sequences differ significantly: according to Taylor and McLennan (1985) volcaniclastic sediments and greywackes are characteristic of greenstone belts, as are chemical sediments, and also the thick coarse-grained

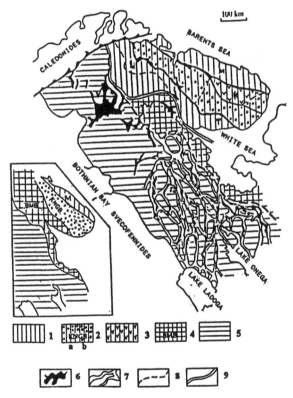

FIGURE 6.2 The main late Archaean structural provinces of the eastern Baltic Shield. Below: a generalized structural scheme. After Sharkov *et al.* (1998).
1 — Kola granulite–gneiss terrane (M — Murmansk block, L — Lotta domain, T — Tersk domain); 2 — Kola-Norwegian granulite belt; 3 — Keivy domain; 4 — Belomorian Mobile Belt; 5 — Karelian granite–greenstone terrane; 6 — fragments of greenstone belts; 7 — reconstructed greenstone belts in Karelia (see Fig. 6.3); 8 — original boundary between the KGGT and the BMB; 9 — the late Palaeoproterozoic Main Lapland Thrust.

and volcaniclastic turbidites overlying the volcanic series. At the same time, mature sediments such as quartzites, marbles and paraschists, (metapelites), all of which suggest formation in large sedimentary basins, are typical of granulite–gneiss terranes.

The relationships of these terranes can be seen on the Baltic Shield. There are four late Archaean (3.0–2.7 Ga ago) structural provinces here: large Karelian GGST, Kola-Norwegian GGT of moderate pressure, and a transitional zone of tectonic flowage (Belomorian Mobile Belt, BMB), made up from partly folded tectonic slices with lithology of adjacent Karelian GGST between them. In the north part of the Kola Peninsula the Murmansk block occurs — a fragment of GGT, divided from the Kola-Norwegian Granulite Belt (KNGB) by a regional fault. In its geological features it is rather close to the South Greenland Shield and, probably, was formerly a part of it (Mitrofanov, 1995).

6.1 BASEMENTGRANITE–GNEISS ("GREY GNEISS") ASSEMBLAGES

Archaean granite–gneiss areas are generally composed of 80–90% peculiar "grey gneisses", which are actually plagiogneisses of tonalite–trondhjemite–granodiorite (TTG) composition. On many ancient cratons their age is estimated as 3.96– 3.6 Ga. The origin of these "grey gneisses" has grown more controversial over the past decade, because it was established that even at the earliest, most reliably dated stages of formation of the Earth's crust, its composition included large volumes of sialic material. This raises the question about the Earth's development as compared with that of the Moon, where rocks from that time period can be correlated with the basic complexes (Ringwood, 1978; Bogatikov et al., 1984).

Summarizing the results of studies on this rock series, Barker (1979) emphasizes that trondhjemites predominate only in the Archaean grey gneiss complexes. The main mineralogical association of grey gneisses is oligoclase + quartz + biotite (K-feldspar and hornblende are minor phases). Archaean grey gneiss complexes have a distinctive geochemistry that does not recur in later formations (Kratz, 1978, 1981; Martin; 1994). In comparison with the younger tonalite–trondhjemite complexes, they are characterized by increased aluminium, magnesium, sodium, nickel, chromium, strontium and REE content and by decreased potassium, rubidium, radiogenic lead and uranium content (Table 6.1). Both mineralogical and chemical compositions indicate that TTG belong to the I-type granitoids of Chappel and White (1974).

According to Martin (1994), one of the more characteristic features of Archaean TTG is their REE patterns, which are strongly fractionated and generally display similar degrees of fractionation for both LREE and HREE. Generally, TTG do not have significant negative or positive Eu anomalies, but in some cases, positive Eu anomalies have been reported, and were interpreted as reflecting the fractionation of REE-rich accessory mineral phases (i.e. allanite).

Table 6.1 Chemical composition of the most ancient TTG-rocks.

Components	1	2	3	4	5	6
SiO_2	64.66	69.84	68.26	68.18	64.4	74.5
TiO_2	0.68	0.27	0.31	0.32	0.35	0.39
Al_2O_3	16.24	16.46	16.34	16.36	15.8	14.2
Fe_2O_3	1.67	0.96	1.04	1.04	1.18	0.36
FeO	2.99	1.48	1.97	2.05	1.79	1.92
MnO	0.06	0.04	0.04	0.05	0.04	0.05
MgO	1.81	0.83	1.17	1.17	1.14	0.45
CaO	2.68	2.45	3.03	3.15	3.37	2.43
Na_2O	3.85	4.86	4.28	4.40	4.68	4.08
K_2O	2.66	1.40	1.80	1.45	1.58	1.95
H_2O	0.14	0.02	0.02	0.12	0.54	0.37
LOI	2.09	1.56	1.22	1.13	0.11	0.03
Total	99.53	100.17	99.48	99.42	98.57	100.73
Rb	97	34	63	101	44	75
Sr	568	468	584	500	460	110
Ba	—	340	470	350	400	420
Zr	—	110	160	160	175	290
Cr	—	10	10	10	12	8
V	—	37	70	65	—	—
Ni	12.0	9.3	13.0	12.0	13	7
Rb/Sr	0.17	0.07	0.11	0.16	0.12	0.68
K/Rb	228	341	237	262	290	216

Columns 1–4 — Archaean oligoclase granites of the Koikar area (Karelian GGST) (Kratz, 1979); 5–6 — Archaean tonalites (Condie, 1981).

They are characterized by extremely low initial $^{87}Sr/^{86}Sr$ ratios (0.699–0.703); $\Sigma_{Nd(T)}$ values range from $+4$ to -3.

The most ancient "grey gneisses" often have a complex polymetamorphic history. As an example, for the Amitsoq gneisses of Western Greenland (Baadsgaard et al., 1976; Barker, 1979; Nutman et al., 1993) a combination of mineral parageneses from at least three stages of deformation and mineral formation, corresponding with ages of 3.87–3.625, 2.6 and 1.6 Ga, has been established. The greatest ages have been attributed to the TTG gneisses cross-cutting the rocks of the Isua supracrustal complex (layered amphibolites, gabbros, ultramafic rocks and minor iron-banded formation). For Early Archaean alone, at least three major phases of tectonomagmatic activity (at >3.7, 3.65 and 3.25 Ga), accompanied by additional injections of TTG, have been shown (Nutman et al., 1993). All this makes the reconstruction of the original lithology of the grey gneisses extremely difficult.

Most investigators tend to favour an initially magmatic (intrusive or volcanic) origin for the grey gneisses (Windley, 1979; Kratz, 1979, 1981; Bogdanova *et al.*, 1987; Condie, 1994; etc.). In some cases, when the grey gneisses are the least deformed (for example, in the Isua area), the polyphase body structure, the injection of later and lighter gneissic phases in homogeneous and darker diorite-gneisses has been revealed. Based on these attributes and on the presence of xenoliths of gabbroid and supracrustal rocks, the Amitsoq and Uyvak "grey gneisses" are considered to be syntectonic intrusions, whose crystallization proceeded under conditions that caused the development of primary gneissosity (Collerson and Bridgwater, 1979). Theories about the volcanic nature of the "grey gneisses" are based on the similarity of their sections over vast areas,

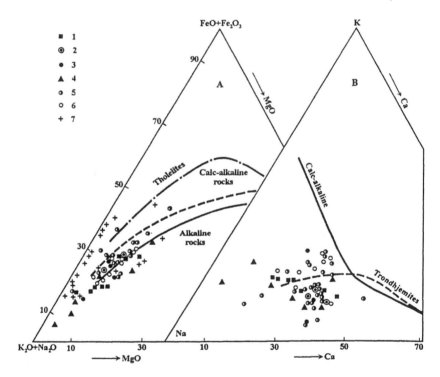

FIGURE 6.3 Location of tonalities and plagiogranites of granite–greenstone areas, on the diagrams ($K_2O + Na_2O$)–FeO–MgO (A) and Na–K–Ca (B). After Bogdanova *et al.* (1987). 1–7 — tonalites and plagiogranites with the age (in Ga): 1 — 3.4–3.2; 2 — 3.1–2.9; 3 — 2.8; 4 — 2.7 and less; 5 — tonalities of Karelia, with age of 3.4–3.1; 6 — tonalities and plagiogranites of Karelia, with ages of 2.9–2.8; 7 — earliest tonalite gneisses (with the ages more than of 3.5 Ga) from granulite–gneiss and granite–gneiss areas.

their great thickness, and quite often the alternation of acid and inter-mediate gneisses with amphibolites and sometimes with metasediments (Rosen et al., 1988).

The genetic and age inhomogeneity of the "grey gneiss" complexes is reflected in their variable petrographic composition. They include the metamorphic equivalents of quartz-diorites, granodiorites, tonalites, trondhjemites, granites and quartz monzonites (Fig. 6.3). The most potassium-rich varieties are often associated with zones of repeated migmatization, or with the injection of later K–Na melts.

6.2 MAGMATISM OF GRANITE–GREENSTONE TERRANES

The granite–greenstone terranes (GGST) of the Archaean have been less altered as compared with granulite-gnesis terranes, and therefore the majority of the igneous rocks of these areas have preserved their initial geological and petrological features (Fig. 6.4). The most ancient GGST were formed over the time interval 3.9–3.4 Ga (e.g. gneisses of the Slave Province of the Canadian Shield — 3.96 Ga, see Bowring et al., 1989; Amitsoq in Western Greenland — 3.82 Ga, Kinny, 1986; metavolcanics of the Pilbara Block, Australia — 3.7 Ga, Gruau et al., 1987; the Barberton belts of Southern Africa — 3.4 Ga, Gruau et al., 1988; Volotsky belts in Karelia, Baltic Shield — 3.5–3.4 Ga, Pukhtel et al., 1991; Sergeev et al., 1991; etc.). The younger (Late Archaean) greenstone belts are 2.9–2.7 Ga old (for example, the Abitibi, Cambalda and Belingwe belts, as well as most of the greenstone belts of the Baltic, Aldan and Brazil shields: Windley, 1992; de Wit and Ashwal, 1997).

In Archaean granite–greenstone areas, two major structural elements have been distinguished: greenstone belts (GSB) and granite–gneiss areas composed of mainly tonalite–trondhjemite–granodiorite gneisses (TTG) (Fig. 6.3). Many attempts have been made to classify greenstone belts; however, no clear-cut classification of these structures has been made as yet. In practice the concept of greenstone belt is applied to a combination of narrow, elongated structures infilled by supracrustal volcanosedimentary complexes, with predominantly basic volcanic rocks occurring among the large, often symmetrical, dome-shaped structures formed by tonalite–trondhjemite series rocks ("herds of domes"). In some belts acid volcanic rocks (leptites) predominate, and often they are associated with the largest banded-iron formations. Most greenstone belts include thick terrigenous clastic units, including few or no volcanic, auto-clastic, or pyroclastic rocks. Metamorphism of granite–greenstone terrane rocks usually did not exceed amphibolite-facies conditions, decreasing to greenschist facies in the central parts of the greenstone belts (de Wit and Ashwal, 1997).

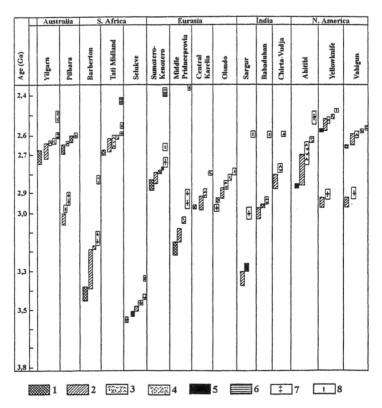

FIGURE 6.4 Schematic succession and stratigraphy of the igneous formation of Archaean greenstone belts. After Bogdanova *et al.* (1987). 1–8 — formations: 1 — high-magnesian volcanic rocks; 2 — basic volcanics; 3 — andesites; 4 — acid pyroclastics and volcanics; 5 — dunite–harzburgites; 6 — layered basic–ultrabasic plutons; 7 — tonalite–trondhjemites; 8 — granitoids and grandiorites; the relative dimensions of the columns, within the time interval for the development of each belt, approximately reflect the quantitative ratios of the assemblages in the section.

Most investigators consider greenstone belts to be extensional structures in Archaean sialic proto-crust. However, the magnitude of this extension and the nature of the basement of the greenstone troughs is still controversial, although the prevailing view is still that it was sialic in nature (Kroner, 1988). Greenstone belts of Archaean, and particularly Early Archaean, age are fairly uniform over relatively large areas (Condie, 1994). This suggests that either they represent the relics of originally unified large-scale sedimentary-volcanic structures that were disrupted in the process of granitoid dome formation, or that the geodynamic conditions under which they were formed were remarkably similar.

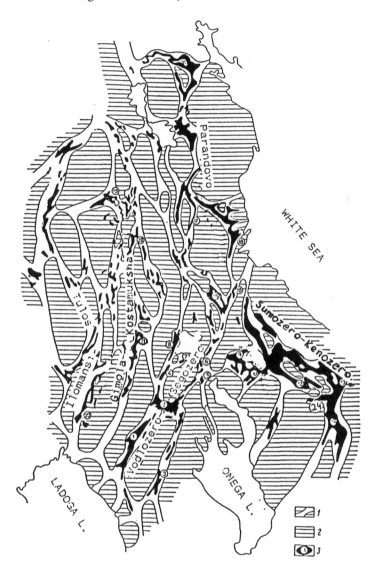

FIGURE 6.5 Schematic reconstruction of Archaean greenstone belts of the Karelian GGST. After Rybakov (1987). 1 — greenstone belts; 2 — basement blocks; 3 — local structures: 1 — Khautvaarsky, 2 — Koikar–Korbozyrsky, 3 — Kindasovo–Manginsky, 4 — Sovdozersky, 5 — Palaselginsky, 6 — Saisozersky, 7 — Bergaulsky, 8 — Luzhmozersky, 9 — Shilossky, 10 — Rybozersky, 11 — Kamennoozersky, 12 — Senegozersky, 13 — Kozhozersky, 14 — Tokshinsky, 15 — Parandovsko–Nadvoitsky, 16 — Vostochno–Idelsky, 17 — Tungudsky, 18 — Pebozersky, 19 — Kostomukshsky, 20 — Hedozero–Bolsheozersky, 21 — Gimolsky, 22 — Yalonvaarsky, 23 — Ilomansi.

Granite–gneiss areas occupy 70–90% of the granite–greenstone terranes and are composed of polygenetic granitoid complexes of varying ages. The majority of them appear to be formed by the tonalitic–trondhjemite gneisses. The rest is considered to be a component of the greenstone belts themselves, in which migmatization, metasomatism and secondary melting of earlier granitoids play an important role. As shown by Samsonov et al. (1993), using the example of the Middle Dnieprov'e area (the Ukrainian Shield), endogenetic activities within the GGST and the granite–gneissic areas nearly coincided in time.

The relationship of the greenstone belts with the surrounding gneiss– migmatite blocks of the infra-complex still remains uncertain. Most investigators believe these belts are younger as compared with the host tonalite matrix (Martin, 1994). However, according to other investigators (Glikson, 1984; Bogatikov et al., 1989), the nature of the distribution of mafic-ultramafite inclusions in Archaean gneisses at the base of the early greenstone belts enables us to consider these inclusions as relics of a once extensive mafic crust, possibly the relics of a primordial crust.

The Karelian GGST of the Baltic Shield, which contains recognizable Early and Late Archaean greenstone belts, is a typical example of an Archaean granite–greenstone area (Figs. 5.1 and 5.2). The Early Archean greenstone belts are exemplified by the Volotsky Belt in the Vodlozersky segment, which is 3.4 Ga in age (Pukhtel et al., 1991) and occurs among tonalitic gneisses 3.54 Ga old (Sergeev et al., 1990). However, the most widely occurring belts here are of late Archaean age (2.95–2.8 Ga: Sergeev et al., 1990), forming a network that is mostly reminiscent of a scattered spreading pattern (Fig. 6.5).

The TTG-series rocks in the vicinity of GGST are the same age or younger, and ancient cores are preserved only at a distance from the latter, between them (Chekulaev et al., 1997).

Greenstone belts are characterized by the following features:

1. A predominance of basic and ultrabasic rocks over volcanics, i.e. 55–80% of the total; intermediate rocks account for 5–40% and acid rocks for 3–15% of the total; however, in some cases the acid volcanics predominate;
2. A predominance of volcanic rocks over plutonic ones. The subvolcanic formations play an essential role here (for example, in the Middle Dnieprov'e, Ukrainian Shield). In some cases, as for NE Karelia, the abundances of plutonic rocks, particularly of acid rocks, are quite similar to those of the volcanics;
3. In the igneous rocks of greenstone belts, the following assemblages of volcanic rocks are represented — komatiite–basaltic, tholeiitic, andesitic and dacite–rhyolitic (leptitic), very rarely low-Ti potassic alkaline types. The komatiite–basalt series is most characteristic of Archaean greenstone belts. The andesitic and leptitic assemblages are subordinate; rocks of the tholeiitic and potassic alkaline assemblages (for example, the leucite-bearing volcanics of Kerkiland Lake (Abitibi Belt) (Condie, 1981)) are met in small amounts in Late Archaean belts;

4. Sedimentary infillings of the greenstone belts are represented by four principal petrologic suites: (a) volcaniclastic and pyroclastic rocks, (b) terrigenous sedimentary units, including mature quartzitic units and craton-derived detritus, (c) orthochemical deposits and (d) biogenic sedimentary rocks (Lowe, 1994). According to him, there are major differences in the sedimentary suites and inferred depositional environments of pre-3 and post-3 Ga greenstone belts: if older belts were formed in broad, flat, shallow-water conditions, in the younger belts sediments often accumulated in deep-water conditions.

The development of typical igneous assemblages in greenstone belts is shown in Fig. 6.4. As seen from the figure, the individual troughs differ significantly in both the completeness of their section and the degree to which the separate assemblages are displayed.

6.2.1 Komatiite–Basaltic Assemblage

The komatiite–basaltic assemblage includes komatiites (peridotite komatites) with a MgO content of $>$ wt. 24% (recounted per dry residue), komatiitic basalts: 18–24 wt.% MgO and basalts: $<$ wt. 18% MgO, and also basic and ultrabasic cumulates. In most cases this series is represented by an alternation of massive flows of ultrabasic and basic pillow lavas (Fig. 6.6), with well-defined chilled zones, slag crusts, and polygonal jointing. Pyroclastic formations are scarce. In some cases (e.g. the granite–greenstone terrane of Middle Dnieprov'e) the komatiites form sill-shaped bodies from a few metres to hundreds of metres thick. In the Barberton, Tati and Olondo belts the komatiitic assemblage accounts for about 25% of the section, whereas in the Sumozersky and Yellow-knife belts it is no more than 1% (Bogdanova et al., 1987).

The structures and textures of peridotite komatiites are highly specific. In the upper part of phenocryst-free lava flows, spinifex textures are formed by the rapid growth of minerals from the top inwards into the flow. Large (2–3 cm long and more) skeletal olivine and pyroxene crystals are typical. They combine to form parallel sheaf-like or radial clusters (Fig. 6.6) which occur in a matrix composed of the same dendritic accumulations of augite, skeletal chromite crystals and devitrified glass. Lava flows containing spinifex structures usually have a dual composition, where the rocks containing these structures occur at the top, and the cumulus varieties are developed at the bottom. In the Abitibi Belt these are 1–20 m thick and are evenly distributed without any notable change in thickness over areas of hundreds of metres to kilometres, which is characteristic of a low-viscosity of magma (Arndt, 1994).

Table 6.2 shows the average compositions of peridotite komatiites and komatiitic basalts from different regions (Bogdanova et al., 1987). They represent high-magnesian rocks with a low content of SiO_2, TiO_2 and alkalies, particularly

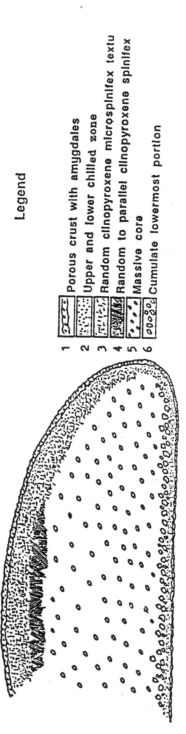

Legend

1 Porous crust with amygdales
2 Upper and lower chilled zone
3 Random clinopyroxene microspinifex textu
4 Random to parallel clinopyroxene spinifex
5 Massive core
6 Cumulate lowermost portion

0   10   20   30   40   50 cm

FIGURE 6.6 Schematic section across a large pillow from the Olondo Greenstone Belt komatiites. After Pukhtel *et al.* (1993).

Table 6.2 Average chemical composition of peridotite komatiites.

Components	1	2	3	4
SiO_2	45.97	45.95	45.26	49.55
TiO_2	0.34	0.38	0.30	0.41
Al_2O_3	2.98	7.41	6.20	7.69
Fe_2O_3	10.50	10.86	11.17	9.46
MnO	0.20	0.18	0.20	0.20
MgO	33.79	26.32	30.75	21.86
CaO	4.73	7.94	5.59	7.57
Na_2O	0.15	0.26	0.41	0.35
K_2O	0.03	0.05	0.04	0.09
P_2O_5	0.03	0.03	0.03	0.05
Number of analyses	24	31	23	16
Cr	2750	3250	3000	2500
Ni	1460	1050	1470	380
Co	170	93	99	91
V	120	160	120	120
Sc	22	20	17	38
Rb	1.8	3.2	3.0	1.2
Sr	33	38	22	43
Ba	11	7.0	5.0	—
Zr	23	18	15	25
Nb	2.0	1.2	1.0	5.0
Y	7.9	10	7.0	9.0
La	1.2	0.44	0.57	0.35
Ce	4.2	1.4	1.7	1.25
Nd	3.0	1.6	1.5	1.16
Sm	0.91	0.74	0.57	0.44
Eu	0.70	0.32	0.21	0.19
Gd	1.3	1.3	0.77	0.66
Dy	1.3	1.7	1.0	0.88
Er	0.75	1.1	0.65	0.59
Yb	0.76	1.0	0.64	0.60
Lu	0.086	0.098	0.098	0.093
Number of analyses	8	25	7	15

Columns 1–4 — greenstone belts of Australia: 1 — Barberton, South Africa, 2 — Abitibi, Canada, 3 — Norsmen–Uiluna, Western Australia, 4 — Olondo, Aldan. Original data of J. Arth *et al.*, N. Arndt *et al.*, S. Barnes, A. Besvik, B. Jahn *et al.*, A. Naldrett and A. Turner, R. Nesbitt, S. Sun, G. Stolts, M. Viljoen and R. Viljoen, Pukhtel *et al.*, etc. have been used for calculations of the average composition of the rocks.

K_2O. The CaO/Al_2O_3 ratio, with the exception of the Barberton Belt komatiites where it reached 1.5–2.5, is normally close to one, i.e., it differs very little from those of chondrites. Komatiites have a relatively high content of some incompatible elements (Ti, Zr, Y and Nb). With the fairly wide variations in the composition of these elements the ratios of pairs for some of them are constant and are close to those of average chondrites. This concerns Ti/Zr, Zr/Y, Ti/Y and Zr/Nb, etc. The peridotite komatiites are also distinguished by high Ni and Cr content. In accordance with the REE distribution pattern, the komatiites are divided into two types: the first are characterized by an unfractionated HREE distribution, whereas the second are strongly depleted in HREE. Two varieties are distinguished according to their LREE content: i.e. depleted and enriched in LREE. The rocks with a chondritic pattern of synchronous distribution of both heavy and light lanthanoids are relatively rare.

Like the preceding group of ultrabasic rocks, the basic volcanics occur mainly as flows and sills with a thickness ranging from a few metres to some hundreds of metres. Occasionally pyroclastics are also encountered. The pillow structure of the flows is highly characteristic. In the external "pillows" zone variolitic and vesicular structures have been recognized. In the more extensive flows a cumulus layering is quite commonly observed.

The high-Mg komatiitic basalts proved to be inhomogeneous, and when this group was mapped in the type region of the Barberton uplands they were subdivided into two varieties, which were later recognized in other regions. A distinguishing feature of both groups is the high CaO/Al_2O_3 ratio (>1), while in the third group of rocks it is less than 0.8. All three groups of rocks are noted for their high (10.5%) to very high (21 wt.%) MgO content, their low TiO_2 content and low LILE content, as well as their unfractionated distribution of HREE. Compositionally, komatiitic basalts are intermediate between the ultrabasic volcanics and the more common, moderately magnesian basalts, whose geochemical characteristics overlap with those of MORB tholeiites (Bogdanova et al., 1987; Arndt, 1994).

The geological and isotope-geochemical features of the komatiite assemblage can be considered using the example of the two greenstone belts of the Olekma granite–greenstone terrane of the Aldan shield: the Tungurcha and Olondo belts (Fig. 6.7) (Pukhtel et al., 1992, 1993b). The first, with an age of 3.4–3.3 Ga, is representative of the Early Archaean belts, whereas the second, with an age of 3.0 Ga, is assigned to the Late Archaean.

The Tungurcha greenstone belt (GSB) represents a fragmented belt, transversed by veined and layered bodies of tonalite–trondhjemite gneiss. Both complexes were strongly affected by the gneiss–plagiogranites, with an age of 2.82 ± 0.14 Ga, which surround the GSB. The age of tonalite–trondhjemite gneisses: 3.21 ± 0.08 Ga, is correlated with the time of tonalite magma intrusion into the upper crust.

In the section of the area studies, three units of komatiites, differing in their Al_2O_3 content (Tables 6.3 and 6.4), have been distinguished. The komatiites of the first and third units have similar concentrations of Al, whereas those of the

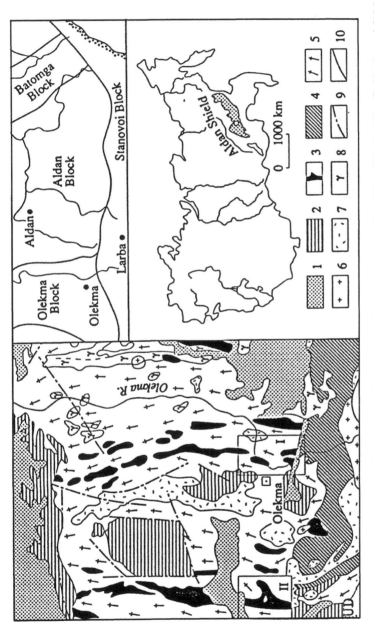

FIGURE 6.7 Schematic diagram of the geological structure of the central part of the Aldan Shield. After Pukhtel *et al.* (1993). 1 — deposits of Vendian–Palaeozoic–Mesozoic cover; 2 — early Proterozoic deposits (UD — Udokan Basin); 3 — Archaean greenstone belts: Tungurcha (1) and Olondo (11); 4 — Archaean schists of the Aldan Complex; 5 — early Archaean Olekma GGST. Intrusive formations: 6 — early Proterozoic syenites of the Tassky Complex; 7 — late Archaean granites; 8 — early Archaean granite–gneisses and subalkaline granites; 9 — major faults; 10 — geological boundaries.

Table 6.3 Major and rare-element content in basic and ultrabasic metavolcanics of the Tungurcha greenstone belt, Aldan Shield (Pukhtel et al., 1993).

Components	1	2	3	4	5	6	7	8
SiO_2	46.2	47.4	44.6	49.2	48.5	49.1	52.7	55.0
TiO_2	0.98	0.56	0.40	0.44	0.40	0.45	0.94	0.81
Al_2O_3	9.56	7.37	5.63	6.00	8.28	8.70	13.7	12.3
FeO^*	13.3	13.4	13.0	12.6	10.9	10.5	12.9	12.1
MnO	0.27	0.32	0.30	0.29	0.18	0.19	0.19	0.20
MgO	19.17	21.9	28.1	24.6	21.2	18.0	6.85	6.31
CaO	8.96	8.16	7.40	7.32	10.3	12.1	10.4	11.7
Na_2O	0.69	0.69	0.50	0.26	0.05	0.75	1.46	1.01
K_2O	0.27	0.07	0.02	0.05	0.12	0.12	0.68	0.37
P_2O_5	0.07	0.15	0.03	0.04	0.07	0.09	0.18	0.23
Total	100.00	100.00	100.00	100.00	100.00	100.00	100.00	100.00
LOI	3.39	2.74	7.26	2.96	3.01	1.13	1.37	1.24
Cr	2,630	2,150	2,515	3,010	2,380	1,810	200	490
V	230	160	120	130	150	165	265	225
Co	100	100	120	100	80	70	60	55
Ni	1,050	1,130	1,680	1,445	1,130	890	145	135
Zr	70	32	23	24	22	24	57	41
Y	15	11	9	9	0	10	21	19
Sr	14	17	30	23	18	9	119	139
Rb	0	2	0	1	1	0	31	11

Columns 1 — Komatiites of unit I; 2–4 — Komatiites of unit II; 5–6 — Komatiites of unit III; 7–8 — basalts.
Samples: 1 — 86156; 2 — 8697; 3 — 8698; 4 — 86171; 5 — 86198; 6 — 86195; 7 — 8696; 8 — 86105.

Table 6.4 REE content in basic and ultrabasic metavolcanics of the Tungurcha greenstone belt (Aldan Shield).

Elements	La	Ce	Nd	Sm	Eu	Gd	Dy	Er	Yb	Lu	$(Gd/Yb)_N$	$(Ce/Sm)_N$	Eu/Eu^*
86156	5.00	12.3	8.10	2.34	0.678	2.68	2.49	1.31	1.10	0.163	1.97	1.30	0.83
8697	1.41	4.36	3.79	1.30	0.374	1.74	1.95	1.18	1.09	1.165	1.29	0.81	0.76
8698	—	2.68	2.51	0.877	0.163	1.18	1.35	0.822	0.778	—	1.23	0.74	0.49
8699	0.884	2.78	2.65	0.923	0.173	1.23	1.40	0.851	0.798	—	1.24	0.73	0.49
86101	0.958	3.11	2.96	1.03	0.220	1.33	1.52	0.901	0.828	0.124	1.30	0.73	0.58
86171	0.771	1.93	1.73	0.734	0.212	1.06	1.37	0.894	0.883	—	0.97	0.64	0.79
86195	0.841	2.37	2.28	0.865	0.219	1.27	1.53	0.935	0.884	—	1.16	0.66	0.64
86198	1.05	2.83	2.56	0.942	0.286	1.33	1.62	1.01	0.983	—	1.09	0.73	0.78
8696	8.07	6.66	2.27	0.816	3.05	—	3.58	2.23	2.15	—	1.14	0.86	0.95

Numbers of samples: see Table 6.3.

second unit (the more voluminous) are clearly depleted in alumina. Moreover, the distinctive features of the REE distribution pattern are outlined. Thus, a sample from the first unit would have a moderately fractionated spectrum, with a high total content of lanthanoids, and would show enrichment in LREE $(Ce/Sm)_N = 1.3$. The second group of komatiites are moderately depleted in both LREE and HREE: $(Gd/Yb)_N = 1.3$; $(Ce/Sm)_N = 0.79$. Finally, the third unit of metavolcanics is strongly depleted in LREE and is distinguished by a virtually unfractionated HREE distribution $(Gd/Yb)_N = 1.1$; $(Ce/Sm)_N = 0.67$ (Fig. 6.8). The presence of a negative Eu anomaly may be accounted for by the leaching of Eu by circulating hydrothermal fluids, since the plagioclase could not be fractionated after crystallization of komatiites, whatever the conditions. The komatiites of the second unit are correlated geochemically with the Al-depleted komatiites, while those of the first and third units are correlated with the Al-undepleted komatiites, according to Nesbitt *et al.* (1982).

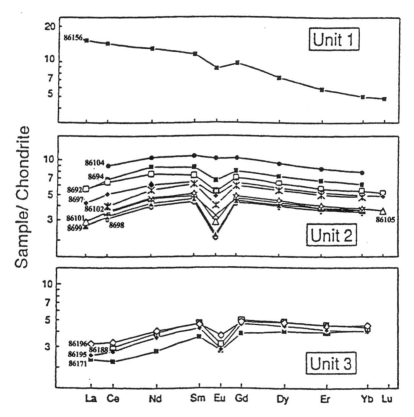

FIGURE 6.8 REE spectra of the earliest komatiites of the early Archaean Tungurcha greenstone belt, from units I, II and III (from bottom to top). After Pukhtel *et al.* (1993). Samples: see Table 6.4.

Intercalated with metakomatiites are metabasalts that can be correlated with high-Mg olivine-normative and low-Mg quartz-normative varieties, characteristic of komatiitic series formations. $\varepsilon_{Nd(T)} = +0.3 + 0.9$ (komatiites) and $+1.9 + 0.2$ (metabasalts).

In the Late Archaean Olondo greenstone belt, among the tonalite–trondhjemitic gneisses, the same two units can be recognized with certainty: a lower one, 500 m thick, composed predominantly of basic and ultrabasic volcanics and an upper one, 600 m thick, composed of intermediate and acid volcanics and a subordinate quantity of sediments. Subvolcanic sill-shaped bodies of basic and ultrabasic composition are present. Komatiites and basalts occur as both massive and differentiated flows and pillow lavas (Fig. 6.6). The results of chemical analyses for typical rock varieties are given in Tables 6.5 and 6.6. As is the case for the Tungurcha greenstone belt, two types of komatiites can be distinguished here: Al-depleted and Al-undepleted; however, the latter predominate. Among these, in turn, there are two distinctive groups of rocks. The first group are enriched in Zr relative to Ti and Y, and in La relative to Sm, that is, in elements accumulated in the Earth's crust (LREE-rich komatiites), suggesting their contamination by crustal material. The second group are strongly depleted in incompatible elements relative to moderately incompatible (komatiites depleted in LREE). The basalts that are associated with the komatiites are represented by olivine- to quartz-normative varieties, as within the Tungurcha belt. In contrast to Early Archaean ones, they are characterized by relatively high concentrations of FeO, Ni, Co and Cr and are distinguished by their severe depletion in highly incompatible elements.

In the northern part of the structure, there are very specific low-Mg olivine-normative tholeiites with unusually high FeO and TiO_2 content (16.4–20.3 and 1.85–2.12 wt.%, respectively). The Cr and Ni contents are relatively low, whereas the V content is high (15–72, 16–70 and 600–1400 ppm, respectively). The REE distribution is not typical for tholeiites (Fig. 6.9). They have a slightly enriched and practically undifferentiated distribution in the lighter part of the spectrum and are strongly depleted in heavy lanthanoids. $\varepsilon_{Nd(T)}$ for komatiites: $+2.24 \pm 0.14$; for komatiitic basalts, northern tholeiites, andesites, dacites and tuffs as well as for gneiss–plagiogranites: $+2.2 \pm 0.10$.

6.2.2 Tholeiite–Basaltic Assemblage

The rocks of this assemblage make up a reasonable proportion of the Late Archaean greenstone belt volcanics. They are particularly widespread in the Canadian greenstone belts (the Abitibi and Yellowknife belts), those of Australia (the Kulgardi and Norsemen belts), of South Africa (the Barberton Belt), of the Aldan Shield (the Olondo belt), and of the Baltic shield (the Kostomuksha and Sumozersko–Kenozersky belts), etc. (Bogdanova *et al.*, 1987).

Table 6.5 Major and rare-element content in basic and ultrabasic metavolcanics of the Olondo greenstone belt (Aldan Shield) (after Pukhtel and Zhuravlev, 1993).

Components	1	2	3	4	5	6	7	8	9	10	11	12
SiO_2	50.2	50.2	50.5	51.3	47.7	44.1	49.1	49.3	49.5	50.4	47.7	48.2
TiO_2	0.77	0.79	0.84	0.32	0.30	0.28	0.53	0.85	1.19	1.14	2.09	1.85
Al_2O_3	7.83	8.04	8.66	7.80	6.98	2.71	8.28	15.2	15.5	15.2	11.8	13.6
FeO^*	12.8	12.4	12.3	9.73	11.7	12.5	12.8	12.0	12.2	12.6	19.8	16.8
MnO	0.20	0.19	0.19	0.18	0.19	0.24	0.20	0.19	0.19	0.19	0.22	0.20
MgO	15.3	14.5	12.6	21.7	27.2	35.0	18.8	8.63	7.60	7.30	6.29	5.65
CaO	12.3	12.6	13.4	8.40	5.69	5.11	9.27	12.2	11.6	11.4	10.5	11.3
Na_2O	0.34	0.94	1.18	0.44	0.08	0.01	0.89	1.18	1.77	1.38	1.06	1.82
K_2O	0.12	0.14	0.10	0.03	0.03	0.02	0.08	0.17	0.26	0.20	0.33	0.36
P_2O_5	0.20	0.23	0.25	0.16	0.10	0.04	0.18	0.25	0.24	0.24	0.22	0.24
LOI	1.84	2.17	1.82	4.28	7.11	8.76	3.53	1.78	1.72	1.74	2.42	2.33
Total	100.00	100.00	100.00	100.00	100.00	100.00	100.00	100.00	100.00	100.00	100.00	100.00
Cr	1,258	1,249	1,180	2,765	3,628	5,072	2,214	420	349	290	68	31
V	204	188	165	124	147	70	115	247	265	264	1,351	876
Co	71	70	61	76	92	173	79	54	1	49	77	69
Ni	363	370	347	938	1,338	1,864	789	148	137	127	64	38
Zr	43	45	50	17	15	17	41	45	62	58	46	51
Nb	2.3	2.6	2.7	2	1	1	1	3	1	3	3	3
Y	14	14	15	7	6	0	9	20	27	26	13	14
Sr	31	74	118	510	12	0	10	92	148	104	35	152
Rb	1	2	3	1	1	1	1	1	2	1	4	6

FeO^* — total iron as FeO, here and elsewhere.

Columns 1-3 — eastern komatiitic metabasalts; 4-5 — western komatiites depleted in LREE; 6 — peridotites of the Porphiritovaya Deposits; 7 — western komatiites enriched in LREE; 8 — western tholeiitic metabasalts; 9-10 — eastern tholeiitic metabasalts; 11-12 — northern tholeiitic metabasalts.

Table 6.6 REE content in basic and ultrabasic volcanics of Olondo GSB, Aldan Sheild (after Pukhtel and Zhuravlev, 1993).

Components	8551	8569	228/18	8550	8550/1	07/2	07/3	07/6	15/1	15/2	86226	231/1	232/2	08/3
	1	2	3	4	5	6	7	8	9	10	11	12	13	14
La	0.456	0.347	0.362	—	0.456	3.29	1.41	2.73	1.84	2.01	1.46	3.68	2.13	3.63
Ce	1.41	1.23	1.12	1.98	1.41	8.89	4.65	7.65	5.17	5.42	3.86	8.91	5.76	9.90
Nd	1.24	1.21	1.04	1.88	1.24	6.98	4.63	6.38	4.40	4.39	3.39	5.96	4.47	7.70
Sm	0.480	0.459	0.389	0.71	0.480	2.19	1.77	2.01	1.51	1.49	1.30	1.80	1.46	2.56
Eu	0.361	0.124	0.090	0.232	0.361	0.805	0.528	0.733	0.56	0.524	0.602	0.70	0.728	0.913
Gd	0.716	0.678	0.577	1.06	0.716	2.68	2.33	2.47	2.17	2.09	1.92	2.25	1.93	3.34
Dy	0.931	0.907	0.795	1.33	0.931	2.81	2.59	2.60	2.63	2.61	2.48	2.49	2.15	4.04
Er	0.626	0.611	0.541	0.857	0.626	1.61	1.49	1.46	1.69	2.64	1.58	1.47	1.29	2.53
Yb	0.634	0.621	0.537	0.828	0.634	1.42	1.30	1.30	1.65	1.60	1.51	1.38	1.19	2.39
Lu	0.099	0.097	0.084	—	0.099	—	—	—	—	—	—	—	—	
$(Ce/Sm)_N$	0.712	0.651	0.695	0.673	0.712	0.980	0.633	0.920	0.828	0.878	0.717	1.20	0.954	0.934
$(Gd/Yb)_N$	0.912	0.882	0.869	1.03	0.912	1.52	1.45	1.53	1.06	1.03	1.03	1.32	1.31	1.13
Eu/Eu*	1.88	0.681	0.582	0.817	1.88	1.02	0.795	1.01	0.948	0.907	1.16	1.07	1.33	0.955

Columns 1–5 — western komatiites: 1–3 — depleted, 4, 5 — enriched in REE; 6–8 — eastern komatiitic basalts: 9–10 — western tholeiitic basalts; 11 — western metagabbroids; 12–13 — northern metatholeiites; 14 — eastern tholeiitic metabasalts.

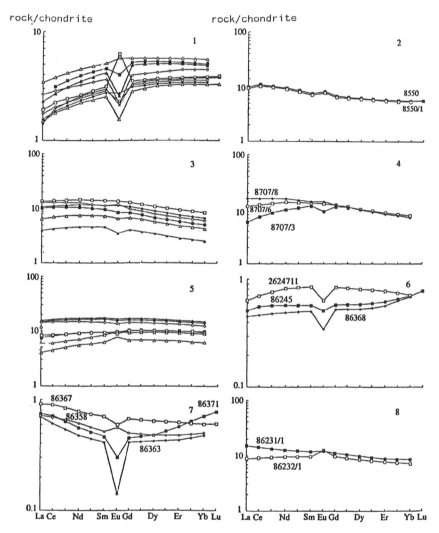

FIGURE 6.9 REE spectra of Olongo GSB basic–ultrabasic rocks. After Pukhtel *et al.* (1993). 1–2 — western komatiites: depleted (1) and enriched (2) in Zr and LREE; eastern komatiitic basalts (represented by rhombs and squares) and peridotites of the Porphirovidny deposit (triangles); 4 — differentiated pillow lavas (Fig. 6.8); 5 — eastern tholeiitic basalts (rhombs and filled triangles) and western tholeiitic basalts (squares) and gabbroids (open triangles); 6–7 — ultra-basic cumulates of Kzasnaya gorka intrusion: 6— dunites, 7 — peridotites; 8 — northern tholeiitic basalts.

The rocks are represented by quartz- or hypersthene-normative tholeiites. They differ from komatiitic basalts in their higher TiO_2, Al_2O_3, Na_2O, K_2O, Ba, Sr, Zr and Y content, their higher Zr/Y and Ti/V ratios and their lower MgO/FeO (<1) ratio. The TiO_2 content in some tholeiitic basalts of the Late Archaean greenstone belts may reach 1.80–2.12%, as has been established in the Abitibi and Olondo belts, but is usually much lower.

The andesitic assemblage is commonly confined to the middle and upper parts of the sections, in a series of Late Archaean greenstone belts, and is unevenly distributed. In some regions: the Baltic Shield (eastern Finland and western Karelia), the Aldan Shield (the Olondo GSB), and in Zimbabwe, the andesites account for 20–40% to 70% of the thickness of the volcanic section. In Central and Eastern Karelia, Kulgari–Norsemen, Barberton and other regions, the andesitic formations make up no more than 5% of the section, and often are completely absent. The most characteristic formations are fine-grained pyroclastics including water-lain tuffs and tuffites. Andesitic breccias and agglomerates are occasionally encountered; lava flows and shallow-depth intrusions are even less common.

Unlike Phanerozoic andesites, Archaean andesites are generally characterized by their low Al content, their increased Mg and Fe content, their high Y content and their high FeO/Fe_2O_3 and Ni/Co ratios. On the other hand, the Archaean andesites are enriched in Ni, Cr, Co and Zn (Bogdanova *et al.*, 1987), which means that they closely resemble the formation of the Proterozoic boninite-like series.

The results of Sm/Nd isotopic study of tonalitic gneisses from the Archaean (3.2 Ga) Aulsky GSB (Middle Dnieprov'e, Ukrainian Shield), which were presumably formed at the expense of andesite dacites, has shown that $\varepsilon_{Nd(T)} = +1.7 \pm 0.7$, MSWD $= 1.9$. This suggests that the source of their parental melts was either the mantle or basic crustal material (Samsonov *et al.*, 1993).

6.2.3 Rhyolite–Dacitic (Leptitic) Assemblage

This assemblage's wide range of compositions cover the interval from dacites to rhyolites, including quartz latites and rhyodacites. In the majority of cases dacites and rhyodacites predominate. Acid lavas are rarely encountered. Coarse-grained volcanic breccias and agglomerates with fragments 10–30 cm in diameter or more form strata up to 300 m thick, and generally grade into other lithologies or wedge out along strike (Bogdanova *et al.*, 1987). Vein rhyolites and porphyritic vein granitoids normally complete the volcanic cycle of greenstone belt development. They are observed in practically all the greenstone belts of the Baltic and Ukrainian shields as well as in complexes of the Canadian, Zimbabwe–Kaapvaal, western Australian and Indian shields.

Within the Late Archaean greenstone belts thick horizons of banded iron formations (iron quartzites) occur frequently; these are the most important source of economic iron ore. In the Baltic Shield they are developed mainly within the Gimola-Kostamuksha greenstone belt of northern Karelia and in the Olenegorsky greenstone belt of the Kola Peninsula, where they are related to large economic deposits of iron quartzites (Kratz *et al.*, 1978).

6.2.4 Granite Plutons

The formation of Archaean greenstone belts was followed by intensive peraluminous granitoid magmatism, often localized at the contacts of the greenstone belts with granite–gneiss areas composed mainly of "grey gneiss" complexes. These granitoids form plutons, domes, diapirs and dykes and show intrusive contacts with greenstone-belt rocks. According to Sylvester (1994), the Archaean granite plutons consist mainly of granodiorite, tonalite and true granite. They were emplaced several million years or more after episodes of greenstone belt volcanism. In general, compressional deformation occurred before and continued through the beginning of granite plutonism but then quickly waned. Calc-alkaline, strongly peraluminous and alkaline granite plutons are each quite abundant. Within a single craton, alkaline plutons tend to be somewhat younger than calc-alkaline and often occur along boundaries of structural domains (Kotov *et al.*, 1993). The Archaean granite plutons were formed by partial melting of crustal rocks, in most cases with little involvement from mantle-derived magmas. Similar granites occur in the Phanerozoic in what are thought to be collision-related orogenic situations (Sylvester, 1994).

The earliest of these granitoids are represented by tonalite–trondhjemite assemblages. This is represented by massifs of variable composition, i.e. quartz-diorites, granodiorites, tonalites and plagiogranites. Their intrusion more frequently preceded the formation of the upper terrigenous parts of the greenstone-belt sections, and, more rarely, completed their formation (Condie, 1981).

Quartz-diorites and granodiorites tend to be spatially separated from tonalites and plagiogranites and are located close to greenstone belts of various types. Tonalites and plagiogranites are widespread close to bimodal volcanic belts (e.g., the Kostomuksha Belt of western Karelia), whereas the most extensive massifs of quartz diorites and granodiorites are confined to belts containing andesites (the Ostersky and Khautovarsky belts of the central Karelia: Kratz, 1978). This suggests that many granitoids of this series were comagmatic and synchronous with greenstone-belt acid and intermediate volcanics. This has been proved in some cases by the similarities in their chemical composition (i.e., in the Abitibi Belt) and their isotope-geochemistry (e.g. Suomosalmi in Finland) (Martin *et al.*, 1984). The tonalites and plagiogranites of greenstone belts correspond to the calc-alkaline magmatic series with a trondhjemite crystallization trend (Fig. 6.10).

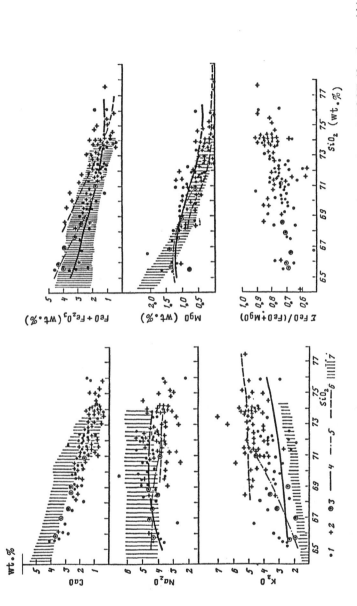

FIGURE 6.10 Relationship between major-element content (in wt.%) and silica in the granites of GGST (Southern Africa, Canada, Western Australia, the Wyoming area and Karelia). After Bogdanova *et al.* (1987). 1–2 — granites: 1 — synkinematic, 2 — post-kinematic; 3 — average for Berens field batholiths (after K. E. Ermanovich); 4–6 — direction of trends: 4 — for synkinematic granites, 5 — for post-kinematic granites, 6 — for Berens field batholith; 7 — compositions of the tonalites and plagiogranites.

Studies on the Sm/Nd isotopic system in tonalites of northern Finland (Jhan *et al.*, 1974) allowed conclusions to be drawn about the multistage crustal history of the tonalites: the data on isotopic contents agree with a model of the remobilization of the acid source at 3.6–3.5 Ga. Tonalites may have been partly derived from melts generated in the lower crust, which by that time had already acquired a fractionated REE distribution. Their melting under granulite-facies conditions led to precipitation of the HREE in garnet–pyroxene restites and a concomitant depletion of HREE in the melts. Considerable variations in the degree of REE fractionation (Fig. 6.11), as well as in the concentrations of other rare elements, could indicate different methods of tonalitic melt formation. They could be formed during partial melting of basic rocks at variable depths, determined by the stability fields of minerals such as hornblende and garnet.

The suggestion that the tonalites are heterogeneous is often confirmed by the relatively high values of $^{87}Sr/^{86}Sr$ obtained for acid volcanics series from green-stone belts (Jahn *et al.*, 1984), which indicate their crustal origin. This may be attributed to some comagmatic plutonic tonalites.

Uniformity in major-elements concentration is certainly associated with the cotectic ratios of rock-forming minerals. There is a definite spatial and temporal relationship between the majority of the TTG and the mafic volcanics, which may be accounted for by assuming that the ascending mantle diapirs and komatiite–basaltic magma underplating served as a heat source for the partial melting of these tonalites.

In all of the granite–greenstone terranes their late-stage development was followed by the formation of migmatite–granites. In a few specific regions, metamorphism, metasomatism and anatexis mainly occurred in the granite-gneiss areas separating the greenstone belts, while in other localities the granitization covered the marginal parts of the granite–greenstone belts as well.

Metasomatic reworking of the marginal parts of greenstone belts also resulted in the formation of a series of granites containing potassium feldspars. The quartz diorites there grade into granodiorites, tonalites and plagiogranites, whereas the terrigenous rocks of the greenstone sections grade into granite and K–Na migmatites. These rocks contain high concentrations of Si, K, Rb and Pb, and low concentrations of Ti, Al, Fe, Mg, Ca, Na, Sr and Zr.

For the synkinematic granites of all of the granite–greenstone belts, there is a trend similar to the tonalitic trend, in changes in the concentrations of aluminium, iron, magnesium and calcium with increases in SiO_2 (Fig. 6.10), although shifted towards a higher iron and potassium content and lower sodium and calcium content. These granites differ even more markedly from the tonalites, as evidenced by their REE content (Fig. 6.11). These are often enriched in HREE, although with a lesser degree of fractionation. All of this confirms the concept of heterogeneity of synkinematic granitoids in both their paragenesis and initial source-rock composition.

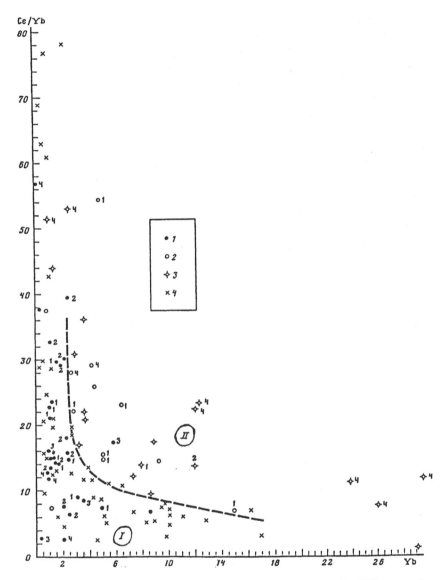

FIGURE 6.11 Degree of fractionation of chondrite-normalized REE, according to the Yb and Ce/Yb ratio. Data from the Amitsoq area (Greenland), Bighorn (USA), Uyvak (Canada), Barberton and Swaziland (Southern Africa), Minnesota and Wyoming (USA), Aldan (Eastern Siberia) and Karelia have been used. 1 — tonalites and plagiogranites; 2 — granite–gneisses and migmatites; 3 — post-kinematic microcline granites; 4 — "grey gneisses"; I — field where rocks of the Na-series predominate; II — K–Na series; numbers near the symbols refer to the age: 1 — 3.4–3.2 Ga; 2 — 3.0–2.9 Ga; 3 — 2.8 Ga and 4 — less than 2.7 Ga ago.

Granitization processes at the final stages of development of Late Archaean granite–greenstone areas were very intensive and the ultrametamorphic granitoid series thus located essentially determined the appearance of granite–greenstone areas and the composition of the granitoids. The evolution of acid magmatism in Late Archaean granite–greenstone terranes was completed by the formation of post-kinematic granitoids richer in iron and potassium. Usually, they make up a small volume of the present-day eroded section. The largest massifs of post-kinematic granites (up to $1,600\,km^2$) are known from the Kaapvaal craton; they have been dated as 3.1 and 2.66 Ga old (Tankard et al., 1982; Sylvester, 1994). Most commonly they form small bodies, stocks and numerous veins confined to the contacts of various complexes, both of the granite–gneiss areas and greenstone belts.

High and widely varying LILE contents are particularly characteristic of post-kinematic granites. The Zr and Y ratios in granites of the Karelian granite–greenstone terrane imply the formation of the melts at different levels in the Earth's crust. For the granites with low Y content and high Th/U ratios, one can postulate formation of melts from granulite-facies complexes containing garnet and pyroxene. In other cases, veined post-kinematic granites with increased K/Rb ratios (for example, in the Lake Tulos area of Karelia) have remained close to the melting level, which is in accordance with the restite nature of the host granitoids as reflected by their high Y concentrations (Bogdanova et al., 1987).

6.3 MAGMATISM OF GRANULITE–GNEISS TERRANES

Granulite–gneiss terranes (GGT) are mainly belts of intense deformations and high-grade metamorphism to granulite or amphibolite facies. They contain valuable information about magmatism in these specific areas. Increasingly the isotope-geochronological data indicate at least three Archaean phases in the formation of these areas. The most ancient granulites (3.87–3.6 Ga) were recognized in West Greenland (Nutman et al., 1993), in West Australia (Page et al., 1984) and in East Antarctica (Krylov et al., 1993). Mid-Archaean (3.5–3.0 Ga) complexes are found on Ukranian Shield (Bibikova et al., 1990) and on the Aldan Shield (Bibikova et al., 1989). However, a key stage in the development of granulitic global-scale metamorphism in the Archaean was the period from 2.85 to 2.6 Ga (Bibikova et al., 1982; Percival, 1994), which essentially coincides with the time of Late Archaean greenstone-belt development. Another important feature of Late Archaean granulitic metamorphism is the substantial time interval between it and previous magmatism and sedimentation (Bibikova et al., 1982; DePaolo et al., 1982). This suggests that granulite formation was with particular global stages in the formation of Archaean lithosphere.

GGT commonly occur as belts among granite–greenstone terranes as illustrated in Fig. 6.2. They are formed mainly of metasediments (metapelites, silicate-carbonate rocks, marbles, iron-banded formations, etc.); among them are found various amphibolites, which are interpreted as metavolcanics of basic and intermediate composition, and TTG-series rocks. All of them metamorphosed under conditions from amphibolite to granulite facies where the latter predominates. Granulite metamorphism was accompanied by the emergence of abundant synkinematic endebite–charnockite magmatism in the form of migmatites and conformed intrusive bodies. This is the only evident type of magmatism within GGT, because the situation with basic, intermediate and TTG-series rocks is not clear yet: they are usually older than enderbite–charnockite rocks, and often their localities resemble tectonic slices of granite–greenstone terranes, involved in tectonic-metamorphic reworking, and linked with the formation of the GGT. As a rule, their formation is rather close to adjacent GGST, for example the age of granulite metamorphism of the Kola-Norwegian belt (2.83–2.76 Ga; Mitrofanov, 1995) is rather close to the age of the main stage of the Karelian GGST formation (see above). It is very important that, between adjacent GGST and GGT, specific transitional zones of tectonic flowage with the same age occurred. In the case of the Baltic Shield it was the Belomorian Belt between the Karelian GGST and the Kola-Norwegian Granulite Belt (Fig. 6.2); in the case of the Limpopo Granulite Belt (South Africa) there were two Marginal zones between the belt and the adjacent Kaapvaal and Zimbabwe cratons (van Reenen et al., 1992); on the Canadian Shield the same zone is found between low-grade Abitibi and high-grade Optica belts (Calvert et al., 1995).

High-grade metamorphism conditions in the Archaean GGT are, as a rule, characterized by moderate pressures (6–8 kbar) and temperatures of 600–800°C; two-pyroxene-plagioclase parageneses are predominant in metabasites; garnet-bearing granulites are more rare but predominate among the Early Precambrian lower crustal formations (Downes, 1993; Persival, 1994; Sharkov et al., 1996).

Magmatism occurs at all stages during the formation of highly metamorphosed areas. However, the sequence of igneous activity is very difficult to reconstruct because of structural and metamorphic reworking of the original sedimentary-volcanic and intrusive complexes.

Studies of the characteristic features of GGT formation have demonstrated that the latter represent areas of compression, within which complex isoclinal fold systems were formed, and thrusts and nappes were generated under conditions of ductile flow. The Archaean areas often display complex "interference"-type folding, reflecting multiphase deformation and tectonic mixing of different sections, in most cases completely destroying the original layering. As a consequence, complexes from various depths have been juxtaposed. The main components of these highly metamorphosed areas were originally volcanic and sedimentary rocks. These are represented by basic two-pyroxene-plagioclase schists and amphibolites (basic volcanites), metagabbroids, metaultrabasites and

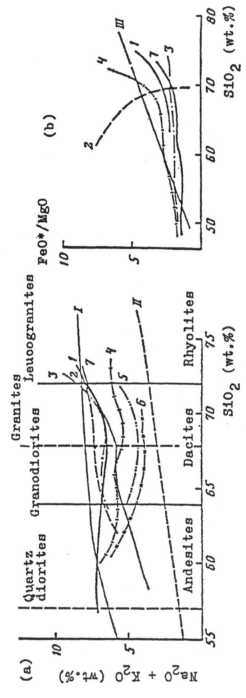

FIGURE 6.12 Petrochemical differences between "grey gneisses" of various ages and enderbites. After Bogdanova *et al.* (1987). (a) on the diagram (Na$_2$O + K$_2$O)–SiO$_2$; (b) on the diagram FeO*/MgO–SiO$_2$; 1,2 — gneisses with an age of 3.8–3.6 Ga: 1 — "grey gneisses" of the North Atlantic granulite-gneiss area, 2 — "augen" gneisses of the same area; 3 — Nuk "grey gneisses", Western Greenland, 3.1–2.8 Ga; 4–7 — enderbitoids: 4 — of the Kola Peninsula, 5 — of the Ukrainian shield, 6 — of the Volga-Urals area, 7 — Enderby Land, Antarctica; I–III — boundaries between the rocks: I — boundaries of normal and alkaline series, II — boundaries of tholeiitic and subalkaline petrochemical series.

various plagiogneisses (intermediate and acid metavolcanites, greywackes), metapelites, marbles, quartzites, etc. The bulk of the sections is occupied by granitoid, especially enderbite–charnockitic and tonalite–trondhjemitic complexes, including pre- and syn- and post-metamorphic ones.

As a rule, they are found in one form of widespread zones of migmatites, intrusive bodies; some of them are interpreted as former lava flows.

6.3.1 Metabasites

Basic and ultrabasic schists locally account for up to 30% of the GGT volume and are extremely diverse in terms of their geological setting and nature. Sometimes they alternate with metasedimentary rocks: highly and moderately aluminous gneisses, calciphyres, ferro-siliceous rocks all presumed to have been derived from sedimentery rocks. Taking into account their petrography and geochemistry the metabasites are attributed to basic and ultrabasic volcanics. As a whole, these segments resemble greenstone belts, but were subjected to granulite-facies metamorphism. In most cases, these metabasites correspond with the formations of the komatiitic series, predominantly basalts, and the situation as a whole resembles greenstone belts.

Intrusive metabasites are always present in these sections, ranging in composition from pyroxenites to gabbro–anorthosites, composing sill-like bodies. Despite significant variations in their content of iron, nickel, chromium, titanium, cobalt, copper and other minor elements, metagabbroids are very close to metabasalts in their chemical composition. Some of the metabasites and ultrabasites could be differentiates of large layered intrusions, including gabbronorite–anorthosite types. Some complexes have been so reworked by high-temperature metamorphism and ductile deformation that they have been transformed into "stratified" pseudovolcanic complexes, conformable with the regional structural trend. Processes of meta-morphogenic granite–charnockite formation and migmatization can completely rework the initial sections.

These massifs are exemplified by the essentially anorthosite Fiskenaesset Complex of West Greenland $(2.86 \pm 0.05 \, \text{Ga}$ old, $\varepsilon_{Nd(T)} = +2.9 \pm 0.04$. $(^{87}\text{Sr}/^{86}\text{Sr})_0 = 0.70108-0.70162$: Ashwal et al., 1989). It is sheet-like and about 500 km at length and up to 2 km in thickness, with a layered and intensely deformed body; it was formed at about the same time as granulite-facies metamorphism. The complex contains significant amounts of dunites, harzburgites, chromitites and norites, and in this respect it is similar to typical layered intrusions of the Stillwater type. Most ancient anorthosite–gabbro–ultramafic intrusions, such as the Manfred complex, which occurs as fragments in the 3.65 Ga old granite–gneisses in the Yilgarn Block of Western Australia, are of analogous structure. In the opinion of Myers (1988) the intrusion and the country rocks underwent two stages of deformation and metamorphism: the earlier one under amphibolite to granulite-facies conditions, and the later one under greenschist-facies conditions. However, in contrast to the Moon, anorthosite massifs are not typical of the Archaean: on the Earth they reached their peak development in Proterozoic times.

According to Ashwal (1993), the Archaean anorthosite complexes rather differ from the Proterozoic ones. They are characterized by megacrysts of equidimensional calcic plagioclase (usually $An_{80\pm5}$), and form extensive sheet-like bodies, mostly emplaced at shallow depths into basaltic volcanic rocks. Original thicknesses of the anorthositic sheets are difficult to determine because of deformation. Archaean anorthosites were derived from fractionated basic magmas, and related basaltic rocks with calcic plagioclase megacrysts also form dykes and flows.

6.3.2 Igneous Rocks of Intermediate and Acid Composition

Pyroxene–plagioclase gneisses or enderbite gneisses of andesite–dacite composition are also widespread in Archaean granulite–gneiss terranes. In a number of cases (the Ukrainian, Anabar and Aldan shields, the Volga–Urals area), enderbite gneisses make up 40–60% of the terranes (Bogdanova et al., 1987). As a rule they are found in the form of widespread zones of migmatites, intrusive bodies; some of them are interpreted as former lava flows.

Thus, the magmatism of the Napier complex of Enderby Land (East Antarctica) has been dated as 3.6–3.5 Ga (Grew, 1984; Krylov et al., 1993) as determined by U–Pb and Sm–Nd methods. The rocks comprise enderbites and charnockites, which have preserved the isotopic composition of their meteoritic-type lead. The originally volcanic origin of the enderbite gneisses has been confirmed by their stratification and alternation with metasediments.

As was shown on the Baltic Shield, early enderbite–charnockite intrusive complexes are synkinematic with deformations and formed of sheet- and lens-like bodies, from tens and hundreds of metres to 5–7 km in size, which are concordant with the country rocks for distances of up to 20 km. It is rare for these intrusions to be discordant xenoliths of the country rocks, and the non-migmatized rocks are generally recognizable (Shemyakin, 1988). These enderbite–charnockite complexes are the most common intrusive complexes of the infrastructure of the Archaean granulite–gneiss terranes.

A pronounced structural feature of these complexes, in which intermediate and acid rocks account for up to 80% of their volume, is their textural inhomogeneity as shown, for example, by the alternation of rocks of different composition. The latter vary from pyroxenites and gabbronorites to quartz diorites, tonalites and plagiogranites, often reworked during late stages of granulitic metamorphism into websterite and two-pyroxene schists, enderbites and charno-enderbites. The metamorphic layering appears to post-date the primary stratification of the bodies, and may indicate that the intrusions crystallized under P–T conditions close to those of granulite-facies metamorphism.

The intermediate and acid rocks have intrusions, with a calc-alkaline character, high Al_2O_3 concentrations and stable K_2O/Na_2O ratios (0.2–0.3), together with significant variations in CaO. The most likely volcanic equivalents of plutonic enderbite–charnockite series could be dacite–rhyolite metavolcanics with age 2.8–2.6 Ga, which infilled the Keivy structure in the eastern part of the KNGB (Fig. 6.2). It follows from Mints et al., (1996) data that in their geochemistry these metarhyolites and metadacites are rather close to enderbites and charnockites.

The early-folded gabbro–tonalitic assemblages are almost identical in many respects to the low-potassium Archaean "grey gneiss" complexes, particularly those of the Late Archaean (Fig. 6.12). Like the "grey gneisses", they are enriched in nickel, chromium and vanadium, and depleted in many lithophiles, for example Rb, U, Th, Y, Li, Nb, Ba, F and REE. They have low initial $^{87}Sr/^{86}Sr$ ratios (0.702–0.704) (Bel'kov et al., 1984). The effect of the host-rock composition cannot be ruled out. There is certainly a tendency for the Fe content of enderbite–gneisses to increase in those cases where the host rocks are ferrosiliceous formations. An increased Al content is also correlated with high-Al composition of the Archaean volcanogenic-sedimentary complexes.

The synkinematic (anatectic and anatectic–metasomatic) igneous rocks of the Archaean are directly related to phases of regional metamorphism and intense metasomatism of earlier-formed complexes. Migmatization, metamorphic differentiation and secondary processes of granite formation occurred during folding and disrupted deformation, and therefore the modes of occurrence of metamorphogenic igneous complexes tend to follow the structural trend of the granulite–gneiss terranes. The true volume of igneous formations of this type is difficult to evaluate because these areas show a composite imbricate-fold structure, and also the newly formed granitoids, especially those from the early stages, differ only slightly in terms of their petrographic composition.

High heat flow in the Archaean crust could have led to the formation of some of the earliest Na-plagiomigmatites (enderbites) by anatexis. These layered and injected migmatites have sharp boundaries with their host rocks and are frequently observed in granulite complexes. However, the products of intensive fluid-metasomatic reworking tend to grade into their parent. These rock types are of much more widespread occurrence; petrographically, they are distinguished by their coarse-grained texture, their more acid composition, and the presence of hydrous minerals such as amphibole and biotite and potassium feldspar, and as a whole they are characterized by less varied mineral composition. In comparison with their source rocks, the plagiomigmatites display higher SiO_2 and alkali content, decreased iron and magnesium content, and a higher degree of iron oxidation. Accordingly, the concentrations of alkali-related Rb, Ba, Pb, REE and Zr increase and the content of the Fe-group elements — Co, Ni, V, Cr and Sc — decreases (Bogdanova et al., 1987).

An even greater degree of metamorphism is exhibited by K-feldspathic migmatites, i.e. rocks of charno-enderbitic and charnockitic composition, which are formed by the reworking of both parental complexes and first-stage pla-giomigmatites. The K-feldspathic migmatites sometimes make up <70–90% of the area of outcrops and are the most variable morphological type. Depending on the intensity of regional migmatization and tectonic activity, migmatite from augen to shadow types are formed. Alaskite–charnockite massifs and fields of shadow migmatites cover areas of hundreds of square kilometres. It is pertinent that orogenic charnockites generally occur associated with rocks of basic composition. These rocks have a very variable mineral composition, which is directly related to that of their source rocks.

Metasomatic and anatectic migmatization at the later stages of metamorphism could have evolved into magmatic melting and could have resulted in the forma-tion of magmatogenic autochthonous charnockitoids intruding the migmatites. Their most common mode of occurrence is as small massifs confined to the nuclei of anticlinal structures, as well as irregular patches inside the intrusive enderbites or plagiomigmatites formed during the early stages of folding. Geochemically, these processes were accompanied by the increasing participa-tion of potassium, and, petrographically, there was an increase in the proportion of the more acid and potassium charnockitoids in the central parts of the autochthonous massifs.

The final stages of Late Archaean granulite metamorphism are succeeded by the injection of dry and K-rich magmas and the formation of fairly large bodies of late-folded charnockitoids, which possess all the attributes of intrusive com-plexes: i.e. the presence of eruption breccia zones, xenoliths of country rocks and quenching zones, and also cut across the earlier-formed fold structures. Characteristically, they tend to be localized in fault zones. The petrographic composition of late-folded charnockitoids is distinguished by a marked differ-entiation: from quartz monzonites to charno-enderbites and monzo-charnockites (Shemyakin, 1988). If these complexes had not undergone metamorphic reworking, the igneous nature of the charnockitoids would have been indicated by their typical trachytoid structures, porphyritic, hypidiomorphic–granular and prismatic–granular textures and the presence of high-temperature minerals, etc. In contrast to the syn-orogenic charnockitoids, these are, as a whole, more acid and less calcium-rich rocks of the subalkaline and calc-alkaline series, with a variable alkali content, particularly of potassium. For the Volga–Urals charnockitoids, a regular relationship has been established between the iron content and the composition of the host Archaean strata: the more Fe-rich charnockitoids are associated with sections containing ferro-siliceous eulysitic rocks (Bogdanova et al., 1987).

One of the famous Late Archaean granulitic terranes is the Anabar Shield in the north of the Siberian platform (Fig. 6.13) formed predominantly by rocks of the granulitic complex (Rosen et al., 1988). In this section of the complex, three

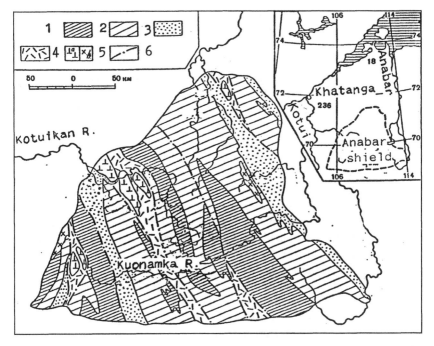

FIGURE 6.13 Schematic diagram of the geological structure of the Anabar Shield geological structure. After Rosen *et al.* (1988). 1 — Daldynsky Series; 2 — Verhneanabarsky Series; 3 — Hapchansky Series; 4 — retrograde metamorphic rocks of deep fault zones; 5 — igneous rocks of the deep fault zones (a — anorthosites, b — K-porphyroblastic granites); 6 — faults.

formations can be distinguished (from the bottom upwards): the Daldynskaya: 3,500–4,500 m thick; the Upper Anabarskaya: 5,000–6,000 m thick; and the Khapchanskaya — 4,500–5,000 m thick. In the Daldynskaya Formation, a dominant role is played by meso-melanocratic two-pyroxene crystalline schists and plagiogneisses (enderbites) in varying quantities, especially in the upper part of the section. Some 8–10% of the volume is accounted for by quartzites, some of which are magnetite-bearing, which are associated with high-Al gneisses containing cordierite and sapphirine. The Upper-Anabarskaya Formation dominates over nearly half of the shield area. Hypersthene and two-pyroxene plagiogneisses (enderbites) predominate here, and associated with these are small amounts of garnet-bearing rocks (biotite–garnet, garnet–biotite–hyperstene, garnet–sillimanite–cordierite, etc.) and occasionally quartzites and calciphyres. The rocks are usually highly migmatized and granitized. The upper Khapchanskaya Formation is composed of various biotite–garnet, garnet–hyperstene and two-pyroxene plagiogneisses and crystalline schists. Thin interlayers and units of marbles, calciphyres and salite-scapolite rocks, sometimes with rhythmical structures, are widely developed. The rocks of this suite were also subjected to migmatization.

It is necessary to note that the rocks of these formations were strongly deformed during granulite metamorphism and now look like irregular alternated tectonic slices (Fig. 6.13). So, their stratigraphic sequence is highly conjectural.

Investigations conducted by Rosen *et al.* (1988) have shown that some rocks of the Daldynskaya Formation are similar in their main and rare element content to the komatiite-basaltic series rocks (Table 6.7). Various plagiogneisses (enderbites and charnockites) correspond compositionally with andesites, dacites and rhyolites, and in some cases a significant amount of potassium, up to $K_2O/Na_2O > 1$, has been observed in the rocks, i.e. judging from this parameter some of the granitoids are similar to latites. A detailed study of the carbonate– gneissic assemblage, using strontium and oxygen isotopic data, has ascertained that from the contents of most of the major- and trace-elements, the carbonate and calc-silicate rocks, as well as the garnet gneisses, were initially sedimentary in origin. It should be emphasized that, in contrast to the greenstone belts, the original textures and structures here are almost completely absent and all the reconstructions are based exclusively on the data on the chemical and isotopic composition of the rocks.

The rocks of all three formations underwent granulitic metamorphism at pressures of 10–12 kbar and temperatures, of 900°C. The age of premetamorphic rocks of the Daldynskaya Formation, from the data of E.V. Bibikova (Rosen *et al.*, 1988), is evaluated as no younger than 3.32 Ga, and that of granulite metamorphism is taken as 2.7 ± 0.1 Ga ago. The Anabar Shield granulitic complex is cross-cut by a number of thick, Early Proterozoic zones of retrograde metamorphism under amphibolite-facies conditions with an age of 1.97 ± 0.02 Ga. These are associated with formations similar in age to the extensive anorthosite–mangerite massifs, which were found to be 2.1 Ga old at a $\varepsilon_{Nd(T)}$ of -6.0 (Sukhanov, 1989).

Thus, in the tectonic history of the shield, two main stages can be distinguished. The fold structure of the Archaean Anabar Shield is marked by granulite-facies parageneses and is characterized by linear, narrow, often isoclinical folds, with a NNW strike. The folds are often overturned, suggesting an intensive tangential compression at about the time of granulitic metamorphism, and a substantial increase in the total thickness of the Archaean complex at about ~ 2.7 Ga ago. The other major tectonothermal event was the formation of c. 2 Ga old zones of intense retrograde metamorphism and the shear zones with which the major anorthosites massifs are related.

The predominant types of the Anabar Shield granulite are various types of enderbites (including hypersthene and two-pyroxene plagiogneisses) which, coupled with the metabasites, make up three-quarters of its volume. In the opinion of Rosen *et al.* (1988), most of them were volcanics and intrusive bodies; from other research the Anabar Shield metabasite–enderbite complex is partly fairly close in composition to the "grey gneisses" complexes: a marked prevalence of intermediate and acid rocks of the Na series and the presence of sheet-like metabasite bodies.

Table 6.7 Chemistry to plagiogneisses, metabasites and associated rocks of the Anabar Complex (after Rozen et al., 1988).

Components	1	2	3	4	5	6	7	8	9	10	11	12	13	14	15	16	17	18	19	20	21	22
SiO_2	48.85	50.67	53.13	58.63	61.33	63.92	67.70	71.11	48.71	48.17	52.23	57.96	61.73	62.73	67.50	70.71	48.13	54.96	58.18	61.48	68.18	71.20
TiO_2	1.07	0.26	0.72	0.65	0.81	0.91	0.36	0.30	0.42	0.52	0.31	0.92	0.72	0.65	0.38	0.42	1.81	0.96	0.74	0.94	0.32	0.14
Al_2O_3	13.70	3.99	15.31	17.35	18.04	11.99	6.26	14.00	5.80	15.32	15.28	16.03	16.82	15.90	15.38	14.15	14.80	19.34	16.39	14.43	14.11	14.61
Fe_2O_3	14.41	5.94	11.34	8.26	5.30	3.79	4.18	2.74	15.04	9.61	9.36	8.80	5.87	6.18	4.57	4.56	12.59	7.74	7.95	9.54	5.10	2.64
FeO	—	10.00	7.60	5.93	4.04	6.00	2.93	2.07	11.72				5.34	4.08			9.90	5.15				
MnO	0.22	0.19	0.15	0.10	0.10	0.06	0.03	0.03	0.19	0.14	0.06	0.11	0.09	0.05	0.06	0.06	0.17	0.08	0.12	0.12	0.05	0.02
MgO	7.24	19.67	5.83	3.90	2.97	2.30	1.22	1.51	18.49	8.86	1.88	3.47	3.43	2.44	1.88	1.36	5.60	2.84	2.75	2.50	1.87	0.45
CaO	11.04	7.75	8.58	5.62	5.60	4.65	4.52	4.68	8.84	12.18	4.17	6.62	4.90	5.17	4.17	3.04	11.90	5.77	5.42	4.12	2.89	2.39
Na_2O	2.40	0.78	3.67	4.10	4.56	4.22	4.05	3.00	0.76	2.18	3.16	4.25	3.98	4.22	3.16	3.00	2.60	4.70	3.90	3.18	3.50	3.30
K_2O	0.43	0.20	0.50	0.65	0.61	0.99	0.79	1.28	0.50	0.95	1.61	1.06	1.53	1.18	1.61	1.76	1.16	2.00	3.18	2.94	2.89	4.25
P_2O_5	0.08	0.05	0.09	0.24	0.21	0.17	0.08	0.11	0.04	0.04	0.12	0.20	0.20	0.16	0.12	0.16	0.25	0.31	0.22	0.26	0.11	
LOI	0.06	not def.	0.18	not def.	—	0.62	0.31	0.74	0.71	0.90	0.74	0.08	—	0.82	0.74	0.15	0.49	0.80	0.65	not def.	0.48	0.50
Total	99.50	99.50	99.50	99.50	99.53	99.50	99.50	99.50	99.50	99.50	99.50	99.50	99.27	99.50	99.57	99.37	99.50	99.50	99.51	99.50	99.50	99.50
Ni	64	180	76	40	23	136	13	19	840	300	330	32	30	25	12	16	100	14	23	23	40	30
Co	62	59	33	19	23	14	8	6	57	45	39	22	19	13	8	12	31	14	15	16	3	4
Cr	62	1000	160	71	39	45	55	90	240	1,600	580	28	80	86	27	82	290	57	66	50	27	292
V	410	170	260	35	120	80	38	30	18	330	180	220	100	36	105	70	130	100	150	72	30	20
Sc	—	—	—	—	13	—	—	—	—	46	22	—	17	—	6	10	—	—	18	—	9	7
Ba	120	70	200	500	280	390	550	200	420	110	170	400	450	490	600	780	400	1900	1100	1470	680	680
Sr	90	60	190	320	300	280	470	300	50	160	220	250	680	360	185	320	410	1300	320	195	220	140
Pb	6	9	10	17	9	16	23	15	25	19	21	14	14	12	18	22	13	20	23	22	16	32
Zn	150	90	120	100	50	80	50	<30	170	30	40	70	60	70	32	—	140	90	30	65	30	<30
Cu	140	9	59	49	30	13	14	10	38	92	22	130	6	20	18	16	74	52	27	17	22	20
Zr	—	50	50	50	110	230	110	120	50	40	30	—	290	190	125	220	480	350	180	215	150	90
Ga	30	8	19	26	15	22	23	20	14	19	24	26	—	19	19	28	27	26	20	24	28	17
Ge	1.8	1.8	1.5	1.2	0.7	0.9	0.7	1.7	1.9	—	—	1.1	1.0	0.9	1.1	—	1.3	1.0	—	—	—	—
Nb	<0.7	<7	<7	<7	<14	7	7	10	7	—	—	<14	<14	<7	<7	—	91	<7	—	—	—	—
Ta	<0.7	0.6	1.1	0.7	0.7	1.2	1.6	0.7	1.9	—	—	<0.6	0.7	0.7	<0.6	—	3.3	1.0	—	—	—	—
Li	31.0	8.5	6.4	8.3	7.3	6.8	7.8	32.0	14.0	26.0	13.0	15.0	6.4	11.0	13.7	12.4	12.0	15.0	14.0	8.6	7.6	4.3
Rb	<5	<5	<5	<5	<5	<5	5	16	19	<5	<5	73	14	18	33	32	25	12	52	57	65	69
B	—	—	—	—	7	—	—	—	—	22	23	8	—	15	12	—	—	12	—	—	12	15
F	520	1300	300	300	680	500	200	200	100	—	—	250	—	700	1,700	1,900	800	400	—	—	—	—

All iron is determined as FeO: Fe_2O_3 content is defined in individual cases and is not included in the total. 1–8 — low-potassium rocks. 9–15 — moderate-potassium rocks. 16–22 — high-potassium rocks. 1 — amphibole-two-pyroxene schist, 2 — clinopyroxene schist, 3 — amphibole–clinopyroxene schist, 4–6 — enderbite, 7 — hypersthene plagiogneiss, 8–10 — amphibole-two-pyroxene schists. 11–14 — hypersthene plagiogneisses, 16–17 — two-pyroxene schists, 18–21 — enderbites, 22 — hypersthenegneiss.

The Kola-Norwegian granulite Belt (KNGB) of the Baltic Shield (Fig. 6.2) is formed by various metasediments: biotite-hyperstene-, biotite-garnet-, garnet-hypesten-cordierite and garnet-cordierite-sillimanite gneisses, eulisites, banded iron formations, carbonate-silicate rocks, etc. In metasediments the thin layering of turbidite type is sometimes observed. Above are orthogneises and amphibolites of unclear age, which were interpreted as metavolcanics (Mirofanov, 1995). All rocks of the KNGB underwent intensive deformations and repeated metamorphism under conditions from high-temperature amphibolite- to granulite-facies grade (T = 650–800°C, P = 5–7 kbar; locally pressure could reach 8 kbar) (Glebovitsky, 1997) with predomination of the latter. The age of the earliest metamorphism is about 2.83 Ga and the latest – 2.6 Ga (Mitrofanov, 1995). Formation of the belt occurred under conditions of bilateral compression and was accompanied by the appearance of crust-derived granitoids of the ender-bite–charnockite series.

Between KNGB and Karelian GGST a transitional Belomorian Mobile Belt (BMB) occurred. Its regional structure is deciphered as a huge package of tectonic slices with a system of late Archaean and Palaeoproterozoic folds; nevertheless the structures retained a dominant NW strike and NE dipping, in the direction of the KNGB. Tectonic slices have the lithology of the Karelian GGST (TTG-gneisses and migmatites, fragments of greenstone belts), and its NE border is defined as an organic element of a single primary subhorizontal tectonic flow (Miller and Milkevich, 1995). Tectonic-metamorphic processes in the BMB began about 2.85 Ga ago; however, the main events occurred 2.8–2.75 Ga ago, when the BMB was a zone of tectonic-metamorphic reworking of the Karelian GGST under conditions of high-pressure amphibolite facies (Glebovitsky, 1997). According to him, the feature of the BMB's metamorphic is lateral zoning: from amphibolite facies of moderate pressure and temperature (T = 600–650°C, P = 5.5 kbar) at the NE edge of the Karelian GGST to a zone of regional migmatization (T = 650–730°C, P = 6–8 kbar) with local zones of granulite metamorphism (T = 725–800°C, P = 7.5–9 kbar) in central parts and in those adjoining the KNGB. Late Archaean magmatism in the BMB is represented by intrusions of gabbronorites, gabbrodiorites, enderbites, charnockites and tonalites (Zinger et al., 1996; Chekulaev et al., 1997).

Because the KNGB was formed as a result of bilateral compression, it suggests that its NE edge was primarily similar to the SW one, i.e. a transitional zone like the BMB occurred here. The same regularities are characteristic of the Limpopo Granulite Belt (see below). A large ductile shear zone occurred between the low-grade Abitibi and high-grade Opatica belts (Superior Province, Canadian Shield: Calvert et al., 1995). Seismic data show that reflections of this zone extended 30 km into the mantle, which was interpreted as a relic of a 2.9 Ga old suture associated with subduction. In both cases (i.e., Belomorian and Superior ones), the investigators suggest that these sutures were formed as a result of the Late Archaean collision; however, the same picture could appear as a result of plume tectonics activity, not linked with subduction (see Chapter 8).

Another well-studied example of a granulitic belt is the 2.7 Ga old Limpopo Belt of South Africa (van Reenen *et al.*, 1992). It is located between two granite–greenstone cratons: the Kaapvaal to the south and the Zimbabwe craton to the north. The granulite belt with a width of ~100km is subdivided into three zones, each characterized by intrinsic geological features and separated from each other and adjacent cratons by thick shear zones: a Central and two marginal zones.

The Central zone is formed by metasediments (predominantly garnet-cordierite-sillimanite gneisses, marbles, carbonate-silicate rocks, quartzites, iron-banded formations, etc.). As in the other granulite belts, abundant enderbite–charnockite magmatism took place. Along the boundaries with the adjacent ancient (Mesoarchaean) Kaapvaal and Zimbabwe cratons there are specific marginal zones 60–80 km in width; they are represented by systems of tectonic slices with the lithology of granite–greenstone material from corresponding cratons: Southern Marginal zone — from the Kaapvaal Craton, and Northern Marginal zone — from the Zimbabwe Craton. In their geological and petrological features, these marginal zones remain close to the BMB of the Baltic Shield, and like the latter, they are characterized by lateral metamorphic zoning, which intensified from cratons to the granulite belt, both in temperature and in pressure.

6.4   GEODYNAMICS IN THE ARCHAEAN

Did plate tectonics operate during the Archaean? Unfortunately this is not a question that can be answered at present (de Wit and Ashwal, 1997). Many geologists now suggest that plate tectonics were active in the Archaean (Windley, 1992; Condie, 1994; etc.). From that point of view, greenstone belts represent the ancient oceanic spreading zones, and almost all Archaean high-grade terranes can be explained in the context of one or two tectonic settings: deep crustal portions of magmatic arcs or collision orogens (Persival, 1994). However, most of the komatiites in greenstone belts underwent crustal contamination. This suggests that they ascended through continental crust and that the Archaean greenstone belts were not formed in an oceanic environment and could not be defined as ancient oceanic crust, since all the more Archaean ophiolites are absent (Arndt, 1994; Bickle *et al.*, 1994).

In the upper parts of the lava sequences of some greenstone belts, especially the Late Archaean ones, more silicic magmas — andesites, dacites and rhyolites, sometimes K-rich ones — are present. It rather resembles the subduction-related magmas at the Phanerozoic active plate margins and in essence provides the main argument for a similar interpretation of the Archaean situation. However, the composition of magmas depends only on the melting source, not the geo-dynamic situation in general. A good example of it is the early Palaeoproterozoic high-magnesian boninite-like series, which did not link with active plate margins at all (see Chapter 5).

As shown above, the geological situation in the Archaean was rather different from that in the Phanerozoic. There were two main types of geological structures in the Archaean: (1) granite–greenstone terranes or nuclei about 500–1000 km in diameter, and (2) granulite–gneiss belts between them. GGST consist of an essentially granitic (tonalitic) matrix with a network of greenstone belts where mantle-derived komatiitic–basaltic magmatism took place. As a whole, the geological features of the GGS terranes closely resemble the Phanerozoic continental rift zones (areas) like Baikal or the Province of Basins and Ranges (Chapter 2); Grachev and Fedorovsky (1980) first paid attention to it. Like the Phanerozoic continental rifts, greenstone belts were elongated depressions infilled by different sediments and mantle-derived volcanics. Such a situation in the Phanerozoic is linked with the spreading of the heads of asthenospheric diapirs beneath such areas, and the same mechanism is suggested for the Archaean GGS terranes, for example for Karelian GGST (Kulikov, 1988). In essence, GGST were the ancient extensional areas.

In contrast, GGT are characterized by strongly deformed and thrusting rocks, and were zones of compression and the descending flow of crustal material; on the Earth's surface above them sedimentary basins were common (Taylor and McLennan, 1985). The lithology of the GGT shows that they were formed of both old material of neighbouring GGSTs and new-formed sediments which underwent granulitic metamorphism (Rosen et al., 1988; van Reenen et al., 1992). Magmatism in GGT was different from GGST — there were mainly sili-cic (composed by andesites and dacites) melts which resembled subduction-related magmas.

As shown above, in relatively well preserved late Archaean areas like the Baltic and South African shields, or the Canadian Shield, it is established that GGST and GGT existed at the time and that specific transitional mobile belts evolved between them. They were particular zones of tectonic flowage of crustal material from the former to the latter and were characterized by structural-meta-morphic zoning which intensified in this direction. This picture does not corre-spond to the plate-tectonics model, but is in good agreement with the plume-tectonics model (see Chapter 8).

6.5 SUMMARY

1. The most ancient formations carrying information on the earliest stages of the Earth's crustal development are Archaean quartzites containing detrital

zircons with ages of 4.28 and some tonalitic gneisses (grey gneisses) of 3.96 Ga. Their age is only 0.3–0.6 Ga younger than the most ancient zircons in the Solar System.

2. The most ancient of the geological structures discussed are granite–greenstone terranes. The most ancient greenstone belts (3.7–3.4 Ga) were made up mostly by the formations of the komatiite series (komatiites, komatiitic basalts and basalts) and, in some instances, by volcanics of transitional and acid composition and the sedimentary rocks: siliceous and carbonate-siliceous sediments, shales, iron quartzites (iron-banded formation), etc. The more common are Late Archaean greenstone belts (3.0–2.7 Ga). Along with the komatiite series formations, there are also low-Ti tholeiitic basalts — in rare cases some of these are titanous; andesites and dacites begin to play an essential role, and turbidites and banded-iron formations become important among the sedimentary rocks. The ancient frame of the greenstone belts is largely formed by plagiogneisses of TTG series ("grey gneisses"). In the Late Archaean the K-granites appeared; on occasion even K-alkaline rocks of the miaskite series were present. Mantle-derived magmatism within the greenstone belts and their granitoid frame occurred almost simultaneously. GGSTs were vast extensional zones which in the geological sense are reminiscent of the Phanerozoic continental rift systems.

3. The formation of granulite–gneiss terranes (belts) began at least 3.65 Ga ago. They are located between cratons formed by granite–greenstone terranes and were large compressional zones. GGTs are formed mainly from metasediments; among them metapelites, iron-banded formations and carbonate-silicate rocks (including marbles) predominated. It suggests that at the time they were large sedimentary basins. GGTs were characterized by synkinematic crustal-derived abundant enderbite–charnockite magmatism.

4. Granite–greenstone and granulite–gneiss terranes existed simultaneously and were linked by specific transitional mobile zones like the Belomorian Belt of the Baltic Shield and the Marginal zones of the Limpopo Granulite Belt (S. Africa).

5. The Archaean formations, excluding some of the Late Archaean sedimentary basins of the Witwatersrand Supergroup type, are characterized by a high degree of migmatization, which dies out towards the central parts of the greenstone belts. This suggests that the upper crust was highly ductile at the early stages of the Earth's development.

6. The transition to Early Palaloproterozoic-type magmatism and geodynamics occurred rather unevenly, from 2.9 Ga in the Kaapvaal Craton to 2.5 Ga ago on the Canadian Shield. This is probably evidence for the non-uniform cooling of the lithosphere which did not completely stabilize until 2.5 Ga ago.

7. It is not clear yet: Did plate tectonics operate during the Archaean? The geological data within GGSTs more closely resemble the situation of continental rifting above extending mantle plume heads; there is no available geological evidence for interpreting these terranes in the same way as the situation on

convergent plate boundaries, only the presence of intermediate and acid volcanics in greenstone belts. Above, located among GGST and coincidently with them, vast granulite belts of moderate pressure with metasedimentary lithology and abundant crustal-derived magmatism have no analogy among the Phanerozoic environments and their origin could not be characterized in plate-tectonic terms.

CHAPTER 7

GENERAL FEATURES OF MAGMATIC EVOLUTION THROUGHOUT THE EARTH'S HISTORY

O.A. BOGATIKOV, V.I. KOVALENKO and E.V. SHARKOV

When discussing the main features of igneous evolution three problems must be considered:

1. To what extent were the relationships between magmatism and geodynamics as established for present-day geodynamic environments, also present in the Phanerozoic and Middle–Late Proterozoic?
2. What is the difference between these relationships and those of Early Precambrian?
3. What are the main stages of tectonic-magmatic evolution in the Earth's history?

7.1 COMPARATIVE ANALYSIS OF RECENT MAGMATISM AND THAT OF PHANEROZOIC FOLD BELTS AND ANOROGENIC ENVIRONMENTS

The magmatism of Phanerozoic orogenic and anorogenic regions is closest to that of the present day. This fact has been emphasized repeatedly by many investigators, although problems of correlation between fold-belt magmatism and present geodynamic environments have not been resolved completely as yet.

The results of the comparison of Phanerozoic magmatism and contemporaneous geodynamic environments, on the basis of schematic formulae of indicator magmatism (only taking into account rock types that make up more than abundances exceeding 10% of the assemblages), are presented in Table 7.1. We have restricted ourselves to discussing magmatism of the main development stages of fold belts and the basic types of geodynamic environment.

7.1.1 The Early Stages of Fold-Belt Development

a) Ophiolitic Assemblages

Ophiolite assemblages (Fig. 7.1) are typical of the earliest stages of fold-belt evolution. The reconstruction of the initial tectonic position of ophiolites is of

Table 7.1 Magmatism of present-day geodynamic environments and their palaeo-analogues.

Type of magmatism	Oceanic	Oceanic–continental	Continental	
			Orogenic	Anorogenic
State of fold-belt development	Early	Early-Orogenic	Orogenic	Anorogenic
Formula of indicator magmatism	T_{MORB}	$T + CA + A_{LTiK-Na, K}$	$CA + A_{LTiK-Na}$, $HA_{LTiK-Na, K}$	$T_{HTi} + A_{K-NaHTi} + HA_{K-Na, KHTi}$
Simple environments				
I Constructive (convergent)				
oceanic rifts (T_{MORB})	++			
continental rifts ($T_{HTi} + A_{K-NaHTi} + HA_{K-Na, KHTi}$)				++
II Destructive (divergent) island arcs				
Young ($T_{LTi} + CA_{HMg+LMg} + A_{K-NaLTi}$)	++	++		
Developed ($T_{LTi} + CA + A_{K-Na, KLTi}$)		++		
Mature ($CA + A_{K-Na, KLTi} + HA_{K-Na, KLTi}$)		++	++	
back-arc seas ($T_{MORB} + CA$)	++			

Active continental
margins (AKO) Andean
type ($CA_{LMg} + A_{K-Na, KLTi}$
$+ HA_{K-NaLTi}$)   ++

Collision Alpine-Himalayan
type ($CA + A_{K-Na, KLTi}$)   ++   ++

III Intra-plate

Continental rifts,
Oceanic islands
($T_{HTi} + A_{K-NaHTi, KHTi}$
$+ HA_{K-Na, KHTi}$)   ++

hot-spots ($T_{HTi} + A_{K-Na+HTi}$
$+ H_{AK-Na, KHTi}$)   ++

Complex environments

Californian or
Mongolian–Okhotsk
($T_{MORB} + CA + A_{K-Na, K}$
$+ HA_{K-Na, K}$)   ++   ++   ++

Collision of
Mediterranean type
($T_{MORB} + CA_{LMi}$
$+ A_{K-Na, K} + HA_{K-Na, K}$)   ++   ++   ++

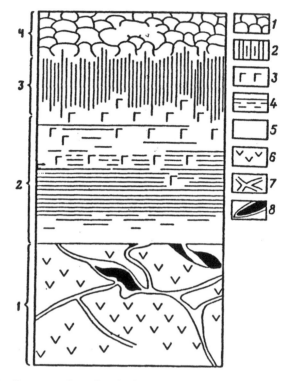

FIGURE 7.1   Cross-section of typical ophiolite sequence (after Laz'ko, 1988a).
1 — pillow-lavas; 2 — sheeted dykes; 3 — gabbroids; 4 — layered complex (alternation of ultramafites and gabbroids); 5 — dunites; 6 — harzburgites and lherzolites; 7 — veined ultramafites and gabbroids; 8 — chromitites.

major importance for the geodynamic interpretation of the early stages of fold-belt formation. At present, the overwhelming majority of researchers agree that ophiolites were once fragments of oceanic crust and the upper mantle, but there is some controversy about the precise region of the palaeo-ocean floor in which they were formed. Firstly, the ophiolites have been interpreted as the analogues of lithosphere of major oceans formed within mid-oceanic rift zones by large-scale spreading (Coleman, 1977). However, subsequent studies of the petro-chemical characteristics of ophiolitic basalts have demonstrated that they are characterized by both tholeiitic and calc-alkaline trends (Fig. 7.2). In this regard, attempts have been made to elaborate the classification of ophiolites (Dobretsov, 1980; Kovalenko 1987; Laz'ko and Sharkov, 1988; Nicolas, 1989; Detrick, 1991; etc.), the ultimate objective being to elucidate the original initial geodynamic environments in which they were formed.

At present, more than ten types of ophiolite have been distinguished by various investigators, on the basis of petrological, geochemical, lithological,

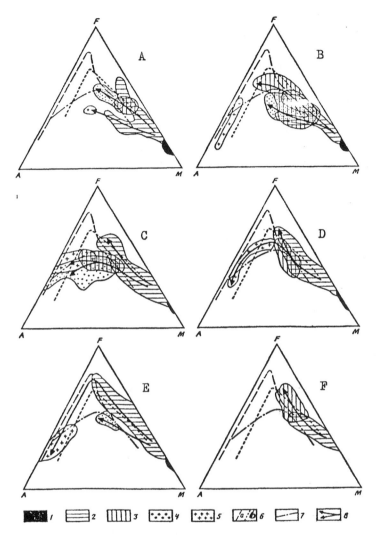

FIGURE 7.2 Fields of igneous rock compositions for some ophiolites on the AFM diagram. After E.E. Laz'ko (Bogatikov *et al.*, 1987). (A) Kamchatsky cape, Kamchatka; (B) Troodos, Cyprus, (C) Samail, Oman; (D) Bay of Island, Newfoundland; (E) Maynitsky Complex, Koryak Highland; (F) Tortuga, Patagonia; 1 — ultrabasic rocks; 2 — gabbroids; 3 — parallel dykes and associated volcanics; 4 — volcanics unrelated to dykes; 5 — intermediate and acid rocks associated with various ophiolite complexes; 6 — igneous differentiation trends (a — Skaergaard Intrusion, after Wager and Brown (1968), b — tholeiitic basalts of the Hawaiian Islands); 7 — boundary between the tholeiitic and the calc-alkaline series, after Irving and Bargar; 8 — possible trends of melt evolution during the formation of ophiolites.

stratigraphic, tectonic and other evidence. Among the possible geodynamic environments of their generation are mid-oceanic ridges, seamounts and oceanic islands, transform faults, back-arc (marginal sea) basins, island arcs at early stages in their development, fore-arc zones and deep-sea trenches, zones of inter-arc spreading, passive continental margins, rift zones on continental crust, etc. As noted by Dobretsov (1980), the number of types of ophiolite may be even greater, because their variations are determined by independent factors and often do not correlate with each other. For that and other reasons, in some cases it is difficult to provide an unambiguous correlation between the types of ophiolites and the particular geodynamic setting, although some analogies can be outlined with a certain degree of confidence.

Certainly, the presence within the ophiolites of sheeted dyke complexes, which indicate consolidation of the assemblage during intense sea-floor spreading, is of crucial importance in the interpretation of ophiolites. On the present-day Earth's surface, these conditions occur in mid-oceanic rifts and back-arc spreading centres in island arc–marginal sea systems.

It may be necessary to point out a difference — repeatedly stressed in the literature — between peridotites of ophiolites and those of mid-oceanic assemblages, i.e. there is a marked predominance of lherzolites in mid-oceanic assemblages, but harzburgites predominate in the ophiolites, along with a nearly complete absence of ultramafic vein and ultramafics of gabbroid complexes in the oceans (Kovalenko, 1987; Laz'ko and Sharkov, 1988; etc). According to the available data (see Chapter 2), in the Atlantic and Indian oceans lherzolites actually occur more frequently than the harzburgites. However, within the zones of transform faults cutting the East Pacific Rise, the peridotite ratios are reversed, at least in those parts where representative material was obtained (the Clarion, Garret and Eltanin faults). In the basement rocks of a number of island arcs (Marianas, Izu–Bonin, Tonga, Central American and Puerto Rican deep-sea trenches), there are harzburgites that are extremely depleted in fusible components, and which contain high-Cr spinels. These rocks are identical to the ultrabasic rocks that dominate the composition of most of the ophiolites studied.

Data on the ultrabasic rocks of spreading zones in the marginal seas are scarce. Evidence obtained from two well-known exploration sites (the Parece–Vela Zone in the Philippine Sea and the Cayman Trough in the Caribbean) suggests the presence of both types of peridotites here (Laz'ko and Sharkov, 1988). Thus, it must be concluded that among the ultrabasic rocks of the ophiolites, analogues of island-arc ultrabasites are predominant, although examples of lherzolite–harzburgitic Alpine-type assemblages are also known (for example: some massifs of the Urals, the Koryak Highlands, the Balkan Mountains, the North American Cordilleras, Japan, etc.) which could have been formed in structures similar to present-day zones of active back-arc spreading or fast-spreading mid-oceanic ridges like the East Pacific Rise. Finally, in at least one region — the western Mediterranean (the western Alps, the Apennines and the

Dinaric Alps) — there are ophiolites with a marked predominance of lherzolites in their ultrabasic complexes. The present-day analogues of these complexes could be the ultrabasic rocks of slow-spreading mid-ocean ridges of Arabia–Indian or Mid-Atlantic types. However, even in this case one cannot rule out the possibility of correlating these peridotites with the rocks of back-arc spreading zones.

Analogous geochemical correlation of oceanic and ophiolitic gabbroids is still difficult to interpret because of wide variations in the chemical composition of the rocks, ascribed to their cumulus character (Bogatikov *et al.*, 1987; Laz'ko, 1988a). At the same time the gabbroid assemblages of the ophiolites show essential differences in the crystallization trends of the primary basic melts associated with varying conditions of differentiation.

According to available data, this may be attributed to three fractionation trends. The first one is characterized by the following sequence of mineral crystallization: olivine, plagioclase–clinopyroxene (orthopyroxene) and Fe–Ti oxides. In the corresponding cumulus series of rocks, dunites are successively replaced by troctolites, olivine gabbros, gabbro and ferrogabbros (gabbronorites are occasionally encountered). A similar trend is characteristic of the so-called high-Ti ophiolites (the name reflects the chemical composition of primarily volcanic and subvolcanic members of the assemblage) of the northern Apennines, Corsica, Newfoundland, Chile, etc. Their present-day equivalents are series of plutonic basic igneous rocks on mid-oceanic ridges (for example, on the West Indian Ridge, see Chapter 2) and back-arc spreading zones in marginal seas (the Philippine Sea, see Chapter 2).

The other trend is represented by a different series of cumulus phases, i.e. olivine–clinopyroxene–plagioclase–orthopyroxene–Fe–Ti oxides–amphibole. In the intrusive complexes of ophiolites this corresponds to the replacement of dunites by wehrlites, clinopyroxenites, gabbronorites, ferrogabbronorites and diorites. The gabbroid assemblages of this type referred to as "low-Ti" constitute the majority of the assemblages studied and correspond with current island-arc igneous activity.

The last type of low-Ti gabbroid complexes which occurs within ophiolites (e.g. Mongolia, New Guinea and Newfoundland) is close to the second type in terms of its geodynamic interpretation. The sequence of mineral crystallization in the third type of series (olivine–orthopyroxene–clinopyroxene–plagioclase–Fe–Ti oxides–quartz–amphibole) leads to the appearance of lherzolites, websterites and gabbronorites at the early stages of fractionation. The relationship between such series and orthopyroxene-bearing basalts and boninites indicates that they are very likely have been formed at the early stages of island-arc evolution. Appreciable differences in the recognized crystallization trends are most likely to have been caused by variations in the water concentrations in magmatic chambers and, in part, probably by the composition of parental magma sources and the dynamics of fractionating systems. It should be pointed out that classification and geodynamic correlation of ophiolite gabbroid assemblages can be hindered in some cases by the presence of various fractionation trends within a single assemblage.

Despite the advantages of using the characteristics of intrusive rocks for palaeotectonic reconstructions, so far ophiolites have been classified according to the chemical composition of their dykes and volcanic rocks.

Extensive geochemical studies of ophiolitic lavas have shown that two main groups of basalts can be distinguished in ophiolites. These correspond with:

1. volcanics of oceanic rifts or those of spreading zones in marginal seas;
2. volcanics of island arcs.

This conclusion is also relevant to basic dykes and is clearly illustrated by the Ti/Cr vs. Ni diagrams (Fig. 7.4) and Ti–Zr diagrams (Fig. 7.3). Moreover, a lot of importance is attached to the discovery of boninite series rocks of various ages within ophiolites (Dobretsov, 1980). The present-day analogues of these rocks are known only in the fore-arc zones of ensimatic island arcs.

Thus, there are a wide range of geodynamic environments in which ophiolites could have been formed, taking into account their intrusive and volcanic members; i.e., in principle, all of the above-mentioned environments of intensive oceanic crustal spreading are feasible. However, statistical analysis has demonstrated that, within fold belts, assemblages of island-arc rock type are commoner than those of intra-oceanic type (Laz'ko, 1988a). Moreover, it is conceivable that various fold belts ended their development at different stages of evolution, which might account for the differences between them. In this respect, it must be remembered that there are some differences between the two main environments of magma genesis in contemporaneous oceans, which may be relevant to the problem of ophiolites:

1. The presence of acid rocks in many ophiolites distinguishes the latter from present-day mid-oceanic assemblages, but makes them closer to the products of island-arc magmatism.
2. In most cases, the lithological compositions and thicknesses of the sediments that occur together with the igneous rocks of ophiolites do not correspond with the sedimentation processes in the central parts of the oceans, but are similar to those observed at its margins.
3. The age gaps between the eruptions of ophiolitic lavas and their consolidation within the fold belts (e.g. the Mediterranean, Pacific and North Atlantic) are out of proportion with the duration of spreading of present-day oceans (hundreds of millions of years) and in general are not more than several tens of million years. This is roughly equivalent to the duration of back-arc basin evolution, although the presence in the fold belts of ophiolites of different ages does suggests that oceans did have a bigger life span in the past.

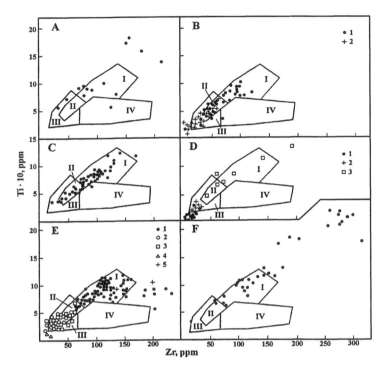

FIGURE 7.3 Distribution of points representing the basalt compositions of volcanic and dyke complexes of some ophiolites, plotted on the Ti/Zr diagram of Pearce and Cann (1973). After Bogatikov *et al.* (1987). (A) massive and pillow lavas of the Sevan–Akerinsky Zone, Lesser Caucasus; (B) Troodos, Cyprus (1 — the lower suite of volcanics and the dyke complex, 2 — upper suite of pillow lavas); (C) pillow lavas of the Bay of Island, Newfoundland); (D) Betts Cove, Newfoundland (1 — lower suite of volcanic dykes, 2,3 — pillow lavas of intermediate and upper suites); (E) Samail, Oman (1 — volcanics of the Geotimes Formation and parallel dykes; 2–5 — volcanics of the following suites: Leceil (2), Elly (3), clinopyroxene-phyric (4) and Salahi (5)); (F) pillow lavas of the Maram Complex, New Guinea; fields I and II correspond with the areas of tholeiitic basalt compositions of spreading zones in oceans and marginal seas; II, III — represent tholeiitic basalts of island arcs; II and IV — represent calc-alkaline basalts.

All of these data indicate that the ophiolites originated on the active margins of palaeo-oceans rather than in their centres. In particular, it should be emphasized that there are no clear-cut criteria for discriminating between the igneous rocks of current mid-oceanic rifts and those of back-arc spreading zones in marginal seas, which prevents reliable classification of their palaeo-analogues according to their respective environments. The available data are still inadequate for recognizing mid-oceanic ophiolites. At the same time, evidence for the

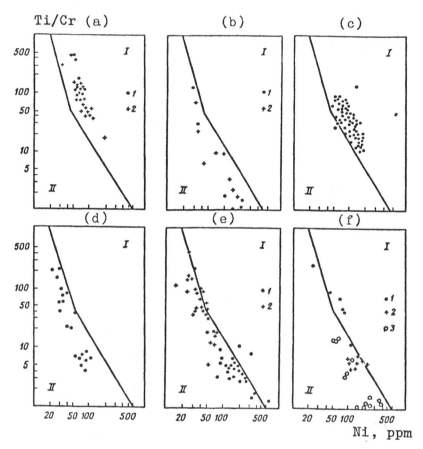

FIGURE 7.4 Distribution of plotted points for basalt compositions of volcanic and dyke complexes for some ophiolites, on the Ti/Cr–Ni diagram of Beccaluva *et al.* (1979). After Bogatikov *et al.* (1987). (a) Mugodjary (1 — pillow lavas, 2 — sills and parallel dykes); (b) Khan–Tayshir, Mongolia (1 — pillow lavas, 2 — parallel dykes); (c) pillow lavas, northern Apennines; (d) massive and pillow lavas, sills and dykes, Vurinos, Greece; (e) Troodos, Cyprus (upper (1) and lower (2) pillow-lavas suites); (f) Betts Cove, Newfoundland (1, 2 — volcanics of the upper (1) and intermediate suites (2), 3 — volcanics of the lower suite and dykes); field I — tholeiitic basalt composition of spreading zones in oceans and marginal seas; field II — tholeiitic basalt composition of island arcs.

island-arc and back-arc nature of ophiolites suggests a genetic relationship between the assemblage and palaeo-analogues of present-day destructive margins between oceanic lithospheric plates. The complete set of the rocks known from ophiolites can be observed along the present western margin of the Pacific Ocean, over an area of several hundred kilometres across the strike of young

Marianas- and Tonga-type island arcs. It is probable that these regions are the best model for the eugeosynclinal zones of fold belts at the early stages of their evolution, although it would be premature to deny that the relics of past mid-oceanic assemblages occur in these regions.

The classification of the active margins of palaeo-oceans as the probable parental structures of ophiolites does not preclude the possibility of a more detailed geodynamic classification (for example, into island-arc and marginal sea assemblages). This appears to be rather simple in cases where all of their members follow the same evolutionary trends. However, ophiolites not infrequently include rocks whose compositions allow varying geodynamic interpretations or suggest different stages of development for the individual structures within the belts. In this respect, the existence of the compositionally complex but tectonically undivided pseudostratified volcanic complexes is particularly representative. One of the most striking examples of this kind was recognized in Oman (Alabaster *et al.*, 1982) where in the northern part of the Samail Complex, five markedly contrasting volcanic suites succeed each other within the section (Fig. 7.3). Judging by their geochemical characteristics the basalts of the lower suite which grade downward into a dyke complex, must have been formed within a spreading centre in a marginal sea. They are overlapped by volcanics characteristic of evolving island-arc environments. The upper suite includes rocks with signs of intra-plate magmatism, consistent with island-arc and passive continental margin collision (Alabaster *et al.*, 1982). A different sequence of events can be deciphered in the Betts Cove ophiolites of Newfoundland (Coish *et al.*, 1982). Here, in the lower suite of the volcanic complex, directly related to the dykes, lavas of the boninite type are widely represented; in the middle suite typical island-arc tholeiites occur and in the upper suite there are basalts that are similar to MORB (Fig. 7.4). The Troodos volcanics are the most typical example of a complicated volcanic complex (Coleman, 1977).

The above-mentioned evolution of volcanic rock composition provides evidence for sequential accretion of the assemblage section under various geodynamic conditions, i.e. large-scale horizontal movements of oceanic plates accompanied by various types of magmatism.

b) Fold Belts as Possible Palaeo-Analogues of Destructive Plate Margins

If the suggestion regarding the similarity between many ophiolites and the crust of marginal seas is correct, then back-arc sea-island-arc systems, as mentioned above, can be an appropriate model for the evolution and the lateral zonation of most fold belts (Fig. 7.5).

It can be seen from Table 7.1 that the early stage of fold region development (inner eugeosynclinal zones) most closely resembles the present-day young

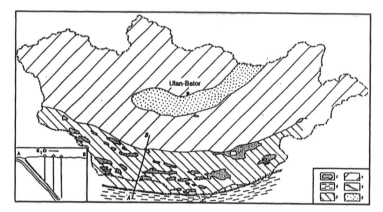

FIGURE 7.5 Location map for the Carboniferous volcanics of Mongolia. (after Kovalenko and Yarmolyuk, 1987). 1 — present outcrops of Carboniferous volcanics; 2 — late Hercynian rocks of Inner Mongolia; 3 — Hercynian rocks of southern Mongolia; 4 — Caledonides of the central and northern Mongolia; 5 — outcrop of the subduction zone on the surface; 6 — intra-continental Khangai–Khenteysky miogeosynclinal trough. A reconstructed section across the boundaries of the late Palaeozoic plates: the continental northern Asian one and the oceanic palaeo–Tethyan one are shown in the inset; 1–3 — locations where volcanics were sampled.

island arcs. Under both conditions, tholeiitic and calc-alkaline magmatism are widely developed and to a lesser extent, subalkaline K–Na magmatism. Although comparisons between the island-arc and early-orogenic (early geosyn-clinal) structures have been made already, at present this conclusion is convinc-ingly supported by statistics: the widespread occurrence of calc-alkaline petrochemical series rocks has been observed within the early-stage (including ophiolitic) assemblages, which here account for about one-third of the total amount of volcanics (Ronov *et al.*, 1990). Certainly, one cannot exclude the pos-sibility of the contribution of oceanic rift magmatism in such assemblages, espe-cially those which originated as a result of the closure of Tethyan-type oceans, when some of the lithospheric plates of the oceanic lithosphere proper may have been involved in obduction (Dobretsov and Kirdyashkin, 1991).

The presence of moderately alkaline K–Na series in early-stage igneous assemblages is probably, to a certain extent, associated with the spilitization of basalts, but the presence of true alkaline igneous rocks is a possibility, as estab-lished for oceanic islands and seamounts and island arcs.

In general, the origin of the spilites remains uncertain. They did not appear until 2 Ga ago, concurrent with the first Fe–Ti basitic rocks, and these two rock types have much in common in terms of their mineral composition and geo-chemistry. Spilites have declined up to the present day. The conventional view is that spilites were derived from the alteration of tholeiitic basalts as a result of their reaction with sea-water (Amshtutz, 1974; Wedepohl *et al.*, 1983), but this is thought to be unlikely, because on the present-day ocean floor where this reaction should have proceeded on a large scale, spilites have not been observed. The hypothesis of their relationship with regional Na-type metamorphism, characteristic of the early stages of orogenic belts, still does not solve the problem. Together with high Na content, they possess all the other attributes of moderately alkaline Fe–Ti basalts, in both the composition of their relict igneous minerals (titanomagnetite, titan-augite, kaersutite, apatite, etc.) and by their major and rare elements. The latter are characterized by high concentrations of Ti, Fe, P, Y, Ba, Be, V, Sr, Zr, etc., contrasting sharply with those of MORB tholeiites.

Therefore the most reasonable hypotheses seem to be those postulating a correlation between the origin of spilites and specific basaltic magmas of moderate and increased alkalinity with relatively high water and carbon dioxide content (Amstutz, 1974; Sharkov, 1983b) and/or with metamorphism of moderately alkaline rather than tholeiitic basalts. From this standpoint, the predominance of spilites among the basic rocks during the early stages of Phanerozoic fold-belt development may suggest the more widespread occurrence of moderately alkaline basaltic magmatism in Palaeozoic oceanic crust than was the case for the Mesozoic and Cenozoic.

Comparisons of early magmatism of fold belts and that of young island arcs conform with general geological data. In particular, structures from the early stages of folding are characterized by thick, deep-sea sedimentary sequences. In terms of the thickness and character of the section, sedimentary formations of young arcs and adjoining marginal seas may be considered as fairly satisfactory equivalents of sedimentary formations from the early stages of fold-belt development.

As in this and other cases, tholeiite-series rocks up the section are usually replaced by calc-alkaline series formations. Alternations of tholeiite- and calc-alkaline-series rocks are observed occasionally, attesting to repeated changes in extensional and compressional conditions, which is also atypical of mid-oceanic rifts.

Paired metamorphic belts are characteristic of many fold regions combining units regionally metamorphosed under low- and high-pressure conditions. The studies of Miyashiro *et al.* (1982) have demonstrated that, in Alpine regions, low-pressure, high-temperature belts, followed by abundant granitoid magmatism, are confined to the back-arc zones of island arcs, for example Japan, whereas high-pressure low-temperature metamorphism (the glaucophane-schist facies) occur in the fore-arc zones of these structures. These blue-schists are thought to have been generated by subduction, and they indicate the presence of a Benioff zone. Therefore the presence of similar units in Phanerozoic and Late Precambrian fold belts (the Palaeozoic Kiyama, Omi and Sangun belts of Japan, the Palaeozoic Kyrento Series in Chile and NW Kamchatka, and, in the Mesozoic–Cenozoic, the Alps and California, and the Sanbagava belt of Japan, as well as those of New Caledonia and Central Kamchatka) provides further evidence that these structures were once island arcs.

7.1.2 The Orogenic Stages of Fold-Belt Development

The magmatism characteristic of the late stages of fold-belt development closely resembles that of the early orogenic stages (Table 7.1). The difference between early- and late-stage magmatism is quantitative rather than qualitative; in early orogenic igneous assemblages, moderate alkaline series are widely represented, but the tholeiitic series are less common. According to indicator formulae for igneous assemblages the rocks of the late stage of fold-belt development and early orogenic magmatism are more similar to the present-day magmatism of developed and mature island arcs. The latter, which have a formula similar to these types of magmatism, are distinguished by a rarer occurrence of tholeiitic magmatism and a more frequent occurrence of alkaline magmatism in the mature arcs. Accordingly, the late-stage magmatism is similar to that of developed arcs, and the magmatism of the early orogenic stages parallels that of mature arcs. However, it is not inconceivable that these two stages are so intimately interrelated that they can intergrade. Therefore these possibilities are also shown in Table 7.1.

In the late orogenic stage of fold-belt development, rocks of the tholeiitic series rocks effectively disappear and the magmatism is similar to that of Andean-type continental margins and of Alpine–Himalayan-type continental collision zone (Table 7.1). It was also a stage of intense primarily granitoid, calc-alkaline and moderately alkaline (K–Na and K) magmatism. Andesites and rhyolites account for about 80% of the total volume of volcanics, and rhyolites may possibly predominate over andesites (Ronov *et al.*, 1990).

The igneous assemblages of anorogenic regions (see Chapter 4) most closely resemble those of present-day continental intra-plate magmatism. They are usually formed under extensional conditions, followed by dome formation, the generation of rift structures, the development of large basaltic lava fields, etc. There appears to be a marked similarity between igneous assemblages in which rocks of the moderate- and high-alkaline series (mostly Ti-rich K–Na and, to a lesser extent, K), tend to predominate. The characteristic feature of Phanerozoic anorogenic magmatism is that, because these areas have been deeply eroded, the intrusive equivalents of the lava series tend to be widely exposed and in larger volumes. Moreover, these former rift environments sometimes contain predominantly intermediate and acid moderately alkaline rocks, which are not characteristic of the Cenozoic.

In addition, among the anorogenic formations there are assemblages with no present-day equivalents, for example the titaniferous lamproites and kimberlites, although this may be due to the fact that their volcanic analogues are very rare and probably have not been reliably identified among Late Cenozoic volcanics as yet.

The formation of volcanic traps is still incompletely understood. They appear to be absent in the Late Cenozoic. As mentioned in Chapter 2, the Columbia River and Snake River basalts are most likely to have resulted from back-arc spreading, and are hardly the equivalents of vast intracontinental basalt fields such as the traps of Siberia, Karoo and Deccan. Neither can they be correlated with the products of Late Cenozoic intra-plate magmatism, which although fairly widespread, occur in relatively small areas rather than forming large regions.

The evolution of magmatism within the trap fields is generally similar to that observed during ocean opening: i.e. in a sequence from Fe–Ti K–Na basalts and picrites of moderate and high alkalinity to MORB-type tholeiitic basalts (see Chapter 4). In some cases the magmatism led to ocean opening (e.g. the Parana–Etendeka traps and the Deccan traps), but in others, it did not (e.g. the Siberian traps). The reasons for these differences are so far not fully understood, particularly in view of the evidence from recent geochronological data that the exceptionally high rates of trap formation coincided with global-scale catastrophic events confined to the boundaries of the major geological periods. This may explain their relative rarity and their different course of development.

Table 7.1 also shows the indicator formulae for the magmatism of complex geodynamic environments, as discussed in Chapter 2, including Californian-, Mongolian–Okhotsk-, and Mediterranean-type collision zones. Their magmatism is characterized by almost all of the known petrochemical series, and igneous assemblages from different stages of fold-belt development and from anorogenic zones (see Table 7.1), are being formed simultaneously. In these cases the most complex, often zonal, tectonomagmatic areas, with the greatest variety of magmatism, are being produced. As in present-day environments the magmatism of these areas may have been formed by the superposition of several simple geodynamic environments.

Many investigators have reconstructed the trends of development of fold belts on the basis of analogies with current geodynamic environments. Of these the most common is the Wilson cycle (Wilson, 1968), according to which an orogenic cycle beings with ocean opening and ends with ocean closure. The scheme was elaborated subsequently by Mitchell and Garson (1981) and Kovalenko (1987), among others. Thus, as follows from Table 7.1, continental rifting or trap formation, evolving into sea-floor spreading, should have been preceded by the formation of fold belts. As a result, zones of oceanic spreading appeared that were similar to those of the Red Sea, the Atlantic or the Tethys. The formation of a fold belt as such could be initiated with the emergence of active continental margins, or island arcs at the margins of oceans. These would have started off as intra-oceanic and later (or concurrently) would have evolved into developed or mature types. The future fold belt would have passed through the following phases: the early (young arc–marginal sea system), late (developed arc), and early orogenic (mature arc) up to the late orogenic stage of Alpine–Himalayan collision type. This was probably the sequence of development of the

Alpine, Urals and Central Asian fold belts. In this scheme of fold-belt development, complex geodynamic environments are formed at almost every stage, due
to the effects of hot-spots, thus supplementing the range of other, simple geodynamic environments.

However, the process of fold-belt formation along the periphery of the Pacific
Ocean proceeded in a rather different manner and did not end in ocean closure.
Here, fold belts were formed at the boundary with an ancient ocean (Puschrovsky
and Melanholina, 1989) and were asymmetrical in character ⌐ from the initiation of active margins which have evolved into island arcs and back-arc seas, as
in the case of the western periphery of the Pacific Ocean — or could have followed the existing active continental margin regime. In the first case, a typical
example of which is the Japan Sea (see Chapter 2), continental rifting occurred
first, in the back-arc part of the structure, and progressively evolved into back-arc
spreading, followed by subsidence and the formation, on the same site, of a back-
arc sea floored by both oceanic and subcontinental crust. Fold-belt formation
occurred here by the suturing of island arcs to continents after the closure of
back-arc seas, as established for the Phanerozoic fold belts along the Pacific rim
(Puscharovsky and Melanholina, 1989). In the second case, which is an active
margin, continental sedimentary basins are formed in the arc-trench gap, fairly
often accompanied by basalt magmatism of the Columbia and Snake River types.
It is conceivable that these basins will evolve further to form fold belts.

As noted above, the selection of a particular model of fold-belt development
depends on whether or not the ophiolites can be interpreted as relics of oceanic
crust or crust of marginal seas.

It is obvious that both the Wilson cycle and other cycles with strictly fixed
sequences of geodynamic environments and associated magmatism are of limited
applicability to actual mobile belts. This has to be taken into account when considering fold-belt development. However, particular fold belts differ mainly in
their sequence of geodynamic environments and their indicator magmatism, the
degree of igneous activity and/or orogenesis or their absence, all of which determine the specific type of magmatism in a particular region and the type of ore
mineralization that might be expected there.

7.2 GENERAL FEATURES OF EARLY PRECAMBRIAN MAGMATISM

The Phanerozoic type of magmatism and geodynamics can be traced with cer-
tainty back to the Late Proterozoic. In effect, Late Proterozoic (Pan-African or
Baikal) fold belts do not differ from the Caledonian and Hercynian fold belts
(Borukaev, 1985). However, in the Middle Proterozoic, before 1.2–1.0 Ga, and

particular in the Early Proterozoic at the 2.2–2.0 Ga boundary, orogenic processes differed markedly from those of the Phanerozoic. At that time, huge anorthosite–mangerite–charnokite intrusions were formed in ancient collision zones. As can be seen from data on the Grenville, Lapland and Stanovoi belts, they are closely associated with thick shear zones and high-pressure zonal amphibolite- and granulite-facies metamorphism. Inverse metamorphic zoning often occurred. As in the Phanerozoic, oceanic- or continental-type crust (the latter with dyke swarms) must have been formed in the back-arc spreading regions, for example the oceanic crust of Pechenga–Varzuga on the Baltic Shield and the continental crust containing dyke swarms such as the McKenzie dyke swarm of the Canadian Shield, and sedimentary sequences (many of which are continental, for example the Udokan Basin on the Aldan Shield). Within these basins the magmatism is predominantly MORB-type basaltic, with smaller amounts of Fe–Ti picrites and the basalts of normal, moderate and high alkalinity — typical of the Phanerozoic intra-plate magmatism.

This type of tectonomagmatic activity is essentially an ancient variation on plate tectonics. The first half of this interval (up to 1.7–1.6 Ga) is characterized by the worldwide acid igneous activity (mostly of the potassiun series), a typical representative being the anorthosite–rapakivi granite assemblage. The volcanic–plutonic formations of this stage form huge belts (the Baltic–Ukrainian, Trans-Siberian and North American belts among others) which were created at the site of ancient intercratonic ocean collapse. A reconstruction of the former position of these belts shows that they formed an interconnected system that provides evidence for the presence of the first supercontinent and the emergence of Panthalassa — a precursor of the Pacific Ocean.

The situation changed abruptly at the 2.2–2.0 Ga boundary, representing a radical change in the development of the Earth's lithosphere. In rocks more ancient than these, obvious geological evidence for the existence of plate tectonics is lacking. Indirect evidence in favour of Archaean plate tectonics is based largely on the geochemical data and on implying that the geodynamic position of ancient igneous activity did not differ from the situation in the Phanerozoic (Taylor and McLennan, 1985; Condie, 1994) and could have a different explanation (see Chapter 10). Titaniferous igneous rocks of moderate and high alkalinity, corresponding in composition with Phanerozoic intra-plate formations, are also absent prior to the 2.2–2.0 Ga boundary. However, rocks typical of plate boundary magmatism in the Phanerozoic were predominant before 2.2–2.0 Ga ago, but occupied an entirely different geodynamic position.

In particular, one of the principal rock types within the 2.5–2.2 Ga interval was the siliceous high-magnesian boninite-like series, which are rarely encountered in the Phanerozoic and are observed only on the active margins of oceans and continents. However, in the Palaeoproterozoic they possessed all the features of intra-plate formations, producing large dyke swarms, giant layered intrusions and infilling graben-like volcanic–plutonic structures. The areas where such rocks were developed consist of large-scale oval structures reminescent of plume-tectonic features on Venus (Phillips *et al.*, 1991). It is obvious from the above-mentioned data that a radically different type of deep petrogenesis was operating at that time, as reflected in both the geodynamics and igneous rock composition, as well as the mechanism of magma generation. There is no available geological evidence for plate tectonics in the Early Palaeoproterozoic and plume tectonics was predominant.

Archaean magmatism is even more distinctive as expressed by the widespread development of komatiite-series formations and tonalite-series plagiogranitoids. These and other rock types can be observed within the two main types of geological structures from that time: granite–greenstone and granulite–gneiss terranes, presumably representing complementary structures.

As shown in Chapter 6, the granite–greenstone terranes form large-scale areas which must have developed within extensional areas. Within the greenstone belts an important role is played by komatiite-series formations: peridotitic komatiites, komatiitic basalts and mildly magnesian basalts often overlapping with MORB in their chemical characteristics. Andesites and rhyolites are occasionally present in subordinate quantities. In Late Archaean belts they are sometimes supplemented by moderately titaniferous tholeiitic basalts and low-Ti alkaline rocks of the K series. In the middle and upper parts of these belts volcano-sedimentary complexes are encountered fairly frequently. A large proportion of these complexes contain specific types of andesties, somewhat similar to those of the high-magnesian Early Palaeoproterozoic calc-alkaline series, as well as acid rocks of the dacite–rhyolite series, associated with ferroquartzites and turbidites. In Late Archaean greenstone belts low-Ti high-K latite series rocks are sometimes found in association with these rocks. The superstructures of all greenstone belts are predominantly gneiss–granites of tonalite–trondhjemite composition ("grey gneisses"). Their origin is controversial; some geologists consider them to be the result of melting and granitization of the primary basic crust, whereas others regard them as the remains of partly reworked primary sialic crust.

In contrast to granite–greenstone terranes, granulite–gneiss terranes appear to represent zones where material must have accumulated. Judging by data from the Kola-Norwegian and Limpopo belts and other analogous structures (see Chapter 6), they were compensated zones of descending movements, where excess crustal material from adjacent GGSTs was accumulated. GGT were vast sedimentary basins, where mature sediments (pelites, limestones, quartz sandstones, etc.) predominated; above, slices of adjacent GGSTs (TTG-rocks and material from GST) were found. All this material was subjected to granulitefacies matamorphism, so that its textural and structural features were barely preserved. Abundant synkinematic enderbite–charnockite magmatism is very characteristic for GGT. The appearance of GGTs in many instances coincided with the peak of GGST development of both Early Archean — 3.6–3.4 Ga and Late Archaean — 2.9–2.6 Ga.

From available geological evidence, it can be inferred that the dominant type of Archaean geodynamics was plume tectonics, characterized by the upwelling of major mantle diapirs, above which the granite–greenstone terranes were formed; the granulite–gneiss terranes were probably formed above descending convection currents.

While the existence of rigid sialic crust in the Palaeoproterozoic is undeniable, it is less certain for the Archaean. The ubiquitous development of migmatization and the ductile flow of crustal material are the most characteristic features of Archaean formations. The stabilization of the Earth's crust appears to have proceeded rather irregularly, and appears to have taken some 300 Ma: from 2.8 Ga in South Africa to 2.5 Ga on the Canadian Shield, when Proterozoic-type sedimentation commenced. Cratonization of crust younger than 2.5 Ga has been recorded on all of the Precambrian shields.

The relatively constant value of Nd (inherent to depleted mantle) for the period 3.8–2.7 Ga and the comparatively lower depletion of Archaean volcanics as compared with MORB basalts could have been a consequence of one of the following events:

1. in the Archaean, severely depleted mantle was repeatedly enriched as a result of the recycling of continental (oceanic) crust,
2. the primary mantle melts were contaminated by ancient crustal material or
3. undepleted material from the lower mantle was involved in melting.

The isotope-geochemical data do not conflict with these suggestions, and they do allow the effect of each of the factors to be readily differentiated. However, the lack of conclusive evidence for Archaean plate tectonics makes the first model seem rather less probable than the other two.

7.3 THE COMPOSITION OF THE PRIMORDIAL EARTH'S CRUST

This topic has been very difficult to study because of a chronic lack of data, i.e. no primordial Earth's crust has been preserved.

There are two opposing hypotheses:

1. that the primordial crust was basic in composition, or
2. that it was sialic tonalitic in composition.

Isotope-geochemical investigations carried out in recent years suggest that depleted mantle reservoirs already existed even in the very early stages of the

Earth's development. The most ancient Archaean volcanics, i.e. from the Pilbara Block of Western Australia (3.7 Ga: Gruau et al., 1987), SW Greenland (3.8 Ga: Hamilton et al., 1983), the Barberton belts of South Africa (3.4 Ga: Gruau et al., 1988) and the Volotsky Belt of East Karelia (3.4 Ga: Pukhtel et al., 1991), are characterized by positive initial $\varepsilon_{Nd(T)}$ values from $+1$ to $+3$. The most ancient tonalites from Mount Narrier, Western Australia (3.73 Ga: Nutman et al., 1991) have $\varepsilon_{Nd(T)}$ values ranging from $+1.7$ to $+1.9$. The enderbites of the Napier Complex, East Antarctica, are of age 3.7–3.8 Ga: Krylov et al. (1993). The tonalitic Amitsoq gneisses (3.88 Ga: Nutman et al., 1993) has $\varepsilon_{Nd} = +2$. Analogous evidence has been obtained from Pb–Pb isotope data, suggesting mean values $M_1 = 8.3$ and $K_1 = 4.2$ for the early crustal rocks. This indicates that the source of their parent melts was mantle that had been depleted in the most incompatible elements, long before the time of their formation (Chase and Patchell, 1988; Bennet et al., 1993; Shirey, 1990). The removal of crustal material of both continental and oceanic type from the upper mantle could well have produced its ubiquitously recognizable depleted composition.

There are arguments for both the first and second suggestions. On the other hand, the trend of continental crustal accretion obtained by Sm–Nd, Lu–Hf and Rb–Sr isotopic modelling of the crust-depleted mantle system presumes that, by 3.8 Ga, at least 40% of the continental Earth's crust was already in existence (Jacobsen, 1988). However, Chase and Patchell (1988) consider that the volume of preserved continental crust older than 3 Ga could not have caused such a marked depletion of the mantle reservoir. This leads us to consider the possibility that primordial continental-type crust could have covered the entire Earth's surface, as predicted from models of primordial sialic crust.

According to Bogatikov et al. (1990), the primary Earth's crust had a predominantly basic composition, i.e. possibly komatiitic–basaltic or anorthosite–norite–troctolitic as on the Moon and, presumably, on Mars. The most ancient rocks on Earth are found to have a relatively low K content, and the ratios of certain characteristic elements, for example, Th/K, are identical to those typical crustal rocks. It is assumed that this proto-crust did not form all at once, and was relatively heterogeneous in terms of its chemical composition, and that it was subsequently destroyed in the upper mantle. According to this opinion, the initial stage of terrestrial magmatism was komatiitic–basaltic in composition. The appearance of essentially granitic Archaean crust is related to the melting of these basic rocks at deep levels in the Earth's crust, under the influence of komatiitic melts.

On the other hand, as mentioned above, zircons with minimum ages of 4.28 Ga were identified in Archaean metasediments from Western Australia. In the opinion of a number of researchers this is a strong argument for the existence of very ancient sialic crust within this region (Compston and Pidgeon, 1986; Kober et al., 1989; Pidgeon et al., 1988). The results of geochemical studies on zircons have shown that the primary melts were characterized by a high La/Sm ratio, were depleted in Eu and had a non-fractionated HREE distribution. Modelling of these melts suggests that they were probably formed by the remelting of more primitive acid crustal material and thus suggests the existence of an extensive, multiply reworked block of granitic continental crust (Maas and McCulloch, 1991).

Other evidence for the existence of very ancient sialic crust has been derived from studies of the Slave Province in the North-west Territories of Canada. The Akasta Gneisses comprise a complex association of tonalitic gneisses, gneissic granites and amphibolites. Uranium-lead (SHRIMP) analyses of zircons from tonalitic gneisses have shown that these rocks must have been formed some 3.96 Ga ago. Neodymium isotopic data indicate a crustal source for these rocks: their model ages T_{CHUR} are less than or equal to 4.1 Ga, and their $\varepsilon_{Nd(T)}$ is -4.8. The Akasta gneisses are the first example of rocks from a highly enriched source, as opposed to depleted Early Archaean mantle reservoir (Bowring et $al.$, 1989a).

Thus, the data currently available do not enable an unambiguous conclusion to be drawn about the nature of the primordial Earth's crust. There are strong argument for both hypotheses and only further study will resolve the problem.

7.4 THE MAIN STAGES OF MAGMATIC EVOLUTION THROUGHOUT THE EARTH'S HISTORY

As seen from the data presented above, the Earth's geological history can be subdivided into four main stages (Kovalenko et $al.$, 1987):

1. the formation of the primordial crust at 4.5–4.1 Ga (the Lunar stage);
2. the formation of the primitive crust at 4.0–2.5 Ga (the nuclearic stage);
3. the formation of the mature crust at 2.5–2.2 Ga (the cratonic stage) and
4. from 2.2 Ga up to the present day: the continental–oceanic stage.

Little is known about the first "pre-geological" stage because these formations were not preserved. As already noted, there are two alternative viewpoints about the composition of the primordial crust: basic and sialic. At the present time the question remains controversial. By analogy with the Moon, where the earliest formations are better preserved, this stage has been termed "Lunar".

The nuclearic stage (of the primitive Earth's crust) includes the Archaean era of our planet's development. The main types of structure are granite–greenstone and granulite–gneiss terranes. The basic types of igneous rocks are komatiitic and tonalite–trondhjemitic assemblages; andesite and dacite–rhyolite assemblages are encountered in subordinate quantities and their geochemical characteristics are close to those of formations of siliceous, high-magnesian,

boninite-like series (SHMS), typical of the early Palaeoproterozoic. By the end
of this stage, in the Late Archaean, moderately titaniferous tholeiites and low-Ti
alkaline rocks of the K-series began to appear in small amounts. During this
period the formation of ancient cratons had begun and, by 2.5 Ga, had resulted in
the total cratonization of all ancient shields. Plume tectonics were the dominant
type of geodynamics at this time.

The subsequent cratonic stage continued from 2.5 to 2.2 Ga. The dominant
rock types were the ancient equivalent of the calc-alkaline series, namely, the
boninite-like SHMS assemblages (low-Ti picrites, magnesian and moderately
magnesian basalts, andesites, dacites and rhyolites) developed in the form of
volcano-sedimentary complexes, dyke swarms and large layered intrusions of
basic and ultrabasic rocks. From their geodynamic environments rather than
their composition, these formations correspond with the Phanerozoic intra-plate
magmatism. The prevailing type of geodynamics was still plume tectonics.

The fourth, continental–oceanic stage began after 2.2–2 Ga ago. It can be
divided into two substages: early — Proterzoic and late — Phanerozoic. The
principal feature of this stage is the development of the Earth's lithosphere by
plate-tectonics processes, the formation of oceans like those of the present day,
i.e. with oceanic-type lithosphere, whose fragments occur in fold belts as ophiolitic complexes and floor of present-day oceans, as evidenced by deep-sea
drilling and dredging. At this stage, distinct differences could be recognized in
the magmatism of divergent (constructive) and convergent (destructive) plate
margins, and geochemically enriched magmas of Fe–Ti type picrites and basalts
of normal, moderate and high alkalinity related to "hot-spots" (intra-plate magmatism) appeared on continents and in the oceans.

The first Proterozoic stage spanned a period of almost 1 Ga. It was characterized by the emergence of gigantic, sometimes arc-like belts of dynamothermal
metamorphism, containing huge anorthosite–mangerite massifs. Here, back-arc
spreading processes generated MORB-type tholeiites and moderately alkaline
Fe–Ti picrites and basalts ($A_{K–NaHTi}$). At the final stages of fold-belt development the alkaline-K granites and monzonites — the intrusive equivalents of the
shoshonite-latite series (A_{KLTi}) — were formed. Anorthosite–mangerite massifs
are characterized by calc-alkaline series geochemistry and occupied the position
of the gabbro–granite assemblage of island arcs and active continental margins.
The relatively short-term opening of oceanic structures occurred concurrently
where the eruption of MORB-type basalts took place. These oceans existed for
approximately 200–300 Ma and finally closed, with the formation of collision
fold belt reminiscent of the mesozoic Tethys. These collision zones were characterized by intense acid, essentially potassic magmatism completed by the formation of major anorthosite–rapakivi granite massifs. By this time, all of the
basement of the most ancient platforms had been formed. According to some
theories, as a result of all these processes, the first supercontinent, Pangaea, and
the first ocean, Panthalassa, were formed.

All this time, continental rifting was occurring continuously, which led to the formation of the most ancient kimberlites and high-Ti lamproites. This type of activity occurred for the last time 1.6–1.0 Ga ago.

The second, Phanerozoic, phase of the continental–oceanic stage started in the Late Proterozoic about 900–800 Ma ago. All of the igneous assemblages encountered up to the present day are characteristic of this stage. Several episodes of intense magmatic activity can be distinguished here (Ronov *et al.*, 1990). These peaks of activity are associated with the intermediate stages of the Caledonian, Hercynian and Alpine cycles (Fig. 7.6). For most of the volcanic

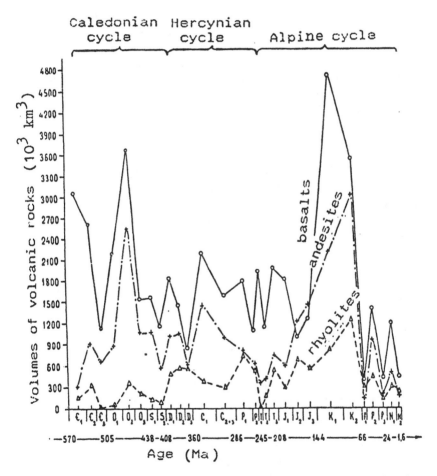

FIGURE 7.6 Stratigraphic distribution of the volume of Phanerozoic basalts, andesites and rhyolites within the sedimentary cover of the continents (after Ronov *et al.*, 1990).

episodes, basalts made up the large volume of erupted material, followed by andesites, and rhyolites were the least numerous. In the Caledonian cycle the eruptive maxima coincide in time for all these types of volcanics, and occurred during the middle Ordovician. In the Hercynian cycle, the eruptive maxima coincided for basalts and andesites and were confined to the early Carboniferous, whereas the most intensive extrusions of rhyolites were shifted to the early Permian. In the Alpine cycle the basalts peaked in the early Cretaceous, while andesites and rhyolites peaked in the late Cretaceous.

It is interesting to provide concrete examples for this rather generalized picture. The type of igneous activity for the Phanerozoic fold belts is shown in Fig. 7.7. It differs markedly from both the magmatism of the orogenic continental regions (Fig. 7.8), which is represented mainly by collision zones dominated by rhyolites, and the "platform" type, which is represented by intra-plate magmatism on the continents, where the basalts predominate (Fig. 7.9). As shown in Figs 7.6–7.8, the peaks of igneous activity in the fold belts and platforms may not coincide. Thus, for example, in the latter two situations the Caledonian cycle was rather poorly displayed, whereas in the Hercynian cycle, these peaks tend to occur less frequently in the fold belts. Only the Alpine cycle is equally broadly represented in all of the geodynamic environments.

A dramatic burst of volcanic activity occurred in the oceans at the end of the Meosozoic-earliest Cenozoic. This produced the seismic layer II in the oceanic lithosphere. Together with the basalts of oceanic islands, submarine plateaux and aseismic ridges, the total volume of oceanic tholeiitic basalts constitutes $522 \times 10^6 \, km^3$ that is, twenty times more than the total volume of synchronous volcanics on the continents ($20 \times 10^6 \, km^3$); nine times more than the total volume of Upper Jurassic–Middle Neogene volcanic rocks in the continents as a whole ($56 \times 10^6 \, km^3$) and almost four times the total volume of volcanics of the whole of the Phanerozoic and late Proterozoic of the continents and their active margins taken together ($135 \times 10^6 \, km^3$: Ronov et al., 1990). As already noted in Chapter 4, this pulse largely coincided with the Cretaceous Superchrone, that is, the 40 Ma long period during which there were no magnetic field reversals.

As can be seen from Fig. 7.10, there have been at least five stages of granite formation on the continental platform, with the greatest igneous activity occurring at: 3.5; 2.6; 1.7; 1.1 and 0.4 Ga ago. The first of these is confined to the end of the Early Archaean, the second to the Late Archaean and the third to the time of the collapse of the first oceans, while the fourth corresponds with the end of the Middle Proterozoic (the Grenville phase of folding) and the fifth with the boundary between the Caledonian and Hercynian phases of folding. However, no inferences can be made about the scale of the Earth's crustal formation. In particular, numerous studies have shown that continental-crustal accretion and growth of the hydrosphere throughout the Phanerozoic were very insignificant and hardly exceeded several per cent of the mass of the Precambrian crust and hydrosphere (Ronov et al., 1990). Moreover, as seen

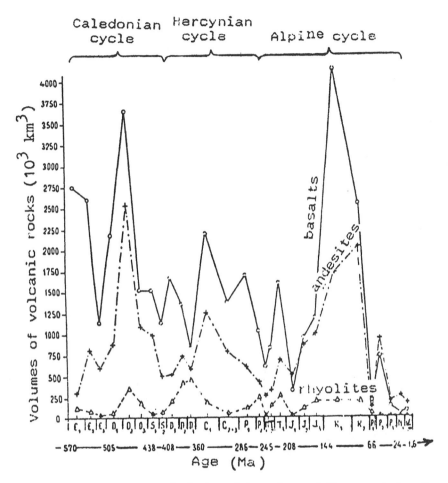

FIGURE 7.7 Stratigraphic distribution of the volume of Phanerozoic basalts, andesites and rhyolites in the sedimentary sequences of continental fold belts (after Ronov et al., 1990).

from Fig. 7.6, at the 0.4 Ga boundary a minimum in total igneous activity is observed, coinciding with an appreciable maximum of acid magmatism in the orogenic regions.

The general scheme of evolution of the terrestrial magmatism is shown in Fig. 7.11. It provides evidence that this activity evolved irreversibly: some types of rocks became extinct, others emerged and evolved very actively; the diversity of igneous rocks abruptly increased after 2 Ga.

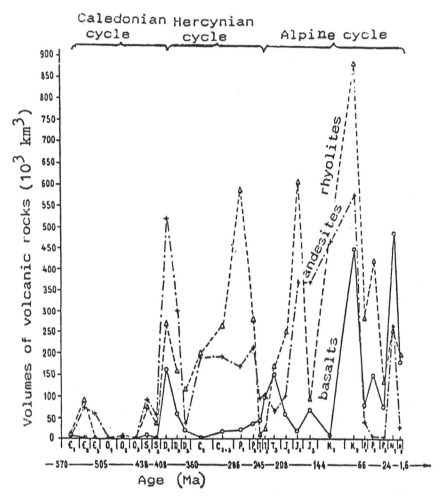

FIGURE 7.8 Stratigraphic distribution of the volume of Phanerozoic basalts, andesites and rhyolites in the sedimentary sequences of continental orogenic belts (after Ronov *et al.*, 1990).

7.5 SUMMARY

1. Not surprisingly, the Phanerozoic type of magmatism and geodynamics is the one most closely related to present-day magmatism. In fold belts and anorogenic regions, one can trace the same igneous assemblages as those that occur in present-day zones of tectonomagmatic activity. The magmatism of

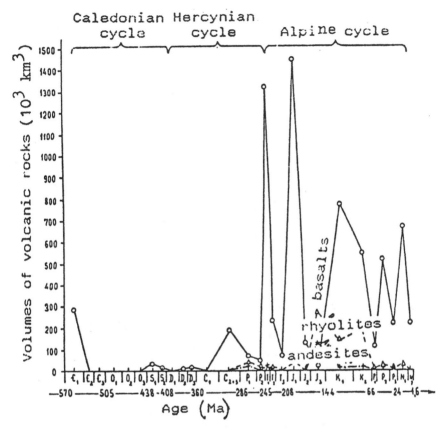

FIGURE 7.9 Stratigraphic distribution of the volume of Phanerozoic basalts, andesites and rhyolites in the sedimentary sequences of continental platform (after Ronov *et al.*, 1990).

fold belts is generally analogous to that observed on active (constructive) plate margins, in volcanic arc–back-arc basin systems and in ophiolitic complexes that largely represent the fragments of the marine back-arc sea lithosphere, although, in some cases, the presence of fragments of the true ocean floor cannot be ruled out.

2. Phanerozoic magmatism of anorogenic regions is very similar to continental intra-plate magmatism, particularly that of rifts and hot-spots. Trap assemblages occupy a special position within the magmatism of this type, having essentially no equivalents among Late Cenozoic formations. The Columbia River and Snake River basaltic fields are sometimes considered as traps, and some researchers also assign them to back-arc spreading assemblages.

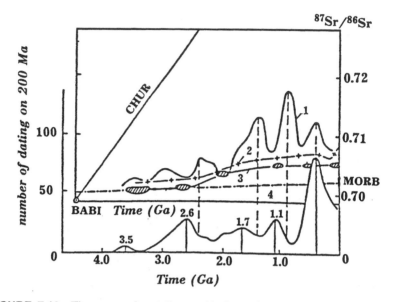

FIGURE 7.10 The curve of variations, with time, of weighted average values of: 1 — initial isotope composition of strontium $(^{87}Sr/^{86}Sr)_0$ of granitoids; 2 — evolution of $(^{87}Sr/^{86}Sr)_0$ in the ocean; 3 — evolution of $(^{87}Sr/^{86}Sr)_0$ in the source of the granitoids according to the maxima of $(^{87}Sr/^{86}Sr)_0$ density distribution with time; 4 — extrapolation of the trend of Sr isotopic composition in a homogeneous reservoir — UR (Balashov, 1985). At the bottom — a histogram of Rb/Sr dates and corresponding $(^{87}Sr/^{86}Sr)_0$, with time, for the above-mentioned granitoids. Elementary time interval at an average $(^{87}Sr/^{86}Sr)_0 = 200$ Ma, CHUR — chondrite homogeneous reservoir; After Pushkarev (1990).

Kimberlitic and titanifer (agpaitic) lamproites have no equivalents among the present-day igneous formations (or cannot be reliably identified).

3. At the end of the Early and, especially, the Middle Proterozoic, the magmatism became even more distinctive, although it was still governed by the same geodynamic processes. The collision zones may be identified here by using shear belts and zones of high-pressure metamorphism. They are confined to huge massifs of anorthosite–mangerite assemblages, which played the same role as gabbroic and granitic batholiths in the Phanerozoic.

    In the rear zones of these structures, basaltic volcanic fields and large dyke swarms made up of MORB-type basalts, Fe–Ti picrites, and basalts of varying alkalinity analogous to Phanerozoic intra-plate basalts have been observed.

4. By the middle of this stage (peaking at 1.7 Ga) large belts of acid, predominantly potassic volcanism related to anorthosite–rapakivi–granite assemblage intrusions were of worldwide occurrence. From their geochemical

Scheme of evolution of magmatism
in the Earth's history

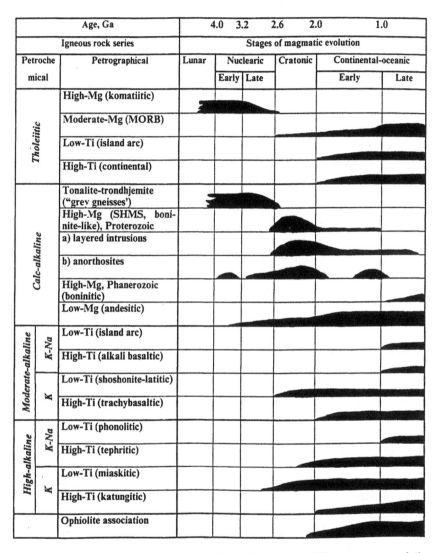

Age, Ga			4.0	3.2	2.6	2.0		1.0	
Igneous rock series			Stages of magmatic evolution						
Petrochemical	Petrographical		Lunar	Nuclearic		Cratonic	Continental-oceanic		
				Early	Late		Early		Late
Tholeiitic	High-Mg (komatiitic)								
	Moderate-Mg (MORB)								
	Low-Ti (island arc)								
	High-Ti (continental)								
Calc-alkaline	Tonalite-trondhjemite ("grey gneisses')								
	High-Mg (SHMS, boninite-like), Proterozoic a) layered intrusions								
	b) anorthosites								
	High-Mg, Phanerozoic (boninitic)								
	Low-Mg (andesitic)								
Moderate-alkaline	**K-Na**	Low-Ti (island arc)							
		High-Ti (alkali basaltic)							
	K	Low-Ti (shoshonite-latitic)							
		High-Ti (trachybasaltic)							
High-alkaline	**K-Na**	Low-Ti (phonolitic)							
		High-Ti (tephritic)							
	K	Low-Ti (miaskitic)							
		High-Ti (katungitic)							
	Ophiolite association								

FIGURE 7.11 Distribution of magmatic series in the different stages of the Earth's evolution.

characteristics the rapakivi granites in some cases resemble the lithium–fluorine granites of the Phanerozoic. The first supercontinent, Pangaea, and the first ocean, Panthalassa, appear to have been formed at about this time.

5. A major change in the Earth's development occurred 2.2–2.0 Ga ago. From this point on, all isotopically and geochemically enriched melts of Fe–Ti basalt type virtually disappear and geological signs of plate-tectonic activity are lacking. Calc-alkaline series are represented by siliceous high-magnesian formations, in contrast to Phanerozoic ones, of an intra-plate nature. They form dyke swarms, large layered intrusions of basic and ultrabasic rocks and volcano sedimentary complexes confined to graben-like structures.

6. Komatiitic-series formations are unique to the Archaean. They are confined to greenstone-belt systems among the tonalite–trondhjemite series granite-gneisses ("grey gneisses"). Among the volcano-sedimentary complexes of the greenstone belt, andesites, dacites and rhyolites are encountered, all of which are associated with iron-banded formation. In the Late Archaean greenstone belts, low-Ti subalkaline and alkaline potassic-series acid rocks occur occasionally. The granulite–gneiss terranes are located between the granite–greenstone terranes, with the granulite–gneiss belts presumably representing ancient compression zones with sedimentary basins in their central parts. They are characterized by synkinematic enderbite–charnockite magmatism. At this stage plume tectonics was the predominant type of geodynamics.

7. At present, the composition of the Earth's primordial crust is not known with certainty: two viewpoints exist on its original basic or sialic composition.

8. Judging from the character of the igneous assemblages, two main periods in the Earth's evolution can be distinguished: an early one, up to the 2.2–2.0 Ga boundary, and a late stage from 2.2–2.0 Ga to the present.

9. In the early period, three stages are delineated: a lunar, or pre-geological stage; a nuclearic, or primitive crustal stage (4.0–2.6 Ga) and a cratonic stage (2.5–2.2 Ga).

    Granite–greenstone terranes and the granulite–gneiss belts located between them are typical of the nuclearic stage. Volcanic structures of intra-plate type are characteristic of the cratonic stage. The predominant type of geodynamics for these stages appears to have been plume tectonics.

10. The late, continental–oceanic period is characterized by the processes of plate tectonics. It may be subdivided into two stages: firstly an ancient one from 2.0 up to 1.0 Ga, characterized by the widespread development of anorthosite assemblages and an independent epoch of stabilization of ancient platform basement, accompanied by large-scale acid K magmatism, including rapakivi granites; and, secondly, a young or Phanerozoic stage where the character of the magmatism was, in principle, indistinguishable from that of present-day magmatism.

11. The most important features of the evolution of igneous processes are their irreversibility and the complication caused by the changes in both the magmatic sources, and, apparently, the mechanisms of magma generation. Changes in the nature of deep petrogenesis occurred concomitantly with changes in the mechanism of geodynamics. The primitive plume tectonics of the early stages have been largely replaced by plate-tectonic processes that continue up to the present day.

CHAPTER 8
TERRESTRIAL AND LUNAR MAGMATISM:
AN EVOLUTIONARY OVERVIEW

E.V. SHARKOV, O.A. BOGATIKOV and V.I. KOVALENKO

As already discussed in previous chapters, igneous processes have changed irreversibly during the Earth's evolution. This has been caused both by changes in the composition of the melt substrates and by an increase in their complexity. In addition, the Earth's thermal regime has also varied over time, with concomitant changes in the type of the geodynamics involved: from plume tectonics in the early Precambrian to plate tectonics in the Phanerozoic. The general scheme of tectonomagmatic evolution is shown in Fig. 8.1.

Four main stages have been distinguished in the Earth's development: lunar, nuclear, cratonic and continental–oceanic, each characterized by its own style of igneous activity and its relationship with geodynamics. Some researchers consider the boundaries between the first three stages as being caused by significant impact events 3.5, 3.0 and 2.7 Ga ago (Glikson, 1993), although the available geological and petrological data rather tend to favour internal reasons for this type of evolution.

Before proceeding to discussions about the stages of magmatic evolution, it is necessary to deal with some general principles arising out of isotopic data. Figure 8.2 shows the isotopic characteristics of the main magma sources (Azbel and Tolstikhin, 1988). From this diagram, it follows that the mantle succession could have arisen from the primitive mantle, as a result of the partitioning of the geochemical reservoir into crust and depleted mantle. As a result, points corresponding with samples of continental Earth's crust vary within a rather wide range, being grouped in two major directions although the averaged-out crustal material forms a continuation of the mantle sequence. This distribution of points can be considered as a strong argument in favour of crustal differentiation accompanying the growth of this reservoir. This differentiation resulted in the formation of a separate upper crust, characterized by a higher Rb/Sr ratio, and the lower crust, in which this ratio is probably close to that of the undepleted mantle reservoir. It is evident that the continental crust grew at the expense of mantle reservoir depletion; during this process the circulation of crustal material occurred: it was added to the mantle during subduction and partly returned to the crust with the products of island-arc magmatism (Azbel and Tolstikhin, 1988).

The study of Rb–Sr isotopic systems has demonstrated that the values of $^{87}Sr/^{86}Sr$ ratios in the continental-crustal rocks suggest that they were most likely to have formed no later than 4.0 Ga ago (Azbel and Tolstikhin, 1988). On the

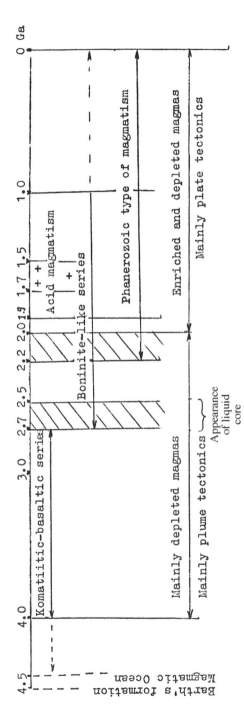

FIGURE 8.1   Schematic representation of the tectonomagmatic evolution of the Earth.

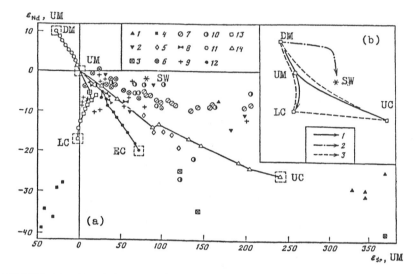

FIGURE 8.2 Neodinium and Sr isotopic ratios for various mantle and crustal rocks (a) and the location of the isotopic sources (b) (after Azbel and Tolstikhin, 1988). (a) 1 — igneous rocks; 2 — sedimentary rocks; 3 — amphibolite-facies gneisses; 4 — granulite-facies gneisses; 5 — continental volcanics; 6 — I-type granites; 7 — S-type granites; 8 — young granitoids of the western USA; 9 — Caledonian granitoids; 10 — continental granitoids; 11–14 — calculated data: 11 — for the depleted mantle (DM), 12 — for the Earth's crust (EC), 13 — for the lower crust (LC), 14 — for the upper crust (UC); SW — sea water; UM — undepleted mantle. (b) 1–3 — processes: 1 — evolution of Earth's sources, 2 — contamination of oceanic crust by Sr and Nd, 3 — mixing of material from various sources.

other hand, these ratios are much higher in young granitoids than in oceanic basalts, which are the representatives of depleted mantle reservoirs. An increase in this ratio began at approximately 2 Ga and seems to have been caused by the onset of subduction processes, during which isotopes fractionated during the melting of intermediate and acid magmas, as well as by the removal of basic restite in the form of eclogite.

8.1 THE PRE-GEOLOGICAL HISTORY OF THE EARTH

8.1.1 The Formation of the Solar System

According to modern views (Ringwood, 1979; Wasserburg, 1984; Safronov, 1987; etc.), the Sun represents a single second-generation star, in a peripheral part of the Galaxy. The Sun itself is no older than 5 Ga, whereas first-generation stars can exceed 10 Ga in age. The distance to the nearest stars is so large that the Solar System can be considered to be effectively isolated in space, its evolution being influenced only by internal factors. The Solar System contains nine planets, with a total of 54 satellites, all orbiting the Sun. Thousands of planetoids occur in the asteroid belt between the orbits of Mars and Jupiter. Comets may have originated from the Oort cloud at the periphery of the Solar System.

The Solar System is believed to have been derived from a gas–dust nebula, as a result of the condensation of material in the spiral arms of the Galaxy. The age of the Solar System is estimated as 4.57 Ga (Safronov, 1987), and the age of the most ancient meteoritic material (form Zircons using the SHRIMP $^{207}Pb/^{206}Pb$ methods) is assessed as 4563 ± 15 Ma (Ireland and Wlotzka, 1992).

According to Safronov (1987) and Vityasev et al. (1990), in general terms the model of the early Solar System is as follows. During the collapse of the gas and dust cloud, a hot central core was formed, which was surrounded by a disc-shaped gas and dust nebula rotating around it. The composition of the core would have been similar to that of the Sun, and its dimensions would have been close to those of the planetary system. The mass of the disc reached 0.01 of the solar mass, that is, one order higher than the present mass of all planets put together. After compression had ceased, the system cooled, and the particles themselves condensed. The particles subsided through the gas, forming a dust layer in the central part of the disc. In this case, the particles coagulated, forming clumps that gradually increased in size. This process of coagulation probably lasted for about a thousand orbits around the Sun. Iron-bearing particles tended to coagulate more rapidly and formed a thin disc. As a result, the first pre-planetary bodies were probably enriched in iron.

As the density of the disc increased, it became gravitationally unstable, which then led to the formation of a ring and then at a density of 6–7 times this, the ring collapsed to form a large number of aggregates, which finally coagulated to form a cluster of pre-planetary bodies (planetesimals), which acted as building blocks in the formation of the planets, their satellites and the parental bodies of meteorites. These bodies grew during collision; the largest ones were in a more favourable situation and became the embryonic planets, which finally used up all the surrounding material. The accumulation process for the terrestrial-group planets continued for about 100 Ma. As shown in Chapter 6, it is the order of magnitude (0.2 Ga) that characterizes the difference between the time of formation of zircons in meteorites and that of the most ancient zircon on the Earth.

8.1.2 The Formation of the Earth and the Moon

Planetesimals are believed to have had a different thermal history and accordingly, a different composition (Ringwood, 1979; Dreibus and Wanke, 1984; etc.). Two types of material can be distinguished: oxidized, volatile-rich and compositionally close to carbonaceous chondrite C1; and a strongly reduced type, containing iron as the metal and some Si, Cr, V and Mn in a reduced state.

The two most popular models for the Earth's formation are: the homogeneous (Safronov, 1987; Ringwood, 1979; Vityazev *et al.*, 1990; etc.) and hetero-geneous (Turekian and Clark, 1969; Dreibus and Wanke, 1984; etc.). According to the first model, from the very beginning the Earth was composed of relatively homogeneous well-mixed material and its layering was formed by later differen-tiation processes. The second model suggests that the accumulation started from the strongly reduced high-ferrous component, to which further progressively oxidized material was added. In this case, the layering of the Earth was to a large extent initiated at the time of its formation.

Currently the majority of researchers support the first view. Core separation is considered to have arisen from melting during the accumulation of colliding planetesimals, which resulted in the rapid sinking of the metallic phases (Taylor and McLennan, 1985), during the heating of protoplanetary material due to the energy of accretion, the gravity condensation of primary material, the radio-active decay of ^{26}Al, ^{244}Pu, U, Th and ^{40}K, and the kinetic energy of meteorite impacts and tidal events (Sorokhtin and Ushakov, 1991) or because of the lower melting temperature of iron fusion as compared to silicates under conditions of high pressure (Ringwood, 1979).

Whatever the accepted hypothesis, it is postulated that the Earth's silicate component — its mantle and crust — to a large extent corresponded with the composition of Cl chondrite meteorites (average wt.%): SiO_2 21.74% TiO_2 0.07; Al_2O_3 1.59; MnO 0.18; FeO 22.86; MgO 15.24; CaO 1.18; Na_2O 0.71; K_2O 0.07; P_2O_5 0.27; H_2O 19.17; Cr_2O_3 0.35; NiO 1.19; CoO 0.06; C 2.99; FeS 3.65; organic matter -6.71 (Mason, 1958); (in ppm): La 0.2446; Ce 0.6379; Pr 0.09637; Nd 0.4738; Sm 0.1540; Eu 0.05802; Gd 0.2043; Tb 0.03745; Dy 0.2541; Ho 0.05670; Er 0.1660; Tm 0.02561; Yb 0.1651; Lu 0.025539 (Evensen *et al.*, 1978).

8.1.3 The Structure and Development of the Moon

By the time of the meteorite bombardment, primary crust with the following composition already existed on the Moon (average wt.%): SiO_2 45.0; TiO_2 0.56; Al_2O_3 24.6; FeO 6.6; MgO 8.6; CaO 14.2; Na_2O 0.45; K_2O 0.075 (Taylor and Bence, 1975).

In the case of the Earth, reconstruction of worldwide igneous and metamorphic processes can be done only by drawing analogies with the Moon, where the events of this period have not been overprinted by later processes. The "lunar" stage of terrestrial evolution first recognized by Academician A.P. Pavlov at the beginning of the twentieth century is now accepted by many researchers (Kovalenko, 1987).

However, when reconstructing the earliest history of the Earth, it has to be taken into account that the proto-lunar material could have differed in composition from that of the Earth; for example, the Moon is depleted in iron (it has a more poorly developed iron core, which accounts for no more than 2% of its mass), whereas it accounts for one-third of the Earth's mass), and lunar rocks are deficient in volatile and low-melting-point components (i.e. Pb, Bi, Tl, In, Hg, Zn, Cd, Cl, Br and other elements). According to Safronov (1987), the deficit of all the above elements occurred *before* the Moon had accumulated all its mass, and was associated with its formation from a secondary circumterrestrial cluster.

The low volatile content of lunar rocks can be inferred from the widespread development of the iron–troilite assemblage and the almost complete absence of hydrated minerals and ferric iron. Nevertheless, detailed study of these rocks has shown that relic minerals retained from impact processes contain minor volatiles that testify to the participation of CO_2, N_2, H_2, H_2S, SO_2, NH_3 and also H_2O and other volatiles in lunar igneous activity (Bogatikov *et al.*, 1985). In this case, the relative abundance of native elements on the lunar surface reflects the reducing conditions for mineral formation on the Moon. These data suggest that the Moon was not completely devoid of volatiles, although, judging from typical mineral parageneses, their contents were rarely high and, on the whole, were lower than for those of the Earth.

According to current views (Ringwood, 1979; Bogatikov and Frikh-Har, 1984; Taylor and McLennan, 1985; etc.) the main structural features of the lunar surface are the ancient *continents* and *mares*, composed by different types of rocks. The continents are formed by earliest anorthositic crust and more younger magnesian-suite magmatic rocks from the highlands of the Moon, represented by high-Al basalts and intrusive rocks of anorthosite-norite-troctolite series (ANT) which intruded the primary anorthositic crust (Snyder *et al.*, 1996). Significantly younger (approximately 0.3–0.4 Ga younger) roughly circular *mare* composed of basalts with elevated Ti content. The average thickness of the latter is relatively low, and their total volume does not exceed 1% of the lunar crust. In their geology, these basalts are reminiscent of flood basalts (trap formations) on the Earth.

The lunar crust exists in a state close to isostatic equilibrium. The exceptions are mascons (accumulations of excess mass) connected with *mare* basins. The crustal thickness varies from 30–35 km (above the mascons) up to 90–110 km in the continental areas of the dark side of the Moon, and averages 60–70 km (Ringwood, 1979), reminiscent of the Earth's range of crustal thicknesses. The boundary between the upper and lower mantle is located at a depth of about 400 km. Some researchers have suggested the presence of a small metallic core with a radius of about 170 km.

The average density of the Moon is about 3.34 g/cm^3 (in contrast to the Earth, where the average density is 5.5 g/cm^3, Bills and Ferrari, 1977), which is very close to the density of the Earth's upper mantle. According to Ringwood (1979), the average composition of the lunar crust and mantle is quite close to the pyrolite (lherzolite) of the Earth's upper mantle, distinguished by a slightly increased content of SiO_2 and the above-mentioned deficiency in volatile and low-melting-point components. Available data suggest that the anorthositic rocks of the lunar upper crust, at a depth of nearly 40 km, give way to melanocratic gabbroids, and that the crustal composition is generally close to that of aluminous basalts, with 18 wt.% Al_2O_3.

The most ancient lunar continental rocks are composed of two main petrochemical types of basaltic rocks:

1. low-Ti (0.3–0.7 wt.% of TiO_2) low-K (0.03–0.11 wt.% of K_2O) high-Al basalts and their plutonic equivalents of ANT series, and
2. medium-Ti (about 1–4 wt.% TiO_2), high-K for the Moon (0.3–1.2 wt.% of K_2O) basalts (KREEP).

All "continental" basalts are distinquished from the "later" *mare* basalts by their increased magnesium and alumina content ($FeO/(FeO + MgO) = 0.4-0.5$; $Al_2O_3/(FeO + Fe_2O_3 + MgO) = 0.8-1.8$). Medium-Ti K-rich varieties of "continental" basalts are enriched in REE and phosphorus (Bogatikov et al., 1985).

The oldest (up to 4.56 ± 0.07 Ga in age: Alibert et al., 1994; the next oldest, the most plagioclase-rich highland samples, do not exceed 4.27 Ga with their youngest age being c. 4.14 Ga, Snyder et al., 1995) are high-Al low-Ti low-K basalts and the rocks of the ANT series. Investigators usually associate the formation of essentially feldspathic (anorthositic) primary lunar crust with processes of gravity differentiation in a planet-wide magma "ocean" that originated shortly after its formation. This magmasphere was 200 km thick. The appearance of anorthositic crust, which concentrated nearly 40% of the lunar Al_2O_3, was commonly associated with plagioclase flotation (Walker and Heys, 1977; Taylor and McLennan, 1985; etc).

However, as was demonstrated by Jeffries (1929) and Magnitsky (1965), the consolidation of the melted planets, and in particular of the magmatic oceans, could have proceeded only from the bottom upwards, because of the difference in values of the adiabatic gradient and the melting-point temperature gradient (see Chapter 1, Fig. 1.1). There are no reasons to believe that the process proceeded any differently during the solidification of the Moon. Therefore less-dense plagioclase crystals, which floated upwards, must have been dissolved in the convecting melt above, which was overheated relative to the liquidus. At most, if flotion took place, this "partial flotation" could only have accelerated the rate of melt evolution towards Al and Ca enrichment.

It seems that a more realistic model of lunar continental crust formation is of graduae solidification of the magma ocean from the bottom to the surface with ANT series rocks as the low-temperature cumulates. From this position, it follows that Al-rich magmatism could have occurred in the same way as the magmatism of the cratonic stage of the Earth, with remelting of ancient lower lunar crust by ascending asthenospheric diapirs.

The early Archean (4.45–4.25 Ga) magnesian-suite magmatism from the highlands of the Moon, which intruded the primary anorthositic crust, is represented by high-Al volcanics and intrusive rocks (Snyder *et al.*, 1995). Pristine intrusive rocks of this suite are represented by dunites, harzburgites, troctolites, norites, gabbronorites and anorthosites (ANT-series). Orthopyroxene and pigeonite play an important role in these rocks; Cr-spinel, orthoclase, quartz, apatite (or whitlockite) and Ti-phases (ilmenite or armalkolit) occur in many varieties. Thus, the petrographic and mineralogical features of these intrusive rocks are rather close to those that characterize the above rocks of the early Palaeoproterozoic layered intrusions and drusite complex of the Baltic Shield; the only difference is a high An content in plagioclase, which ranges from 84 to 98% and could be linked with the depletion of the Moon interior during the formation of the primordial anorthositic crust. In their major and rare elements pattern they are also close to the rocks of the drusite complex (and large layered intrusions) and differ mainly in their low Na_2O content. What's more, the value of $\varepsilon_{Nd(T)}$ in the rocks of the lunar magnesian-suite plutonic rocks is about -1, which correlates with the value of $\varepsilon_{Nd(T)} = -1$ to -2 in the studied SHMS rocks of the Baltic Shield. There are also some similarities in the origin of these lunar and earth's series. As in the case of the Earth's SHMS magmas, where assimilation of the lower crustal basic material (commonly garnet granulites) is very likely, it is suggested that the origin of magnesian-suite melts could be the products of primitive melts of the lunar interior which have assimilated the ancient plagioclase-rich lunar crust.

Such similarities of the early Archaean lunar highlands magnesian-suite magmatism and the early Palaeoproterozoic siliceous high-Mg magmatism of the Earth suggest that both planet bodies passed through this specific stage during their evolution. The main features of this stage were availability of highly depleted mantle sources in the bodies' interiors as a result of previous melting events, the appearance of a rigid lithosphere (after magma ocean solidification, and after Archaean stabilization of the Earth's crust), and large-scale involvement of crustal (mainly lower crustal) material in magma-forming processes during the ascent of high-temperature magmas to the surface in stable tectonic conditions.

The medium-Ti KREEP basalts, enriched in incompatible elements (Zr, Nb, U and Th, etc.), are considered to be the final product of magma ocean crystallization (Snyder *et al.*, 1995). According to their data, the parental KREEP basalt magmatism was not a unique event, but was an important process possibly repeated several times throughout the first 600 to 700 Ma of lunar history. It is in good agreement with the proposed model of a "cratonic-like" development of crust of the early Moon.

The eruption of the main mass of *mare* basalts occurred between 3.8 and 3.6 Ga in at least three phases (Snyder *et al.*, 1994) and in a very few cases even as late as 2.4–2.3 Ga (Shanin *et al.*, 1981). They differ from continental types in their higher titanium and iron content and their lower aluminium and calcium content. Among the *mare*, medium-Ti (2–4 wt.% TiO_2) and high-Ti (5–12 wt.% TiO_2) rocks have distinctive titanium and potassium content.

Judging by the presence of a well-defined Eu anomaly and depletion in REE, the initial material of these basalts had undergone differentiation by cumulative processes 4.56–4.14 Ga ago. Characteristically for basalts with high Ti contents, LREE depletion is significantly lower, as is the Eu anomaly, and the Sm/Eu ratio is close to chondritic. From this, it has been concluded that both upper-mantle cumulates and the initial upper mantle material were involved in the melting process.

This succession of events is connected with the fact that the accumulation of the Moon occurred relatively rapidly and that the energy released led to heating of the upper layers, whereas the interior remained relatively cool. The melted upper layers then cooled, and progressively thicker and more solid lithosphere began to form. The temperature distribution was affected by radioactive heating, from the decay of U, Th and K, and this heating was more efficient at great depths where heat was not lost by conductivity (Ringwood, 1979). This heat was probably the cause of the continuation of continental magmatism and the appearance of the *mare* volcanism, which began well after the formation of the lunar continental regions.

According to another hypothesis, the formation of the lunar *maria* and their underlying mascons was associated with the impacts of large meteorites penetrating the thin crust and causing large-scale volcanic eruptions. The mascons themselves are considered to be the residues of similar bodies. However, the assymetrical position of the *maria* (mainly at the visible side of the Moon) and their incomplete correspondence with the ancient collisional structures tend to suggest internal causes for this process (Zharkov and Trubitsyn, 1975); for example, the mascons could be regarded as solidified asthenospheric diapirs, which supplied the *mare* volcanism.

8.1.4 The Lunar Stage of the Earth

There are good reasons to believe that, in general, the sequence and the results of primary crust generation were similar for both the Moon and the Earth. In the opinion of many researchers, accretion, collisional, tidal and radiogenic heating of the Earth and the melting of its outer layers must have resulted in the formation of a magmatic ocean (Anderson, 1989; Sorokhtin and Ushakov, 1991; etc.). Obviously, this ocean must have had a mainly chondritic ultrabasic composition. It suggests that, as in the case of the Moon, the crystallization of the terrestrial magma ocean could only have occurred upwards: from the bottom to the surface, according to the fraction crystallization model. The primordial upper mantle and Earth's crust must have formed as a consequence of this. The Earth's crust must have concentrated the majority of incompatible elements because it contained the original material from the magmatic ocean, as suggested by experiments on zone refinement (Vinogradov, 1961). Results of recent experimental work on trace element partioning at high pressures (33, 70 and 217 kbar) in the case of differentiation of a terrestrial magma ocean also support this idea (Hauri *et al.*, 1996). They show that such crystallization led to enrichment of residual melt by Si, Al, incompatible elements, LREE, etc.

The composition of the primordial crust could have been either basic or sialic (see Chapter 7). It largely depends on the distance covered by the solidification front — the greater the distance, the more efficient the acceleration. In this case the distance was determined from the depth of the magmatic ocean, and the composition of the final product was affected by that of the initial material, the presence within it of volatile components, etc. Since in any specific case the scale of planetary melting and initial composition could not have been identical, it is difficult to envisage that the primordial crusts of the Earth and Moon would also be analogous. On Earth, as compared with the Moon, the depth of the magmatic ocean was appreciably larger and the presence of significant amounts of volatile components, especially water, must have shifted the evolutionary trend towards the appearance of acid igneous rocks (Ryabchikov, 1988) as the final products of crystallization. In any case there are no physical and physicochemical constraints for this model.

In general, it is probably not very important what particular composition the primordial crust had, since both basic and acid rocks could have originated only during deep differentiation of primary ultrabasic material.

The important role of sialic material in the formation of primary Earth's crust is evidenced by the widespread development of the most ancient granitoids of the TTG, comprising the upper 20–30 km of the Precambrian continental crustal section. Up to the present day, these granitoids have been and continue to be the main source of acid melts.

The continental crust, 70–80% of which is composed of Archaean rocks (Taylor and McLennan, 1985), presently accounts for less than 1% of the mantle mass. The extraction of this quantity of present-day upper crust must have caused minimal changes in upper-mantle composition. However, study of the degree of crustal depletion in micro-elements (Cs, Rb, K, Ba, U, Th, La and Ce) relative to the primitive mantle suggests that reworking of large volumes of mantle was necessary to form the crust. According to different assessments, the extraction of crust rich in LILE required 30–50% of the primitive mantle (or one-third, according to Jackobsen and Wasserburg (1979)) to have been reworked (Ringwood, 1978; O'Nions *et al.*, 1979; Allegre 1987; etc.). In fact, it means that all of the upper mantle, down to depths of about 670 km, must have been involved in reworking to a certain degree, in order to produce the observed ratio of incompatible elements and the observed isotopic characteristics of the rocks. In this case, the crust is enriched in lithophile elements to the extent that their radii and valencies differ from those of typical mantle cations (Mg and Fe) in six-fold co-ordination. This regularity signifies that crystal-melt fractionation was a leading cause of the generation of crustal material from the mantle (Taylor and Mclennan, 1985). Based on these data, the depth of magmatic ocean can be evaluated as approximately 700 km.

Evidence in favour of this model also comes from studies on $\varepsilon_{Nd(T)}$ variations in Archaean rocks indicating the separation of the primitive mantle into a mantle reservoir depleted in incompatible elements and ancient crust, enriched in incompatible elements, at the earliest stage of the Earth's differentiation, and, in any event, by 4.2 Ga (Balashov, 1990; Goldstein and Galer, 1993).

Following on from this, it is necessary to consider the question of eclogites, which play an important role in the ancient continental lithosphere. For some eclogites, found as xenoliths in kimberlites, very high magnitudes of $\varepsilon_{Nd(T)} = +(19.8–36.7)$ were obtained, which could be interpreted to indicate sharp fractionation of the upper mantle 4.5–4.4 Ga ago (McCulloch and Wasserburg, 1978). These rocks are most often considered as possible analogues of the buried primary Earth's crust.

On the other hand, it follows from the experiments of Green and Ringwood (1967) that, at pressures of more than 20 kbar, crystallization of garnet and clinopyroxene from the ultrabasic and basic melts is feasible (i.e. "igneous eclogites"). Therefore it is not improbable that similar eclogites represented cumulates formed from the melts of basic composition at the solification front of the magmatic ocean at moderate pressures. At lower pressures, i.e. 10–15 kbar, the parageneses with Pl corresponding with garnet granulite can be separated from melts of basic composition. At still lower pressures, garnet-free parageneses of enderbite type are formed from rocks of intermediate composition. And finally, the acidic melts at low pressure can be crystallized as tonalites and trondhjemites. This is consistent with the data of Wyllie et al. (1976) and Rapp and Shimizu (1996) who have demonstrated the essential role of basic igneous rocks in the formation of the ancient "grey gneisses". It was supposed that the formation of the tonalite–trondhjemite–granodiorite assemblage could be connected with the partial melting of the earlier basic igneous rocks, or the latter could be their residuals under upwards crystallization of the magma ocean.

From this point of view, a section of the upper primordial lithosphere could be characterized by the transition from diamond-facies ultramafic rocks to eclogites, garnet- and spinel-facies granulites and gabbroids, to be completed by enderbites and tonalites that formed at moderate and low pressures. All of these formations are widely represented by xenoliths from kimberlites erupted from ancient continental lithosphere, and the tonalites and trondhjemites came to the surface directly within the Archaean terranes.

Also noteworthy is the high Al content of the many lower-crustal granulites of the ancient platforms where Al_2O_3 content often reaches 20–25%, and even garnet-bearing granulites with anorthosite composition are encountered here (Sharkov and Pukhtel, 1986; Kempton et al., 1995; Rudnick and Fountain, 1995).

This appears to have led to the wide development of anorthositic complexes among the igneous rocks of the Proterozoic. They represent the crystallization products of mantle–crust melts, which presumably originated as a result of the assimilation of lower-crustal material by deep-seated melts (see Chapter 7). This suggests that the Earth's primitive lower crust could be analogous to ancient anorthositic lunar crust, although it would have been an intermediate rather than a final product of primary melt differentiation. In other words, the formation of the high-Al layer could represent one of the stages of terrestrial-group planet formation, but its location in the section would have depended on both the composition of the primary planet-forming material and the degree of its differentiation.

A particular role seems to have been played by the different volatility of oxides, during their evaporation as a result of collisional processes at the final stages of planetary accumulation. As was demonstrated by Dikov *et al.* (1990), among the rock-forming elements CaO and Al_2O_3 are the least volatile, being second only to Na, K and Si. Owing to this, in the upper layers of planetary bodies, at the final stages of their formation, when collisions reached their maximum intensity, primary layering could have arisen, associated with the so called exotic type of differentiation of material during collisional processes. This probably predetermined the primary composition of the upper layers of the planets, which was further modified to some extent during crystallization of the magma ocean.

Intensive mantle degassing must have been an important consequence of directed upward crystallization of the magma ocean, due to the high solubility of the volatile components and especially water in magmatic melts at high pressures (Kadik, 1991). Although a proportion of the volatiles could have remained in the lithosphere as water-bearing minerals, carbonates, etc., the bulk of them, at the final stages of magmatic ocean solidification (already at low pressures), must have escaped and formed a thick atmosphere. This hypothesis agrees with data on the isotopes of noble gases, which have demonstrated that the main stage of mantle degassing occurred synchronously with formation of the primordial Earth's crust at 4.48–4.40 Ga (Taylor and McLennan, 1985; Azbel and Tolstikhin, 1988; etc.). From this it follows that by at least c. 4.4 Ga the first atmosphere had appeared, and, after surface cooling, divided into the hydrosphere (the primary ocean) and the atmosphere itself, the volume of which did not essentially change over time. Thus, the atmosphere and hydrosphere are in essence ancient "geological" formations retained from the earliest stages of the Earth's formation.

As on the Moon, at this stage, the deeper layers of the primary mantle apparently were not involved in reworking, and the primary material from which the Earth was formed was preserved there. The survival of this material up to the present day has been confirmed by the discovery, in Recent magmatic fluids, of the ^3He isotope, which could have arisen only from nuclear fusion during star formation (Mamyrin *et al.*, 1968; Azbel and Tolstikhin, 1988).

8.2 MAGMATIC AND GEODYNAMIC PROCESSES IN THE EARLY PRECAMBRIAN

8.2.1 The Nuclearic Stage

The formation of granite–greenstone areas in the Early Precambrian appears to have begun at least 3.96 Ga ago, but mainly at 3.8–3.4 Ga. At present there are no data available on the processes that occurred between 4.5 and 4.0 Ga. This suggests that here, as on the Moon, there must have been a gap between the time of primary crust formation and the renewal of igneous activity associated with the accumulation of heat in the mantle. According to isotopic data, the Archaean lithosphere was stabilized before at least 3.5 Ga ago and its thickness was about 200 km; the recent type of geodynamics was not active this time (Pearson et al., 1995).

The generation of magma at the nuclearic stage could be attributed to the ascent of asthenospheric diapirs from the base of the upper mantle. This may be inferred from the great depth at which it occurred and the almost chondritic, or weakly depleted, character of the geochemical spectra of ancient komatiites, as emphasized in Chapter 6. The upwelling of these diapirs appears to have occurred rather quickly within still heated ductile lithosphere as komatiitic melts, in some cases retaining signs of their extremely deep origin (for example, the involvement of majorite in their genesis), although they could have separated from the matrix at depths of no more than 150–200 km — probably the depth up to which the ancient diapirs (plumes) rose. The ascent of diapirs was responsible for the formation of komatiites, because, in any other case, i.e. with high geothermal gradient, the considerably lower-temperature crustal rocks would have melted first.

As shown in Chapter 6, there is no geological evidence of plate tectonics in the Archaean. Moreover, geological and petrological data shown that plume tectonics are active during this time. The data presented above are evidence that asthenospheric diapirs did not reach the base of the Earth's crust and that their heads spread out within the upper mantle like a mushroom cap (Fig. 8.3). This suggests that granite–greenstone nuclei formed above the spread-out plume heads, and that granulitic belts formed among them, in some places above the descending flows of the mantle material. Specific nappe-folded belts of the Belomorian Belt type appeared between low- and high-grade terranes. Andesites, dacites and rhyolites in the upper parts of the lava sequences of some greenstone belts are mainly high magnesian and rather similar to the early Palaeoproterozoic boninite-like series. Their origin was unlikely to be related to active plate margins as in the Phanerozoic.

This spreading occurred at substantially greater depths than the base of the Earth's crust, and therefore could have resulted in the formation of a series of weakened zones, into which mantle-derived melts penetrated, causing the

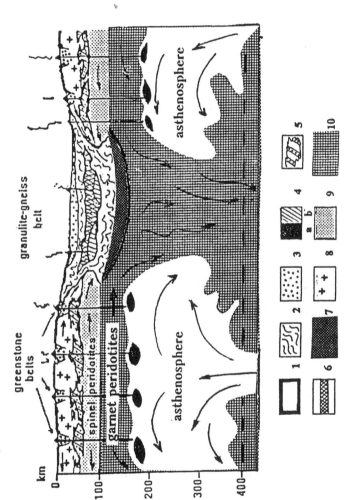

FIGURE 8.3 Suggested scheme of the Early Precambrian plume tectonics. 1 — extended asthenosphere diapirs; 2 — areas of sagging movements where excess crustal material was accumulated and granulite belts formed; 3 — sedimentary basins; 4 — magma-generation zones: in the mantle (a), in granulite belts (b); 5 — places of underplating beneath greenstone belts; 6 — places of newly-formed lower crust in granulite belts; 7 — garnetized spinel peridotites beneath granulite belts; 8 — ancient continental crust; 9 — ancient lithospheric spinel peridotites; 10 — ancient garnet peridotites.

formation of a network of greenstone belts without rupturing the sialic crust. One would expect that underplating of komatiitic melts occurred beneath these terranes and that loss of their heat resulted in widespread migmatization within them and the appearance of large granitic bodies in nearby greenstone belts. These processes led to mixing of material and changes in the isotopic characteristics of granitoids from different areas.

Evidently, the complementary zones, to which the granulite–gneiss belts were confined, must have developed around such areas. Under these conditions, the excess crustal material from the neighbouring granite–greenstone areas must have flowed and accumulated above the zones of descending convection currents that formed the granulite–gneiss areas. The main problem in the formation of granulite belts is the source of a heat supply, especially if we take account of their geological position. Instead of GGST, where mantle-derived magmatism was widespread simultaneously with extension, within GGT only crustal enderbite–charnockite magmatism, synkinematic with deformation and metamorphism, occurred. From this it follows that there were no mantle diapirs or asthenospheric lenses beneath these terranes, and the source of the heat had another origin, not linked with mantle heat. We suggest that it could be expected that, in conditions of large-scale ductile-viscous rock flow, shear heating would be released which can contribute as much as 200°C to temperature (Molnar and England, 1995) and would favour the appearance of granulite-facies metamorphism. As they were located above areas of descending currents, they must have represented the sedimentary basins characteristic of these structures (Taylor and McLennan, 1985). As a result, two types of material occurred within GGT — ancient from neighbouring GGST and newly formed sediments (van Reednen *et al.*, 1992). So, in the early Precambrian, GGT in some sense played the role of the Phanerozoic subduction zones, where excess crustal material from extensional areas has been absorbed.

From this it follows that within high-grade terranes, crustal material underwent sealing and warming up processes during the flow, which led to the partial melting of descending material and the wasting away of newly formed acid melts. As a result, restites became basic in composition but with specific features: high-iron minerals, not typical of basic igneous rocks but typical of enderbites and charnokites, are widespread here (Sharkov *et al.*, 1996). They are represented by high-density garnet granulites and eclogites, which could descend into the ultramafic mantle to the necessary depths. Since they are associated with garnet peridotites (see Chapters 4 and 9), it suggests that the natural lower limit of penetration of crustal material into the upper mantle was the discontinuity at a depth of 100–120 km, where the phase boundary between the spinel and garnet facies in ultramafic rocks occurs and a change of density takes place (Willye, 1988). In this case, both the lower crust and the upper part of the upper mantle must be formed if similar material, but metamorphosed under different conditions. The crust–mantle boundary here was probably consistent with a limit of plagioclase stability in the basic rocks. Spinel peridotites beneath these structures could be transformed into garnet varieties, as evidenceal by the xenoliths of spinel–garnet peridotites in kimberlites (Sobolev, 1974; Laz'ko, 1988c).

The interface at a depth of 650–670 km is considered to be associated with the transition of olivine ($MgSiO_4$) from a spinel to a perovskite-type structure (Ringwood, 1979; Ringwood and Irifune, 1988). This interface is easily passed through by ascending heated material, but can be an insuperable barrier for descending currents which will then spread out over its surface. The situation in the Archaean is still insufficiently understood; however, judging by the predominance of depleted mantle substrate throughout the Earth's geological history, if mass exchange between the upper and lower mantle did exist, it would have been restricted.

This type of tectonomagmatic activity was maintained throughout the whole of the Archaean. The temperature of the diapirs apparently decreased as adiabatic melting occurred at depths of 100–150 km and the Al-depleted komatiites became scarce or disappeared completely. In Late Archaean greenstone belts, in some instances the emergence of moderately Ti-rich tholeiites (although in small amounts), and K-low-Ti alkaline rocks lacking in the Early Archaean formations, has been noted. This could be evidence of the beginning of the mantle metasomatism activity, however, on a small scale. The appearance of liquid core coincided with this time (Hale, 1987). However, these rocks are atypical of Archaean magmatism, which is characterized by the presence of the melting products of variably depleted mantle substrates.

Locally, in the Late Archaean, stable areas appeared with Proterozoic-type sedimentation (e.g. the Witwatersrand Group of South Africa). Continental sediments are also typical of these areas.

8.2.2 The Cratonic Stage

By 2.5 Ga, further cooling of the lithosphere had resulted in widespread stabilization of the Earth's crust and the development of brittle deformation (see Chapter 5). Nevertheless, plume tectonics still continued. There are vast areas of spreading with the igneous activity characteristics of major dyke swarms, layered intrusions, and volcanic-sedimentary complexes in graben-like structures that are often inherited by Late Archaean greenstone belts. Granulitic belts occur among them, and specific transitional zones with dispersed magmatism (e.g. the drusite complex within the Belomorian Belt, the Baltic Shield) took place. The K-granites of normal and increased alkalinity are confined to compression zones.

The generation of mantle magma occurred in strongly depleted substrates, i.e. depleted at the expense of the melting of Archaean komatiite substrates, at depths of 60–90 km below the base of Earth's crust. The high-magnesian boninite-type mantle–crustal magmas began to increase in importance. Their origin is associated with crustal melting caused by the upwelling of magma bodies of high-temperature high-Mg melts, in accordance with the principle of zone refinement.

In rocks form that time, the earliest magnetic field reversal in the history of the Earth with an age of 2.45 Ga, was recorded from the Matachevan dyke swarm (Halls, 1992).

The number of continental sedimentary basins dating from that time had increased, but the erosion of sialic rocks was still insignificant and their contribution to the total sedimentation was negligible, as suggested by Sr isotope data from sediments (Taylor and McLennan, 1985). As these isotopic data are usually treated as evidence for the relatively weak development of sialic crust, it should be borne in mind that actually this could reflect the lack of erosion of sialic material, which resulted from the relatively low topography of the Earth's surface.

8.2.3 The Beginning of the Continental–Oceanic Stage

The situation changed drastically at the 2.2–2.0 Ga boundary when the first subduction zones, and the first back-arc basins, appeared synchronously all over the Earth, mantle metasomatism began, and the first geochemically enriched Fe–Ti basalts and picrites of normal, moderate and high alkalinity and their associated Ti-alkaline rocks of the K–Na and K-series appeared. As a result of subduction, an active influx of crustal material to the mantle began — earlier the mass exchange was of a predominantly one-sided character caused mainly by the transport of deep-seated material up to the Earth's surface, reflected in Rb–Sr (Azbel and Tolstikhin, 1988) and Re–Os (Cohen et al., 1996; Shirey and Walker, 1996) isotope systematics. At the same time, the first complex geodynamic environments emerged, which were associated with the superposition of intra-plate magmatism on back-arc spreading zones (for example, the Pechenga–Varzuga zone, see Chapter 5).

Iron–titanium picrites and basalts, as well as associated alkaline rocks, are typical products of intra-plate magmatism, and represent qualitatively new material that did not participate in earlier mass exchange in the outer layers of the Earth. Before this point, the depleted sources predominated to varying degrees, with the maximum depletion occurring at the beginning of the Early Proterozoic. The first emergence of this type of substance occurred in the Late Archaean but was only completed during Early Proterozoic.

It is significant that model isotopic dates for MORB and oceanic-island basalts vary from 2 to 1.5 Ga (Olsen et al., 1990), i.e. this event had global consequences for the entire upper-mantle reservoir. As shown in Chapter 2, a similar picture has been observed in mantle xenoliths from the Late Cenozoic basalts of Central Asia (Kovalenko *et al.*, 1990). Furthermore, apart from the model age of 2 Ga, relic dates of 600, 250 and <60 Ma have been determined there, coinciding with major tectonic events in the region. This coincidence is hardly accidental and this region hardly represents the exception to the rule. It suggests that the most important geodynamic events after 2 Ga were, in some form, associated with the influx of independent portions of this deep-seated material into the upper mantle, and in terms of petrology the phenomenon of mantle reactivation is connected with similar events.

In this regard, there is a similarity in the major geochemical parameters between various magmatic melts in the Early Precambrian (up to 2.2 Ga) and igneous rocks from Phanerozoic active margins (of both the divergent and the convergent types). Obviously, this testifies to the similarity of melt substrates for all those formations where a leading role was played by the upper layers of the Earth, represented mainly by variably depleted upper mantle and crustal material. Attention is drawn to the fact that the evolution of the Archaean magmatism of the Moon is rather close to the Earth's Palaeoproterozoic magmatism. It began from depleted low-Ti melts of the magnesian suite of the highlands, rather close in the composition of rocks, their mineralogy and geochemistry to the Earth's early Palaeoproterozoic magmatism of the siliceous high-Mg series, and, after a long break, was ended by high-Ti *mare* basalts, which resembled Fe-Ti basalts of the late Palaeoproterozoic of the Earth. The two-stage evolution of magmatism from the earlier type, specific to the highlands, usually high-Al and siliceous to the younger magmatism of *mares* or plains, is very typical for the solid planets of the Earth's group: on Venus it is represented by the ancient magmatism of tesseras and the more recent magmatism of plains (Basilevsky, 1997, pers. com.), and similarly on Mars and Mercury(Kobinson and Lucey, 1997).

The similarity in the evolution of endogenous processes in these planetary bodies can hardly be regarded as accidental, and is more likely to point to general regularities in the development of the terrestrial group. We suggest that it could be linked with the formation of liquid cores. On Earth this seems to have occurred at the 2.7–2.6 Ga boundary, and on the Moon, because of the Moon's considerably smaller size, it might have occurred even earlier.

Obviously, the formation of the liquid core must have played a decisive role in changing the character of deep-seated processes in the planetary bodies, because the situation at the upper–lower mantle boundary did not change significantly. Thus, for the Earth, it took some 400–500 Ma for the results of liquid core formation to affect the upper mantle. Apparently, this time was needed for the lowest-melting-point components that had separated from the liquid core to reach the upper mantle as a fluid: the main agent of mantle metasomatism. Up to this point, deep-seated mantle material effectively had not participated in worldwide mass exchange: in any case traces of it have not been found (apart from minor events in some late Archaean greenstone belts). By 2.2–2.0 Ga magma was being generated mainly from material derived from the outer layers of the Earth.

The emergence of heated mantle fluids — in addition to decreasing the solidus temperature of already depleted upper mantle material — actually changed the situation significantly. Renewed activity led to low-density, newly formed asthenospheric material upwelling to higher levels, reaching the base of the Earth's crust and higher, i.e. 40–30 km in depth. Therefore, during the spreading out of the heads of these asthenospheric diapirs, fracturing could have arisen that could have led to the development of plate tectonics. Up of this time, diapiric spreading occurred considerably below the base of the crust; therefore the stresses in the crust appeared to be insufficient to cause fracturing.

Rifting and subduction of crustal material within the collision zones resulted in the formation of the first basins on oceanic crust, and in MORB-type magmatism, i.e. in geological terms, formation of the first oceans.

From this time onwards the quantity of radiogenic strontium in sedimentary material and in sea-water increased sharply, suggesting the involvment of essential amounts of ancient sialic crust material in the sedimentary cycle. This is a direct indication that collisional events were followed by the appearance of a rugged topography, suggesting that large massifs of sialic crust were subjected to erosion.

At the 1.7–1.6 Ga boundary, the closure of the most ancient oceans occurred and huge belts of sialic magmatism were formed at the sites of their collapse. As a result of these processes the first Pangaea and, apparently, the first Panthalassa were formed.

The final stage of this kind of tectonomagmatic activity, in many respects analogous to that developed 2 Ga ago, took place at the end of the late Proterozoic at 1.2–1.0 Ga.

8.3 THE CONTINUATION OF THE CONTINENTAL–OCEANIC STAGE IN THE PHANEROZOIC

Phanerozoic-type tectonomagmatic activity started 850–750 Ma ago. The main tendencies outlined in the Early and Middle Proterozoic here developed further. Subsequent cooling of the outer layers of the Earth and the greater degree of asthenospheric diapiric upwelling resulted in a nearly complete predominance of plate-tectonics processes. Zones of pervasive schistosity at convergent (destructive) plate margins now occur in the low-temperature conditions of the glaucophane (blue) schist facies (Dobretsov, 1980). Anorthosite–mangerite complexes disappeared and were replaced by gabbro–granite series and Andean batholith-type granitoids. Calc-alkaline and low-Ti K-series formations of moderate and high alkalinity were widespread. In the back-arc regions, large oceanic- and continental-type sedimentary basins, associated with the upwelling of asthenospheric diapirs, were developed, often together with strong basaltic volcanism, and new oceanic crust was generated.

Large-scale upper-mantle convection was obviously the basis for the evolution of tectonomagmatic processes in the Phanerozoic and probably throughout the whole of the Earth's history. According to plate-tectonic models this convection can be expressed by the upwelling of asthenospheric superplumes, above which vast areas of extension (oceanic spreading zones) and complementary zones of descending currents (subduction zones) have emerged. However, advances made in seismic studies and in isotope geology and petrology may require this simple scheme to be reconsidered.

At present, it is established that the adiabatic curve in the outer core is significantly higher than in the lower mantle (Olsen et al., 1990). Therefore at the core–mantle boundary (CMB), the superadiabatic thermal layer (D"), which is similar in its characteristics to the asthenosphere, is located at the base of the lower mantle (Artyushkov, 1983). The boundary between the solid and the liquid core should correspond with the liquidus of the core material. The overall picture is reminiscent of the zone refinement model, with melting events at the CMB and crystallization of high-melting-point material at the base of, and at the surface of, the solid core. According to these researchers, the conductive heat flow passing through the CMB is a basic energy source for the initiation of thermal convection in the Earth and the principal energy source for all of the planetary geodynamics.

Judging by seismic tomography data (Dziewonski and Anderson, 1981; Olsen et al., 1990), the structures of mantle convection currents in the lower and the upper mantle differ significantly, in that they are appreciably simpler in the former and more complex in the latter. Moreover, the main descending currents can be very long-lived, in contrast with the major upwelling ones. Structural changes in the currents are associated with the boundary between the upper and lower mantle at depths of about 670 km.

A major element of the descending current of the convective structure of the lower mantle is a huge ring of high-velocity cold material located around the Pacific Ocean (Fig. 8.4). This structure divides two vast areas of low-velocity (heated) material into one located beneath the African continent and the Atlantic, and another located beneath the central Pacific Ocean. It is unclear so far whether the subducted plate material passes through the intermediate zone or remains on its surface as a megalith (Ringwood and Irifune, 1988), with the descending currents continuing to carry upper-mantle material cooled from above.

It is significant that ascending currents are not localized beneath mid-oceanic ridges, i.e. the structure of their distribution is different in the upper and lower mantle. Nevertheless, areas of heated lower mantle are characterized by the geoid topographic "highs" and, what is more, represent large clusters of intra-plate

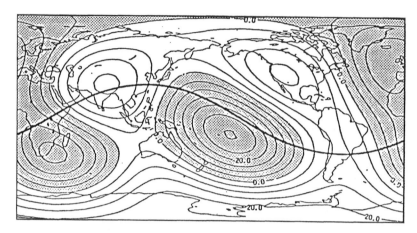

FIGURE 8.4 Schematic representation of the model of seismic-wave velocity distribution in the lower mantle (Dzievonski, 1984). The relatively slow-velocity areas are dashed. The two predominant slow-velocity areas correspond to the entire Pacific basin and a region centred over South Africa; they are interpreted as areas of ascending lower-mantle currents. The distribution of both hot-spots and geoid topography can be well correlated with these data. The interval between the isoseismic lines is 5 m/sec.

magmatism (Fig. 2.1). In some instances, isotopic anomalies are associated with these major upper-mantle upwellings. Among these, the most well-documented is the so-called Dupal anomaly in the southern parts of the Atlantic and Indian oceans, where $^{87}Sr/^{86}Sr$ ratios in the basalts increase as compared with the other oceans. The same anomaly is less traceable in the south of the central Pacific. Hot-spots with high initial $^{3}He/^{4}He$ ratios — Reunion, Afar, Iceland, Hawaii and Samoa — are related mainly to these areas (Azbel and Tolstikhin, 1988). The exception is the Yellowstone hot-spot located in the back-arc zone of the North American active continental margin.

The source of this primary helium seems to be connected with the processes at the CMB. It is not clear how it passes through the lower mantle but in the upper mantle it migrates with the intra-plate magmatic melts (Polyak, 1988). In so far this type of magmatism is closely connected with upper mantle metasomatism, there are reasons of believing that ^{3}He, together with other agents of metasomatism, passed through the lower mantle as a fluid flow. Judging by the helium anomalies associated only with the strongest hot-spots, at least some of these flows had local characteristics.

Thus, the data currently available tend to suggest that the upper and lower mantle have independent convection systems, although they may interact with each other. It is suggested that the main source of mantle convection energy seems to be located at the CMB. Juvenile material from this boundary probably passes through the lower mantle as a fluid. Since the lower mantle is most likely to be composed of relatively weakly depleted silicate material, the fluids separated at the CMB must have been oxidized, must have lost part of their material in the inevitable metasomatic reactions and must have had a proportion of their components removed by leaching. When reaching the upper mantle, this adiabatic overheated fluid must have substantially heated its base, causing a decrease in rock density, leading to a reduction in the solidus temperature and thereby causing mantle diapirism and different geodynamic processes in the Earth's crust. Accordingly, one can get some idea of the character of upper-mantle convection on the basis of specific geological, geophysical and petrological data.

8.3.1 Tectonomagmatic Processes in Global Extensional Zones

In the Phanerozoic, there must have been at least three types of tectonomagmatic activity in the extensional areas. Two of them developed while the Earth's crust was completely unfractured, and the third develops when a rupture in the Earth's crust occurs and a zone of oceanic spreading appears.

Intra-plate magmatic zones and continental rifts, whose development is not completed by ocean opening, for example the Dnieper–Donets and Oslo grabens, are attributed to the first type. They are obviously related to the relatively small sizes of the diapirs, with their correspondingly restricted energy capacities.

The second type is probably reminiscent of the regime of Early Precambrian plume tectonics (in their geodynamic character rather than their rock composition). It appears to occur in those instances where asthenospheric diapiric material was heated for some reason beyond the average for the Phanerozoic. In this case, its upper part began melting and spreading mainly below the base of the crust, not leading to its rupturing but forming extensive basalt areas like the Siberian traps.

The third type of tectonomagmatic activity, when the top of a major diapir reaches the base of the Earth's crust and even intrudes into it, produces a break in continuity and the formation of oceanic spreading zones (as occurred in the case of the Red Sea, the North Atlantic, which originated at the site of the Palaeogene British–Arctic Province, the South Atlantic (Mesozoic traps of the Parana-Etendeka), the western Indian ocean (the Deccan traps), and others). Owing to be decompression arising from the thinning and break-up of the Earth's crust, viscous asthenospheric material rises and almost reaches the surface, thereby occurring immediately beneath the ocean-floor basalts and becoming accessible to observation and sampling in the deep trenches and transform-fault zones (Laz'ko and Sharkov, 1988; Wilson, 1989; etc.). Only such an ascent of the asthenosphere could have led to the emergence of a global system of oceanic-spreading zones (mid-oceanic ridges) and the separation of huge masses of continental lithosphere into different blocks (continents), moving in opposite directions.

The presence of high-Ti picrobasalts and basalts of moderate and high alkalinity, usually of K–Na and rarely of the K-series (see Chapter 7), is characteristic of the early stages of melting of fresh asthenosphere material that was ubiquitously established in the areas of intra-plate magmatism. As the diapir rose and the processes of asthenospheric melting were enhanced under conditions of adiabatic temperature distribution at 15–20% melting, magmatic melts of this type gradually evolved into the MORB-type tholeiites characteristic of mid-oceanic ridges and back-arc seas.

The nature of intra-plate magmatism hardly seems to vary between the continents and the oceans, indicating the high homogeneity of the asthenospheric material supplied by mantle convection, as well as the relatively high degree of upwelling of material, thus keeping its interaction with the host lithospheric rocks to a minimum.

In this context, hot-spots can be treated as portions of consistent upwelling currents of convective asthenosphere, in contrast to mid-oceanic ridges, particularly the slow-spreading types, where the thickness of the asthenosphere usually does not exceed 200–300 km (Olsen et al., 1990; Anderson et al., 1992) and therefore the inflow of fresh material from a common mantle reservoir is restricted. This is more important in fast-spreading ridges with deep roots (for example, the East Pacific Rise), where the basaltic composition, according to some parameters, approaches that of intra-plate basalt (see Chapter 2). The stability of hot-spots in time might be explained by their confinement to definite zones of increased permeability, which could be dictated by the convection system within the upper mantle (Fig. 8.5). At the same time the possibility of their association with stable zones of extension and decompression cannot be ruled out.

As indicated above, one of the problems associated with hot-spots is the fact that in the rocks of the most intensely active hot-spots (Hawaii, Iceland, Reunion, Afar, Samoa and Yellowstone), high $^3He/^4He$ ratios have been observed, which testifies to the significant contribution of juvenile material to their formation, whereas in the majority of other hot-spots, crustal radiogenic helium predominates (Tristan da Kunha, Goph and the Azores, etc., Polyak, 1988). This possibly favours the concept of recycling of the subducted crustal material being accumulated at the surface of the transitional zone, gradually heated and re-implicated in magma generation processes (see below). On the other hand, it is possible that in the mantle there must have existed local zones of increased fluid discharge, influencing the scale of magma generation, whereas in areas where the moderate discharge of these fluids occurs the ^3He content decreases. It is most likely that both of these factors exist in nature.

In these terms, the creation of complex geodynamic environments in oceans and back-arc seas, caused by the coincidence of mid-oceanic ridges and intra-plate magmatism, is obviously associated with the various depths of these

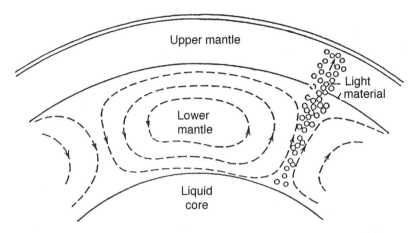

FIGURE 8.5 Possible forms of convection currents transporting the light material generated at the core–mantle boundary into the upper mantle, for adiabatically homogeneous lower mantle (after Artyushkov, 1983).

processes. Under the mid-oceanic ridge, the lens of anomalous mantle that was formed as the result of advanced melting of the asthenosphere has been developing according to its own laws, spreading at the level of its buoyancy. It is evident that this lens is fed continuously by heat from, and jets of, hot-spot material. However, a substantional proportion of it, particularly in the case of the intensively active "jets", has had enough time to pass through such a lens and to form independent eruption centres, such as oceanic islands and seamounts.

8.3.2 Tectonomagmatic Processes in Global Compressional Zones

Global compressional structures complementary to global extensional areas are expressed as destructive plate margins. The characteristic feature of these structures is the presence of descending convection zones of lithospheric material — i.e. subduction zones — localized mainly along the margins of the Pacific Ocean. The systems of andesite–latite volcanic arcs and back-arc basins, with basaltic magmatism and thinned crust of intermediate or of oceanic type, emerged simultaneously. Along the back-arc basin–arc boundary, magmas of various origins can be encountered, and their mixing often occurs within a single intermediate chamber. In the opinion of most researchers, only oceanic crust could have been involved in the subduction process (Dobretsov, 1980; Wilson, 1989). However, as was shown in Chapter 2, analogous structures with the analogous magmatism are also developed during continental plate collision.

Therefore, the involvement of continental-crustal material in the subduction process, both as the immediate subducted plate and from the back-arc zones in the process of back-arc asthenosphere diapir spreading, appears to be the most probable. The lower-crustal eclogites and garnet granulites, whose density can exceed even that of oceanic crust, have to be involved firstly here. When continental plates collide, the sialic substance is obviously accumulated mainly at the top of the crust, thus creating a complex picture of "two-storey plate tectonics" (Lobkovsky, 1988).

The situation in back-arc spreading zones is slightly different (Fig. 9.25). Here, owing to the one-sided spreading of the diapir, the sialic material of the back-arc area turns out to occur between two subsiding plates (oceanic and lower crustal), and can also be involved in general circulation, which at the usual subduction rates of 7–10 cm/year contributes to the movement of these rocks that accumulate as accretionary prisms above zones of subduction (Sharkov and Svalova, 1991). Finally, sooner or later, oceanic-type lithosphere with MORB-type magmatism is formed in the back-arc zone, and the ancient continental crust located there is largely subducted by the subduction zone.

Accordingly, both types of crust can participate in the generation of magma at active plate margins. In addition, since a large amount of volatile components, mainly water, is separated by the dehydration of the subducted plate, melting here occurs under the unique P_{H_2O}, contributing to a cotectic shift into the area of melts enriched in SiO_2. The primary melts here seem to have been andesites and latites (see Chapter 7). Another consequence of high water saturation of newly formed melts is their high capacity to extract elements such as Au, Ag, Hg, W, Mo, Sb, Pb, etc. from the rocks. As a consequence, areas of active plate interaction become intensive ore-forming systems. Moreover, as they pass through the mantle wedge, these water-saturated melts leach out some of its components, which probably gave rise to the basic members of the calc-alkaline series, including the boninites.

The important feature of magmatism at convergent plate margins is the presence of subalkaline and alkaline formation, often with a K specificity at a low Ti content, along with rocks of normal alkalintiy. It appears to be a common feature of compressional zone magmatism, traceable from the end of the nuclearic stage and associated with the inflow of K-rich fluids in these areas. The nature of these fluids is not yet ascertained; however, the distinct tendency for an increase in the significance of these K-rich melts from island arcs to active continental margins and, further, to zones of continental-plate collision suggests the significant role of the continental lithosphere in their generation. This is also confirmed by the high $^{87}Sr/^{86}Sr$ ratios in rocks of the shoshonite–latite series, up to 0.710 (Tsvetkov et al., 1993).

When discussing the problem of subduction, it should be borne in mind that a subducted plate is not a plate in the true sense of the word. As observations of ancient subduction zones have shown (Dobretsov, 1980; Priyatkina and Sharkov,

1979; Kovalenko, 1987; etc.), they are more likely to be large-scale zones of schistosity and metamorphism, along which crustal material penetrates deep into the mantle. As these rocks subside, they are metamorphosed to form various amphibolites, garnet granulites and eclogites (Dobretsov, 1980; Wyllie, 1988), while rocks of acid composition are metamorphosed to form quartz eclogites and various high-density garnet–coesite–kyanite schists (Chopin, 1984; Wyllie, 1988). Owing to this metamorphism, the presence of those rock types does not distorb the general pattern.

At present, there is no unanimity of opinion regarding the fate of the plates. As has been demonstrated by Artyushkov (1983), only a small amount of the crustal material that has subsided into the mantle actually ascends to the surface. According to his calculations, the volume of volcanic material accumulated in the arcs in an order of magnitude less than that of subducted material. From this it is inferred that the majority of the material irreversibly descends into the mantle. However, even in this case it is not clear what happens to it later. Some investigators (Silver *et al.*, 1988) believe that the subducted plate penetrates the transitional layer and reaches the lower mantle. In the opinion of other researchers, after reaching the interface the plate can spread over it forming megaliths, whose material is subsequently involved in the recycling process, thus forming intra-plate magmas (Ringwood and Irifune, 1988).

Figure 8.6 shows the situation arising from this. As can be seen from the scheme, the back-arc diapir can eventually be cross-cut by the subducted plate and from this moment onwards can exist independently, congealing gradually. The top of the diapir remains heated for a long time, which can contribute to the continuation of the formation of oceanic crust which started during diapiric spreading.

In the case of continental-plate collision, subduction of continental crust can occur from both sides, at the expense of subducted plate and the involvement of material from the back-arc zone in the process. Currently, the same situation appears to take place in the Caucasus region of the Alpine Belt (see Chapter 2), and, in the early Precambrian, in the eastern part of the Baltic Shield (see Chapter 5).

From this it follows that ancient continental lithosphere appears to be continuously subducted in subduction zones at active plate margins. Andesitic and shoshonite–latite magmas melting out during subduction will partly compensate for the loss of ancient continental lithospheric material; however, the majority of it is most likely to have irreversibly disappeared, as far as the geodynamic processes of the outer layers of the Earth are concerned.

This is indicated particularly by the present distribution of different types of crust at the Earth's surface. Thus, Archaean sialic crust comprising ancient platform basement accounts for 30% of the area of the solid surface; the newly formed continental crust originated in late Proterozoic and Phanerozoic fold belts and accounts for only 10%, i.e. about a quarter of the present continental

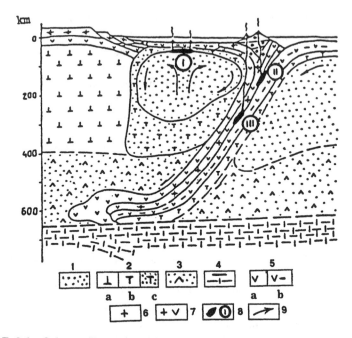

FIGURE 8.6 Scheme illustrating the late stages of back-arc spreading development (Sharkov and Svalova, 1991). Key: 1 — asthenosphere material; 2 — lithospheric mantle material: (a) continental, (b) newly formed upper mantle, (c) oceanic; 3 — upper mantle material (below 400–450 km below the interface); 4 — lower-mantle material (650–670 km below the interface); 5 — lower-crustal material: (a) of the continents, (b) of the oceans; 6 — sialic crust; 7 — tectonic mixture of the formations of sialic and lower-crustal material; 8 — magma-generation zones: I — tholiite series, II — boninite series, III — K moderate and high alkaline series; 9 — the direction of movement of mantle material.

crust (Taylor and McLennan, 1985); the remainder is newly formed oceanic crust.

One more type of complex geodynamic environment, caused by the overlapping of asthenospheric sources of oceanic spreading by continental plates, may be associated with zones of continental-plate collision. Examples of similar Cenozoic structures are found in the eastern Pacific and in the Alpine–Himalayan Belt (see Chapter 2). In the latter case, the asthenosphere probably consists of relatively small residual lenses; in any case, it is missing on schemes of global seismic tomography. In igneous formations, only the location of radiogenic helium is fixed, which testifies to the lack of deep processes associated with the interaction between the upper and the lower mantle.

8.3.3 Zones of Global Extension and Compression in the Present-day Tectonosphere

It should be noted that present-day active plate margins, which are marked by belts of andesite–latite magmatism, are grouped into two huge rings contiguous along the Alpine–Himalayan Belt and broken by the Atlantic spreading zone (Fig. 2.1, 8.7). The first structure is traced by the North American active margin, through the Aleutians, the Kurile–Kamchatka arc, Japan, the Philippines and the Izu–Bonin island arcs, and through the Burma–Indonesian arc and the eastern end of the Alpine–Himalayan Belt traceable up to the Atlantic Ocean (Fig. 8.8). An equivalent of the Alpine–Himalayan Belt on the American continent is the system of volcanic arcs and basins of the Gulf of Mexico,

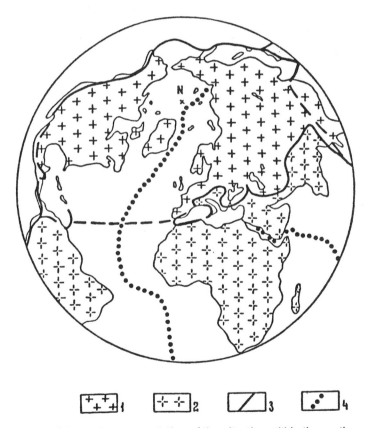

FIGURE 8.7 Schematic representation of the situation within the northern "ring" (the post-Laurasian ensemble). 1 — Laurasian fragments; 2 — Gondwana fragments; 3 — active plate margins; 4 — mid-oceanic ridges.

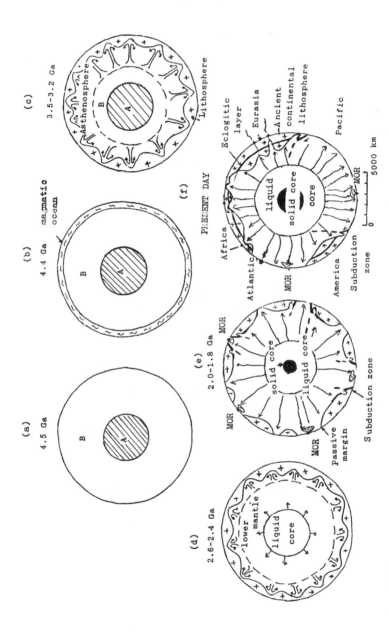

FIGURE 8.8 Schematic representation of the main stages in the Earth's evolution. a — primordial Earth (primordial mantle material: A — reduced essential iron phase, B — oxidized silicate phase); b — lunar stage; c — nuclearic stage; d — cratonic stage; e — beginning of the continental-oceanic stage; f — present-day situation.

the Caribbean Sea and Central America (Gorodnitsky and Zonenshain, 1979; Khain, 1985).

An analogous ring can also be traced in the Southern Hemisphere. Here it is displaced east wards and is formed by the volcanic arcs of Australasia, Tonga, Kermadec and New Zealand, and runs as far as the submarine Macquarie Ridge region, to be broken by the Indian–Pacific spreading ridge. The western branch of the Southern ring is connected with the Northern one in the Central American region, and then can be observed along the western edge of South America and through the South Sandwich arc and the Antarctic Cape, and then runs on to Antarctica. A belt of Cretaceous granites in Marie Byrd Land in West Antarctica (Adams *et al.*, 1995) could be a continuation of it. The Transantarctic Ridge was developed here in the Mesozoic. At present, it separates the area of intensive reactivation of western Antarctica and a large Late Cenozoic plateau-basalt volcanic belt, which extended over the 2,000 km from Marie Byrd Land to Elsworth Land, west of the Antarctic Shield. In general, the situation is reminiscent of that in the Himalayan segment of the Alpine–Himalayan Belt, and this is the site where the closure of the Southern ring appears to occur.

As shown in Fig. 8.7, each of these rings contains fragments of only one supercontinent: Laurasia or Gondwana, formed at the beginning of the Mesozoic as a result of the Late Palaeozoic breakup of Pangaea (Khian, 1985). It is characteristic that, inside these rings, representing "post-Laurasia" and "post-Gondwana", only passive plate margins are developed. Only the nature of the western Antarctic terrane located beyond these rings is still obscure.

It is known that the Mesozoic essentially saw the break-up of these continents, and their separation by newly formed zones of oceanic spreading into a series of fragments. Some of them moved northwards, leading to the closure of the Tethyan Ocean at the end of the Cretaceous/beginning of the Palaegene and the formation of a massive collision zone — the Alpine–Himalayan Belt and the new supercontinent of Eurasia. However, the fragments that are moving apart have remained within the contours delineated by zones of active plate interaction. Nowadays these structures are easily reconstructed from the presence of andesite–latite volcanic arcs.

Apart from the Alpine–Himalayan Belt, all of the other structures of active plate interaction are mainly developed along the periphery of the Pacific Ocean. From the Mesozoic onwards, thrusting of the post-Gondwanan and post-Laurasian areas on to the Pacific Plate has been universally observed (Khain, 1985). This suggests that the present geodynamic setting is made up of three major ensembles of structures, or "superstructures" — two spreading supercontinents and the ancient Pacific Ocean, which is now decreasing in area.

It should be emphasized that the post-Gondwanan and post-Laurasian areas appear to be different, and probably their break-up occurred in different ways. Therefore Gondwana broke up into at least five large fragments, which dispersed from a single centre, Africa. Laurasia was split into two large fragments by the Atlantic Ocean — Northern Eurasia, and North America plus Greenland. The nature of the differences in their geophysical fields is also interesting. In contrast to the other continents, Africa is located above an African–Atlantic lower-mantle upwelling, with a thermal centre occurring beneath the SE tip of this continent, as indicated by seismic tomography data (Fig. 8.4). Geologically, it is expressed in the wide development of intra-plate magmatism; in the presence of the largest isotopic Dupal anomaly (Wilson, 1989); in a geoid "high" (Le Pichon and Huchon, 1984; Olsen et al., 1990); and also in the higher heat flow in the Southern Hemisphere as compared with the Northern Hemisphere (Anderson et al., 1992). Africa is the largest fragment of former Gondwanaland and is surrounded on practically all sides, except the north, by oceanic spreading zones — the Mid-Atlantic and West Indian ridges and the currently forming East African–Red Sea spreading zone. These ridges, as well as the East Indian Ridge, are responsible for the destruction and fragmentation of Gondwanaland.

Unlike the post-Gondwanan continents, the post-Laurasian continents are generally underlain by more strongly cooled lower mantle and form geoid "lows", especially in the area of the Siberian Platform and Canadian Shield. Only those segments where geological activity, accompanied by intra-plate basalt and collisional andesite–latite magmatism, can be currently observed are above the average level of the geoid. These are mainly the Alpine–Himalayan area and the eastern periphery of the Eurasian continent contiguous with the Pacific Ocean. In North America, geological activity is concentrated along the edge of the Pacific Ocean. From this standpoint, the role of the Pacific Ring of Fire appears to be different. It is most likely that the Pacific Ocean is actually a fragment of ancient ocean — Panthalassa (Khain, 1985), at whose expense the spreading of supercontinents occurred and is still occurring.

Interrelationships between the active margins of spreading supercontinents and mid-oceanic ridges suggest that they represent contemporaneous structures, although during their development the young spreading ridges could have operated autonomously, breaking up the earlier-formed active margin structures. Thus, in the Cenozoic, the northern Mid-Atlantic Ridge splits the structure of the Alpine–Himalayan Belt of Eurasia and Central America into two parts, and the Early Cenozoic Indian–Pacific Ridge truncated the Tonga–Kermadec–New Zealand arc. It is true that in the case where the structures of the ridge and active margin are similar in their direction of strike, as observed in eastern North America, the continental plate overlaps the East Pacific spreading zone, here creating a complex Californian-type geodynamic environment.

It is significant that the largest oceanic basins of the Pacific and the Atlantic and the majority of the Indian ocean basins are located above the most intense ascending lower-mantle convection currents (Fig. 8.4). At the same time, the continents (with the exception of Africa) occur within zones of descending convection currents.

According to the classical concept of plate tectonics, the geodynamics of the modern Earth's surface can be explained by ocean-floor spreading, with further subduction of the surplus material in subduction zones. However, this is not so much new crust accreting in oceanic spreading centres, as the uprising and spreading of huge masses of heated low-density ultramafic material moving apart major blocks of ancient, cooled continental lithosphere towards the less viscous oceanic lithosphere of the Pacific Ocean. From this standpoint, mid-oceanic ridges appear to represent zones for the release of the main thermal–elestic stresses below large asthenospheric rises. Within their boundaries, oceanic-floor spreading actually reflects the main stages in the evolution of oceanic lithosphere, but is not a basis for processes occurring there.

From this viewpoint, the formation of the post-Gondwana ensemble was evidently connected with the appearance — from the earliest Mesozoic — in the south of the present-day South Africa, of a major upwelling of lower-mantle material, as has been discussed above. A result of this must have been a rise in the level of the geoid surface and the formation of a large thermal field, which has been observed until recently (Fig. 8.4). Eventually, this must have given rise to extensional stresses and the initiation of the extensive rift zones of the present-day Red Sea–East African type, as well as the splitting of the supercontinent into a number of fragments with deep roots (Olsen et al., 1990). The African continent is of particular importance as it represents the centre of the thermal region. It appears to truncate the flow of heated material from the lower mantle (Ballard and Pollack, 1987, 1988), which led to the formation of a mid-oceanic ridge system encircling Africa and causing post-Gondwanan spreading. In part, this material penetrates the interior of the continental plate, which is expressed here in the widespread development of intra-plate magmatism.

Within Laurasia the situation was completely different. This supercontinent had been active in the Palaeozoic, and, by the Mesozoic, it represented a stable cooled lithospheric block. Its fragmentation became possible only in the latest Triassic–earliest Jurassic, due to the opening of the Central Atlantic Ocean. However, it is mainly associated with the latest Cretaceous–earliest Palaeogene, when the Tethys had closed and the North Atlantic and the Arctic Ocean were opening. This led, on the one hand, to the spreading of the Laurasian fragments towards the Pacific Ocean (which is still occurring), and, on the other, to the emergence of a large area of intra-plate magmatism in Central Asia.

Probably the main reason for all of these events rests on changes in the lower-mantle convection system; rather than E–W, near-equatorial convection, leading to the development of the Tethys, a N–S thermal structure are initiated. The active processes of Arctic Ocean opening are occurring here even today. In this regard, the Pacific upwelling is notable. Its position, judging by the stability of the Pacific ocean, was noted for its surprising stability during the whole of the Phanerozoic and, possibly, the Late Precambrian (Puscharovsky and Melanholina, 1989).

However, as has been shown above, these spreading processes do not produce, but only maintain, the existing active margins. Their origin obviously required particular conditions and may be due to thermal–elastic stresses accumulated above areas of lower mantle upwelling. As a result of their origin on these geoid "highs", the disintegration of supercontinents and the gravity slippage of their fragments towards topographic lows (Anderson, 1989) tend to occur. The initiation of subduction zones seems to be associated with just such sudden events in the development of the limiting shear stresses, which apparently "start" a complex mechanism of processes operating at active plate margins. This immediately channels zones of excessive stress release in the outer layers of the Earth, thus limiting them to the periphery of the supercontinents.

Thus, huge asthenospheric diapirs, above which areas of mantle magmatism are developed both at constructive plate margins and within tectonic plates, were of major importance to the character of igneous processes in the Phanerozoic. The active margins (destructive plate margins) with intensive mantle–crustal magmatism are complementary zones of compression and descending convection currents, arising along the periphery of mantle upwellings. At the same time, portions of the collisional structure can become quiescent, being retained only as an intracontinental fold belt such as the Urals. This global reconstruction of the Earth's surface may be related to the periodic transformation of the system of world-wide lower-mantle convection.

8.4 THE ROLE OF MAGMATIC PROCESSES IN THE EARTH'S EVOLUTION

The Earth is virtually a closed system (the contribution of solar radiation to the energy balance is insignificant: Magnitsky, 1965), continously losing energy to outer space. Evidently, according to the second law of thermodynamics, it may become a tectonically "dead" planet. However, this energy loss has its own peculiar features. As follows from the thermodynamics of non-equilibrium systems (the Le Chatelier–Brown principle: Glensdorf and Prigozhin, 1973), in such thermogradiental (dissipative) systems, the processes of energy release should operate to compensate for the deficit. These processes can be of different character (radioactive decay, tidal energy, the energy of phase changes (transitions), gravitation energy, compaction, etc.) and, possibly, have been of variable significance at the different stages of the Earth's evolution. This is evident, if only from the fact that, in thermogradiental dissipative systems, their outer parts must primarily be involved in compensatory processes, and, as their energy capacities are exhausted, the significance of their inner regions gradually increases (Gladyshev, 1988). In other words, in terms of general thermodynamics it would be expected that the energy sources which would have compensated for energy losses would have to be displaced inwards, i.e. a "front of energy mobilization" has to move from the surface to the planetary core.

It seems reasonable to suppose that since the main energy losses occur through heat loss, it is processes of heat generation and heat exchange that play a leading role in the Earth's evolution. The essence of this development, in thermodynamic terms, appears to be the evolution of the above-mentioned compensation processes, and the specific mechanisms of this evolution which define the type of evolution. Thermal conductivity is barely significant because of its very low rate. The most probable mechanism is convection, which is much more efficient. Decompression partial melting of mantle material in upwelling parts of this convective system is the leading mechanism of magma generation.

It is known that the rate of convection and, hence, its efficiency is directly related to viscosity, with viscosity melt being several orders of magnitude higher than in that of heated rigid bodies. From this it is inferred that igneous processes alone must have been the most efficient mechanism for the Earth's evolution and obviously for the other planets of the Solar System. Moreover, due to their mobility, melts can easily migrate towards the surface, which essentially increases the possibilities of heat transfer. Thus, in terms of non-equilibrium system thermodynamics, melt formation is the most efficient way for compensation processes to be realized in thermogradiental systems of the cooling planets, and igneous processes, along with mantle convection, represent the leading mechanisms for energy transfer from the interior of the planet to the surface. Accordingly, all energy sources ultimately result in heating of the host rocks to produce melts.

Upward migration of magmas escaping from partially molten matrix is the principal cause of global geochemical differentiation, resulting in the depletion of the upper mantle in components concentrating in the melt. It is obvious that at different stages of the Earth's evolution, the scale of melting would have varied depending on the composition of the melting substrate, the heat supply, the scale and depth of the heating, and the lithospheric ductility that determined the mechanics of convection. Evidently, with the decay of radioactive elements, particularly short-lived decay, the contribution of this heat source to the total heat balance decreased continuously and the contribution of the tidal energy was also reduced (due to the Moon's gravitation), because of the increase in lithospheric viscosity (Sorokhtin and Ushakov, 1991). This must have caused changes in the heat-balance pattern, leading to a progressively increasing role for the deep-seated energy sources associated with gravity condensation, high-pressure phase transitions, etc.

This hypothesis follow on from the main postulates of non-equilibrium thermodynamics, and if it is true, during the Earth's evolution, material from even deeper in the Earth must have been gradually involved in planetary mass exchange. Moreover, it is not inconceivable that the peak of this process has already passed, which is verified by the formation of a liquid outer core as far back as the Early Precambrian.

Magmatism is closely associated with mantle convection, and in this respect, along with geodynamics, they represent two sides of energy dissipation. This factor enables us to unravel various detailed features of present-day mantle convection and to reconstruct ancient deep-seated convection systems, based on evidence from magmatism. Along with this, mantle convection itself is inseparable from the processes that occur throughout the Earth, including its core. Thus, all of the problems connected with both the development of the Earth's tectonosphere and the planet as a whole can be traced back to magmatic evolution.

Following on from the concepts of non-equilibrium thermodynamics, immediately after the Earth's formation, energy-release compensation processes must have begun, working from the surface towards the centre. The rate of these processes must have been determined by the rate of energy loss into outer space, which resulted in a self-regulating process characteristic of all dissipative systems (Glensdorf and Prigozhin, 1973). From this it is inferred that "the energy mobilization front" will not move any deeper until the energy at the previous level is exhausted, and considering that the energy sources are different and time lags vary, that does take a certain amount of time. Accordingly, one can except that, as in all dissipative systems, this process must have been non-linear and wave-like in character. In fact, as has been shown already in this book, the definite large-scale stages reflected in the character of the magmatism and the tectonosphere are outlined in the Earth's evolution. In these terms, the sequence of events could be as follows (Fig. 8.8).

Immediately after the Earth's formation, its outer layers must have been involved in energy generation. This stage (the lunar stage) was completed by the formation of the primary upper mantle, the crust, and the atmosphere and hydrosphere. Apparently, once the energy store of the upper layer had been exhausted, the "energy mobilization front" moved into the upper mantle, against a background of progressive surface cooling.

This might account for the transition to the nuclearic stage of evolution at 4.2–4.0 Ga. Rising asthenospheric plumes (diapirs) supplied undepleted or weakly depleted material, whose melting resulted in the formation of komatiite–basalt series. The end of the cratonic stage (2.7–2.6 Ga ago) coincides with the formation of the Earth's magnetic field. It probably indicates the appearance of a liquid core, and that the "energy mobilization front" reached the innermost regions of the planet, considerably exhausting the energy capacity of the lower mantle. At the initial stages, the process was fairly small-scale and therefore the old energy sources continued to operate, supplying what was, by now, strongly depleted mantle material to the surface. By that time the crust had already lost its ductility and was responding to endogenous processes by brittle deformation, and the cratonic stage began. Mantle–crustal magmatism, associated with upwards transfer of deep-seated magma chambers of high-temperature melts in the lithosphere by zone refinement, became very important. In essence, this stage is a logical continuation of the previous one, which took place under the conditions of a more rigid lithosphere and extremely depleted ancient upper mantle.

Only up until the beginning of the continental–oceanic stage at 2.2–2.0 Ga did the scale of energy release in the central part of Earth reach a level that affected processes in the tectonosphere, in terms of both rock composition and geodynamics. Evidently, form that time onwards, the general reorganization of the internal structure of the Earth was initiated. As a whole this was determined by a moving inwards of the energy mobilization front. By the time that the liquid core had been formed, the mobilization of the internal energy of the system seems to have been mainly completed. A process of final energy release began, which was realized by the mechanism of zone refinement along the boundary of the outer core, which gradually moved upwards.

The mantle fluids have played an essential role in mixing with the depleted material of the ancient upper mantle, making it lighter, decreasing the temperature of its solidus, and probably heating it as a result of various chemical reactions. Finally this must have resulted in the reactivation of deep-seated petrogenetic processes in the upper mantle and the appearance of a lower-temperature asthenosphere capable of ascending to shallow depths and thus leading to plate-tectonic activity.

Under the influence of these processes, a large-scale reorganization of the Earth's previous structure began, which occurred during the internal mobilization of energy. The remains of this structure were preserved only in ancient lithospheric blocks beneath the Precambrian platforms. As this stage still continues up to the present, it is interesting to discuss the main tendencies of its development, particularly the problem of the Earth's crustal evolution. As shown in Chapter 2, at the moment two processes are proceeding simultaneously:

1. generation of new oceanic crust in mid-oceanic ridges and back-arc spreading zones at the expense of the melting of ultrabasic mantle material;
2. new continental crust formation in island arcs due to reworking of oceanic crust.

The question is: which of the processes is more important, and, correspondingly, what is happening at the present day: "ocean-to-continent conversion" or "continent-to-ocean conversion" of the lithosphere?

There are currently two viewpoints about it. According to the first, continental crust is being formed continuously from the oceanic crust in island arcs, which has passed through several stages of development — from earlier to developed and mature (see Chapter 2). In this case, the magmatic evolution of island arcs is reflected by the successive change, over time, of tholeiite series to calc-alkaline, subalkaline and alkaline series against a background of regularly increasing amounts of intermediate and acid igneous rocks and their meta-morphic equivalents. The final stage of this evolution involves the joining of a mature island arc (and accordingly, newly generated crust) to the continent and, as a consequence, an increase in its mass (Bogatikov and Tsvetkov, 1986; Kovalenko, 1987). Comparison of geological cross-sections from a large number of present-day arcs worldwide has failed to reveal a distinct and unambiguous correlation between their degree of maturity and their age, which indicates that the continental crust formed at different rates. The reason for this difference probably lies in the details of the island-arc geodynamic regime: the rates of subduction of lithospheric plates, the ratios of extensional to compressional zones within a subduction zone, the relative thickness and composition of the Earth's crust and asthenosphere, etc.

An investigation of the magmatic evolution of island arcs has allowed us to draw some conclusions concerning the way in which they change their positions relative to the continents. In particular, it has been shown that there has been a lateral migration of the tholeiitic and calc-alkaline magmatic fronts towards the back-arc zone, over time. Intensive disruption and erosion of the ocean-ward side of the island arc and the dragging down of clastic material in the subduction zone tend to contribute to this process. In combination with the intense lateral pressure by the subducted oceanic plate, this process can result in the gradual closure of the back-arc sea, the crushing of its basement, and, eventually, the joining of a mature island arc to the continental margin. Something similar to this may have occurred in the very south of the South American continent in Jurassic times, western North America (California, Oregon and Washington) in the Palaeozoic and the Mesozoic, and some parts of the Alpine–Himalayan Belt in the Mesozoic. As an illustration of the role that crustal accretion plays in con-tinental development, let us consider north-east Asia, where, since the Palaeozoic, successive alteration of volcanic belts has been correlated with palaeo-island arcs by the character of the magmatism (Figs. 8.9 and 8.10).

According to another viewpoint (first suggested by Belousov, 1989), up to the present day, development of oceanic crust at the expense of continental crust has been occurring concurrently with this process. Supporters of this view have emphasized that the average chemical composition of rocks of the fold belts, according to Ronov et al. (1990), is much more basic (basalts, andesites, dacites and rhyolites are present there in ratios of 6 : 2 : 0.5 : 0.5) than the average composi-tion of the Precambrian shields, i.e. sialic crust is not fully restored. As indicated above, the majority of subducted material, including that of ancient continental crust, could sink irreversibly into the mantle. Only a small proportion of it returns to the surface as the magmatism of destructive plate margins. Proceeding from the established fact that the proportion of continental crust generated in the Phanerozoic was much less than that generated in the Early Precambrian, one might infer that its formation occurred at a lower rate than the destructive processes. From these standpoints, in the Phanerozoic, lithospheric development

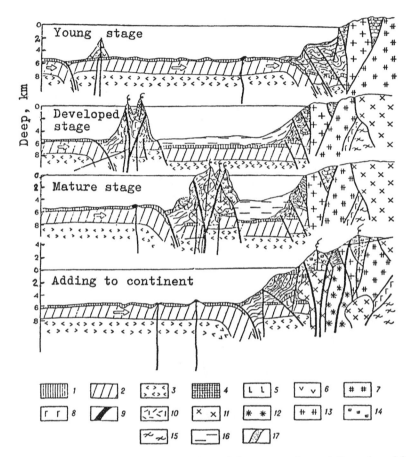

FIGURE 8.9 Schematic representation of the magmatic evolution of an island arc, from its initiation to when it became joined to a continent. From Bogatikov and Tsvetkov, 1986. 1 — sediments; 2 — oceanic crust; 3 — upper mantle; 4 — deep trough sediments; 5 — boninite–marianite assemblage; tholeiitic series: 6 — basaltic, andesitic, plagiorhyolitic lavas, 7 — plagiogranites, 8 — gabbroids; K–Na subalkaline and alkaline igneous series: 9 — moderately alkaline basalts, nephelinites; calc-alkaline series: 10 — andesite, dacite and rhyolite lavas, 11 — grandiorites and granites; 12 — syenites; K moderately alkaline series: monzonites, 14 — shoshonites, latites, K granites; 15 — metamorphic rocks; 16 — terrigeneous sediments; 17 — volcanic–sedimentary rocks.

proceeds towards "continent-to-ocean conversion". At this evolutionary stage the ancient continental lithosphere was gradually destroyed and at present appears to represent a relic structure (Sharkov and Svalova, 1993).

Simple mass-balance calculations show evidence of such a sequence of events. As demonstrated above, the Early Precambrian basement of the ancient

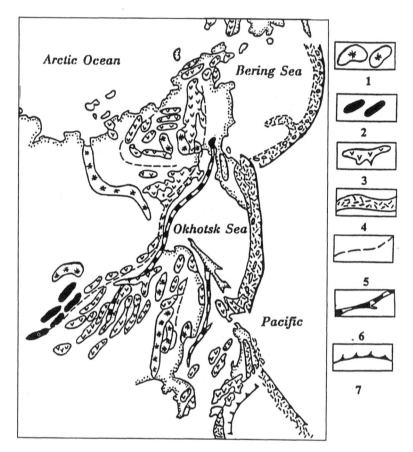

FIGURE 8.10 Schematic representation of volcanic belts (palaeo-island arcs) of north-east Asia (Scheglov, 1984). Palaeo-island arcs: 1 — of the middle and late Palaeozoic, 2 — of Triassic and Early Jurassic, 3 — late Jurassic–Cretaceous, 4 — Cenozoic; the position of subduction zones: 5 — in the Palaeozoic and early Mesozoic, 6 — in the late Mesozoic, 7 — in the Cenozoic.

platforms currently accounts for 30% of Earth's surface. The majority of the basement (including the shields) was metamorphosed under amphibolite- and granulite-facies conditions, i.e. it experienced metamorphism at pressures of 5 to 7–10 kbar (Taylor and McLennan, 1985; Rundquist and Mitrofanov, 1993). These parameters imply that these rocks were formed at depths of 15–30 km (on average 20 km), since high-pressure granulite parageneses are relatively rare. According to Ronov et al. (1990), the volume of the granite–metamorphic layer of the present continental block accounts for about 2,995 km^3 if the average thickness is 14 km. From this it follows that during the Late Precambrian–Phanerozoic about 3,280 km^2 of the continental crust was destroyed by erosion whereas the total Earth's sedimentary cover (taking into account the sedimentary cover of the fold belts) accounts for only 1,130 km^2, i.e. approximately three times less. The question arises: where has this material gone? Evidently, part of it could return to the surface as mantle–crustal and crustal volcanics; however, their contribution to the total balance of orogenic belts is not as great as that of the basalts, and these areas play a subordinate role in the formation of the structure of the continental lithosphere. Of course, more precise mass-balance calculations are needed; however, they will not significantly modify the general pattern, as this requires the continued withdrawal of most of the continental crust from the tectonosphere. This is consistent with the calculations of McDonough (1992), who has demonstrated that residual eclogites, which were generated after the dehydration and melting of subducted plates, form an independent mantle reservoir, containing c. 2% of the Earth's silicate material, which is at least twice as great as the present mass of the continental lithosphere, constituting 1% of the mantle mass.

If this scenario of the Earth's evolution is correct, the logical completion of this process will be a complete reworking of both the interior and surface of the planet, which then will be composed predominantly of oceanic crust, after which it will become a lifeless body.

Thus, currently there are two alternative approaches to understanding the evolution of the Earth's lithosphere. One of these approaches considers the continued growth of continental crust at the expense of the oceanic crust, while the other is based on a belief in the progressive "continent-to-ocean conversion" of the surface. This problem should be solved by further investigation.

8.5 SUMMARY

1. Igneous processes which continuously drain the melting zones in the Earth's depths and supply melted material to the surface are an important source of information about the nature and dynamics of deep-seated processes, as well as about the composition of melting substrates at a particular time and beneath a particular geological structure. They are closely connected with mantle convection, which affects the geodynamics at the Earth's surface. In many instances, this circumstance enables us to solve the reverse problem — the reconstruction of similar structures of the distant past, in the basis of the composition of the igneous rocks.

2. At the early stages of their development, the Earth and the Moon presumably passed through the magmatic ocean stage, in which entire planetary surface, down to depth of several hundred kilometres, was melted. The directed upwards solidification of these oceans must have predetermined the primary

layering of the outer layers of the planets and must have specified some of the basic features of their composition, in particular the primary depletion of upper mantle material in incompatible elements and the accumulation of LILE in the primordial Earth's crust and the gabbro–anorthositic (continental) lunar crust. The composition of this primordial Earth's crust is unknown, and there are various hypotheses concerning either its basic or sialic chemistry. The primary hydrosphere and atmosphere were formed synchronously on the Earth, and their volume did not change significantly at a later date.

3. The geological stage of the Earth's development began at approximately 4 Ga (the nuclearic stage). The predominant type of geodynamics — plume tectonics — must have been determined by the spreading of asthenospheric diapir heads in the mantle, far below the base of the Earth's crust, with formation of large granite–greenstone nuclei above them. The excess of sialic crustal material was accumulated above descending mantle currents with the formation of granulite–gneissic areas.

4. The early Palaeoproterozoic (the cratonic stage) (2.5–2.3 Ga ago) was characterized by siliceous high-Mg (boninite-like) series of igneous rocks. In contrast to the calc-alkaline series of the Phanerozoic, these formations have an intra-plate environment and their dominant lithology is magnesian and moderately magnesian basalts rather than andesites. The geodynamics was also plume tectonics. In essence, this stage is a logical continuation of the previous one, but developed under rigid conditions.

5. The critical change in the Earth's geological evolution occurred at 2.2–2.0 Ga when the continental–oceanic stage began. The formation of geochemically and isotopically enriched mantle substrates, presumably associated with mantle metasomatism, was characteristic of this stage. Iron–titanium picrites and basalts of normal, moderate and increased alkalinity, and also associated high-Ti alkaline rocks of K–Na and K-series, appeared for the first time in large quantities. MORB-type basalts began to play an important role.

6. At the same time, the type of geodynamic activity changed, i.e. the plume tectonics of the earlier stages gave way to plate tectonics, which appeared to be associated with the fact that the less dense asthenospheric diapirs had already reached the base of the Earth's crust, leading to its break-up and the formation of the oceanic crust. From this time on, the relationship, typical of the Phanerozoic, between the basic types of magmatism and the main types of geodynamic structures was established.

7. These breaks in the Earth's development seem to be due to the heating of its interior, the formation of a liquid core and the involvement of qualitatively new deep-seated material in a process of global mass-exchange. Evidently, these data favour the heterogeneous accumulation of the planets, with the initial presence, in their central regions, of reduced metallic material that was mobilized in a process probably related to gradual inward warming of the Earth as a result of its energy dissipation.

8. On the Moon a change in the nature of endogenic activity was followed by the formation of Ti-enriched *mare* basalts, which occurred much earlier, at about 3.8–3.6 Ga. The development of most of the terrestrial group of planets was probably completed at this stage and their further evolution took various different paths.

9. The period from 1.7 to 1.6 Ga saw the collapse of the most primitive oceanic structures, with the formation in their place of huge belts of acid magmatism with K affinities, including anorthosite–rapakivi granite massifs. Their formation can be correlated with the temporary preservation of thermal sources of oceanic spreading zones beneath overlapping sialic plates.

10. Starting from 0.85–0.75 Ga, and throughout the Phanerozoic up to now, the development of the lithosphere has proceeded by the mechanism of plate tectonics. Igneous processes are concentrated predominantly along plate margins — both constructive (the tholeiitic series) and destructive (calc-alkaline and low-Ti tholeiitic, K–Na and K moderate and high-alkaline series). Intra-plate magmatism is of little importance, although in some cases (trap formation) it was very extensive. The subduction of ancient continental crust and newly generated oceanic crust persists in subduction zones. Only part of this material appears to return to the surface in the formation of calc-alkaline and shoshonite–latite magmas and fluids, and the remainder appears to have accumulated at the base of the upper mantle as megaliths.

11. The present-day geodynamics of the Earth are characterized by three major ensembles of geological structures: the pushed — apart Mesozoic Gondwanan and Laurasian supercontinents' and the Pacific super-ocean. Mantle magmatism is most typical of the inner parts of these superstructures, whereas their outer regions, to which active plate margins are confined, are mantle–crustal. These superstructures must have been formed at the beginning of the Mesozoic.

12. The important feature of the Earth's mantle-derived magmatism, over the whole of its history, is the gradual rising of the magma generation front and apparently of the top of the asthenospheric diapirs, from depths of about 200–150 km in the Early Archaean to 60–20 km in the Phanerozoic. The tops of the diapirs gradually reached the base of the Earth's crust, and this probably led to a change in the type of geodynamics.

13. The igneous melts of the first stage (the plume tectonic stage) have much in common with those from active plate margins (both divergent and convergent) of the second (plate-tectonic) stage. They are likely to be associated with the mantle substrates, being represented in both cases mainly by the material from the Earth's outer layers: the upper mantle and the Earth's crust are depleted to different degrees.

14. From the Archaean to the Proterozoic the scale of crustal (sialic) magmatism has decreased consistently. Mantle–crustal magmatism has increased in

importance, starting from the cratonic stage, in the early Palaeoproterozoic. At its earlier stages, it was similar in character to intra-plate magmatism. Only during the continental–oceanic stage, after 2.2–2.0 Ga, was it associated mainly with destructive plate margins, and with subduction zone processes. After this stage the scale of mantle magmatism increased rapidly and changed in character.

15. Presently available data are in better agreement with hypotheses concerning the existence of independent convection systems in the upper and the lower mantle and a special convection system in the liquid outer core. The major source of thermochemical convection in the mantle appears to be located at the core–lower mantle boundary. Conductive heat flow and most low-melting-point fractions, including Fe, Ti, alkalies, ^{3}He, etc., pass through the mantle, probably in liquid–fluid form, and are an important component of intra-plate magmatism, both in the context of its initiation and in its composition.

16. The present-day geodynamics of the Earth, and associated igneous events, are defined by the two largest extensional zones: the Pacific and the African–Atlantic, located above extensive areas of heated lower mantle, which coincide with zones of ascending convection currents. They are characterized by intense basaltic magmatism. Global compressional zones are situated between them, especially around the periphery of the Pacific Ocean and along the Alpine–Himalayan Belt above the descending convection currents. Mantle–crustal magmatism is typical of these zones. Judging from geological data, the upwelling of the Central Pacific lower mantle is the most stable in time. The position of the second branch of ascending convection currents appears to have changed with time — from E–W (the multiple openings and closures of the Tethys) to N–S, which took place in the Cenozoic and possibly the Palaeozoic when the Urals fold system was formed.

17. It is likely that the Earth's tectonomagmatic evolution occurred in two stages. Firstly, that of the Early Precambrian is related to gradual cooling of the outer layers of the planet, heating of the inner layers down to the Earth's core and their involvement in the Earth's differentiation process. The remains of this ancient structure are preserved only within ancient continental lithosphere beneath the Precambrian platforms. The beginning of the second stage was probably connected with the formation of a liquid core at the boundary between the Archaean and the Proterozoic, with a stage involving the final release of energy into outer space. Starting form this time, the global reconstruction of the internal structure of the planet began. At a planetary scale, this occurred in the outer layers of the Earth, 2.2–2.0 Ga ago, expressed both in a change in the character of the geodynamic activity and the formation of new types of magmatic melts. This renewal of the Earth's surface continues to the present day.

CHAPTER 9
THE SOURCES AND ORIGIN OF MAGMAS

E.V. SHARKOV, O.A. BOGATIKOV and V.I. KOVALENKO

Recent evidence suggests that magmas are generated from both mantle and crustal material. Thus both basic and ultrabasic magmas are derived from ultrabasic mantle material, acid magmas are derived from crustal material, and magmas of intermediate composition may have a mixed mantle and crustal source. During the Earth's evolution, both the composition of magmatic melts, and, in some instances, the character of the magma-generating processes, have changed fundamentally, which has led to irreversible changes in the tectonomagmatic activity of our planet.

9.1 PROCESSES OF MAGMA GENERATION IN THE EARTH'S MANTLE

9.1.1 Mantle Substrates: Their Geochemical Classification and their Role in the Formation of Subcrustal Magmas

In recent years, two major trends have been outlined in the study of the petrology and geochemistry of subcrustal zones. The first is based on study of the composition and equilibrium P–T conditions of deep-seated rocks that are solid when they reach the surface (predominantly as xenoliths in basalts and kimberlites). These data allow the reconstruction of the position of subcrustal rocks within the Earth. The collation of these data has led to the development of concepts of vertical and lateral heterogeneity of the upper mantle, and refinement of the rather generalized picture of changes in facies with depth and construction of summary vertical sections of subcrustal zones for the major tectonic structures of the Earth (Sobolev, 1974; Dobretsov, 1980).

The basis of another concept is the generally accepted theory that the mantle is the source region for various types of magma. Therefore, deep magmas may serve as an important source of information about the composition of parental melts. The intensive, detailed study of the geochemistry and isotopic composition of volcanics that are presumed to have a subcrustal origin, as carried out over the past few years, has led to the idea of several different mantle sources for the magmas. This new evidence has allowed us to postulate an undifferentiated "primitive" mantle and its derivatives — magma sources geochemically and isotopically depleted or enriched as compared with primary mantle. These magma sources are the mantle reservoirs, i.e. deep-seated zones with corresponding isotope-geochemical composition are implied (Balashov, 1985; Hart, 1988; Allegre, 1987; etc.).

It is perfectly obvious that the most important results can be obtained only from combined study of both volcanic rocks and mantle formations of known origin, as found in xenoliths within high-alkali volcanics and kimberlites, ultramafic rocks from high-pressure complexes, ophiolites and oceanic assemblages. They provide different information: the former concerning the total result of melting and the latter about the restites. Data accumulated up to the present time allow us to outline the main connections among them, which are summarized in Table 9.1 (Laz'ko et al., 1993).

9.1.2 Volumetrically Significant Mantle Rock Types and the Main Features of their Composition

It is currently believed that the Earth's upper mantle has an essentially peridotitic composition. This hypothesis rests upon independent bodies of evidence from geophysical, petrological and geochemical data (Ringwood, 1975; Dobretsov, 1980). However, in addition to peridotites, dunites, various pyroxenites, eclogites, etc. are also of common occurrence in the mantle. Taking into account their quantitative ratios in deep-seated rock series from particular igneous areas, the subcrustal assemblages can first be subdivided into widely developed peridotite (main) assemblages and sporadic petrographically heterogeneous (secondary) assemblages. Three types of rocks are distinguished in each of them, using a complex of petrochemical, micro-elemental and isotopic characteristics (Table 9.1). Simultaneously, all of them are classified as mantle substrates: the source rocks of the Earth's subcrustal magmatism.

a) The Main Types of Mantle Rocks

Peridotite type I is conventionally termed an undepleted undifferentiated ("chondritic") peridotite. Although the present-day Earth's lithosphere is heterogeneous, there are good reasons to believe that a large variety of subcrustal rocks were formed from roughly homogeneous primordial mantle material. A hypothetical "primitive" peridotite, corresponding in its mineral composition with spinel or garnet lherzolite with close to chondritic geochemical characteristics, serves as a model for this unfractionated primary substrate (Jagoutz et al., 1979; Hart and Zindler, 1986; McDonough and Frey, 1989). Petrochemically, it corresponds with primary mantle silicate material, calculated on the basis of the chondritic model (Table 9.2). "Flat" REE spectra (Fig. 9.1) and Sr and Nd isotopic

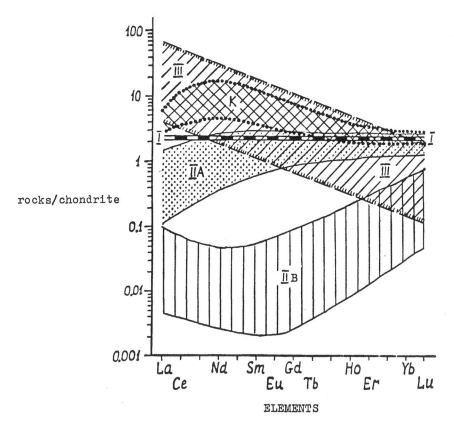

FIGURE 9.1 REE distribution in natural peridotites — representatives of the main types of mantle substrate (after Laz'ko *et al.*, 1993). The types of substrates indicated by the Roman lettering are identical to those in Table 9.1; K — amphibole-bearing lherzolites of the Kossu Massif, with an anomalous bell-gaussian REE distribution.

composition close to those expected in currently undifferentiated mantle should be characteristic of this type of peridotite (Fig. 9.2). Type I is conventionally distinguished because the peridotites that fit this theoretical model have not been found in nature as yet.

The other main types, II and III (Table 9.1), are a combination of well-known natural peridotites based on substrate I. These rocks are either depleted or enriched relative to model "primitive" lherzolite in a set of magmaphilic components, firstly in incompatible lithophile elements (ILE) with low partition coefficients (K, Rb, Sr, Ba, LREE, etc.) and also in radiogenic isotopes. The depletion in ILE observed in low melting-point oxides (CaO, Al_2O_3, TiO_2 and alkalies) is associated with the loss of these components, predominantly on partial melting,

Table 9.1 Types of substrates in the Earth's mantle and some features of their compositions.

Type of substrate	Typical features of the rocks		
	Petrochemical	Geochemical	Isotopic
	Major		
1. Peridotite: undepleted and undifferentiated ("chondritic")	$MgO/SiO_2 = 0.85$; $f = 18\%$; $Al_2O_3 = 4$, $CaO = 3.5\%$, $TiO_2 = 0.20\%$, $K_2O = 0.03–0.05\%$	$La(N) = Sm(N) = Yb(N)$; moderate concentrations of the ILE at 2–3 × chondritic Cl	$\varepsilon_{Sr} = 0$ $\varepsilon_{Nd} = 0$
2. Periodotite: depleted			
(a) "Mid-oceanic" subjected to weak depletion at 1–3 Ga	$MgO/SiO_2 = 0.8–0.9$; $f = 16–18\%$; $Al_2O_3 = 3–5$, $CaO = 3–4\%$, $TiO_2 = 0.10–0.20$, $K_2O < 0.03\%$	$La(N) < Sm(N) < Yb(N)$; low concentrations of ILE	$\varepsilon_{Sr} < 0$ $\varepsilon_{Nd} > 0$
(b) "Marginal-oceanic", extremely depleted	$MgO/SiO_2 = 0.9–1.1$; $f = 14–16\%$; $Al_2O_3 = 1–3$, $CaO = 0.5–2$, $TiO_2 < 0.10$, $K_2O < 0.01\%$	$La(N) < Sm(N) < Yb(N)$; extremely low concentrations of ILE	$\varepsilon_{Sr} < 0$ $\varepsilon_{Nd} > 0$
3. Peridotite: enriched			
(a) Long (>0.5–1 Ga) developed as enriched	$MgO/SiO_2 = 0.9–1.1$; $f = 14–16\%$; $Al_2O_3 = 1–3$, $CaO = 0.5–2$, $TiO_2 = 0.05–0.15\%$ and more, K_2O up to 0.10% and more	$La(N) > Sm(N) > Yb(N)$; increased concentrations of ILE	$\varepsilon_{Sr} > 0$ $\varepsilon_{Nd} < 0$
(b) Long developed as depleted, subjected to enrichment at <0.5 Ga ago	$MgO/SiO_2 = 0.85–1.0$; $f = 15–18\%$; $Al_2O_3 = 2.5–5$, $CaO = 1–4$, $TiO_2 = 0.10–0.20\%$, sometimes more, K_2O of up to 0.10% and more	$La(N) > Sm(N) > Yb(N)$; increased concentration of ILE	$\varepsilon_{Sr} < 0$ $\varepsilon_{Nd} < 0$
	Secondary		
4. Dunite–pyroxenite (low-Ti magnesian "green" series)			
(a) Low-Al	$MgO/SiO_2 = 0.5–1.2$; $f = 15–25\%$; $Al_2O_3 < CaO$, $TiO_2 = 0.10–0.50$, $K_2O < 0.05\%$	$La(N) < Sm(N) < Yb(N)$; low concentrations of ILE	$\varepsilon_{Sr} < 0$ $\varepsilon_{Nd} < 0$

Typical representatives	Potential genetic links with volcanics
Not reliably established, possibly some xenoliths of "primitive" lherzolites from kimberlites and basalts	Intra-plate alkaline and tholeiitic basalts of oceans and continents (?); basaltoid kimberlites (?); peridotite komatiites (?)
Lherzolites of high-pressure complexes rich in Cpx and aluminous phases, oceanic, ophiolitic assemblages, xenoliths of Cpx-rich lherzolites in basalts and kimberlites	Tholeiitic basalts of MOR, peridotite komatiities and low-alkaline picrites
Oceanic peridotites, low in Cpx (lherzolites and harzburgites) ophiolitic assemblage and high-pressure complexes	Volcanics of the -boninite series; island-arc tholeiitic basalts (?)
Garnet and spinel peridotites (low in Cpx, which includes those with small amounts of Am and Phl) in kimberlites and similar rocks	Micaceous kimberlites; ultrabasic lamproites; intra-plate continental tholeiitic and subalkaline basalts
Xenoliths of garnet and spinel peridotites in kimberlites and basalts (including those with Am and Phl hornblende peridotites of high-pressure complexes)	Basaltoid kimberlites, intra-plate alkaline and tholeiitic basalts of oceans and continents, high-alkaline volcanites
Vein ultramafic (dunites and pyroxenites) in the peridotites of ophiolites and high-pressure complexes. Xenoliths of ultramafic rocks, low in spinel and garnet (mainly websterite) of the chromediopside series in basalts and kimberlites	?

Table 9.1 (*contd.*)

Type of substrate	Typical features of the rocks		
	Petrochemical	Geochemical	Isotopic
(b) High-Al	$MgO/SiO_2 = 0.5-0.7$; $f = 15-25\%$; $Al_2O_3 \ll$ CaO, $TiO_2 = 0.10-0.50$, $K_2O < 0.05\%$	$La(N) < Sm(N) <$ $Yb(N)$; low concentration of ILE	$\varepsilon_{Sr} < 0$ $\varepsilon_{Nd} > 0$
5. Wehrlite–pyroxenite–hornblendite (high-Ti ferrous "black" series)			
(a) Low-alkaline	$MgO/SiO_2 = 0.3-0.7$; $f = 20-60\%$; $Al_2O_3 =$ CaO, $TiO_2 = 0.5-1.5$, $K_2O = 0.05-0.50\%$	$La(N) < Sm(N) <$ $Yb(N)$; moderate and high concentrations of ILE	$\varepsilon_{Sr} > 0$ $\varepsilon_{Nd} > 0$
(b) Alkaline	$MgO/SiO_2 = 0.3-0.7$; $f = 20-60\%$; $Al_2O_3 =$ CaO, $TiO_2 > 1$, $K_2O > 0.5\%$	$La(N) > Sm(N) >$ $Yb(N)$; very high concentration of ILE	$\varepsilon_{Sr} > 0$ $\varepsilon_{Nd} < 0$
6. Eclogitic			
(a) Magnesian	$SiO_2 = 40-48\%$; $f < 50\%$ $MgO = 12-24$, $Al_2O_3 =$ 8–20, CaO $= 5-13\%$, $TiO_2 = 0.1-2.0\%$, $TiO_2 = 0.1-2.0\%$	$La(N) < Sm(N) <$ $Yb(N)$; ILE concentrations are higher than chondritic Cl	$\varepsilon_{Sr} > 0$ $\varepsilon_{Nd} < 0$
(b) Ferrous	$SiO_2 = 40-48\%$; $f > 50\%$ $MgO = 9-16$, $Al_2O_3 =$ 5–16, CaO $= 5-15$, $TiO_2 = 1.5\%$	$La(N) < Sm(N) <$ $Yb(N)$; ILE concentrations are higher than chondritic Cl	$\varepsilon_{Sr} > 0$ $\varepsilon_{Nd} > 0$
(c) Aluminous	$SiO_2 = 38-40\%$; $f > 50\%$; $MgO = 8-12$, $Al_2O_3 > 20$, CaO $= 10-15$, $TiO_2 = 0-1.5\%$	$La(N) < Sm(N) <$ $Yb(N)$; positive Eu-anomaly. ILE concentrations are higher than or equal to chondritic Cl	$\varepsilon_{Sr} > 0$ $\varepsilon_{Nd} < 0$

Typical representatives	Potential genetic links with volcanics
Vein spinel and garnet pyroxenites (mainly websterites) in peridotites of high-pressure complexes; xenoliths of analogous pyroxenites in basalts and kimberlites	Low-Ti rocks
Vein ultramafites (mainly spinel and garnet clinopyroxenites and websterites) in peridotites of high-pressure complexes, xenoliths of ultramafic rocks (mainly of olivine clino-pyroxenites, wehrlites and websterites with spinel and garnet) of Al–Ti augite series basalts; xenoliths of garnet websterites in kimberlites	Low-alkaline picrites (?); participation of intra-plate tholeiitic basalts and ultrabasic volcanics of normal alkalinity in parental magma generation
Vein hornblende pyroxenites and hornblendites in peridotites of high-pressure complexes; xenoliths of phlogopite–amphibole–clinopyroxene ultramafic rocks with Iem, Rut, Ap, etc. (MARID) ilmenite–phlogopite and hornblendite pyroxenites, glimmerites, and hornblendites, etc. in kimberlites	Fe–Ti alkaline and alkaline picrites (?); participation of intra-plate subalkaline and alkaline basalts and alkaline ultrabasic volcanites in parental magma generation
Xenoliths of high-magnesian eclogites in kimberlites	Low-Ti basalts and Picrites (?)
Xenoliths of Fe–Ti eclogites in kimberlites and subalkaline basalts	Tholeiitic basalts (?), participation of tholeiitic basalts (?) in parental magma generation (?)
Xenoliths of kyanite and corundum eclogites, grospydites in kimberlites	Participation of calc-alkaline series (?) rocks in parental magma generation

ILE — incompatible lithophile elements,
La(N), Sm(N), Yb(N) — Cl chondrite-normalized REE; concentrations Cpx — clinopyroxene,
Am — amphibole, Phl — phlogopite, Sp — spinel, Gr — garnet, Ilm — ilmenite, Rut — rutite,
Ap — apatite; $f = FeO^*/(FeO^* + MgO)$, where $FeO^* = 0.9Fe_2O_3$ (wt.%).

Table 9.2 Chemical composition of natural peridotites — representatives of the main types of mantle rocks.

Components	1	2	3	4	5	6	7	8	9	10	11	12	13	14	15	16
SiO_2	45.06	45.96	44.55	45.04	45.36	44.33	45.53	46.63	43.54	44.70	43.27	46.27	44.25	44.92	43.80	42.51
TiO_2	0.22	0.18	0.20	0.17	0.27	0.06	0.10	0.07	0.05	0.03	0.03	0.08	0.01	0.32	0.11	0.03
Al_2O_3	3.96	4.06	4.02	3.92	4.81	2.94	3.24	3.14	2.25	2.09	0.98	2.61	0.46	4.70	3.07	0.60
Cr_2O_3	0.46	0.47	0.33	0.34	0.31	0.29	0.37	0.38	0.40	0.43	0.42	0.38	0.36	0.53	0.28	0.34
FeO	7.81	7.54	8.51	8.29	8.76	8.47	7.90	8.09	7.49	7.68	8.06	6.14	7.43	7.47	8.40	7.75
MnO	0.13	0.13	0.14	0.13	0.13	0.13	0.15	0.12	0.13	0.13	0.13	0.15	0.15	0.14	0.13	0.15
NiO	0.27	0.28	0.24	0.24	0.23	0.27	0.25	0.29	0.28	0.30	0.31	0.23	0.31	0.29	0.30	0.31
MgO	38.23	37.78	38.32	38.03	35.61	40.53	39.09	37.92	43.75	42.81	45.95	40.82	46.24	36.83	41.53	47.68
CaO	3.49	3.21	3.30	3.51	4.05	2.71	3.06	2.98	2.01	1.72	0.82	1.92	0.66	4.17	2.16	0.06
Na_2O	0.33	0.33	0.34	0.31	0.45	0.27	0.29	0.32	0.10	0.10	0.03	0.27	0.08	0.46	0.16	0.06
K_2O	0.03	0.03	0.03	0.01	0.00	0.00	0.02	0.06	0.00	0.00	0.00	0.58	0.01	0.12	0.03	0.02
P_2O_5	—	0.02	0.01	0.01	0.01	0.00	—	0.00	—	—	0.00	0.05	0.04	0.05	0.03	—
Co	105	100	107	102	97	110	100	110	112	105	110	99	—	105	115	124
V	77	—	—	73	76	54	22	38	—	42	26	35	16	51	—	—
Sc	17	—	14	15	16	11	11	—	11	11	—	11.5	6.7	—	13	5.2
Rb	0.81	—	—	0.4	0.4	0.1	0.22	0.10	—	0.10	<0.15	17.4	0.17	2.6	0.082	—
Sr	28	19.6	—	22	22	2.0	11	3.6	—	1.8	1.5	166	34.5	17	8.5	—
La	0.63	—	0.24	0.22	0.40	0.059	0.012	0.076	0.079	0.023	0.0035	12.1	3.6	5.9	0.84	2.2
Ce	—	—	1.2	0.83	1.2	0.18	0.055	0.21	0.205	—	0.010	19.7	7.4	18	2.0	4.4
Nd	—	1.17	1.2	0.88	1.2	0.30	0.17	0.17	0.16	0.080	0.013	9.87	2.04	6.9	1.5	2.2
Sm	0.38	0.38	0.44	0.33	0.53	0.12	0.12	0.092	0.070	0.030	0.0087	1.34	0.30	1.7	0.34	0.40
Yb	0.42	0.42	0.42	0.42	0.54	0.32	0.32	0.33	0.17	0.14	0.071	0.216	0.060	1.1	0.27	0.068
Type	I	I	IIA	IIA	IIA	IIAB	IIAB	IIAA	IIAA	IIB	IIB	IIIA	IIIA	IIIB	IIIB	IIIB

Here and in Table 9.4: all iron occurs as FeO; all rock analyses have been recalculated for dry residue and adjusted to 100%. 1, 2 — model "primitive" lherzolites: 1 — Jagoutz et al. (1979), 2 — (Hart, 1988), 3 — spinel lherzolite with plagioclase and amphibole, Zabargad Island Massif (Bonatti et al., 1986); 4–6 — spinel-, garnet- and plagioclase-lherzolites, Ronda Massif (Frey et al., 1985); 7 — plagioclase lherzolite, Kraka Massif (Laz'ko, 1988); 8 — garnet lherzolite, xenolith in Mir Pipe kimberlite (Zhuravlev et al., 1991); 9 — spinel lherzolite, xenoliths in basanite from the Dreiser–Weiher Volcano (Stosch and Lugmaier, 1986; Stosch and Seck, 1980); 10 — the same, Ronda Massif (Frey et al., 1985); 11 — harzburgite, Voykar-Syninsky massif (Laz'ko, 1988); 12 — phlogopite-garnet lherzolite, xenolith in kimberlite from the Bultfontein Pipe (Menzies, 1987); 13 — amphibole harzburgite with apatite, xenolith from the basanite of the Gnotuk Volcano (O'Reilly, 1988; Griffin et al., 1988); 14 — hornblende peridotite, St Paul Island (Roden et al., 1984); 15 — spinel lherzolite, xenolith in San Carlos basanite (Frey et al., 1978; Zindler and Jagoutz, 1988); 16 — amphibole-bearing harzburgite, xenolith in basanite from the Dreiser–Weiher Volcano (Stosch et al., 1986; Stosch and Zeck, 1980).

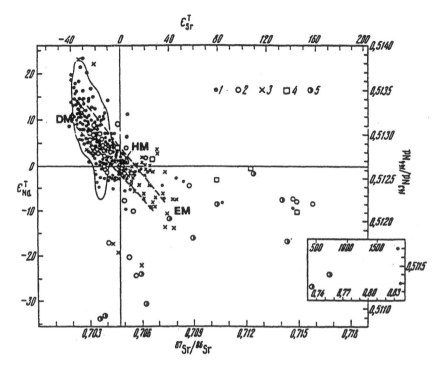

FIGURE 9.2 The Sr and Nd isotopic composition in ultrabasic mantle rocks and their minerals (after Laz'ko et al., 1993). 1 — xenoliths of the "green" Cr-diopside series in basalts and their clinopyroxenes; 2 — the same, "black" Al-titanaugite series; 3 — xenoliths of high-magnesian ("green") ultramafic rocks in kimberlites; 4 — xenoliths of Fe-rich ("black") ulramafic rocks; 5 — knorringite garnets of the xenoliths of diamond-bearing dunite–harzburgite from kimberlites and garnet inclusions in diamonds. Here and in Fig. 9.3: DM — isotopically depleted mantle, NHM — non-differentiated "primitive" mantle, EM — isotopically enriched mantle. The solid line outlines the field of ultramafic rocks of high-pressure peridotite complexes and ophiolites; the broken lines indicate the boundaries of the mantle correlation area.

and enrichment due to the influx of deep melts or fluids, which in a broad sense can be considered as mantle metasomatism (Menzies and Hawkesworth, 1987; Ryabchikov, 1988; Laz'ko and Sharkov, 1988; McDonough and Frey, 1989). The time of the redistribution of radiogenic isotope in rocks is of great importance, i.e. the duration of their isotopic systems development, which appreciably affects the isotopic characteristics. The distinctions between the main mantle substrates are caused mainly by two processes — selective melting and metasomatism, which were repeatedly manifested and superimposed throughout the Earth's history.

The depleted and enriched peridotites in turn may be subdivided into a series of subtypes. Among the depleted substrates, slightly depleted (IIA) and highly depleted (IIB) subtypes are distinguished, with gradations between them. Judging by their major characteristics, peridotites of IIA subtype experienced slight depletion at 1–3 Ga ago (Balashov, 1985; Menzies and Hawkesworth, 1987). It is reflected primarily in the fractionation of a number of micro-elements: all such rocks are depleted in LREE relative to HREE (Fig. 9.1); their Rb/Sr and Nd/Sm ratios are reduced as compared with those expected for undifferentiated mantle. Long-term depletion is reflected by Sr and Nd isotopes: $\varepsilon_{Sr} > 0$, $\varepsilon_{Nd} > 0$ (Fig. 9.3). However, petrochemically, the subtype IIA representatives are close to "primitive" type I (Table 9.1), which suggests a negligible initial depletion.

Subtype IIA is conventionally called "mid-oceanic" because the associated rocks are an important component of the material from which modern oceanic-ridge basalts are partially melted (see below). This subtype includes phases rich in clinopyroxene and alumina phases (plagioclase, spinel and garnet) lherzolites of the World ocean, ophiolites and high-pressure complexes. The analogous spinel and garnet lherzolites are widely developed as xenoliths in basalts and kimberlites (Dobretsov, 1980; Kornprobst, 1984; Laz'ko and Sharkov, 1988). All these rocks can be termed "dry" since they are practically devoid of minerals containing volatile components.

The more intensive depletion of lherzolites is reflected not only in their micro-element composition, but also in their petrochemical make-up. Rocks noticeably depleted in low melting-point oxides and ILE relative to model "primitive" lherzolite (modification IIAB is conventionally classified as transitional between subtypes IIA and IIB, see Table 9.2) are encountered in all mantle peridotite assemblages. However, this depletion does not usually affect the isotopic characteristics of the rocks, which in most cases remain analogous to "pure" subtype IIA (i.e. $\varepsilon_{Sr} < 0$, $\varepsilon_{Nd} < 15$) (Stosch and Lugmair, 1986; Stosch et al., 1986; Reisberg et al., 1989). This implies that "transitional" varieties are most likely to have originated from the same ancient weakly depleted mantle as did the petrochemically "primitive" lherzolites, and their obvious depletion relative to the first depletion cycle was repeated and relatively recent (younger than c. 0.5 Ga), because the rock isotope system had no time to respond to it.

However, the appreciable depletion in mantle material with the initiation of "transitional" lherzolites could be also be relatively ancient. In this instance, the long-term development of closed isotopic systems should give rise to peridotites with more radiogenic Nd (Stosch and Lugmair, 1986) than for lherzolites of subtype IIA. In Table 9.1 such rocks are conventionally classified as a particular modification of subtype IIA, i.e. IIAA.

The strongly depleted subtype IIB is represented by "dry" peridotites low in clinopyroxene or completely devoid of it: i.e. harzburgites, or, more rarely, lherzolites. These are widely developed in the basement of modern island arcs and active continental margins (Laz'ko and Sharkov, 1988), and so this subtype is

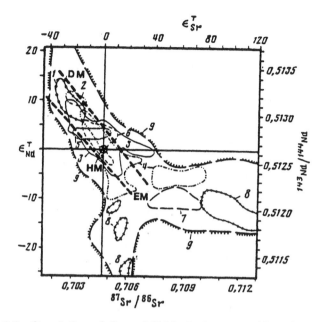

FIGURE 9.3 Correlation of Sr and Nd isotopic compositions in mantle ultra-mafic rocks and some types of volcanics of subcrustal origin (after Laz'ko *et al.*, 1993). 1–8 — fields of composition of volcanic rocks: 1 — of mid-oceanic ridge basalts, 2 — of picrites in Gorgona Island ophiolites, 3 — intra-plate basalts of seamounts and oceanic island, 4 — of plateau basalts (traps) of the continents, 5 — of high-alkaline lavas of continental rifts (melilitites, olivine nephelinites, etc.), 6 — of basaltoid kimberlites, 7 — of micaceous kimberlites, 8 — of lamproites; 9 — contour of composition area of mantle ultramafic rocks (see Fig. 9.2); mixing of isotopically contrasting substrates and their melts in widely varying proportions, with the generation of magmas of intermediate isotopic composition, is possible within the mantle ultramafic field.

known as "marginal-oceanic". The analogous peridotites dominate within ophio-lites, and have been encountered in the World ocean and high-pressure complex-es. They are extremely depleted in magmaphile petrogenic components and ILE (Table 9.2), and are strongly depleted in LREE relative to HREE. However, a proportion of these peridotites have V- and U-shaped spectra, with La > Sm (Fig. 9.1), attesting to the secondary influx of LREE or the complex history of rock melting (Prinzhofer and Allegre, 1985; Navon and Stolper, 1987; Laz'ko and Sharkov, 1988; McDonough and Frey, 1989).

Despite their extreme depletion in ILE the isotopic characteristics of subtype IIB peridotites hardly differ in ε_{Sr} and ε_{Nd} from less-depleted rocks. Hence it follows that these peridotites were derived from mantle of the same composition as the isotopically related lherzolites of subtype IIA, and the process of very intensive depletion occurred relatively recently (less than c. 0.5 Ga ago).

Theoretically, strongly depleted peridotites that underwent advanced melting at the early stages of the Earth's history could be preserved in the mantle. Their existence can be postulated from the unusual character of early Precambrian magmatism, with the widespread occurrence of komatiitic and particularly boninite-like melts. Such peridotites have to be identical to harzburgites of subtype IIB in the content of major oxides and ILE, but will be distinguished by their extremely low concentrations of radiogenic Sr and non-radiogenic Nd ($\varepsilon_{Sr} < -40$–50, $\varepsilon_{Nd} > 20$–25). Rocks with these characteristics (which conventionally could be referred to as the IIBB variety) have not yet been recognized in nature. Petrochemically, they are consistent with xenoliths from unusual diamond-bearing dunite–harzburgites and are deficient in clinopyroxene garnet peridotites from kimberlites (Sobolev, 1974; Sobolev et al., 1975; Pokhilenko, 1989; Laz'ko, 1988c). However, in terms of their geochemistry all these rocks are enriched in ILE (Shimizu and Richardson, 1987).

The concentrations of major oxides and ILE in enriched peridotites vary considerably, but are not functionally related. Therefore it is more convenient to classify the rocks of subtype III depending on the time of ILE influx, judging by the Sr and Nd isotopic evidence. A long-term geochemically enriched subtype IIIA and the relatively recently enriched subtype IIIB have been distinguished (Table 9.3).

Some peridotites of subtype III that are enriched in amphibole, phlogopite and ilmenite (metasomatized) have a characteristic gaussian REE distribution with a maximum of normalized concentrations in the Ce–Sm region (Fig. 9.1). Petrochemically these rocks are somewhat depleted in low melting-point oxides relative to "primitive" lherzolite or closely approach its composition (Stosch et al., 1986; Basaltic volcanism, 1981; Bodinier et al., 1988). Their isotopic characteristics are usually similar to those expected for modern undifferentiated mantle.

Thus, mantle peridotites are highly diversified, first of all in terms of their geochemistry, which testifies to their long-term and complex evolution. Their basic characteristics, normalized to a hypothetical "primitive" mantle composition, are summarized in Table 9.3.

The secondary mantle rock types, in contrast to the main types, differ significantly not only in their geochemistry, but also in their mineral composition. The dunite–pyroxenite (IV), wehrlite–pyroxenite–hornblendite (V) and eclogite (VI) types have been distinguished each with their subtypes (Table 9.4).

Dunite–pyroxenite type IV includes rocks corresponding broadly with low-Ti magnesian ultramafics of the "green" (Cr-diopside) series, originally described from deep-seated xenoliths in basalts (Whilshire and Shervais, 1975). Among these there are two distinct subtypes.

The low-Al subtype IVA includes a variety of formations. These occur as xenoliths of pyroxenites low in spinel and garnet, i.e. predominantly olivine websterite

Table 9.3 Characteristics of real assemblages from natural mantle peridotites as compared with hypothetical undifferentiated peridotites of type I.

Modification	Characteristics of composition and evolution as related to "primitive" peridotite		
	Geochemical (content of low melting oxides)	Geochemical (concentrations of ILE)	Isotopic (duration and character of isotopic systems development)
IIA	Not depleted or slightly depleted	Insignificantly depleted, Nd < Sm	Long-term initial weak depletion (1–3 Ga) without the following changes; ε_{Sr} and ε_{Nd} of MORB type
IIAB	Markedly depleted	Markedly depleted, Nd < Sm	Long-term initial weak depletion (1–3 Ga) and additional recent depletion (< 0.5 Ga); ε_{Sr} and ε_{Nd} of MORB type
IIB	Strongly depleted	Extremely depleted, Nd < Sm	The same
IIAA	Markedly depleted	Markedly depleted, Nd < Sm	Long-term marked depletion (> 1 Ga); ε_{Sr} of the rocks > ε_{Nd} of MORB
IIIA	Depleted to a variable degree	Enriched, Nd > Sm	Long-term enrichment (> 0.5–1 Ga)
IIIB	Not depleted or depleted to a variable degree	Not depleted or depleted to a variable degree	Long-term initial weak depletion (1–3 Ga) and recent enrichment (< 0.2–0.5 Ga)

and clinopyroxenites and rarer orthopyroxenites and dunites in basalts and kimberlites. The analogous vein ultramafic rocks that occur in the peridotites of high-pressure complexes and ophiolites can also be attributed to this subtype. Their petrochemical characteristics vary widely (Table 9.2). The distinctive features of these rocks are constantly low concentrations of Al_2O_3 and a marked predominance of CaO over Al_2O_3 in essentially clinopyroxene ultramafic rocks (3–10 times or more). The geochemical characteristics of these rocks, as a rule, are similar to those observed in the peridotites with which they occur in natural assemblages

Table 9.4 Chemical composition of natural ultramafic rocks and eclogites — representatives of secondary types of mantle rocks.

Components	1	2	3	4	5	6	7	8	9	10	11	12	13	14	15	16
SiO_2	40.91	53.83	54.43	47.26	49.6	46.18	45.99	46.99	45.13	42.95	37.29	48.82	46.93	46.63	49.32	47.38
TiO_2	0.01	0.00	0.28	0.11	0.21	0.25	1.24	0.68	1.78	2.86	4.06	5.10	0.34	0.28	1.13	0.09
Al_2O_3	0.90	1.34	4.04	13.59	8.50	12.90	5.57	13.03	13.78	13.41	17.51	1.60	14.45	14.75	13.49	25.02
Cr_2O_3	1.07	0.62	0.73	0.24	0.97	0.80	0.20	0.05	0.05	0.11	0.02	0.10	0.09	0.05	0.04	0.04
FeO	10.05	3.96	7.35	7.12	5.83	4.56	12.03	13.57	12.39	12.40	11.58	6.80	6.71	11.89	10.48	3.89
MnO	0.15	0.11	0.14	0.17	0.11	0.08	0.15	0.27	—	0.17	0.13	0.07	0.16	0.21	0.13	0.07
NiO	0.31	0.05	0.12	0.33	0.24	0.10	0.06	0.01	0.03	0.20	0.06	0.10	0.04	0.03	0.03	0.00
MgO	46.50	19.09	31.21	16.26	26.41	23.36	21.84	10.86	13.51	14.64	6.60	21.81	14.92	12.59	9.12	7.80
CaO	0.06	20.54	1.12	14.10	7.55	10.93	11.70	12.74	11.57	10.76	13.54	6.63	13.61	11.56	12.93	12.22
Na_2O	0.02	0.42	0.31	0.76	0.52	0.81	1.02	1.73	1.71	1.81	3.92	2.36	1.84	1.28	3.08	2.05
K_2O	0.01	0.01	0.06	0.02	0.02	0.00	0.15	0.03	0.02	0.58	0.81	4.42	0.82	0.62	0.16	1.10
P_2O_5	0.01	0.02	—	0.03	0.04	0.02	0.05	0.03	0.02	0.10	2.69	0.03	0.09	0.11	0.08	0.02
Co	130	30	65	58	62	—	—	—	75	69	105	73	56	60	52	45
V	10	120	—	278	223	224	182	615	323	325	335	229	210	400	275	300
Sc	—	—	25	30	27	50	25	69	23	34	—	37	—	68	35	—
Rb	0.3	1.8	—	2.0	1.0	0.26	1.4	1.6	0.4	1.0	4.7	65	22	20	4.0	17
Sr	1.2	10	—	27	26	151	105	30	203	258	1293	11400	153	61	105	2700
La	0.0050	0.017	0.15	0.070	0.67	16	2.9	1.58	0.97	4.05	59	4.93	1.9	5.11	1.63	2.5
Ce	0.012	0.058	0.41	0.24	2.18	42	9.7	5.52	4.3	13.6	102	11.6	4.0	10.0	4.4	5.2
Nd	0.0060	0.097	0.35	—	—	16	8.8	3.65	4.9	12.0	56	5.62	2.2	2.90	3.9	3.0
Sm	0.0017	0.058	0.11	0.17	0.64	2.0	2.7	1.15	2.09	4.07	13	0.833	0.62	0.56	1.26	0.41
Yb	0.0016	0.15	0.26	1.4	0.66	1.3	0.60	4.88	1.4	2.03	2.3	0.081	0.46	9.40	1.32	0.11
Type	IVA	IVA	IVA	IVB	IVB	IVB	VA	VA	VA	VB	VB	VB	VIA	VIA	VIB	VIC

In sums of analyses the following have been included (in wt.%): in sample 11 — 1.50 Cl, 0.15 SrO, 0.14, BaO; in sample 12 — 1.35 SrO, 0.38 BaO; in sample 16 — 0.32 SrO. 1, 2 — dunite and clinopyroxenite, Voykar-Syninsky Massif (Laz'ko, 1988); 3 — spinel orthopyroxenite, xenoliths in San Carlos basanite (Frey et al., 1978); 4, 5 — garnet clinopyroxenite and spinel websterite, Lherz Massif (Bodinier et al., 1987); 6, 7 — spinel websterite and wehrlite, xenoliths in basanite of Lake Bullenmeri maar (Griffin et al., 1988); 8 — garnet websterite, Ronda Massif (Suen et al., 1987); 9 — garnet clinopyroxenite, xenoliths in olivine nephelinite of the Salt Lake Crater volcano (Frey, 1980); 10 — garnet-amphibole clinopyroxenite, Lherz Massif (Bodinier et al., 1987); 11 — hornblendite, St Paul Island (Roden et al., 1984); 12 — phlogopite–amphibole–rutile–ilmenite–diopside rock (MARID Series), xenolith in kimberlite of the Bultfontein pipe; 13–16 — xenoliths of eclogites in kimberlites: 13 — diamond-bearing eclogite from the Udachnaya Pipe (after Zhuravlev), 14 — two-mineral eclogite from the Bellsbank Pipe (Taylor et al., 1989), 15 — the same, from the Kao Pipe (Shervais et al., 1988), 16 — kianite eclogite from the Udachnaya Pipe (after Zhuravlev and Ponomarenko).

and therefore are very dissimilar. The ultramafic rocks of subtype IVA are most commonly depleted in LREE relative to the middle and heavy REE.

The spinel and garnet ultramafic xenoliths, devoid of amphiboles and mica or poor in them (wehrlites, websterites and clinopyroxenites), occur in mantle-derived volcanics and analogous high-pressure vein rocks are classified as low-alkaline subtype VA. They resemble "green" high-Al ultramafic rocks in their petrochemical and geochemical properties, but are more enriched in Fe, Ti, K and REE (Table 9.2). These low-alkaline "black" ultramafic rocks can be either depleted or enriched in light lanthanoids, but they are characterized particularly by gaussian distribution spectra, with the normalized concentrations maximum located in the middle REE area (Frey, 1980; Suen et al., 1987; Griffin et al., 1988; Bodinier et al., 1988).

A high-alkali subtype VB of the "black series" encompasses vein hornblende pyroxenites and hornblendites of high-pressure complexes and xenoliths of phlogopite–amphibole–clinopyroxene ultramafic rocks containing ilmenite, rutile and apatite (MARID), found in kimberlites and basalts. Petrochemically and geochemically they are similar to rocks of subtype VA, but in general they are even more enriched in Ti, K, P and particularly ILE (Table 9.2).

Among the eclogites, the magnesian (VIA), ferrous (VIB) and aluminous (VIC) subtypes have been distinguished. These occur mainly as xenoliths in kimberlites. Apart from their generally variable composition, the distinguishing feature of magnesian eclogites is their low $FeO^*/(FeO^* + MgO)$ ratio (less than 50%); the ferrous ones, by contrast, have a higher ratio, accompanied by a marked enrichment in TiO_2, and the aluminous eclogites are characterized by a marked accumulation of Al_2O_3 (Table 9.2).

The geochemical characteristics of eclogites also vary widely, although all of them are enriched in ILE. The varied fractionation of REE, with the enrichment or depletion in light lanthanoids relative to heavy ones, is inherent in the magnesian and ferrous eclogites. In some cases the U-shaped spectra of REE distribution and distinctive Eu maxima have been established for these eclogites. The aluminous eclogites are more often depleted in light REE, and typically have a positive Eu anomaly. However, the published values of the total ILE concentrations, particularly of Rb, K, Ba and LREE, in eclogites are not always reliable because of the continuous contamination of the xenoliths by host kimberlite material. The Sr and Nd isotopic data for many eclogites, irrespective of their composition, are extremely unusual for mantle rocks: they have extremely high "unbalanced" values of ε_{Sr} and ε_{Nd} at any particular time (Shevrais et al., 1988; Neal et al., 1990).

b) Mantle Substrates and Subcrustal Magma Genesis

In contrast to the classification of mantle material, which can be done reliably enough on the basis of available data, the genetic correlation of particular volcanics to particular assemblages of subcrustal rocks that may constitute the sources of parental deep-seated melts is not always unambiguous, and so far is valid only in general terms (Laz'ko et al., 1993). The most obvious potential links have been tentatively outlined in Table 9.1, but a number of aspects of this problem require further discussion.

Firstly, most of the deep-seated rocks studied in laboratories are certainly not "pure" direct sources of widespread subcrustal magmas. The real mantle rocks are more frequently complementary to the melts, i.e. represent restites that are depleted to a certain degree. At the same time all are capable of further melting to participate in the generation of fresh parental magmas. The other complication rests on the fact that different liquids can be formed from the same material, depending on the melting conditions, and, vice versa, compositionally similar magmas can be derived from various sources. Finally, they can be formed due to two or more sources.

The above-mentioned difficulties, which arise during the attempt to match mantle rocks with a direct source of volcanics, have led to a broader concept of mantle substrates. In contrast to the abstract mantle sources already suggested (Allegre, 1987; Hart, 1988), these are implied to be real subcrustal rocks or their assemblages as distinguished and characterized above (Table 9.1), which, according to their specific compositions, may be either the main source (in individual cases the single source) of some other melts, or have a definite bearing on the characteristics of the source, or, finally, they may be complementary to restites (Laz'ko et al., 1993). In other words, mantle substrates represent a generalized series of chemically related subcrustal ultramafic rocks and eclogites, whose material can be subjected to, or earlier has been subjected to, melting.

The main conclusion from the foregoing is that available data do not yet allow us unambiguously to interpret the genetic relationships between mantle substrates and subcrustal magmas. A researcher is forced to judge these relationships on case-by-case basis. At the same time it is clear that mantle substrates and melts from them appear regularly during the evolution of subcrustal material. The controlling factors for the latter are the depletion and enrichment of particular zones in the Earth's interior, in low melting-point oxides and ILE, i.e. partial melting and mantle metasomatism. With regard to these processes Fig. 9.4 illustrates, in a generalized form, the suggested genetic relationships of different types of ultramafic substrates and their development throughout the irreversible evolution of the Earth. Eclogite substrates are not included in the simplified scheme presented here since their composition could be initially crustal.

c) The Distribution of Mantle Substrates in the Earth's Interior

Undoubtedly, particular mantle substrates make up isolated zones and segments in the lithosphere, but their actual distribution remains obscure. Within the

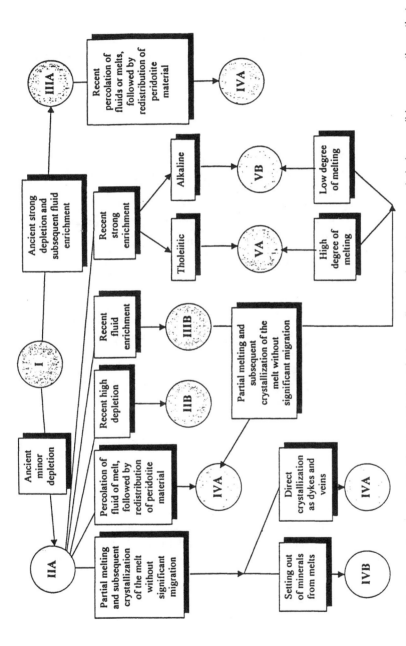

FIGURE 9.4 Generalized scheme of the evolution of ultramafic mantle sustrates and their possible connections that are caused by subcrustal petrogenetic processes.

framework of the popular layered model of upper-mantle structure, it is suggested
that the convecting layer (asthenosphere) is represented by depleted peridotites;
the underlying zone is close to "primitive" lherzolite in composition; and the
overlying lithosphere is composed of a wide range of ultramafic rocks both
enriched and depleted. Much importance has been attached to the eclogitization
of the Earth's crust and its descent into the mantle at subduction zones. Figure 9.5
presents a schematic version of the process, which also relates geodynamics to
present-day subcrustal magmatism.

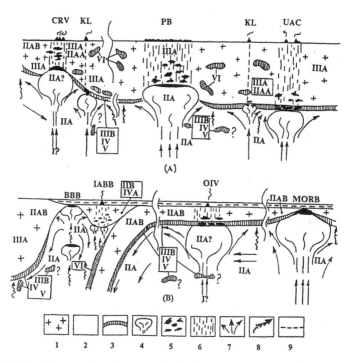

FIGURE 9.5 The mantle substrate distribution in the tectonosphere (A) of continental segments and passive ocean–continent transitional zones; (B) oceanic
segments and active continent–ocean transitional zones; 1 — lithosphere; 2 —
sublithosphere mantle; 3 — boundary layer between convecting asthenosphere
and conductive lithosphere; 4 — ductile material rising from the depths of the
mantle (mantle diapirs); 5 — melting sources; 6 — zones of increased permeability in the lithosphere; 7 — suggested convective currents in the mantle; 8 —
directions of fluid transfer; 9: CRV — volcanics of continental rifts, PB — plateau
basalts, KL — kimberlites and lamproites, UAC — ultra-alkaline volcanic–plutonic
complexes with carbonatites, MORB — mid-oceanic rift basalts, OIV — intra-
plate volcanics of seamounts and oceanic islands, BBB — basalts of back-
arc basins.

Commenting on the proposed scheme, it should be noted that one can argue with certainty in favour of the position of ancient, geochemically enriched but petrochemically strongly depleted, peridotite substrate IIIA. Judging from the isotopic studies of deep-seated xenoliths in kimberlites (Cohen et al., 1984; Richardson et al., 1985; Nixon, 1987; Menzies and Kornprobst, 1987; Laz'ko, 1988c), such rocks underlie the ancient platforms forming the thick geodynamically "dead" subcratonic lithosphere. These rocks can be considered as restites, which have lost their low melting-point fraction during the differentiation of the early Earth. The consolidated peridotites were then affected by fluids and fluidized melts seeping from below (from the underlying asthenosphere). Because of this, unusual "contradictory" characteristics are present.

Depleted peridotite substrates appear to be particularly abundant in the upper mantle. In accordance with modern concepts, convecting asthenosphere serves as a source for oceanic rift basalts (Allegre, 1987; Hart, 1988), and its main component is mildly depleted lherzolite IIA. After the melt separates from it, a strongly depleted "transitional" peridotite IIAB is produced, which constitutes the majority of the young spreading oceanic lithosphere. Extremely depleted harzburgites IIB are widespread in the basement of island arcs and active continental margins in the present-day continent–ocean transitional zones and in their fossil equivalents in the continental fold belts as well. The presence of depleted peridotites in subcratonic lithosphere has also been confirmed (Ionov and Jagoutz, 1988; Zhuravlev et al., 1991).

The position of the source of isotopically depleted intra-plate volcanics with high ILE content (substrate IIIB) is less clearly defined. It is usually believed that the substance from which their parental magmas are melted arose as a plume from subasthenosphere mantle levels or even deeper zones. However, in the local series of deep-seated xenoliths and peridotites of the high-pressure Ronda Massif, wide isotopic variations were discovered which overlap most of the corresponding characteristics for the basalts of the World ocean in general (Nixon, 1987, Griffin et al., 1988; Roden et al., 1988; Reisberg et al., 1989). When such a "hybrid" substrate is melted, one would also expect intra-plate magmas to be produced. In this regard, both weakly differentiated upper mantle material and the so-called marginal layer between the consolidated lithosphere and the convecting asthenosphere may be regarded as sources of intra-plate magmas (McKenzie and Bickle, 1988).

Judging from studies of ophiolites and high-pressure complexes and the discovery of xenoliths of complex structures, the minor ultramafic substrates form small bodies, interlayers, dykes and vein series in mantle peridotites. It is most likely that they originate from the local redistribution of major substrate material, episodes of intra-mantle magmatism and fluid mass transfer (Laz'ko and Sharkov, 1988; Kovalenko et al., 1989). Eclogites are likewise sporadically developed in the mantle.

d) Lower Crustal Substrates

Whereas the upper crust is accessible to geological sampling and measurements, the deep portions of the crust are relatively inaccessible. There are two main types of lower crust: continental and oceanic. The oceanic type is now better understood due to numerous detailed works on oceanic floor and ophiolite complexes. It is accepted that the lower crust in an oceanic environment is represented by layer 3 of the oceanic crust (Dobretsov, 1980; etc.). In ophiolite assemblages it is represented by a layered complex, composed mainly of gabbro and gabbronorites, which in the low part of sections alternates with pyroxenites, wehrlites and dunites (Laz'ko and Sharkov, 1988; Nicolas, 1989). In contrast, the continental lower crust remains a poorly known region of the Earth. Nevertheless, these portions of the crust contain important information related to the bulk composition of the continental crust as well as its forms.

The continental lower crust (below ~20–25 km depth) is believed to consist of metamorphic rocks of granulite facies, which are accessible as tectonic blocks and xenoliths in diatrems. However, the composition of the deep crust remains the greatest unknown because of the large compositional differences between granulites that occur in surface tracts (granulite terranes, in which felsic rocks dominate) and those that are carried as xenoliths, which are dominated by mafic rocks (Downes, 1993; Rudnick and Fountain, 1995).

The most common xenolith lithologies considered to be representative of the lower crust are mafic granulites, composed largely of clinopyroxene, orthopyroxene and plagioclase, and only in the Early Precambrian do garnet granulites and granulite eclogites predominate (Downes, 1993; Sharkov et al., 1996). Most of them were formed as cumulates from basaltic magmas and this origin suggests that they are a result of basaltic underplating, i.e. intrusion of magmas along the crust–mantle boundary. In general, these xenoliths have equilibration pressures between 6 and 9 kbar (approximately 20–30 km depth). However, mafic garnet granulites form the thick cold crust of ancient platforms have experienced higher pressure (12–22 kbar: Kempton et al., 1995; Sharkov et al., 1996). Metasedimentary granulites and felsic meta-igneous granulites are much rarer than mafic granulites. For available data about the geochemistry of lower-crustal rocks from different occurrences see Rudnick and Fountain (1995).

As shown above (see Chapters 5 and 6), the Early Precambrian lower crustal rocks could play a significant role in the petrogenesis of some magmatic series at that time. So it is necessary to discuss this problem in more detail. We studied two locations of early Precambrian lower crustal rocks from major occurrences of the ancient lithosphere of Eurasia: the Russian and Siberian platforms. They are represented by xenoliths from a diatrem in the Kola Peninsula on the Baltic Shield, and by a tectonic block of high-pressure rocks in the Southern Prisayanian, Eastern Siberia (Sharkov et al., 1996).

On the Baltic Shield the lower crustal rocks are represented by deep-seated xenoliths of garnet granulites and eclogites in the Elovy Island diatrem (the Kola

Peninsula) within the Belomorian belt — the eastern continuation of the Archaean. Karelian granite–greenstone terrane which was strongly reworked in the Palaeoproterozoic (see Chapter 5). Their chemical composition occupies an intermediate position between tholeiitic and island-arc basalts (Table 9.5) which is typical for SHMS rocks. The rocks were formed with $P = 12–19$ kbar and $T = 750–930°C$. Their model Sm–Nd age varies from 2.9 to 2.4 Ga (Kempton *et al.*, 1995).

In the Southern Prisaynian (Eastern Siberia, 150 km to NW from Lake Baikal), lower crustal formations are represented by a tectonic block (the Arbansky massif) within the Late Archaean Onotsky granite–greenstone terrane. The massif is formed of garnet granulites and eclogites with interlayers of garnet–kyanite schists (metapelites) and thick (about 100 m) bodies of Gr-bearing enderbite–gneisses. In its turn, the Onotsky terrane is a large tectonic block in the early Palaeoproterozoic (c. 2.45 Ga: Aftalion *et al.*, 1991) Sharyzhalgay granulite belt. Tectonic emplacement of the Arbansky massif took place c. 2 Ga ago. These rocks, probably, were primarily a volcanic-sedimentary sequence, and enderbite–gneisses with relics of magmatic textures were intrusive bodies, as were ultramafics. They differ from the Baltic rocks in that the chemical composition of the garnet granulites is unusual for igneous rocks (Table 9.5), and suggests their residual origin, after the melting of acid rocks. (Sharkov *et al.*, 1996).

The rocks were formed under different P–T conditions: garnet-bearing ones at high pressures (eclogite at $P = 21–24$ kbar, Gr granulites at $P = 15–20$ kbar, $T = 840–960°C$). This probably means that the lower crust here was formed by tectonic slices of rocks from different depths which were "soldered" together before c. 2 Ga ago, when they were thrust out as a uniform body.

The complexes studied represent the two main types of lower crustal rocks which were formed during the Archaean and the Early Proterozoic. Our studies suggest that the origin of the Kola type was linked with underplating beneath granite–greenstone terranes, whereas the origin of the Prisaynian one was probably related to zones of "drawing" beneath granulite belts.

9.1.3 Mantle Magma Generation in the Phanerozoic

Mantle magma generation here is generally associated with oceanic and back-arc spreading zones, as well as with intra-plate magmatism. The mantle substrates considered above are involved in melting and hence we will dwell mainly on melting mechanisms, based on data from petrological and geological studies.

a) Magma Generation in the Oceanic Spreading Zones

In accordance with numerous studies summarized in the works of Kovalenko (1987), Wilson (1989), etc., the origin of MORB is related to the 20–30% partial

Table 9.5 Composition of the lower-crustal rocks from the Early Precambrian terranes and the present Earth.

Components	1	2	3	4	5	6	7	8	9	10	11
SiO_2	49.69	48.84	43.41	48.25	52.40	57.80	55.20	52.30	60.60	66.00	59.10
TiO_2	0.65	0.55	1.03	2.60	0.80	0.70	0.70	0.80	0.70	0.50	0.70
Al_2O_3	22.02	12.68	10.97	36.72	16.50	15.80	15.40	16.60	15.50	15.20	15.80
Fe_2O_3	6.60	14.79	3.38	1.66	0.10	0.10	0.10	0.10	0.10	0.08	0.11
MnO	0.08	0.22	0.27	0.05	0.10	0.10	0.10	7.10	3.40	2.20	4.40
MgO	2.33	9.59	8.85	4.01	7.10	4.70	4.90	9.40	5.10	4.20	6.40
CaO	10.05	11.27	8.00	0.89	9.50	6.70	6.90	2.60	3.20	3.90	3.20
Na_2O	4.70	1.73	0.34	0.15	2.70	3.10	3.10	0.60	2.01	3.40	1.88
K_2O	1.10	0.26	0.05	0.06	0.60	1.70	1.20	0.10	0.10	0.40	0.20
P_2O_5	0.10	0.04	—	—	0.10	0.20	0.20	0.10	—	—	0.20
LOI	3.77	0.42	1.13	0.10	—	—	—	—	—	—	—
Total	101.09	100.39	99.36	100.16	100.00	100.00	100.00	100.00	100.00	100.00	100.00
Mg*	.45	.60	—	—	61	55	54	60	48	47	54
Nb	7	1	2	—	5	11	7	5	8	25	12
Zr	47	78	56	—	69	119	105	68	125	190	123
Y	13	21	18	—	16	19	16	16	22	22	20
Sr	681	131	51	—	349	331	317	348	281	350	325

	1	2	3	4	5	6	7	8	9	10	11
Rb	29.8	9.0	2	—	12	52	29	11	62	112	58
Ba	336	57	56	—	263	379	339	259	402	550	390
Zn	59	151	—	—	77	75	69	78	70	71	73
Cu	25	88	—	—	26	24	22	26	20	25	24
V	146	221	297	—	194	141	162	196	118	60	131
Sc	17	41	—	—	30	22	20	31	22	11	22
Ni	27	132	23	—	88	55	77	88	33	20	51
Cr	41	956	259	—	213	128	162	215	83	35	119
La	10	11	—	—	9	17	18	8	17	30	18
Nd	17	16	—	—	11.4	19.0	15.4	11	24	26	20
Ce	35	34	—	—	21	40	36	20	45	64	42

Columns 1–2 — lower-crustal garnet granulites from Elovyi Island diatrem (Kempton et al., 1995); 3–4 — lower-crustal rocks from Arbansky massif; 3 — garnet granulite, 4 — garnet kyanite rock (Sharkov et al., 1996); 5–7 — average composition of continental crust of shields; 5 — lower crust, 6 — total; 7 — Archaean total (Rudnick and Fountain, 1995); 8–11 — major and trace element composition of the continental crust; 8 — lower, 9 — middle, 10 — upper, 11 — total (Rudnick and Fountain, 1995).

melting of a depleted lherzolite source of IIA type and harzburgites of IIB type at pressures lower than 15–20 kbar (depths of less than 50–60 km). This is consistent with the results of seismic studies on the structures of mid-oceanic ridges, particularly the East Pacific Rise, where the decondensing zone along the ridge axis is located at depths from 20 to 60–70 km (Hekinian, 1982). However, magma eruption on to the surface occurs from less deep-seated magma chambers, where melts separated from mantle sources at different pressures — from 10 to 5 kbar (Sobolev and Shimizu, 1992) — accumulate. Relatively small magma chambers of this type, i.e. 1–2 km thick and tens of km wide, have been revealed by seismic methods on the East Pacific Rise, at depths of 8–4 km beneath the ridge axes (see Chapter 2).

However, the actual situation is often more complex. Thus, according to Sharma *et al.* (1995), detailed Nd isotopic studies of Voykar ophiolite have shown that the oceanic crustal rocks and harzburgites are complementary and come from a MORB source. But one sample of harzburgite is far removed from the array, indicating the presence of older, depleted mantle material unrelated to the crustal section. The major-element and low-REE composition of the Voikar harzburgites can be formed by progressive extraction of melt from an undepleted mantle protolith with the requirement that melt separation began in the garnet lherzolite stability field. The extremely chemical and isotopic characteristics of the Voikar harzburgite demonstrate that extraction of partial melt can generate an upper mantle with highly variable ε_{Nd} and very low Nd concentrations. The potential presence of such material in the upper mantle poses an intriguing geochemical problem in that its bulk isotopic signature may not be detectably expressed in basalts, which are dominated by sources with higher average Nd concentration and lower ε_{Nd} and $f_{Sm/Nd}$.

This poses a definite problem for the interpretation of the layered complexes of ophiolite assemblages, which require large magma chambers for their formation. In an endeavour to solve this contradiction, Quick and Delinger (1993) suggested the "gabbro glacier" model. The essence of this model is a combination of continuous feeding of the magmatic reservoir by the magma generation zone below it, and its crystallization under conditions of continued spreading of the chamber floor, as a consequence of which the newly generated cumulates are removed from the zones of crystallization, thus freeing up a large area of the stratified ophiolite complex. Similar processes also appear to take place in back-arc spreading zones.

b) Magma Generation in Intra-Plate Environments

The most characteristic example is the picrite–basalt series of normal, moderate and high alkalinity (K–Na and K); kimberlites and lamproites are rare.

As has been shown above for the petrogenesis of intra-plate basalts, it is important to note that often their source was metasomatized not long before the generation of mantle magma (IIIB type substrate), producing increased concentrations of titanium, alkalies, barium and incompatible elements. In the opinion of most researchers, intra-plate magmatism is associated with the ascent of asthenospheric diapirs in lower-pressure areas, where decompression can result in the decrease of the solidus temperature of ultrabasic material, which in turn leads to its melting. Judging from the results of experimental studies and seismic data, the areas of intra-plate magma generation occur at depths of about 40–70 km (Kovalenko, 1987; Wilson, 1989; etc.).

The geological evidence indicates that Fe–Ti picrite–basaltic magmas of moderately alkaline and rarely alkaline (K–Na and K) composition, observed at the base of the sections of continental rift zones and oceanic islands, usually have melted first. This type of melt could have existed throughout the entire duration of "hot-spot" activity, but more frequently it is replaced by tholeiitic series formations ("intermediate tholeiites"), for example many continental rift zones, trap fields and Hawaiian-type oceanic islands or even MORB-type basalt, as we can see in the Red Sea Rift or Siberian Traps (see Chapter 4). According to Gast (1968) this could be due to subsequent depletion of the mantle source as a result of igneous processes; the latest melts are known to contain very few incompatible elements. However, in trap regions, continental-lithospheric contamination could contribute to melt petrogenesis.

The essential feature of K–Na alkali-basalt magmatism is often the presence of mantle xenoliths in lavas and pyroclastics. It is very important that, irrespective of whether this area is located within a continental plate, a back-arc basin or an ocean, the composition of xenoliths is practically identical — the most common here are green spinel, and more rarely garnet–spinel lherzolites and websterites (the "green" or Cr-diopside series), black wehrlites, clinopyroxenites, hornblendites and glimmerites as well as megacrysts of Al-titanaugite, kaersutite, phlogopite and sanidine (the "black" or Al-titanaugite series). The detailed characteristics of these formations are described in numerous publications (Kornprobst, 1987; Laz'ko and Sharkov, 1988; etc.), where these rocks have been shown to be generally similar to those observed in the ultramafic complexes of ophiolite assemblages, oceanic peridotites and tectonic blocks of high-pressure ultramafic rocks of fold belts.

The Cr-diopside series formations are attributed to type IIIB major mantle substrates, and type IV of minor mantle substrates (Table 9.1), and most often these occur in combination, i.e. in the same place one can find xenoliths of both depleted and enriched (metasomatized) mantle. In the opinion of many researchers, all of these represent fragments of asthenospheric material.

The situation with the "black" series xenoliths, referred to as minor substrates of type V, is more complicated. There are various viewpoints concerning their origin; however, recent researchers tend to be of the opinion that these xenoliths, including the megacrysts of kaersutite and Al-titanaugite, represent the crystallization products of high-density fluids or severely fluidized melts that entered the upper mantle from deeper zones within the Earth (Kovalenko et al., 1985c; Ryabchikov, 1988; Sharkov et al., 1990). The specific feature of these essentially carbon dioxide fluids is their high saturation with titanium, iron, alkalies (especially potassium), barium and phosphorus.

As may be inferred from observations on xenoliths, e.g. fragments of druses of minerals from the "black" series or growths of similar minerals on the relict flat surfaces of spinel lherzolite xenoliths, these formations were formed in fracture-shaped hollows within upper mantle material. It is most likely that they have been formed in the upper, cooled parts of the spread heads of asthenosphere diapirs, the fragments of which represent the "green" series formations.

Judging from the composition of the xenoliths, the marginal zones of these diapirs were composed partially of depleted ultrabasic rocks, ranging from lherzolites to harzburgites and even dunites, as has been established for the Quaternary Tell-Danun volcano in western Syria (Sharkov et al., 1989). This seems to indicate active mechanical mixing of mantle material during previous upwelling and spreading of the asthenosphere and its rather high viscosity, preventing homogenization of the material. As a consequence, rocks that earlier represented the material from zones of melting could also be present here.

Thus, from the presented data, two important conclusions can be drawn:

1. observed mantle xenoliths appear to represent fragments of the upper cooled margin of asthenospheric diapirs rather than material from the zone where host basalts were generated, and
2. the magma chambers from which the eruptions have just occurred in many cases are located within the upper margins of part of these diapirs.

A special study of xenolith-bearing basalts from the Baikal rift zone has shown that the formation of these magma chambers occurred at pressures ranging from 5 to 8 kbar, i.e. at depths of 16–25 km (Sharkov and Bindeman, 1991). The belief that the upper edge of asthenospheric diapirs could reach the upper crust is also attested by the virtual absence of lower-crustal garnet granulites among the xenoliths observed in the overwhelming majority of intra-plate basaltic volcanoes where only fragments of mantle and upper crustal material were recorded. At the same time, lower crustal formations are widespread in the kimberlites draining the continental lithosphere section. Moreover, in some cases ultrabasic mantle material even comes to the surface, as on the island of Zabargad in the Red Sea (Bonatti et al., 1986) and St. Paul's Rocks in the Atlantic Ocean (Roden et al., 1984).

Hence, from the available data, it can be inferred that at least part of these chambers was located within individual asthenospheric highs beneath basaltic areas, although their identification by geophysical methods is hindered by the screening effect of overlying lava plateaux. Certainly it does not follow that basalts were formed at these depths, it was just the magmatic chambers that were located here. The areas of magma generation could have been much deeper, at depths of 40–70 km, where, for example, in the Baikal Rift Zone, a layer with increased electric conductivity has been located found (Zorin *et al.*, 1989). It is obvious that mantle sections containing such chambers were distinguished by reduced viscosity and density and, accordingly, had to rise upwards, forming the above-mentioned local rises on the surface of asthenospheric diapirs. It is natural to suppose that this process was often followed by fracturing of the overlying lithosphere, which enabled the melt to reach the surface. A schematic diagram of this process is shown in Fig. 9.6.

It is clear that, in addition to the chambers described above, intermediate crustal chambers must have existed, and, in some cases, there must have been magma chambers in the underlying subcrustal mantle. It is evident from the results of isotope-geochemical studies of the basic rocks of trap areas (see Chapter 4), which testify to the large-scale contamination of similar rocks by basaltic magmas. From this it follows that the contours of the surface of asthenospheric diapirs were complex, and intermediate chambers could have been formed under various conditions, depending on the particular tectonomagmatic setting.

The fact that eruptions can be localized in the same region for several millions of years suggests that the process of magma generation was fairly prolonged, and devastated chambers were periodically replenished by fresh melt. In any case, once magma conduits between zones of magma generation and the

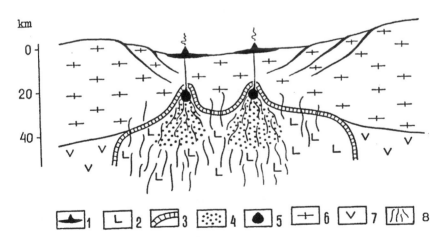

FIGURE 9.6 Suggested topography of the top of the asthenospheric diapir under the Baikal Rift. After Sharkov and Bindeman (1991). 1 — plateau basalt volcanic areas; 2 — asthenospheric diapir; 3 — the chilled zone of this diapir; 4 — melting zones; 5 — igneous sources; 6 — continental crust; 7 — ancient subcrustal lithosphere; 8 — paths of deep fluid penetration.

Earth's surface had developed they remained in position for long periods, enabling the formation of large lava plateaux, including oceanic islands.

Data on isotope distribution in alkaline K–Na rocks has allowed us to draw the following inferences:

1. There are wide variations in ε_{Nd} and ε_{Sr}, including all mantle sources within the main mantle trend such as primitive and depleted mantle, and also beyond the limits of this trend to the right and the left of it (crustal assimilation and mantle metasomatism).

2. For carbonatites most often ε_{Sr} is most commonly greater than zero, although, as with nepheline syenites, wide variations in $(^{87}Sr/^{86}Sr)_0$ have been outlined (Fig. 9.7). The appearance of carbonatites and nepheline syenites with high values of $(^{87}Sr/^{86}Sr)_0$ is connected with the influence of crustal material that also accounts for wide variations in $\delta^{18}O$ (6.7–14.2‰) and ^{13}C (from -6 to 2.5‰). Rocks with high $(^{87}Sr/^{86}Sr)_0$ were formed at about 1.5–1.0 Ga.

High-Ti alkaline series igneous rocks were formed no earlier than 2.2–2.0 Ga, and therefore their sources must have arisen at approximately that time. According to the experimental data, basanitic magmas can arise from a low degree of melting of primitive mantle. At pressures of approximately 30 kbar and a high activity of CO_2, the high-potassium magmas of nephelinitic and melilititic type are melted from garnet lherzolite (at about 50 kbar this is possible without volatiles), the fractional crystallization differentiation of which leads to the formation of agpaitic nepheline syenite magmas (Ryabchikov, 1988).

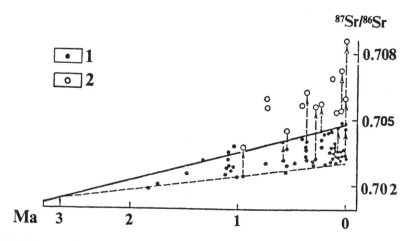

FIGURE 9.7 Evolution of Sr isotope composition for carbonatites (Balashov, 1985). 1 — depleted mantle compositions; 2 — compositions with impurities of crustal material.

New data have been obtained by Kogarko and Ryabchikov (1995) on petrogeneic and REE element distribution in the rocks of meimechite series (Polar Siberia). They show that meimechites, alkaline picrites and katungites have a co-magmatic nature, with meimechites as a primitive melt of the series. It has been supposed that the primary meimechite magmas were formed during the melting of subcrustal lithospheric harzburgites enriched in incompatible elements due to the infiltration of diapiric melts of a low grade of partial melting.

According to Ryabchikov (1995), the REE content observed in kimberlite (see Chapter 4) requires that the primary melt be equilibrated with harzburgitic residual mineralogy. This is consistent with experimental data on the melting of relations in carbonated peridotites and also with the suggested role of refractory harzburgites as traps for the near solidus melt transporting incompatible elements. The selective trapping of low fraction melts transporting incompatible elements from the asthenospheric diapir to the overlying lithosphere by harzburgites may be due to the significantly higher solidus temperatures of these refractory rocks. The balance of CO_2 during melting implies that primary kimberlite magmas are likely to be generated as carbonate-saturated melts; in this case their CO_2 content may approach 30% (Brey et al., 1991). After rising to depths with pressures below c. 50 kbar, kimberlitic melts with 15–30% CO_2 should involve gas, and this may trigger fast transportation of the magma to the surface due to hydraulic fracturing and crack propagation. Because in the case of relatively CaO-poor kimberlites this should happen in the depth range where diamond is still stable, it may be that diamond survives because of the very rapid ascent. By contrast, CaO-rich lherzolite-derived magmas will retain dissolved CO_2 until decompression well beyond the diamond stability field, so they are not likely to be transporting agents for this mineral. However, such melts may be parent magmas for crustal carbonatites.

Wide variations in isotopic parameters mean that the K-alkaline rocks are most often shifted towards $\varepsilon_{Sr} > 0$ and $\varepsilon_{Nd} < 0$ (Fig. 9.8), i.e. their sources are commonly enriched in lithophile elements. This is partly attributed to the crustal influence when a positive correlation is observed between $(^{87}Sr/^{86}Sr)_0$ and $\delta^{18}O$. As for mantle values, however, the $\delta^{18}O$ and $(^{87}Sr/^{86}Sr)_0$ in potassium alkaline rocks are frequently high, which confirms the significant role of mantle metasomatism in their igneous sources. Most petrogenetic models assume that lamproite melts form by a low degree of partial melting (i.e. several per cent of the metasomatized mantle source, enriched by incompatible elements). The composition of mantle material is assumed to vary from phlogopitized harzburgites (Foley et al., 1986) to phlogopitized wehrlites and lherzolites (Dawson and Smith, 1988; Ryabchikov et al., 1991). The Australian lamproites can be used as an example of a fairly long multi-stage (at least three-stage) process of mantle metasomatism. The characteristics of the metasomatizing fluids seem to have been different in the intra-plate and collision settings; in the latter they had a low Ti, Nb and Zr content.

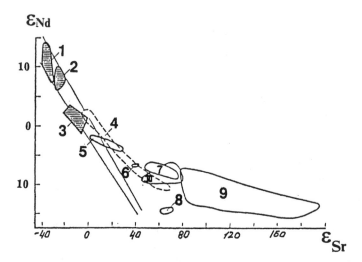

FIGURE 9.8 The ε_{Nd} and ε_{Sr} variations for K-alkaline rocks (Balashov, 1985). The fields: 1, 2 — tholeiitic basalts of MOR N — (1) and E — (2) types; 3 — mica-free kimberlites, southern Africa; 4 — olivine and leucite Verunga nephelinites of Africa; 5 — ugandites and leucitites, eastern Africa; 6 — Precambrian carbonatite, Finland; 7 — leucite rocks, Italy; 8 — leucitites, Antartica; 9 — micaceous kimberlites and lamproites, western Australia; 10 — micaceous kimberlites, southern Africa.

c) The Main Features of Mantle Magma Generation in the Early Precambrian

Of particular interest are komatiitic melts typical of the Archaean. From the experimental data of a number of researchers, particularly Girnis and Ryabchikov (1988), the eruption temperature of komatiite lavas reached 1,600°C. According to their data, the separation of magmas from restites occurred at a pressures of 35–40 kbar and a temperature of 1,750–1,800°C. The temperatures of the initiation melting processes in deep-seated diapirs reached 2,000°C, that is, higher by 600°C than the temperatures of the MORB sources (McKenzie, 1985). It follows that magma sources were located much deeper than was the case for the basaltic melts of the Phanerozoic.

Another important piece of experimental data is the change in the composition of the komatiite liquidus phases. In accordance with Ohtani's data (Ohtani, 1990), for the pressure increase from 10 to 250 kbar, they are modified in the following order: $Ol + Cpx \rightarrow Ol \rightarrow Mj$ (majorite) $\rightarrow Pv$ (Mg perovskite: i.e. olivine with a perovskite structure). From this it will be obvious that majorite (a garnet plus orthopyroxene solid solution) plays a significant role in the formation of high-magnesium magmas at depths of 450–650 km, i.e. at the base of the upper mantle. Below this level, perovskite becomes stable. The majorite concentrates

Al, Sc, Y and HREE, which should cause corresponding changes in the chemistry of the rocks and lead to the formation of Al-depleted magmas (Kato *et al.*, 1988).

The rather similar komatiites are highly characteristic of early Archaean greenstone belts. In accordance with Pukhtel and Zhuravlev's data (1993), they have higher Gd/Yb and Al/Ti ratios as compared with chondrites, whereas in the late Archaean units, volcanics with ratios close to chondrites are the norm. The third, rarer type of komatiites is also restricted to the late Archaean and is characterized by a relative enrichment in Al and HREE. The petrogenetic models for komatiitic melt formation are presented in Fig. 9.9.

Another, more significant achievement of ulta-high-pressure experimental petrology was the discovery of the higher compressibility of the melts relative to restite minerals (Nisbet and Walker, 1982; Rigden *et al.*, 1984; etc.). Hence, at depths of more than 250–300 km, the density of ultrabasic magmas is higher than that of olivine and pyroxene, but lower than that of majorite (Fig. 9.9b). Accordingly, only the latter can be fractionated from the melt, whereas Ol and Px should form a suspension in the magma generation zone. This should create different variants when uprising mantle diapirs, which led to the production of both Al-depleted and Al-enriched magmas, are melted. In diapirs originating at shallower depths (< 400–300 km), the olivine represents the liquidus phase and the melts are not depleted in Al (Ohtani, 1990).

All these data concerning very high pressures during the generation of komatiitic magmas indicate that, despite the higher geothermal gradient in the Archaean (approximately 100–150°C higher than at present; Ryabchikov, 1988), magma generation occurred at great depths. The higher temperature of the crust affected mainly crustal magma generation, but extra heat was also needed for this to happen (see below).

Two point of view exist regarding the origin of Archaean komatiitic high-magnesian and moderately magnesian basalts. In the opinion of McKenzie (1985), Arndt (1994), etc., they were formed by melting of mantle diapirs, which reached depths of c. 50–100 km. However, based on experimental data, Girnis *et al.* (1987) have demostrated that it is more probable that they originated from fractional crystallization of primary komatiitic melts in intermediate-level chambers.

The unusual features of mantle magmas generation at the nuclearic stage in the Archaean were considered by Pukhtel *et al.* (1993a and c), as exemplified by the Olekma granite–greenstone terrane (see Chapter 6). They demonstrated that among the komatiites of early Archaean assemblages (3.4–3.3 Ga in age), Al-depleted high-magnesian rocks predominate. The generation of their parental melts was feasible only under the extreme conditions (P > 150 kbar at T > 1,800°C) that existed in the early Archaean mantle. Here, isotope-geochemical data suggest the presence of vertical isotopic and geochemical inhomogeneity (Fig. 9.10). The upper-mantle horizons were characterized by depletion in LREE and other

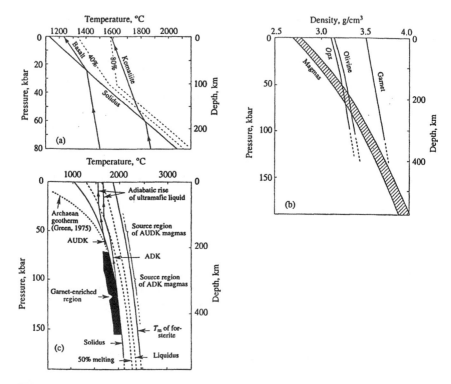

FIGURE 9.9 Petrogenetic models of parental melt formation for Al-undepleted and Al-depleted komatiites. After Pukhtel *et al.* (1993a). See p. 445 for more detailed explanation. (a) — model based on the assumption that the high eruption temperatures of komatiic magmas are due to their generation at far greater depths compared to basaltic magma (McKenzie, 1984); (b) — relationship between the density of mafic-ultramafic magmas (shaded area), some typical liquidus mineral phases and pressure (Rigden *et al.*, 1984); (c) — model of komatiite magma generation by partial melting in rising mantle diapirs (after Ohtani, 1984; Arndt, 1986)

highly incompatible elements, long before the formation of the supracrustal units of the Tungurcha greenstone belt ($\varepsilon_{Nd(T)} = +1.9$). In the deeper levels (>200 km) the following coexisted: reservoirs that were both depleted and enriched in moderately and highly incompatible elements, and reservoirs that existed before the melting processes was established. It is not inconceivable that a reservoir enriched in LREE, which represented ancient crustal material, was assimilated by the ascending komatiitic melt. The degree of depletion of the early Archaean mantle actually decreased with depth.

The late Archaean (3.0 Ga) is characterized mainly by the high-magnesian rocks of Al-undepleted type formed at moderately high P and T (P < 70 kbar,

Early Archaean: 3.3 - 3.4 Ga
Ancient mafic association

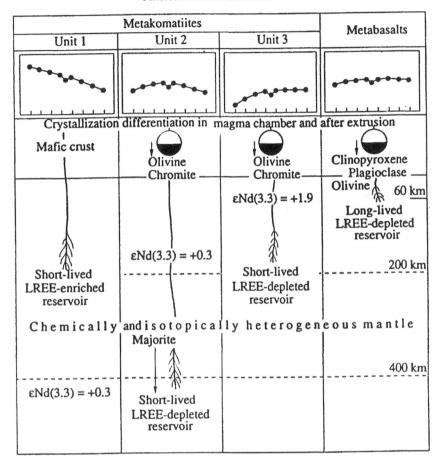

FIGURE 9.10 Scheme illustrating the mechanisms of melting and subsequent evolution of basic–ultrabasic melts and isotope-geochemical characteristics of the upper mantle of the Olekma granite–greenstone in early Archaean. After Pukhtel *et al.* (1993a).

$T < 1,800°C$). Aluminium-depleted rocks are rare and are noted for their low magnesium content. The transition from Early to Late Archaean is characterized by a decrease in the depth range of magma generation and, as a consequence, a change in the liquidus assemblage (majorite → olivine). Isotope-geochronological data provide evidence that the sources of basic–ultrabasic and acid magmas had an identical Nd-isotopic composition (Fig. 9.11). From this it follows that

the late Archaean mantle of the Olekma granite–greenstone terrane was chemically heterogeneous and was characterized by isotopic inhomogeneity caused by
long-term depletion in highly incompatible elements with $\varepsilon_{Nd(T)} = +2.2$ and
Th/U = 2.6. The start of the suggested inhomogeneity occurred during the formation of supracrustal units, owing to the influx of components enriched in
moderately and highly incompatible elements. Isotope-geochemical data suggest
that some komatiites of the Olondinsky greenstone belt were mixed with early
tonalitic–trondhjemitic gneisses of the greenstone belt basement during the

THE EARLY PRECAMBRIAN CRUST-MANTLE EVOLUTION

Late Archaean: 3.0 Ga
Olondo greenstone belt

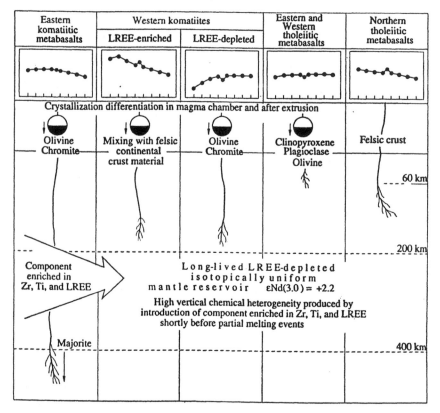

FIGURE 9.11   Scheme illustrating the mechanisms of melting and subsequent
evolution of basic–ultrabasic melts and isotope-geochemical characteristics
of the upper mantle of the Olekma GGST in the late Archaean. After Pukhtel
et al. (1993a).

ascent of highly magnesian melts along the magmatic conduit and/or in intermediate chambers. In the opinion of Pukhtel and Zhuravlev (1993), the acid rocks of greenstone belts most likely derived from melting of basic metavolcanics of the lower parts of the section, and ancient acid crustal material was not involved in their formation. Judging by the data of U–Pb (SHRIMP) dating of zircons from different parts of the section, the duration of the formation of volcano-sedimentary units did not exceed 17 ± 7 Ma.

The Early Proterozoic ultrabasic melts of the cratonic stage have a lower magnesium content, of about 20–22% MgO. They were separated from mantle diapirs (IIBB type substrate) at depths of 60–90 km. As a rule, the Early Proterozoic basic–ultrabasic rocks are enriched in LREE and SiO_2 and have an increased Al_2O_3/TiO_2 ratio that makes them closer to boninites. The $\varepsilon_{Nd(T)}$ varies from near 0 to approximately -3 and $+1$, which on the one hand points to a highly depleted source, and, on the other, to the contamination of the rocks by crustal material.

Beginning with the continental–oceanic stage, i.e. after 2.2 Ga ago, a marked enrichment in moderately and highly incompatible elements (Ti, Nb, Zr and LREE) is observed in the picrites, which suggests the occurrence of anomalous sources of IIIA-type mantle, presumably associated with mantle metasomatism (Fig. 9.12). This tendency first emerged as early as the Late Archaean; however, only from 2.2 Ga did it develop on a large scale. On the other hand, the positive value of $\varepsilon_{Nd(T)} = +1.6$ (in the Olondo Belt) points to the presence at that time of long-lived light REE-depleted reservoirs in the mantle. The process of parental magma melting in the Early Proterozoic was controlled by the fractionation of garnet from picrobasaltic melts at depths of more than 75 km (Pukhtel and Zhuravlev, 1993).

9.2 MAGMA GENERATION IN THE EARTH'S CRUST

In the opinion of most petrologists, acid igneous rocks, including various granitoids, have derived from crustal substrates (Yarmolyuk and Kovalenko, 1987; Ryabchikov, 1988; Wyllie, 1988; Ronov et al., 1990). From the experiments of Wyllie et al. (1976) and Rapp and Watson (1995), tonalites and trondhjemites constituting the "grey gneiss" assemblage at low pressures is plagioclase; at moderate (10 kbar) and high pressures (30 kbar, depths more than 90 km) it is pyroxene and garnet, respectively. These minerals are similar to those from basic magmas but not to mantle ultrabasic magmas and, accordingly, these melts could not be melted out directly from the mantle, but had to pass through the intermediate stage of basic substrates. For K–Na granites, the dominant liquidus minerals under mantle pressure conditions are quartz and coesite, rather than mantle

Early Proterozoic: 2.2 Ga

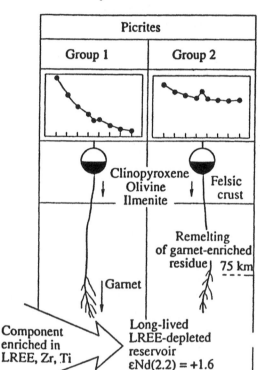

FIGURE 9.12 Scheme illustrating the mechanisms of melting and subsequent evolution of basic–ultrabasic melts and isotope-geochemical characteristics of the upper mantle of the Olekma granite–greenstone belt in the early Proterozoic. After Pukhtel *et al.* (1993).

minerals (magnesian olivine, ortho- and clinopyroxene) or even basic rocks (plagioclase, amphibole, pyroxene). The direct melting of granite magma is ruled out at mantle pressures, even though quartz-feldspatic sediments may reach these depths. In the system albite–orthoclase–quartz at pressures greater than 10 kbar, the composition of the liquid coexisting with quartz and alkaline feldspar (or jadeite, sanidine and coesite) is shifted to alkali feldspar; i.e. under the conditions being studied, melting should result in the production of either trachytic or syenitic melts rather than granitic ones.

The $(^{87}Sr/^{86}Sr)_0$ type isotopic characteristics of granitic magma depend not only on the value of the Rb/Sr ratio of source but also on the time interval between the formation of the source with any given Rb/Sr ratio, and the melting from this source of granite magma (Kovalenko and Bogatikov, 1987). Figure 9.13 shows the complete correspondence between the $({}^{87}Sr/{}^{86}Sr)_0$ variations in granites, gneisses and sedimentary rocks, all of which are possible sources of granitic magmas. Hence, among gneisses and sedimentary rocks there is a sufficient number of compositions with random $({}^{87}Sr/{}^{86}Sr)_0$ values, including "mantle" ones which could have been recorded as being generated from these granites. The tonalitic, trondhjemitic and plagiogranitic magmas with "mantle"

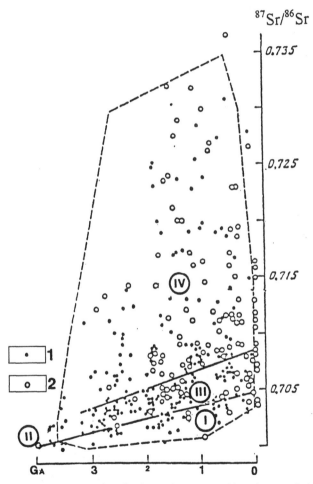

FIGURE 9.13 Evolution of the Sr isotopic composition in granitoids (dotted shading), gneisses (1) and sedimentary rocks (2) (Balashov, 1985). I–III fields — the depleted, primitive and enriched mantle sources, accordingly; IV — crustal sources.

isotopic characteristics could have been derived from sedimentary rocks of "basic" composition and their metamorphosed analogues, as a result of partial melting and further granitization.

Potassium–sodic granite magmas with "mantle" isotopic characteristics can also be formed from the same substrate. These two extreme situations may be visualized (see Fig. 9.14). In the first instance, the Rb influx to the system (source) is assumed to occur long before the generation of granite magma, and the time interval between rubidium influx and magma generation has to be sufficient for the marked accumulation of radiogenic strontium and the increase in $(^{87}Sr/^{86}Sr)_0$ that is also recorded in the granites. This is very similar to the process of crustal granite formation on basic substrate that has been granitized for a substantial period of time. In the second instance the time between the influx of rubidium and substrate melting is negligible, and hence Sr^{87} accumulation fails to occur and the granite magma fixes the low "mantle" $(^{87}Sr/^{86}Sr)_0$ value. This case models the process of basic substrate granitization, with its fast melting. In reality, in terms of total composition the source for the magmas is the same in both instances, but in the first case it is much more ancient relative to the time of magma generation, and in the second case it is very young. It is obvious that, in both cases, granites with crustal and mantle characteristics are typically crustal.

Of course, rubidium influx to the system could be associated not only with granitization but with repeated regional metamorphism and other processes that increase the Rb/Sr ratio in rocks.

Oxygen isotopic data indicate the rarity of granite sources with truly mantle isotopic characteristics (Balashov, 1985). The majority of them have increased $\delta^{18}O$ ($+7.5$–$+8.0$‰) values, suggesting the significant role of sedimentary

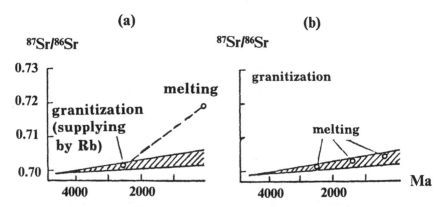

FIGURE 9.14 Theoretical schemes of granitoids with high "crustal" (a) and low "mantle" (b) $(^{87}Sr/^{86}Sr)_0$ (Bogatikov and Kovalenko, 1987).

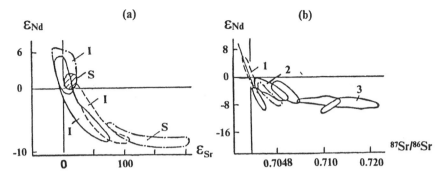

FIGURE 9.15 The variations of ε_{Nd} and ε_{Sr}, $^{87}Sr/^{86}Sr$ in granitoids of the I and S types from batholiths of eastern Australia (a), Malaysia and California (b) (after Balashov, 1985). 1 — California; 2, 3 — East and West Coast Provinces of Malaysia.

rocks as the source of granitoids. The $(^{87}Sr/^{86}Sr)_0$, $\delta^{18}O$ values increase and I_{Nd}^0 decreases to a lesser degree as the role of the latter in the granitoids increases. The extreme (in terms their sources) granitoids are attributed to type I (from the crustal material of primary mantle genesis) and to type S (from sedimentary material). Tonalites, plagiogranites, trondhjemites, granodiorites and related granites are closest to the former, while leucogranites, two-mica and muscovite granites, lithium–fluorine granites and ongonites (Fig. 9.15) most resemble the latter.

Figures 9.16–9.18 demonstrate the evolution of strontium isotope composition in major granitoid varieties. Throughout the Earth's history, crustal rocks of primary mantle origin (basic rocks, amphibolites, etc.), possibly involving sedimentary material as well as the younger rocks, could have served as a source of granitoid of tonalitic–trondhjemitic assemblage (Fig. 9.16). For the granodiorites and monzonites, sedimentary crustal material plays an increased role, which is confirmed by the synchronous rise of $(^{87}Sr/^{86}Sr)_0$ and $\delta^{18}O$ especially for the younger ones ($<0.8\,Ga$) (Fig. 9.17). For the two-mica and muscovite granites, the sources are heterogeneous, but with the predominance of sedimentary rocks (Fig. 9.18), and the $(^{87}Sr/^{86}Sr)_0$ variations increase for younger rocks. For the alkaline granitoids and their associated volcanics the heterogeneous sources are important (Fig. 9.18b): with $\varepsilon_{Sr}<0$ and $\varepsilon_{Sr}>0$. The first can be related to the differentiation of primary mantle magma, the second have been connected with the participation of crustal material (Kovalenko et al., 1985). The role of crustal material increases in rocks younger than 1500 Ma.

As has been shown for mantle magma sources, crustal sources also become more complex in the course of geological history, and are composed of a greater variety of rocks: first basic rocks, then progressively more acid rocks.

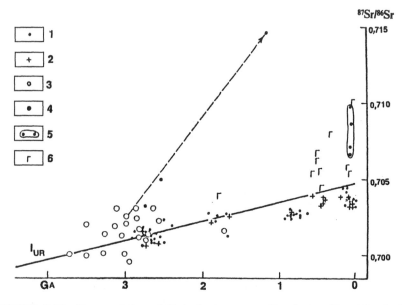

FIGURE 9.16 The evolution of Sr isotopic composition in tonalites (1), plagio-granites and trondhjemites (2), Precambrian "grey gneisses" (3), repeatedly meta-morphosed rocks (4), tonalites of the Adamello Massif, with indications of contamination by crustal material (5), tonalites from granodiorite–granite plutons (6); dotted line — evolution of the tonalite gneisses of the Grenville Province (after Balashov, 1985).

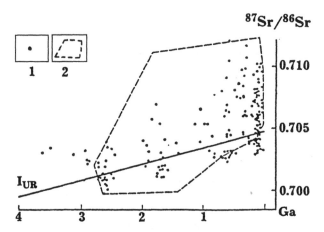

FIGURE 9.17 The evolution of Sr isotope composition in granodiorites (1) and quartz monzonites (2) (after Balashov, 1985).

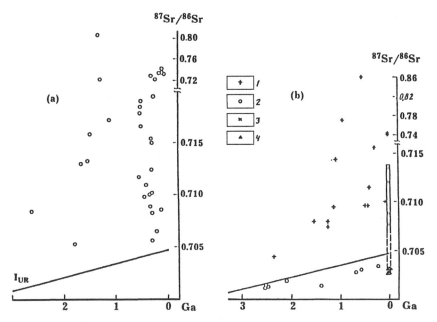

FIGURE 9.18 Evolution of Sr isotope composition in various rocks. (after Balashov, 1985). (a) in two-mica and muscovite granites; (b) in alkaline granites with increased I_{Sr}^0; 2 — the same with I_{Sr}^0 close to alkali basaltoids; 3 — comendites and alkaline granites of Ascension Island; 4 — alkali basaltoids.

9.2.1 Crustal Magma Generation in the Phanerozoic

In contrast to the Archaean when migmatization and granitization processes operated over a very wide area, and the Early Palaeoproterozoic, when they also occurred over extensive areas, in the Phanerozoic crustal magma generation played a secondary role. Here it was associated with areas of intra-plate magmatism and with the processes at destructive plate boundaries.

Volcanic edifices constructed of alkali ignimbrites and rhyolites, characteristic of the initial stages of development of some continental rifts and ensialic back-arc seas, as well as acid alkali igneous rocks (ongonites, pantellerites and comendites), that are connected with some intra-plate magmatic centres, may need to be attributed to the products of crustal magma generation associated with intra-plate environments. Their formation can be explained by the heating of sialic crust above asthenospheric diapirs or individual magma chambers. The increased alkalinity of newly formed acid melts seems to be connected with the formation and crystallization of predominantly moderately alkaline intra-plate magmas.

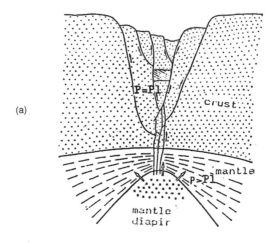

(a)

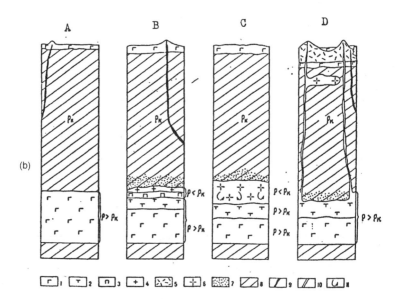

(b)

FIGURE 9.19 The scheme of crustal melt formation during intra-plate magmatism. After Yarmolyuk *et al.* (1990). (a) Scheme of intra-crustal magma chamber (hatched) formation in the rift zone. P — pressure; Pl — lithostatic pressure; double arrows — vectors of excessive pressure caused by mantle diapirism; single arrows mark the routes of upward rising basaltic melt. (b) A model of acid melt formation, in connection with basalt magma differentiation in an intra-crustal magma chamber. 1 — basalts, basaltic melts and their derivatives of basic composition; 2 — trachyte melts; 3 — comendite and pantellerite melts; 4 — anatectic melts; 5 — volcanic acid rocks; 6 — melts formed as a result of the mixing of residual basalt and anatectic melts; 7 — metasomatically altered crustal rocks;

The bimodal basalt–trachyrhyolite–comendite assemblages of Northern Mongolia, studied by Yarmolyuk *et al.* (1990), provide an example of such a process. These authors have shown that these assemblages represent genetically related series in which alkali–sialic rocks originated from a complex process, including the fractionation of alkali-basaltic magmas and the separation of their residual melts; the alkaline metasomatism of crustal rocks and their subsequent melting; and the mixing of residual and anatectic melts. A schematic representation of this process is shown in Fig. 9.19. Judging from the results of mass-balance calculations and Sr isotopic data, the contribution of the crustal component to the formation of alkaline–sialic melts is comparable with that of residual differentiates of alkaline-basaltic magmas or even exceeds it.

As has been demonstrated by Ryabchikov and Kovalenko (1987), the hyperaluminous character of granite magmas services as an indicator of the prevailing crustal sialic source. This stems from the fact that in the system CaO–Na_2O–K_2O–Al_2O_3–SiO_2, the plane with the atomic ratio $(Na + K + 2Ca/Al) = 1$ (quartz-feldspathic section $CaAl_2Si_2O_8$–$NaAlSi_3O_8$–$KAlSi_3O_8$–SiO_2) represents a thermal barrier preventing the transition from normal to hyper-aluminous melts as a result of fractional crystallization. As far as primary plumasite basaltic magmas are unknown, it can be inferred that hyper-aluminous granites must have arisen from partial melting of metapelites or assimilation of a significant quantity of metapelitic material.

With a high fluorine content the quartz–feldspar barrier can be intersected by crystallization trends, because of the appearance of a very wide field of fluorite crystallization. This is apparent from the possibility of simultaneous coexistence of both fluorite and topaz in the granite melt. Consequently, in some instances, topaz-bearing hyper-aluminous igneous rocks can be normal-series magma derivatives. A representative example that may solve the problems of lithium–fluorine granite genesis is given below.

a) The Origin of Lithium–Fluorine Granites and Ongonites

The entire plumasite rock series, from leucogranites to albite–Epidolitic granites with increased content of fluorine, trace alkaline elements, tin, niobium and tantalum, and decreased content of Sr, Ba, REE and Y as compared with the average

FIGURE 9.19 (*continued*)   8 — Earth's crust; 9–10 — the ways by which melt rises to the surface: 9 — basalt melts, 10 — for acid melts; 11 — currents of melt mixing, ρ_σ — rock and melt density, δ_κ — density of upper-crustal rocks. A — the origin of the basalt magma chamber; B — melt crystallization accompanied by contamination of the rocks overlying the magma chamber; C — emergence of anatectic alkaline-salic melts as a result of the melting of overlying metasomatic rocks; D — migration of lighter acid melts to the surface, with the formation of hypabissal intrusions and lava series.

content in acid intrusive rocks, is classified with the lithium–fluorine geochemical type of rare-metal granites. Granites of this type have an age of Precambrian
to Tertiary and are generally associated with the post-orogenic zones of fold belts
or zones of intra-plate magmatism (see Chapter 4). Their volcanic equivalents
are ongonites (topas zhyolites) (Kovalenko, 1977; Ryabchikov and Kovalenko,
1987).

The most differentiated granite massifs considered here have a persistent
zonal structure, both vertically and in plan (Fig. 9.20). Their lower zones are
composed of medium-grained leucocratic biotite granites, which are succeeded
up section by microcline–albite, amazonite–albite and albite–lepidolite granites —
the youngest in these massifs.

The rare-metal lithium–fluorine granites and ongonites (Emmerman *et al.*,
1977; Kovalenko, 1977) are characterized by relatively symmetrical, nearly
horizontal distribution curves of chondrite-normalized REE with a deep Eu
low. There is a sharp decrease in the content of all REE, especially Eu and
LREE between the parental leucogranites and lithium–fluorine types. The available Sr isotopic data on lithium–fluorine granites, which vary from
0.7066 ± 0.0013 to 0.729 ± 0.009, indicate that their magmas derived from a
crustal source. According to Kovalenko (1977; Kovalenko and Kovalenko,
1984), these granites could be a residual eutectic melt, which appeared during
the simultaneous crystallization from the melt and fluid, under conditions of

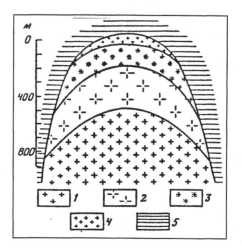

FIGURE 9.20 Generalized zonation scheme of differentiated massifs. After
V.I. Kovalenko (Ryabchikov and Kovalenko, 1987). 1 — biotite leucogranites;
2 — microcline–albite granites with protolithionite; 3 — amazonite–albite granites
with zinnwaldite; 4 — albite–lepidolite granites; 5 — host rocks.

increased fluorine activity. Evidence in favour of this hypothesis is provided by the following facts:

1. the discovery of ongonites representing the subvolcanic and volcanic analogues of rare-metal lithium–fluorine granites;
2. the discovery of the intrusive relationships of rare-metal granite massifs with host rocks;
3. the marked contrast between rare-metal granites and typical metasomatic rocks.

The results of the study of melt inclusions from the minerals of these rocks also support a magmatic origin for these granites (Ryabchikov and Kovalenko, 1987). According to these data, the homogenization temperature of the inclusions was 1200–540°C, the water content in them ranges from 0.2 to 8.6%, and its partial pressure varies within the limits 0.09 to 4.2 kbar; the melt viscosity decreases with decreasing temperature, which is connected with the accumulation of volatile components.

Available experimental data and the results of geological–petrological studies allow us to propose two models for the formation of similar melts from the leucogranitic magma with 0.2–0.5% F: crystallization differentiation and emanation (under gas influence) differentiation.

As experimental investigation have demonstrated (Kovalenko, 1979; Manning, 1981), the presence of fluorine in granitoid systems leads to the lowering of both liquidus and solidus temperatures. In this regard the crystallization of fluorine-bearing granite magma is not completed by formation of a regular granite minimum, but rather extends into the area of lower temperatures, and the crystallization interval broadens by no less than 200°C; the quartz crystallization field also widens causing alternation in the minimum point composition in the system Ab–Ort–Q-F from $Q_{37}Ab_{34}Ort_{29}$ to $Q_{15}Ab_{53}Ort_{27}$ (Fig. 9.21); the compositions of differentiated massifs of lithium–fluorine granites and ongonites are fairly consistent with this trend (Fig. 9.21). In contrast to conventional granites, in which the solidus is close to the equilibrium $Pl_{30-20} + Fsp$, in lithium–fluorine granites the solidus is significantly reduced, which results in the crystallization of virtually pure albite with K-feldspar. It is not inconceivable that albite crystallization was promoted by the release of fluorine from the melt in the mica stability field.

Thus, the fractionation of the main rock-forming minerals (feldspars and quartz) from a granite magma with increased fluorine content may well have led to the formation of fluorine-rich lithium–fluorine granites. In accordance with this model, considering the available estimates of the combined coefficients of element distribution in ongonites, Li, Be, Sn, W, Zn, F, Nb and Ta, and not infrequently, Rb, Cs and B should accumulate in residual magmas. The presence

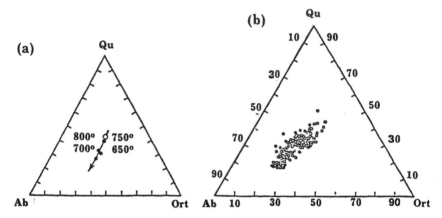

FIGURE 9.21 Evolution of lithium–fluorine granite melts. After V.I. Kovalenko (Ryabchikov and Kovalenko, 1987). (a) compositions of residual melts in the granite–H_2O–HF system within the temperature range 800–650°C. (b) compositions of lithium–fluorine granites from different regions.

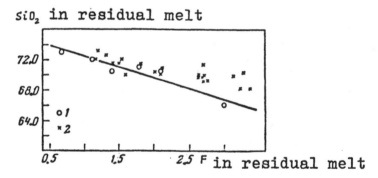

FIGURE 9.22 The SiO_2–fluorine content relationship (in wt.%) in the melts: ongonite–experimental composition (1) and ongonite–natural aphyric varieties (2). After N.I. Kovalenko (1979).

of a Eu anomaly in the lithium–fluorine granites and ongonites is consistent with feldspar fractionation.

The physical probability of emanation differentiation, i.e. the formation of ongonite magma by fluid transfer of fluorine and other components from deep parts of the magmatic chamber to the top of the chamber, has been confirmed by both experimental data and the observations of the compositional variations of natural ongonites. For the latter, an inverse correlation between the fluorine and silica concentrations was established (Fig. 9.22), i.e. there is an obvious relationship between the fluorine content in the melt and its total composition. After crystallization of this type of magma, the surplus F appears in the residual melt, which, together with fluid bubbles, can move into zones of reduced pressure, for example into the apical parts of magma chambers. In accordance with the correlation function (Fig. 9.22), the increase in the F activity of the system should change the composition of the granite magma to an ongonite one. It is significant that, in ongonites, depending on their fluorine concentration, not only the silica content, but also the REE concentrations, were affected. The quantities of Li, Rb, Be, Ta, Tl and Hf increase, while those of Ba, Sr, REE, Y and Pb (Fig. 9.23), and the values of Nb/Ta and Zr/Hf ratios, decrease, i.e. all the characteristic geochemical features of rare-metal lithium–fluorine granites are found to be enhanced (Ryabchikov and Kovalenko, 1987).

Both models of the formation of rare-metal lithium–fluorine granites are founded on the evolution of granite magma with slightly increased fluorine content (several tenths of one per cent). Such primary magmas are being generated within developed continental crust, as confirmed by data on Sr-isotopic composition. Two variants of this magma generation are feasible:

1. The differentiation of common granite magma, which is indicated by the presence of all of the transitional varieties from granites to lithium–fluorine granites and from rhyolites to ongonites. However, in the area of fluorine abundance ratios the distribution coefficients of this element vary widely, and the presence of this type of granite must be much rarer than the formation of granite magma in general.
2. Palingenic melting of magma with increased fluorine content. In crystalline rocks of the lower parts of the Earth's crust fluorine- and REE-rich rocks are known from the zones of metasomatites associated with faults (Apeltsin et al., 1967). In the same zones intensive granitization processes are developed and paligenic evidence is available. The partial melting of these rare-metal metasomatites, which are enriched in fluorine and other rare elements, can lead, according to distribution coefficients between crystals and melts (Fig. 9.24), to the formation of the primary magmas of lithium–fluorine granites. In this case, the zones of rare-metal metasomatic zones around faults may provide a substrate for the primary magmas of lithium–fluorine granites. Another way is the inflowing of such fluorine- and REE-rich fluids into the magma generation zone.

9.2.2 Crustal Magma Generation in the Early Precambrian

As shown in Chapter 6, 80–90% of the granite–gneiss high-grade metamorphosed Archaean areas are on occasions composed of particular plagiogneisses of tonalitic, trondhjemitic and granodioritic composition (TTG: "grey gneisses", see

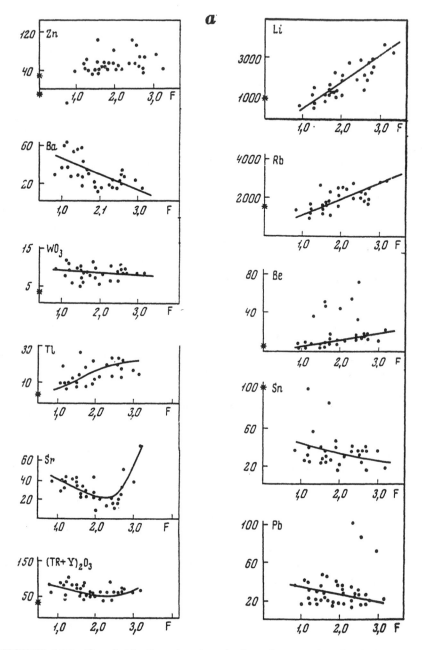

FIGURE 9.23 The distribution of series of micro-elements (ppm) relative to fluorine (in wt.%) in aphyric ongonites. After V.I. Kovalenko (1987).

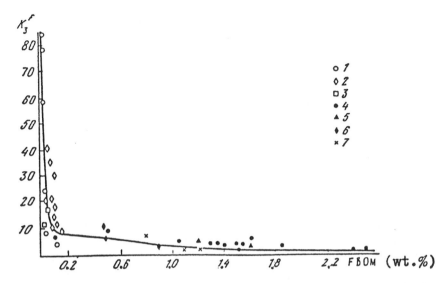

FIGURE 9.24 The relationship between the coefficient of fluorine distribution (K_3^F) for micas and their content in the bulk of igneous rocks. After V.I. Kovalenko (1987). 1 — trachytes and latites; 2 — rhyolites; 3 — andesites; 4–6 — ongonites of Mongolia: 4 — Ongon–Hairkhan, 5 — Baldgigol, 6 — Baga–Gazryn; 7 — Arabylak ongonites, Zabaikalye.

Chapter 6). At the moment, there are two hypotheses of the origin of the sialic crustal magmatism at the nuclearic stage of the Earth's evolution, which are affected to a major degree by ideas on the composition of the Earth's primordial crust:

1. melting of the earlier basic substrate, and
2. remelting of earlier sialic crust.

According to Bogatikov *et al.* (1989, 1990), Samsonov *et al.* (1993) and Martin (1994), based mainly on the results of detailed isotope-geochemical studies, ancient tonalite–trondhjemite gneisses had a volcanic and/or sedimentary-volcanic pre-migmatic origin. The protoliths of these gneisses are represented by crustal metabasites of mantle origin. The tonalitic–trondhjemitic granitoids are associated with the melting of these substrates at lower levels of the Earth's crust. Supporters from the remelting of primordial sialic crust formations (Bel'kov *et al.*, 1984).

Irrespective of these different viewpoints, the guiding mechanism of crustal generation is believed to be anatexis caused by low-temperature eutectic melting of granites from rocks of intermediate and acid composition, including metamorphosed sedimentary rocks (Mehnert, 1968; Mitrofanov, 1985; etc.). Detailed

studies conducted under the guidance of Glebovitsky and Mitrofanov
(Mitrofanov, 1985) have demonstrated that the process (cycle) of granite formation proceeded in a definite sequence, including the following events:

1. the formation of early Na-granitoids;
2. multi-stage anatectic migmatization concomitant with continuous overall
   deformation and resulting in the formation of acidic magma sources;
3. the formation of late K–Na or K-granitoid complexes, usually with indications of intrusive origin due to the deep-seated crustal sources or mature differentiation of anatectic or diatectic magmas. This cycle of granite formation
   generally continued for some 5–10 Ma.

Of particular relevance when discussing crustal magma generation is the
thermodynamic and fluid regime of granitization and the formation of acidic
crustal magma sources by ultrametamorphism. From numerous field studies and
experimental research, it has been established that, due to the high reducibility
of juvenile fluid flows, their water fugacities are low at high pressure. This causes a shift in the dehydration reactions, which are normally marginal for metamorphic mineral facies, towards lower temperatures, and tends to increase the
temperature of granitoid magma formation. During the interaction of reduced
fluids and metamorphosing strata, they are oxidized, the water fugacity increases
and conditions suitable for intensive granite melting are created. The most
favourable location for these processes is at depths equivalent to P of 5–7 kbar
(Glebovitsky and Bushmin, 1983). The relative dryness of granite magmas
(c. 2% H_2O), being generated under granulite-facies conditions at 7–9 kbar,
means that they tend to rise to a higher level. This is characteristic of the
K-granites, which complete the cycle of granite formation. However, water-
saturated (up to 5–8% H_2O), essentially Na-granite melts, which develop under
amphibolite-facies conditions ($P = 4$–4.5 kbar, $T = 600$–750°C are dominant in
the ultrametamorphic zones).

There are different points of view concerning the possibility of the latter rising to higher levels. In the opinion of Glebovitsky and Mitrofanov, these Na-
granites have a low mobility in the Earth's crust. This is accounted for by the
fact that with a decrease in pressure the solubility of the volatiles (mainly H_2O)
in the melt also decreases, which must elevate its liquidus temperature and lead
to rapid solidification of the magma, and hence, to the cessation of its further
advance. However, according to the experimental data of Kadik *et al.* (1991),
the high water solubility in melts of acid composition remains practically
unchanged within the pressure range from 20 to 1 kbar, after which it is drastically reduced. From this it is inferred that all the effects associated with water
solubility in this type of melt operate only at pressures lower than 1 kbar (depths
3–4 km), and that at the greater depths nothing could restrict their vertical
mobility. It is depth that appears to be critical for the acid melts. "Dry" magmas

generated under granulite-facies conditions pass through this barrier without difficulty, but water-saturated ones may overcome it only through catastrophic volcanic eruptions caused by rapid degassing of these melts in shallow magma chambers (Sharkov, 1992). Since water-saturated varieties dominate in acid melts, in most cases these melts fail to reach the surface, solidifying at depth as relatively high-level granite intrusions. It is highly unlikely that the presence of a specific physical–chemical barrier, together with the higher viscosity of acid melts, is responsible for the predominance of plutonic facies in acid rocks in contrast to basic ones. Acid volcanics are generally rare and are most often represented by tuffs and ignimbrites. This holds true for both the Precambrian and the Phanerozoic.

According to Glebovitsky *et al.* (Mitrofanov, 1985), genetic diversity is characteristic of the parental magmas of early plagiogranite complexes, particularly those of the Late Archaean and Early Proterozoic. As in the Phanerozoic, here the gabbro–diorite–trondhjemite assemblages of the intrusive rocks are widely developed. In particular, in the Svekofennian Geoblock of the Baltic Shield they are represented by enderbites associated with gabbroids. Together with basic dykes this is evidence for the existence of a mantle diapir at the base of the crust at that time, which generated a positive thermal anomaly, promoting rapid heating of the metamorphosed rocks. In their view, earlier granitoids in the cycle were derived from deeper igneous sources which presumably are related to melting of the TTG ("grey gneissic") basement above mantle diapirs.

As ultrametamorphic processes proceeded against a background of constant interaction of rising magmas and juvenile fluids, this affected the openness of the granite systems to a number of components, primarily H_2O, Na_2O and K_2O. The increase in water content (probably owing to hydrogen oxidation) resulted in a drop in the temperature of the melting substrate solidus and the enrichment of the melt in normative albite (Luth, 1969), which is characteristic of the main stage of migmatization. As a consequence, the increase in potassium chemical potential should have been unique to the late stages of granite formation.

Studies of the Sm–Nd isotopic ratio demonstrated that the more ancient granites ($>2\,$Ga) have initial Nd ratios very close to the chondritic evolutionary trend, and the ratios in the younger granitoids point to the involvement of increasing amounts of more ancient continental crust (Allegre and Ben Othman, 1980). As shown in Chapter 5, the widespread participation of Archaean sialic crustal material was also established for most Proterozoic granitoids, particularly anorthosite–rapakivi granite assemblages. They related to huge belts of acid magmatism 1.8–1.5 Ga in age, which appeared on all Precambrian shields at the locations of ancient oceans just after their closure, and probably originated due to partial melting of overlapping continental plates above asthenospheric highs.

According to Ramo (1991) and Neymark *et al.* (1994), the basic rocks (diabase dykes, gabbronorites and anorthosites) and the related rapakivi granites were derived from different sources. The former were generated by partial melting of ascending mantle diapirs; this ascent resulted in the appearance of considerable volumes of basic magmas at the base of the crust, causing partial melting of the lower crust and mixing, and leading to the appearance of such rocks as anorthosites, monzonites, diorites, etc. Contamination of the melts with Archaean and Proterozoic upper-crustal material may have caused the scatter of Pb, Sr and isotopic ratios. This anorogenic magmatism could have been related to the formation of the Early Proterozoic Laurentia–Baltica supercontinent.

As is evident from the foregoing, during the stage under consideration the composition of mantle melts did not differ markedly from Phanerozoic continental intra-plate magmatism. The difference consists essentially of the opposing relationships of basic and acid rocks. This may be related to differences in the type of interaction between the asthenospheric diapirs and the sialic crust, which in this case did not form rifts, but actually underwent intense melting, accompanied by the formation of gigantic magmatic centres. The inflow of potassic fluids from the mantle may have played a substantial role in the process. The possibility of inflow is probable, suggested by the discovery of garnet granulites and metasomatic biotite–orthopyroxene–garnet rocks, granitized in the lower crust, in the xenoliths of lower-crustal rocks from the diatrem on Elovy Island, Kola Peninsula, with an age of about 1.7 Ga (Sharkov and Pukhtel, 1986; Bindeman et al., 1990; Kempton et al., 1995).

Thus, the scale of crustal magma generation was reduced with time, from the nuclearic stage to the continental–oceanic stage. The granite–migmatitic assemblages of the tonalitic–trondhjemitic series are the most characteristic of the early stages of the Earth's evolution, and, as the planet evolved, their compositions became more complex, eventually leading to the formation of various alkaline and rare-metal granitoids while at the same time there was a decrease in the amount of crustal melting.

9.3 MANTLE–CRUSTAL MAGMA GENERATION PROCESSES

The problem of the genesis of calc-alkaline series formations, and particularly andesites, is one of the key problems in modern petrology. Andesites and their intrusive analogues are a fundamental component of igneous rocks within zones of active plate interaction: island arcs, active continental margins and collision zones, where huge amounts of material are transferred from the mantle to the crust and from the crust to the mantle, which is characteristic of the continental–oceanic stage. Therefore the problem of the ratios of crust to mantle components in magmatic melts of convergent plate boundaries is important for understanding the processes of evolution and growth of continents in the Phanerozoic. This problem is of no less importance for the cratonic stage, where similar rocks were also widespread but were developed in a different geochemical environment.

Calc-alkaline series volcanics are noted for their wide variations in primary $I^{87}Sr/^{86}Sr$ and $I^{143}Nd/^{144}Nd$ ratios indicating the involvement of both mantle and crustal material in the magma source (Bogatikov and Kovalenko, 1987). The mixed source is proved not only by increased $(^{87}Sr/^{86}Sr)_0$ (> 0.7047) and decreased $\varepsilon_{Nd(T)}$ values (up to -12) but also by the positive correlation of $(^{87}Sr/^{86}Sr)_0$, $\delta^{18}O$ and increased ratios of $^{207}Pb/^{204}Pb$ and $^{208}Pb/^{204}Pb$. The extreme case of crustal participation (metapelites) occurs in the source of cordierite-bearing acid magmas, the $(^{87}Sr/^{86}Sr)_0$ of which increases up to 0.7157–0.7175 at low $\varepsilon_{Nd} = -11$–12.

The isotopic data for the Crater Lake volcano area (see Chapter 2) require sources of primitive magmas to consist of depleted mantle and a subduction component, introduced in variable quantity to the depleted mantle wedge (Bacon et al., 1994). Variable degrees of melting of this heterogeneous mantle, possibly at different depths, produced the diversity of isotopic compositions and LILE abundances in primitive magmas. Trace element ratios do not indicate the presence of an ocean island basalt source component.

9.3.1 Mantle–Crustal Magma Generation in the Phanerozoic

As shown in Chapter 2, the island-arc volcanic rocks are characterized by the following distinctive features:

1. enrichment in different incompatible elements, particularly K, Sr, Ba and Pb;
2. the specific Sr and Pb isotopic characteristics and ε_{Sr}–ε_{Nd} ratios in some arcs;
3. the low values of $^{143}Nd/^{144}Nd$ and $^{176}Hf/^{177}Hf$;
4. the high concentrations of the ^{10}Be isotope; this short-lived isotope is formed as a result of the interaction of cosmic radiation with atoms of the atmosphere and the Earth's surface, and must have reached the magma generation zone only via subduction.

Many researchers associate these features with sedimentary rocks involved in the petrogenesis of island-arc magma series. This association is extremely important as supporting evidence for subduction, i.e. the burial of sediments to depths of some 100–160 km and their participation in melting processes. However, their role in calc-alkaline magma petrogenesis should not be overestimated, according to Taylor and McLennan (1985), who showed that the contribution of the sedimentary component to island-arc volcanics barely exceeded 3%.

The Pb isotope composition of island-arc volcanics practically coincides with that of E-MORB; accordingly, these volcanics could have been derived from the same sources.. The arc volcanics often have higher $^{87}Sr/^{86}Sr$ radiogenic ratios and usually lower $^{143}Nd/^{144}Nd$ ratios as compared with MORB (Tsvetkov, 1990). All of this suggests a multitude of sources for the island-arc melts.

From seismic data, formation of melts took place at depths of about 100–200 km (Fedotov *et al.*, 1988; Wilson, 1989). The presence of high-magnesian olivine and orthopyroxene and also chromian spinels in the rocks, particularly those of the boninite series, as well as some tholeiitic arc basalts, definitely points to the involvement of mantle peridotites in magma generation. It is also evident from the $^3He/^4He$ ratios in the rocks of many island arcs, such as the Aleutian, the Kurile–Kamchatka, the Japanese and the Marianas arcs (Craig *et al.*, 1978; Polyak, 1988). Finally, the third source is the crust actually undergoing subduction. As was shown by M. Kay and R. Kay (1990), Tsvetkov (1990) and many others, the micro-elemental concentrations in island-arc basalts differ essentially from those in MORB, with the former being characterized by enrichment in large-ion lithophile elements, and values of Ba, Rb and Sr/LREE ratios more than twice as high as those of oceanic island basalts. This suggests the presence in the arc melt of Sr-bearing phases not associated with the fractionation of conventional mantle minerals, such as garnet, orthopyroxene, clinopyroxene. It has been suggested that enrichment in Sr, which is easily transported by a fluid, is most likely to indicate that the fluid phase was generated by dehydration of the subducted plate. This fluid could also be a source of large-ion lithophile elements in the melt.

However, as has already been mentioned, the ILE content in most island-arc basalts, and particularly in boninites and andesites, is significantly higher than in oceanic basalts and seems to need its own source material. Moreover, essentially andesite series are widespread in collision zones, where subduction of continental crust rich in these components occurs, and collision-zone andesites are characterized by their enrichment as compared with island-arc types. This suggests that enriched material was implicated in the generation of calc-alkaline series magmas. It could be mantle plume material (Sobolev and Danyushevsky, 1994) and/or continental sialic crust, whose important role in the petrogenesis of calc-alkaline magmas was pointed out by Arculus and Johnson (1981), Babansky *et al.* (1983) and Karig *et al.* (1987). This material could have entered the subduction zone from the back-arc zone during back-arc diapiric extension (Fig. 9.25). In this case the low Sr isotope ratios are most likely to be related to the passage of newly generated magmas through the giant "ion-exchanger" of the mantle wedge where they acquire new characteristics (Yeroshenko and Sharkov, 1993).

Analysis of magma generation processes at active margins presents severe problems, owing to the lack of reliable data on the reasons for and conditions of the melting material in subduction zones and the significant fractionation that most of these magmas have undergone during their ascent to the Earth's surface. Therefore, at present, there is no generally accepted model of magma generation in subduction zones. The view held by the majority of researchers is that tholeiitic basalts, which are usually associated with andesites and are transitional with them, i.e. they tend to grade into andesites within a particular geodynamic

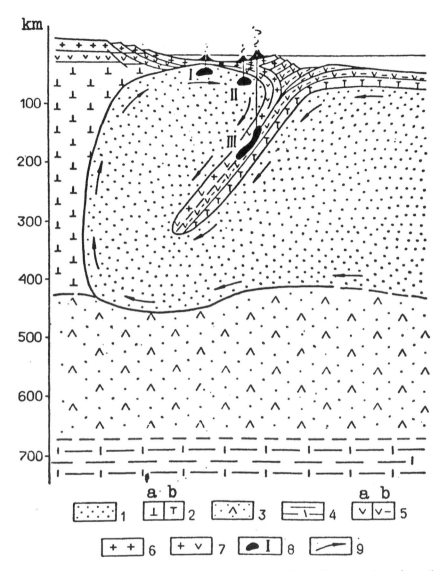

FIGURE 9.25 Scheme of magma generation at the active margins of continents and oceans. After V.A. Yeroshenko and E.V. Sharkov (1993). 1 — asthenospheric material; 2 — lithospheric mantle material: (a) continental, (b) oceanic; 3 — upper-mantle material (400–450 km below the interface); 4 — lower–mantle material (650–670 km below the interface); 5 — "lower crustal" layer material: a — of the continents, b — of the oceans; 6 — sialic crust ("granitic" layer); 7 — tectonic mixture of the formations of "granitic" and "lower crustal" layers; 8 — magma generation zones (I — MORB, II — K–Na moderate alkaline series, III — calcalkaline series); 9 — the direction of movement of mantle material.

environment such as an active margin or a collision zone, represent the initial melts for the formation of the calc-alkaline series (Kay and Kay, 1990; Tsvetkov, 1990). The supporters of this viewpoint believe that andesites form in close proximity to the Benioff zone, as a result of the melting of mantle wedge peridotites and pelagic sediments, under the influence of aqueous fluids derived from the dehydration of the subducted plate.

According to the other view, the reverse process can take place here: the process of partial melting of highly depleted peridotites of the mantle wedge (IIB type substrate) by overheated water-saturated andesite melts (Solovova *et al.*, 1992; Sobolev and Shimizu, 1993) that were generated directly within the subducted plate, leading to the appearance of hybrid melts of varying basicity. In principle, this melting mechanism is much more effective than simply a "fluid effect" because of the high heat capacity of the melts. From this standpoint, the initial melts of the series must have been andesitic as this composition forms the lowest-temperature liquids at pressures of more than 30 kbar, i.e. under conditions where calc-alkaline series magmas are being generated (Green and Ringwood, 1968).

Recently Shiano *et al.* (1995) supported this idea; they showed that hydrous, silica-rich (close to andesitedacite) melts migrating through the mantle wedge are preserved as glass inclusions in mantle minerals in xenoliths from Philippine arc lavas. According their data, these melts, with chemistry that indicates an origin by very low degrees of melting of the subducted oceanic crust, have altered their host peridotites, yielding a metasomatized mantle. Such mantle is the source region of the arc magmas, which share continuous chemical trends with the melt inclusion. These observations provide direct evidence for the importance of slab–mantle interactions in the genesis of island-arc magmas. In this case, basalts (and boninites) are secondary formations and their quantity in a particular case is related directly to the degree of ultrabasic matrix melting under the influence of fluidized andesitic melt (Yeroshenko and Sharkov, 1993; Rapp and Shimizu, 1996). Owing to the complete mutual solubility of basalt and andesite according to the degree of melting, a wide spectrum of intermediate compositions can and does arise, ranging from marianites and boninites to standard island-arc tholeiites having similar REE spectra and isotopic characteristics (Fryer *et al.*, 1992). On the one hand, with this mechanism the isotopic characteristics of these melts equilibrate with those of the mantle wedge rocks, but on the other hand, the essentially andesite-series composition was maintained. Otherwise, in order to produce the observed quantity of andesites, it would be necessary to remelt practically the entire mantle wedge, which turns out to be inconsistent with the energy potential of the process.

It is obvious that an analogous mechanism operated during the formation of the K-subalkaline series, where the dominant variety is represented by latites: the potassic analogues of andesites. The generation of latite melts seems to have occurred under conditions of potassium fluid inflow in the area of magma generation (Pepier, 1994). Probably, material of subducted continental rocks has played an essential role in the generation of this fluid, which is confirmed by the high $^{87}Sr/^{86}Sr$ ratio in rocks of shoshonite–latite series (Tsvetkov et al., 1993). In contrast to the intra-plate environment, there was no titanium in the composition of these fluids, which is generally uncharacteristic of magma generation processes at active plate margins.

The acid members of the calc-alkaline series appear to have originated by both fractional crystallization of andesite and contamination by crustal material. This may be indicated by the results of a study of geological, geochemical and isotopic features on the granitoids of Phanerozoic fold belts, which are presumed to be the intrusive analogues of these formations (Kovalenko, 1987; Taylor and McLennan, 1985; Mueller et al., 1995; etc.).

The involvement of crustal material in the melting processes is most large-scale at the active continental margins and in the zones of continental plate collision (Kovalenko et al., 1987; Wilson, 1989; Brandon and Lambert, 1994; etc.). Specifically this is indicated by a change in the character of the volcanics at active Andean-type continental margins, depending on the type of continental crustal blocks adjacent to the subduction zone (see Chapter 2), and also data on the isotopic composition of the rocks. It is suggested that here melting of crustal sialic gneisses occurs above the large crustal sources of subducted magmas that produce the acid melts. The same is true of zones of continental plate collision, where acid melts (leucogranites and rhyolites) prevail, whose participation in igneous activity not infrequently accounts for more than 50% (Ronov et al., 1990).

9.3.2 Mantle–Crustal Magma Generation in the Early Precambrian

As shown above, calc-alkaline series-type formations are of widespread occurrence in Early Precambrian as well as the Late Precambrian and Phanerozoic, especially in the Palaeoproterozoic, at the cratonic stage. Of particular importance is the time interval from 2.5 to 2.2 Ga where evidence of plate tectonics appears to be lacking, and the magmatism is intra-plate in character. Siliceous high-magnesian (boninite-like) series (assemblages, consisting of low-Ti picrites, high- and moderately magnesian basalts, andesites, dacites and rhyolites) all having features of mantle–crustal generation, but unconnected with subduction zones (see Chapter 5), are highly characteristic of this period.

The intra-plate character of the magmatism, by analogy with the Phanerozoic, suggests a relationship to the upwelling of an asthenospheric diapir. However, judging by the magma composition, diapiric material differed significantly from material supplying intra-plate magmatism in the Phanerozoic. For the Early Proterozoic the mainly depleted substrates of IIB type were endemic (Girnis and

Ryabchikov, 1988; Crawford, 1989; Sivell and McCulloch, 1991; etc.). Picrite with a MgO content of c. 20% appears to have primary parental melt for the boninite-like series, as inferred from the high magnesium content (mg# c. 92) of olivines in the picrites and high-temperature cumulates of layered intrusions. The parental magmas of the series could have formed by varying degrees of melting, and within a particular pressure range, starting from approximately 25 to 5 kbar (Girnis and Ryabchikov, 1988). This yields a specific series of melts, the further differentiation of which can result in the formation of rock series from picrites to low-Mg basalts. Separation of olivine could occur during ascent to the surface and also within intermediate magma chambers. The closing stages of differentiation must have occurred at low pressures, leading to the stabilization of olivine, on the liquidus, for melts with 12% MgO (at less than 5 kbar), thus leading to the appearance of basalts.

In relation to Archaean komatiites, the important feature of the high-magnesian volcanics under study is their increased content of lithophile elements, which have a high degree of incompatibility (e.g. Ba, Zr, Sr and LREE), the degree of enrichment falling off in the series $Ba > Ce > Sr > Zr$. The moderately incompatible elements are characterized by the same chondritic relationship as Archaean komatiites (Girnis and Ryabchikov, 1988). This seems to be evidence in favour of contamination of melts by crustal rocks, which is especially noticeable in low-magnesian rocks (< 12 wt.% MgO).

The temperature of the solidus of the strongly depleted mantle substrate of Proterozoic high-magnesian series must have been fairly high. According to Girnis and Ryabchikov, the temperature of the primary melt here reached 1420°C at atmospheric pressure, approximately corresponding with the temperature of the parental melts of Phanerozoic boninites (Sobolev and Danyushevsky, 1994). The separation of such melts from the mantle substrate occurred at depths of c. 60–90 km. Taking into account the melting-point temperature gradient, the initial temperature of the melt in this instance should reach 1600–1700°C. It is evident that such a high-temperature melt could not be formed at lesser depths, since in this condition — of steep geothermal gradient — far lower temperature rocks would be involved in melting, giving rise to hybrid melts with entirely different characteristics.

In addition to this, similar melts are of common occurrence in the high Mg-series formations of the early Palaeoproterozoic, as follows from the isotopic characteristics of the rocks (see Chapter 5). A possible explanation for this involves the following: upon reaching the base of the Earth's crust — an important interface in the lithosphere — here the melt could form an intermediate magma chamber. As its temperature was 300–400°C higher than that of the solidus of lower-crustal basic rocks, inevitably this would lead to their melting, initiating the formation of major hybrid magma chambers (Sparks, 1986; Huppert and Sparks, 1985). Owing to the difference between the values of both the adiabatic gradient and the melting-point temperature gradient (Jeffries, 1929;

Woster *et al.*, 1990), the crystallization in this type of source must have proceeded from the bottom upwards, with the main heat losses through the top, thus leading to further melting. As a result, this magma body should ascend within the Earth's crust, according to the principle of zone refinement (Yaroshevsky, 1964).

Owing to this mechanism of chamber evolution, the highest-temperature phases were continuously removed from the melt and enrichment in low-temperature components occurred, due to both fractional crystallization and the addition of newly formed melt as a result of the melting of the rocks at the top of the magma body. This obviously led to rapid changes in the compositions of these hybrid magma chambers as they rose within the crust (see Chapter 5). Its composition will rapidly approach that of the host rocks and the low-melting-point components of the primary melt will mostly remain. Thus, in essence, the development of the magma chamber is essentially a process of heat transfer vertically within the Earth's crust. As it rises it is gradually attenuated, and the degree of melting in the upper part of the magma chamber could decrease. Probably, the zone involved in this heat transfer will take the form of a cone (Fig. 9.26). Depending on the scale of the initial heat pulse, the vertex of this cone will either remain within the lower crust or will reach the base of the upper crust, as evidenced from the appearance of rocks of intermediate and acid composition in the upper section of some intrusives, for example the Bushveld and Koilismaa complexes (see Chapter 5).

In this case, the intrusions represented peripheral sources located above the main sources, in the rigid, relatively cold, upper crust. Apparently, these intrusions, which represent very extended lenses of melt, were a type of "trap" for the next set of batches of uprising melt, due to the large areas of the magma chamber (Fig. 9.26). Only magma conduits passing beyond the top of the chamber could have reached the surface, resulting in the formation of volcanic structures or dyke swarms. When these large intra-crustal lenses failed to rise any further for some reason, dyke swarms were generated above these magma chambers, and volcanic zones of Pechenga–Varzuga type emerged at the surface.

Andesites, as a whole, are not typical of the Archaean. They have been observed in some predominantly Late Archaean greenstone belts and occur after komatiites and basalts. In their geochemical characteristics they approximate the high-magnesian series of Early Proterozoic rocks of andesitic composition, and they appear to have a similar origin.

As was shown in Chapter 5, according to available data on the geochemical and isotopic features of rocks of the mesoproterozoic, anorthosite–rapakivi granite complexes, their magmas were derived mainly from crustal material (Ramo, 1991; Neymark *et al.*, 1994) with the addition of specific elements: Ti, alkalies, mainly K, P, Ba, Zr, incompatible elements, etc., which are typical for the fluids responsible for mantle metasomatism. Their inner structure is also unusual: they are huge cone-like trans-crustal magmatic systems, which have been traced to

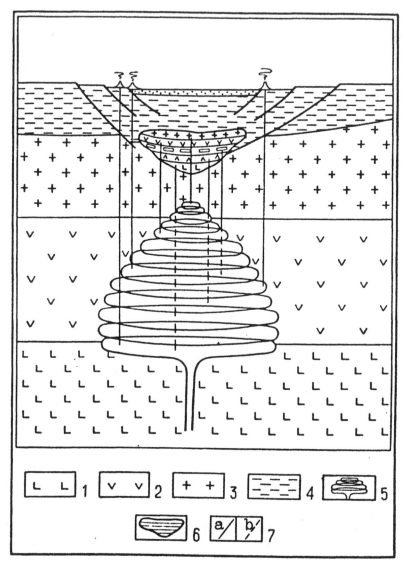

FIGURE 9.26 Schematic diagram illustrating the process of buoyant magma source material moving upwards within the Earth's crust. After E.V. Sharkov (1993). 1 — upper mantle; 2 — lower-crustal formation ("basalt" layer); 3–4 — upper-crustal formations: 3 — "granitic" layer, 4 — "sedimentary" layer; 5 — trend of magma body moving upward; 6 — a large layered intrusion (from bottom to top): ultrabasic rocks, norites and gabbronorites, anorthosites, gabbronorites and diorites; 7 — paths of magma migration from the source: a — maintained, b — reworked during subsequent processes.

depths of 30–40 km (Orovetsky, 1990; Elo and Korja, 1993). Probably, the processes of zone refinement also played an essential role in their formation. In other words, they were similar to the early Palaeoproterozoic magmatic systems, only formed mainly by expence of crustal material.

These complexes are related to huge belts of acid magmatism, which appeared on all Precambrian shields at the locations of ancient oceans just after their closure (see above). Within these belts anorthosite–rapakivi granite complexes appeared at the final stages of their evolution and formed as large magmatic systems which were probably located above asthenosphere highs related to places of mantle fluid relief in conditions of very thick crust (c.50–70 km).

In their geological setting and geochemical features, these complexes were intermediate between granite batholiths of active continental margins and continental collision zones on the one hand, and areas of acid magmatism with Li–F granites and ongonites, which are observed in intra-plate situations (for example, in the Central Asia), as the other. Probably, such a combination of geological conditions was unique in the Earth's history and took place only after the collapse of the first oceans, while plume tectonics were being replaced by plate tectonics. The absence of anorthosite–rapakivi granite complexes is further evidence that such situations were not repeated.

9.4 SUMMARY

1. For the whole of the Earth's history, three major types of magma genesis, differing in the character of their source material, can be distinguished: mantle, crustal and mantle–crustal. The character, scale and sometimes the mechanism of magma generation for each type could have varied significantly at every stage of the planet's evolution.

2. Mantle magma genesis in the Early Archaean nuclearic stage occurred at extreme depths. The dominant Al-depleted komatiites were formed with the participation of majorite at depths of c. 300–400 km. The upper parts of Early Archaean upper mantle were depleted in LREE and other incompatible elements, the degree of depletion decreasing with depth. In the Late Archaean, Al-undepleted komatiites were predominant, testifying to the presence of magma generation zones occurring at depths less than 200 km. It has been noted that small amounts of components enriched in incompatible elements were supplied to the magma generation zones. At the beginning of the Early Proterozoic the magma generation zones rose to depths of 60–90 km and the mantle substrates were severely depleted. At the 2.2–2.0 Ga boundary, the substrates highly enriched in incompatible elements appeared. These

were supposedly associated with mantle metasomatism. Thereafter mantle substrates were characterized by a high degree of heterogeneity, varying in their time of enrichment and depletion of material in different parts of the mantle. The areas of magma generation are located at depths of 60–20 km, reaching 10 km beneath the mid-oceanic ridges.

3. The crustal magma generation of the nuclearic stage was related mainly to the melting of granitic magma from basic and/or sialic material from the continental crust. These processes were most widespread in the Archaean, where this melting involved fluids under granulite- and amphibolite-facies conditions and resulted in the appearance of large migmatite fields. The migmatization cycle, in the initial stages of granite formation, included the melting of plagiogranite (Na)-series melts, in the main stages Na- and K-series melts and in the final stages predominantly K-series melts. The beginning of melting coincided in time with the manifestations of basic magmatism, which probably indicates the essential contribution of mantle diapirs to the heat balance of the process. The role of crustal magma generation *per se* gradually decreased with time — in the Early Proterozoic, at the cratonic stage, it is associated mainly with compressional zones, and from the outset of the continental–oceanic stage at 2.2–2.0 Ga — at the final stages of fold-belt development — with zones of ancient inter-block ocean closure.

In the Phanerozoic, crustal magma generation of minor importance, forming local fields in areas of continental intra-plate magmatism and playing a more essential role at active continental margins and in continental plate-collision zones. As the scale of crustal magma generation decreases, the melt composition becomes more complicated, with the appearance of various alkaline, rare-metal and lithium–fluorine granites. This is characteristic of the whole continental–oceanic stage, including the Precambrian.

4. Mantle–crust magma generation has played an increasingly important role since the cratonic stage, but its mechanism has changed considerably with time. The cratonic stage apparently was characterized by the formation of calc-alkaline series rocks due to zonal melting processes — the rising of buoyant magma chambers containing high-temperature mantle melts in the Earth's crust owing to the melting of the top and crystallization of the highest-temperature phases near the base. From the onset of the continental–oceanic stage 2.2–2.0 Ga ago, mantle–crustal magma genesis is more often associated with subduction zones, and in the Phanerozoic it was the main method of generating calc-alkaline series magma, and forming the low-Ti potassic moderately alkaline (shoshonite–latite) series.

CONCLUSIONS

O.A. BOGATIKOV, V.I. KOVALENKO, E.V. SHARKOV, and
V.V. JARMOLYNK

This book summarizes data on the geology and mineral composition of igneous rocks for the whole of the Earth's history. This systematic overview has enabled the identification of specific features of magmatism at different phases in the Earth's evolution and the pinpointing of their origin, their mode of occurrence and their relationships with various geodynamic environments from the near and distant past. It has been shown that the evolution of igneous processes mainly led to an irreversible increase in the diversity of the igneous rocks.

Four major phases in the history of the Earth: lunar, nuclearic, cratonic and continental–oceanic, are marked by different types of tectonomagmatic activity, pinpointing the main trends of igneous evolution. However, they cannot answer all of the relevant questions. Nevertheless, we have used all new geological, geochemical and isotopic data available; here we are referring to the relatively recent past, when the main concepts of plate tectonics were being formulated, which led to the reappraisal of many well-established geological theories. Undoubtedly, during further study we will obtain unexpected results and make new discoveries. However, the present data, when taken together, provide us with a clear picture of current hypotheses concerning the evolution of magmatic processes, as well as their advantages and drawbacks.

To summarize, it should be noted that there were obvious differences in the nature of magmatism and geodynamics at the early stages of the Earth's history — in the Archaean and the early Proterozoic, and in the succeeding stages up to the present day. If the first time interval was characterized mainly by melting substrates, depleted to varying degrees, and plume tectonics was the dominant type of geodynamics, then, starting from 2.2–2.0 Ga, geochemically enriched mantle melts related to the influx of deep-seated fluids increased in importance, and plume tectonics gave way to plate tectonics. This is confirmed by the statement of Taylor and McLennan (1985), based on their study of continental crustal evolution, that present-day geological activity can be extrapolated in time no further back than the early Proterozoic. The pre-geological phase in the Earth's history, which we identified as a lunar one, still remains vague. The existence of a magmatic ocean and the relevant problem of the primordial Earth's crust is a widely debated topic. The two alternative hypotheses of a primary sialic or a primary basic composition for the Earth's crust require further study using geological, geochemical and isotope-geochronological data.

The nuclearic, Archaean phase is not as controversial. Most researchers agree that granite–greenstone terranes and the granulite–gneiss belts separating them, apparently large complementary extensional and compressional structures, were typical geological structures. This phase was marked by fairly deep-seated mantle-derived komatiite–basalt magmatism in GGST, which could have been caused by the ascent of mantle plumes. In contrast, for GGT synkinematic enderbite-charnockite magmatism was typical. Speculation about the possible existence of plate tectonics at that time still persists, although the available data tend to suggest widespread plume tectonics of the type found on Venus.

As has been shown earlier in this book, the early Proterozoic was a period of dramatic change in the type of deep-seated processes within the Earth, which finally resulted in the replacement of plume tectonics by plate tectonics. Stabilization of the Earth's crust had set in by the beginning of the early Proterozoic; this cratonic phase logically succeeded and terminated the first stage in the Earth's evolution. It was characterized by widespread siliceous high-magnesian series, that were never so common before or after. The origin of these series seems to be related to the continuing development of the outer layers of the Earth, following the principle of plume tectonics, but, unlike the Archaean, under rigid crustal conditions.

It was at that time that the main volume of the continental sialic crust was formed. This phenomenon is unique to the terrestial-type planets: nothing of this type has been recorded on the surface of the Moon, Venus, Mars or Mercury.

An abrupt change in the development of deep-seated petrogenesis and geo-dynamics of the Earth's crust and mantle occurred about 2.2–2.0 Ga ago. At the same time, the widespread occurrence of geochemically enriched mantle melts, with high Ti, Fe, and alkali content, was recorded on all the ancient shields. These melts had earlier been completely absent. The diversity of the igneous rock types increased spasmodically and formations typical of the early Precambrian gradually disappeared. At the same time, linear geological structures were formed, associated with specific tectonomagmatic processes, allowing reliable correlation of the onset of plate tectonics. We believe that this marks the inception of the continental–oceanic stage of the Earth's evolution which differs greatly from the cratonic phase, and continues up to the present day. After that time, the paths of evolution of the Earth and those of the other planets, where plate tectonics was not established, start to diverge. We suggest that the formation of a new type of melt was due to the involvement of material derived from the deepest parts of the Earth, in processes of planetary mass exchange. There is evidence for the formation, at the Archaean–Proterozoic boundary, of a liquid core from which material gradually "percolated" through the lower mantle and reached the upper layers of the Earth by that time.

Apparently, this scenario is feasible only if two conditions are met:

1. the Earth warmed up gradually, from the surface to the deeper layers, and
2. the inner regions of the planet were initially composed of a material different from that of the outer layers.

A similar trend in the composition of magmatic melts on the Moon (from the oldest continental series to the younger titanium *mare* basalts) suggests that the accumulation of at least these two planets was heterogeneous.

The onset of the continental–oceanic phase in the early Proterozoic changed the pattern of geological processes on Earth: they became similar to those operating at present. Its Precambrian interval (about 2.2 to 1.0 Ga) was quite distinct. It can be attributed to a higher heat flow and to the widespread involvement of Archaean crustal rocks in deep-seated petrogenesis, as compared with the Phanerozoic, when subduction of oceanic crust played a major role in planetary mass exchange. The initial suture zones were characterized by high-temperature, high-pressure granulite metamorphism rather than the low-temperature, high-pressure blueschists common in similar Phanerozoic structures. However, the opened oceans, judging from the occurrence of typical ophiolitic assemblages of this age, already contained oceanic lithosphere. The magmatism of that time was marked by widespread anorthosite assemblages, and the formation, at the sites of closed oceans, of very extensive belts composed of potassium-dominated acid igneous rocks, and related structurally complex anorthosite–rapakivi granite massifs.

The continental crust formed in this phase differs greatly from the earliest, Precambrian crust in many respects, primarily in its more basic mineral composition, and it is less extensive, amounting to no more than 25% of its volume. Apparently, the crust from the early phases of evolution (prior to the continental–oceanic stage) is a relict formation that still persists in the basement of ancient platforms. As has been inferred from recent seismic data, large blocks of ancient lithosphere have very deep roots, in contrast to younger oceanic lithosphere. This suggests that the upper layers of the modern Earth consists of two different types of lithosophere which were formed at different times, and which also differ in their structure and mineral composition and probably in their origin. The available data may imply that formation of the earliest continental-type lithosphere took place mainly during the first phase of the Earth's history, and the oceanic-type lithosphere could have been formed during the second phase. The fundamental differences were apparently caused by the different processes of lithospheric formation during these phases. This statement is in good agreement with geological and petrological data on the predominance of various forms of geodynamics and igneous activity during their formation. These differences could be related to the different directions of the processes within the Earth: during the first phase, deep-seated petrogenetic processes were caused by a gradual warming up of the planet, from the surface downwards, and a cooling of the outer layers, whereas during the second phase it involved the displacement of the molten zone (outer core) towards the surface, due to a final release of energy of the Earth into space.

To conclude, it should be noted that the study of the igneous and geodynamic processes discussed in this book represents a complex challenge. Undoubtedly further study is needed, this time using a multidisciplinary approach including consideration of the general problems of petrology, geodynamics and Earth sciences as a whole.

REFERENCES

Adams, C.J., Seward, D. and Weaver S.D. (1995) *Antarctic Science*, **7**(3), 265–276

Aftalion, M. *et al.* (1991) *J. Geology*, **99**, 851–862

Aguirre, L. (1983) *Geol. Soc. Am.*, **159**, 293–316

Akhmedov, A.M., Orlova, M.G. and Yakobson, K.E. (1992) *Doklady RAN*, **326**(2), 305–308 (in Russian)

Alabaster, T., Pearce, J.A. and Malpas, J. (1982) *Contrub. Mineral. Petrol.*, **81**(3), 168–183

Alapieti, T.T. (1982) *Geol. Surv. Finland Bull.*, **319**, 116 pp.

Alapieti, T.T., Filen, B.A. and Lathinen, J.J. *et al.* (1990) *Mineral. Petrol.*, **42**(1–4), 1–22

Alibert, C., Norman, M.D. and McCulloch, M.T. (1994) *Geochim. Cosmochin. Acta*, **58**, 2,921–2,926

Allegre, C.J. (1987) *Earth Planet. Sci. Lett.*, **86**(2/4), 175–203

Allegre, C.J. and Ben Othman, D. (1980) *Nature*, **286**(5771), 335–342

Almond, D.C. (1986) *Tectonophysics*, **131**, 301–322

Almukhamedov, A.I., Kashintsev, V.V. and Matveenkov, V.V. (1985) *Evolutia basaltoidnogo vulcanisma Krasnomorskogo regiona.* (*Evolution of basaltic volcanism of the Red Sea region.*), Nauka, Novosibirsk, 191 pp. (in Russian)

Almukhamedov, A.I., Plyushkin, G.S., Almukhamedov, E.A. *et al.* (1992) *Russian Geology and Geophysics*, **7**, 48–59

Amelin, Yu.A., Heaman, L.M. and Semenov, V.S. (1995) *Precambrian Res.*, **75**(1–2), 31–46

Amstutz, G.C. (Ed.) (1974) *Spilites and Spilitic Rocks*. Intern. Union of Geol. Sci., ser A, 7, Springer Verlag, New York, 387 pp.

Anderson, D.L. (1989) *Theory of the Earth*, Blackwell Scientific Publications, Boston, 366 pp.

Anderson, D.L., Tanimoto, T. and Shang, Y. (1992) *Science*, **265**, 1,645–1,650

Apeltsin, F.R., Skorobogatova, L.N., Yakushin, L.N. (1967) *Geticheskie Cherty Granitoidov Urala i Usloviya ih Redkometalnoy Metallogeneticheskoy Spetcializatcii.* (*Genetic Features of the Ural Granitoids and their Rare Metal Metallogenic Specialization*), Nedra, Moscow, 199 pp. (in Russian)

Arestova, N., Alexejev, N., Bogomolov, E. *et al.* (1995) In: *Precambrian of Europe*: *Stratigraphy, Structure, Evolution and Mineralization. MAEGS 9.* Abstracts. St. Petersburg, p. 3

Arnaud, N.O., Vidal, Ph., Tapponier, P., Matte, Ph. and Deng, W.M. (1992) *Earth & Planet. Sci. Lett.*, **111**, 351–367

Arndt, N.T. (1986) *Terra Cognita*, **6**, 59–66.

Arndt, N. (1994) Archean komatiites. In: K.C. Condie (Ed.) *Archean crustal evolution*, Elsevier, Amsterdam, pp. 11–44

Arndt, N. and Jenner, G.A. (1986) *Chem. Geol.*, **56**, 229–255

Aro, K. and Laitakari, I. (Eds.) (1987) *Geol. Surv. of Finland*, Espoo, **76**, 254 pp.

Arth, J.G. and Hanson, G.N. (1972) *Contrib. Mineral Petrol.*, **37**(2), 161–174

Artyushkov, E.V. (1983) *Geodynamics*. Developments of Geotectonics, 18, Elsevier, 312 pp.

Arzamastsev, A.A., Kaverina, V.A. and Polezhaeva, L.I. (1988) *Daikovye Porody Hibinskogo massiva i ego Obramleniya.* (*Dyke Rocks of the Hibiny Massif and the*

Surrounding Area), Apatity, Kola Scientific Centre of Russian Academy of Sciences, 86 pp. (in Russian)

Ashwal, L.D., Jacobsen, S.B., Myers, J.S. *et al.* (1989) *Earth Planet. Sci. Lett.*, **91**(3/4), 261–270

Ashwal, L.D. (1993) *Anorthosites*, Springer-Verlag, New York, 422 pp.

Azbel, I.Ya. and Tolstikhin, I.N. (1988) *Radiogennye Izotopy i Evolutsiya Mantii Zemli, Kory i Atmosphery. (Radiogenic Isotopes and Evolution of the Mantle, Crust and Atmosphere of the Earth*), Kola Scientific Centre an Academy of Sciences, Apatity, 140 pp. (in Russian)

Baadsgaard, H., Lambert, R.S.J. and Krupicka (1976) *Geochem. Cosmochim. Acta*, **40**(5), 513–527

Babansky, A.D., Ryabchikov, I.D. and Bogatikov, O.A. (1983) *Evolutsiya Schelochnozemelnyh Magm (Evolution of Calc-Alkali Magmas*), Nauka, Moscow, 120 pp. (in Russian)

Balagansky, V.V., Glasnev, V.N. and Osipenko, L.G. (1996). In *Proterozoic Evolution in the North Atlantic Realm. Abstracts vol.*, 16–17.

Balashov, Yu.A. (1985) *Izotopno-Geokhimicheskaya Evolutsiya Mantii i Kory Zemli. (Isotopic-geochemical Evolution of the Mantle and Crust of the Earth*), Nauka, Moscow, 221 pp. (in Russian)

Balashov, Yu.A. (1990) *Izvestiya AN SSSR, Ser. Geol.*, **12**, 126–128 (in Russian)

Balashov, Yu.A., Bayanova, T.B. and Mitrofanov, F.P. (1993) *Precambrian Res.*, **64**(1–4), 197–205

Balashov, Yu.A., Mitrofanov, F.P. and Balagansky, V.V. (1992) New Geochronogical Data on the Archaean Rocks of the Kola Peninsula. In: V.V. Balagansky and F.P. Mitrofanov (Eds.). *Correlation of Precambrian Formations of the Kola–Karelian Region and Finland*, Kola Science Center RAS, Apatity, pp. 13–34

Ballard, S. and Pollack, N.M. (1987) *Earth Planet. Sci. Lett.*, **85**(1/3), 253–264

Ballard, S. and Pollack, N.M. (1988) *Earth & Planet. Sci. Lett.*, **88**(1/2), 132–142

Barberi, F., Ferrary, G., Santacroce, R. *et al.* (1975) *J. Petrol.*, **16**(1), 22–56

Barbey, P., Convert, J., Moreau, B. *et al.* (1984) *Bull. Geol. Soc. Finland*, **56**, 161–188

Barbey, P. and Raith, M. (1990) The granulite Belt of Lapland. In: D. Veilzent and Ph. Vidal (Eds.) *Granulites and Crustal Evolution*. Kluwer, Dardrecht, pp. 111–132

Barker, F. (Ed.) (1979) *Trondhjemites, Dacites and Related Rocks*. Elsevier, Amsterdam, 418 pp.

Barovich, K.M. *et al.* (1989) *Bull. Geol. Soc. Am.*, **101**(3), 333–338

Barnes, S.J. (1989) *Contrib. Mineral. Petrol.*, **101**(4), 447–457

Basaltic Volcanism on the Terrestrial Planets (1981). Pergamon Press, New York, 1,286 pp.

Batiyeva, I.D. (1976) *Petrologiya Shelochnyh Granitov Kolskogo Poluostrova. (Petrology of the Alkali Granites of the Kola Peninsula.*), Nauka, Leningrad, 224 pp. (in Russian)

Basu, A.R., Rubury, E., Mehnert, H. and Tatsumoto, M. (1984). *Contrib. Mineral. Petrol.*, **86**(1)

Basu, A.R., Poreda, R.J., Renne, P.R. *et al.* (1995) *Science*, **269**(5225), 822–825

Batiza, R. (1982) *Earth & Planet. Sci. Lett.*, **60**(2), 195–206

Beccaluva, L., Ohnenstetter, D. and Ohnenstatter M. (1979) *Can. J. Earth Sci.*, **16**(9), 1,874–1,882

Beckinsale, R.D., Sanchez-Fernandez, A.W., Brook, M. *et al.* (1985) *Pitcher, op. cit.*, 177–202

Bel'kov, I.V., Batiyeva, I.D. and Vetrin, V.R. (1984) Drevneyshaya Kora Baltyiskogo Shchita: Sostav, vozrast i Geneticheskie osobennosti. (The Ancient Crust of the Baltic shield: composition, age and genetic features.) In: *Geology of the Precambrian, 27th IGC*, Vol. 5, pp. 92–99 (in Russian)

Belousov, V.V. (1989) Endogennye reshmy: vzaimodeystviye verhney mantii i kory. (Endogenic regimes: interactions between the upper mantle and the crust.) In: O.A. Bogatikov (Ed.) *Crystalline Crust in Space and Time. Magmatism.* Nauka, Moscow, pp. 36–45 (in Russian)

Belousov, V.V., Gerasimovsky, V.I., Polyakov, A.I. *et al.* (1974) *Vostochno–Africanskaya Riftovaya Sistema.* (*The East African Rift System.*), Nauka, Moscow, **3**, 264 pp. (in Russian)

Bennet, V.C., Nutman, A.P. and McCulloch, M.T. (1993) *Earth Planet. Sci. Lett.*, **119**, 299–317

Berezhnaya, N.G., Bibikova, E.V., Sochava, A.V. *et al.* (1988) *Doklady AN SSSR*, **302**(5), 1,209–1,212 (in Russian)

Bergman, S.C. (1987) *Geol. Soc. Spec. Publ.*, **30**, 103–190

Bernard-Griffiths, I. Peucat, J.J., Vidal, Ph. *et al.* (1984) *Precambr. Res.*, **23**, 325–348

Bertrand, H., Boivin, P. and Robin, C. (1990) Petrology and geochemistry of basalts from the Vavilov basin (Tyrrenian Sea), Ocean drilling program Leg 107, Holes 107, 651A, and 654. In: K.A. Kastens, J. Mascle *et al.*, *Proc. ODP, Sci. Results, 107: Colledge Station, TX (Ocean Drilling Program)*, pp. 75–92

Bibikova, E.V., Gracheva, T.V., Drugova, G.M. *et al.* (1989) *Doklady AN SSSR*, **305**, 949–952 (in Russian)

Bibikova, E.V., Grinenko, V.A., Kiselevsky, M.A. and Shukolyukov, Yu.A. (1982) *Geokhimiya*, **12**, 1,718–1,728 (in Russian)

Bibikova, E.V., Mel'nikov, V.P. and Avakyan, K.Kh. (1993b) *Petrologiya*, **1**(2), 181–198 (in Russian)

Bibikova, E.V., Sheld, T., Bogdanova, S.V. *et al.* (1993a) *Geokhimiya*, **10**, 1,393–1,411 (in Russian)

Bibikova, E.V. and Williams, I.S. (1990) *Precambr. Res.*, **48**, 203–221

Bickle, M.J., Nisbet, E.G. and Martin, A. (1994) *J. Geol.*, **102**, 121–138

Bills, B.G. and Ferrari, A.J. (1977) *J. Geophys. Res.*, **82**(8), 1,306–1,314

Bindeman, I.N., Sharkov, E.V. and Ionov, D.A. (1990) *Internat. Geol. Rev.*, **32**(9), 905–915

Biryukov, V.M. *et al.* (1991) *Doklady AN SSSR*, **321**(2), p. 362 (in Russian)

Blaxland, A.B., van Breemen, O. and Steenfelt, A. (1976) *Lithos*, **9**(1), 31–38

Bodinier, J.L., Guiraud, M., Febries, J. *et al.* (1987a) *Geochim. et Cosmochim. Acta*, **51**(2), p. 279

Bodinier, J.L., Fabries, J., Lorand, J.-P. *et al.* (1987b) *Bull. Mineral.*, **110**(4), 561–568

Bodinier, J.L., Dupuy, C. and Dostal, J. (1988) *Geochim. Cosmochim. Acta*, **52**(12), 2,893–2,907

Bogatikov, O.A. (1966) *Petrologiya i Metallogeniya Gabbro-Sienitovyh Kompleksov Altaye-Sayanskoy Oblasti.* (*Petrology and metallogeny of gabbro-syenite complexes of the Altae-Sayany Area.*) Moscow, 240 pp. (in Russian)

Bogatikov, O.A. (1979) *Anortozity* (*Anorthosites*), Nauka, Moscow, 232 pp. (in Russian)

Bogatikov, O.A. (Ed.) (1988) *Komatiity i Visokomagnezialnye Vulkanity Rannego Dokembriya Baltiyskogo Shchita.* (*Komatiites and High-Magnesian Volcanites of the Early Precambrian of the Baltic Shield.*) Nauka, Leningrad, 185 pp. (in Russian)

Bogatikov, O.A. and Frikh-Khar, D.I. (1984) Ob obrazovanii anortozitov Luny. (On the formation of lunar anorthosites.) In: M.S. Markov and O.A. Bogatikov (Eds.) *Anorthosites of the Earth and the Moon,* Nauka, Moscow, pp. 189–198 (in Russian)

Bogatikov, O.A., Frikh-Khar, D.I., Ashihmina, N.A. *et al.* (1979) *Doklady AN SSSR,* **247**(2), 450–453 (in Russian)

Bogatikov, O.A., Karpenko, S.F., Sukhanov, M.K. and Spiridonov, V.G. (1988) *Doklady AN SSSR,* **301**(2), 430–433 (in Russian)

Bogatikov, O.A. and Kononova, V.A. (Eds.) (1991) *Lamproity (Lamproites),* Nauka, Moscow, 302 pp. (in Russian)

Bogatikov, O.A., Kovakenko, V.I., Tsvetkov, A.A. *et al.* (1987a). Isvestia of the USSR Academy of Sciences, *ser.geol.,* **3**, 3–12 (in Russian)

Bogatikov, O.A., Kovalenko, V.I., Yarmolyuk, V.V. *et al.* (1987b) Sravnitelnyi analiz magmatizma sovremennyh geodinamicheskih obstanovok i ih paleoanalogov. (Comparative analysis of present-day geodynamic environments and their palaeo-analogues.) In: V.I. Kovalenko (Ed.), *Evolution of Magmatism over the Earth's History,* Nauka, Moscow, pp. 237–258 (in Russian)

Bogatikov, O.A., Simon, A.K. and Pukhtel, I.S. (1990) Rannyaya kora Zemli: geologiya, petrologiya, geokhimiya. (The Early Earth's crust: geology, petrology, geochemistry.) In: *The Early Earth's crust: its age and composition,* Nauka, Moscow, pp. 15–26 (in Russian)

Bogatikov, O.A. and Tsvetkov, A.A. (1986) *Vestnik AN SSSR,* **6**, 71–82 (in Russian)

Bogatikov, O.A. and Tsvetkov, A.A. (1988) *Magmaticheskaya Evolutsiya Ostrovnyh Dug.* (*Magmatic Evolution of Island Arcs.*) Nauka, Moscow, 248 pp. (in Russian)

Bogdanov, N.A. (1988) *Tektonika Glubokovodnyh Vpadin Okrainnyh Morey.* (*Tectonics of Deep-Sea Basins of Marginal Seas.*) Nedra, Moscow, 221 pp. (in Russian)

Bogdanova, M.N. and Kaulina, T.V. (1995) In: *Precambrian of Europe: Stratigraphy, Structure, Evolution and Mineralization.* MAEGS 9. Abstracts. St. Petersburg, p. 14

Bogdanova, S.V., Lobach-Zhuchenko, S.V., Markov, M.S., Simon, A.K. and Bogatikov, O.A. (1987) Magmatizm i geodinamika drevneishih struktur Zemli. (Magmatism and geodynamics of the oldest structures of the Earth.) In: V.I. Kovalenko (Ed.) *Magmatic Rocks. Evolution of Magmatism over the Earth's History,* Nauka, Moscow, pp. 146–172 (in Russian)

Boily, M., Ludden, J.M. *et al.* (1991) *Can. J. Earth Sci.,* **28**(1), 28–36

Bonatti, E., Ottonello, G. and Hamlyn, P.R. (1986) *J. Geophys. Res.,* **B91**(1), 599–631

Borsuk, A.M. (1979) *Mezozoyskiye i Kainozoyskiye Magmaticheskiye Formatsii Bolshogo Kavkaza.* (*Mesozoic and Cenozoic Formations of the Great Caucasus.*) Nauka, Moscow, 300 pp. (in Russian)

Borukaev, Ch.B. (1985) *Struktura Dokembriya i tektonika Plit.* (*Precambrian Structure and Plate Tectonics.*) Nauka, Novosibirsk, 200 pp. (in Russian)

Bowring, S.A., Houst, T.B. and Isachsen, C.E. (1990) The Acasta gneisses: remnants of Earth's early crust. In: Newsom, H.E. and Jones (Eds.) *Origin of the Earth.* Oxford Univ. Pr, Houston, pp. 319–343

Bowring, S.A., Williams, I.S. and Compston, W. (1989a) *Geology,* **17**(11), 971–975

Bowring, S.A., King, J.E., Housh, T.B., Isachsen, S.E. and Podosek, F.A. (1989b) *Nature*, **340**, 222–225

Boudreau, A.E., Mather, E.A. and McCallum, I.S. (1986) *J. Petrol.*, **27**(4), 627–645

Brey, G.P., Kogarko, L.N. and Ryabchikov, I.D. (1991) *N. Jb. Mineralogie. Mh.* H4, 159–168

Brooks, C., Ludden, J. and Pigeon, Y. (1982) *Can. J. Earth Sci.*, **19**(1), 55–67

Brotzu, P., Morbidelli, L., Peccirillo, E.M. *et al.* (1974) *Bull. Volcanol.*, **38**(1), 206–234

Brown, L., Klein, J., Middleton, R. *et al.* (1982) *Nature*, **299**(5885), 718–722

Buddington, A.F. (1937) *Geol. Soc. Am. Mem.*, **7**, 430 pp.

Buldakov, I.V. and Kotova, I.K. (1991) *Rogovoobmankovye Bazity i Ultrabazity Yuzhnoy Yakutii.* (*Hornblende Basites and Ultrabasites of Southern Yakutia.*) LGU, Leningrad, 159 p. (in Russian)

Bullard, F.M. (1980) *Volcanoes of the Earth.* University of Texas Press Austin.

Burtman, V.S. (1989) Geotektonika, **2**, 67–75 (in Russian)

Bussen, I.V. and Sakharov, A.S. (1972) *Petrologia Lovozerskogo schelochnogo massiva* (*Petrology of the Lovozero alkaline massif*). Leningrad, Nauka Publ., 296 pp. (in Russian)

Byamba, Zh. (1991) *Tektonicheskoe Razvitie Mongolii v Pozdnem Proterozoe-Rannem Paleozoe i Uslobiya Obrazovaniya Drevnyh Fosforitov.* (*Tectonic evolution of Mongolia in the late Proterozoic–early Palaeozoic and conditions of ancient phosphorites forming.*). Doctoral thesis, Moscow (in Russian)

Bykhover, N.A. (1984) *Raspredelenie Mirovyh Resursov Mineralnogo Syr'ya po Epoham Rudoobrazovaniya.* (*Distribution of the World ore Deposits Resources Throughout of the ore-forming epochs.*), Nedra, Moscow, 230 p. (in Russian)

Calvert, A.J., Sawuer, E.M., Davis, M.J. and Ludden, J.N. (1995) *Nature*, **375**, 670–674

Campbell, I.H. (1977) *J. Petrol.*, **18**, 183–215

Chappell, B.W. and White, A.J.R. (1974) *Pacific Geol.*, **8**, 173–174

Chase, C.G. and Patchell, P.J. (1988) *Earth Planet. Sci. Lett.*, **91**, 66–72

Chayes, F. (1963) *J. Geophys. Res.*, **68**(5)

Chekulaev, V.P., Lobach-Zhuchenko, S.B. and Levsky L.K. (1997) *Geochemistry International*, **8**, 805–816.

Chernyshov, N.M., Pereslavtsev, A.V., Molotkov, S.P. and Chernyshova, M.N. (1991) *Izvestiya AN SSSR, Ser. Geol.*, **9**, 11–24 (in Russian)

Chopin, C. (1987) *Phil. Trans. Roy. Soc. London*, **A321**(1557), 183–197

Christiansen, R.L. and Lipman, R.W. (1972) *Phil. Trans. Roy. Soc. London*, **271**, 249–284

Clark, S.P., Turekian, K. and Grossman, L. (1972) Model for the early history of the Earth. In: E.S. Robertson (Ed.) *The Nature of the Solid Earth*, New York, McGraw-Hill, pp. 3–18

Cohen, A.S., O'Nions, R.K. and Dawson, J.B. (1984) *Earth Planet. Sci. Lett.*, **68**(2), 209–220

Cohen, A.S., Burnham, O.M., Rogers, N.W. and Martin, K. (1996) In: *1996 V.M. Goldschmidt Conference, Heidelberg, Germany. J. Conference Abstracts*, Cambridge Publ., **1**(1), 116

Coish, R.A., Hickey, R. and Frey, F.A. (1982) *Geochim. Cosmochim. Acta*, **46**(11), 2,117–2,134

Coleman, R.G. (1977) *Ophiolites. Ancient Oceanic Lithosphere?* Springer-Verlag, Berlin, 229 pp.

Collerson, K.D. and Bridwater, D. (1979) Metamorphic evolution of Early Archaean tonalitic and trondjemitic gneisses of the Saglek area, Labrador. In: F. Barker (Ed.) *Trondjemites, Dacites and Related Rocks*. Elsevier, Amsterdam

Compston, W. and Pidgeon, R.T. (1986) *Nature*, **321**, 766–769

Condie, K.C. (1981) *Archaean Greenstone Belts*. Elsevier, Amsterdam, 434 pp.

Condie, K.C. (Ed.) (1994) *Archaean Crustal Evolution*. Elsevier, Amsterdam, 528 pp.

Courtillot, V., Feraud, G. and Maluski, H. *et al.* (1988) *Nature*, **333**, 543–846.

Cox. K.G. and Jamieson, B.G. (1984) *J. Petrology*, **15**(2), 269–302

Craig, H., Lupton, J.E. and Horible, Y. (1978) *Earth Planet. Sci. Lett*, **3**, 13–16

Crawford, A.J. (Ed.) (1989) *Boninites*, Unwin Hyman, London, 445 pp.

Davis, G.L., Sobolev, N.V. and Kharkiv, A.D. (1980) *Doklady AN SSSR*, **254**(1), 175–179 (in Russian)

Dawson, J.B. (1980) *Kimberlites and their Xenoliths*. Springer-Verlag, Berlin

Dawson, J.B. and Smith, J.V. (1988) *Contrib. Mineral. Petrol.*, **100**, 510–527

Decker, R.W., Wright, T.L. and Stauffer, P.H. (Eds.) (1987) Volcanism in Hawaii. *US Geol. Surv. Prof. Pap.*, **1350**, 839 p.

De Paolo, D.J., Manton, W., Grew, E. and Halpein, M. (1982) *Nature*, **298**(5875), 614–618

De Paolo, D.J. and Johnson, R.W. (1979) *Contrib. Mineral. Petrol.*, **70**(4), 367–379

De Paolo, D.J. and Wasserburg (1979) *Geochim. Cosmochim. Acta*, **43**, 999–1,008

Dermott, S.F. (Ed.) (1976) *The Origin of the Solar System*. Wiley, Chichester, 668 pp.

Deruelle, B. (1982) *J. Volcan. & Geotherm. Res.*, **14**(1/2), 77–124.

Detrick, R.S. (1991) *Ridge Crest Magma Chambers: a Review of Results from Marine Seismic Experiments at the East Pacific Rise. Ophiolite Genesis and Evolution of the Ocean Lithosphere*. Kluwer Academic Publishers, Dordrecht, pp. 7–20

de Wit, M.J. and Ashwal, L.D. (Eds.) (1997) *Greenstone Belts*. Claredon Press, Oxford, 809 pp.

Dietz, R.S. (1961) *Nature*, **190**, 854–857

Dikov, Yu.P., Gerasimov, M.V. and Yakovlev, O.I. (1990) *Geokhimiya*, **5**, 752–756 (in Russian)

Distler V.V. and Kunilov V.E. (Eds.) (1994) *Geology and Ore deposits of the Noril'sk region*. Guidebook of VII Intern. Platinum Symposium. Moscow-Noril'sk, pp. 67

Dobretsov, N.L. (1980) *Vvedeniye v Globalnuyu Petrologiyu. (Introduction to Global Petrology.)* Nauka, Novosibirsk, 230 pp. (in Russian)

Dobretsov, N.L. (1994) *Russian Geology and Geophysics*, **35**(3), 3–19

Dobretsov, N.L., Aschepkov, I.V. and Ionov, D.A. (1989) Evolutsiya verhney mantii i bazaltovogo vulkanizma Baikalskoi riftovoi zony. (Evolution of upper mantle and basaltic volcanism of the Baikal rift zone.) In: O.A. Bogatikov (Ed.) *Crystalline Crust in space and Time (Magmatism)*, Nauka, Moscow, pp. 5–16 (in Russian)

Dobretsov, N.L. and Kirdyashkin, A.G. (1991) *Russian Geology and Geophysics*, **3**, 4–20

Dolenko, G.N., Lyashkevich, Z.M., Alyehina, M.N., Malyuk, B.I. and Zavyalova, T.V. (1991) Geologiya i neftegazonosnost Dneprovo-Donetskoi vpadyny. Endogennye protsessy i neftegazonosnost. (Geology and oil–gas potential of the Dnepr-Donetsk basin.) In: G.N. Dolenko (Ed.) *Endogenic Processes and Oil–Gas Potential*, Naukova Dumka, Kiev, 97 pp. (in Russian)

Downes, H. (1993) *Phys. Earth Planet. Inter.*, **79**, 195–218

Downes, H. and Dupuy, C. (1987) *Earth & Planet. Sci. Lett.*, **82**(1/2), 121–135

Dreibus, G. and Wanke, H. (1984) *Accretion of the Earth and Terrestial Planets*, 27th IGC, Moscow, Vol. 11, pp. 3–11

Duchesne, J.C., Maqnil, R. and Demaiffe, D. (1985) The Rugoland anorthosites: facts and speculations. In: *Deep Proterozoic Crust North Atlantic Province*. Proc. NATO Adv. Study Inst., May 16–30 July, 1984, Dordrecht, pp. 449–476

Duk, G.G., Kol'tsova, T.V., Bibikova, E.V. *et al.* (1989) Problemy glubinnogo petrogenezisa i vozrast porod Kolskoi sverhglubokoi skvazhiny. (Problems of Deep-Seated Petrogenesis and Age of Rocks of the Kola Superdeep Well. In: *Isotopic Geochronology of the Precambrian*, Nauka, Leningrad, 72–86 (in Russian)

Dyuzhikov, O.A., Distler, V.V., Strunin, B.M. *et al.* (1988) *Geologiya i rudonosnost Norilskogo Rayona*. (*Geology and metallogeny of the Noril'sk Region.*), Nauka, Moscow, 280 pp. (in Russian)

Dzievonski, A.M. (1984) *J. Geophys. Res.*, **89**, 5,929–5,952

Dzievonki, A.M. and Anderson, D.A. (1981) *Phys. Earth & Planet. Inter.*, **25**, 297–356

Egorov, A.S., Kostyuchenko, S.L., Mukhin, V.N. and Solodilov, L.N. (1994) Trehmernaya Model Litosfery v Polosovoi Zone Geotraversa Kolskiy Poluostrov-Ural. (Three dimension model of the Lithosphere within geotraverse Kola Peninsula-Urals.). In: *Tectonics and magmatism of East-European Platform*. KMK Ltd., Moscow, pp. 161–168 (in Russian)

Ein, A.S. (1984) Daiki basitov Severo-Zapadnoy Karelii. (Basic dykes of NW Karelia.) In: A.I. Bogachev (Ed.) *Intrusive Basites and Hyperbasites of Karelia.*, Karelian Research Centre, 30–41 (in Russian)

Ellam, R.M. and Cox, K.G. (1989) *Earth & Planet. Sci. Lett.*, **92**(2), 207–218

Elo, S. and Korja, A. (1993) *Precambrian Res.*, **64**(1–4), 273–288

Emmerman, R.A. *et al.* (1977) *Neues Jahrb. Miner. Abh.*, **128**(3), 129–136

Emslie, R.F. and Hunt, P.A. (1989) *Pap. Geol. Surv. Can.*, N89-1c, 11–17

Engel, A.E.J., Engel, C.G. and Havens, R.G. (1965) *Bull. Geol. Soc. Am.*, **76**(7), 387–398

Erickson, R.L. and Blade, L. (1963) *Geochemistry and Petrology of the Aklaloc Igneous Complex at Magnet Cove, Arkansas*. US Government Printing Office, Washington, 95 pp.

Esin, S.V., Prusevich, A.A. and Kutolin V.A. (1992) *Pozdnekainozoiskiy Vulcanism i Glubinnoe Stroenie Vostochnogo Sihote-Alinya*. (*Late Cenosoic Volcanism and Deep-Seated Structure of East Sikhote-Alin.*), Nauka, Novosibirsk, 163 pp. (in Russian)

Evarts, R.C. (1977) *The Geology and Petrology of the Puerto Ophiolite, Diablo Range, Central California Coast Ranges. North American Ophiolites*. Portland, 95

Evensen, N.M., Hamilton, P.J. and O'Nions, R.K. (1978) *Geochim. et Cosmochim. Acta*, **42**, 1,199–1,212

Extended Abstracts. 4th International Kimberlite Conference (1986) *Geol. Soc. Austral. Abstr.*, No. 16.

Fau, Q. and Hooper, P.R. (1991) *J. Petrology*, **32**(4), 765–810

Fedorenko, V.A. and Dyuzhikov, O.A. (1981) *Sovetskaya Geologiya*, **9**, 98–104 (in Russian)

Fedorova, M.E. (1978) *Geologicheskoye Polozheniye i Petrologiya Granitoidov Hangaiskogo Nagorya*. (*Geological Position and Petrology of the Hangai Highlands Granitoids.*), Nauka, Moscow, 140 pp. (in Russian)

Fedotov, S.A. (Ed.) (1984) *Bolshoe Treshchinnoe Tolbachinskoe Izverzhenie*. (*Great Tolbachinskoe Fracture Eruption.*), Nauka, Moscow, 520 pp. (in Russian)

Fedotov, S.A., Bagdasarova, A.M., Kuzin, I.P. and Tarakanov, R.Z. (1968) *Zemletryaseniya i Glubinnoye Stroenie Yuga Kurilskoi Ostrovnoi Dugi.* (*Earthquakes and Deep Structure of the South of the Kurile Island Arc.*), Nauka, Moscow, 212 pp. (in Russian)

Fedotov, S.A., Zharinov, N.A. and Gorelchik, V.I. (1988) *Vulkanologiya i Seismologiya,* **2,** 32–42 (in Russian)

First International Conference of Geochemistry, Cosmochemistry and Geochronology (1982) Tokyo, 582 pp.

Fletcher, I.R., Libby, W.G. and Rosman, K.J.R. (1987) *Austral. J. Earth Sci.,* **34,** 523–525

Foley, S.F., Taylor, W.R. and Green, D.H. (1986) *Contrib. Mineral. & Petrol.,* **94,** 183–192

Forster, R. (1975) Geological history of the sedimentary basin of southern Mozambique and some aspects of the origin of the Mozambique Channel. In: *Palaeogeogr., Palaeochimatol., Palaeoecol.,* **17,** 267–287

Fountain, D.M., Arculus, R. and Key, R.M. (Eds.) (1992) Continental lower crust. *Developments of Geotectonics,* 23, Elsevier, 485 pp.

Francis, D. (1985) *Contrib. Mineral. & Petrol.,* **89**(2/3), 144–154

Fraser, K.J., Hawkesworth, C.J., Erlank, A.J. *et al.* (1985) *Earth & Planet. Sci. Lett.,* **76,** 57–70

Freeman, R. and Knorring von, M. *et al.* (1989) *Tectonophysics,* **150,** 253–318

Frey, F.A. (1980) *Am. J. Sci.,* 280A, 427–449

Frey, F.A. and Prinz, M. *et al.* (1978) *Earth & Planet. Sci. Lett.,* **38,** 129–176

Frey, F.A., Suen, C.J. and Stockman, H.W. (1985) *Geochim. et Cosmochim. Acta,* **49**(11), 2,469–2,491

Frolova, T.I., Perchuk, L.L. and Burikova, I.A. (1989) *Magmatizm i Preobrazovaniye Kontinentalnoi Kory Aktivnyh Okrain.* (*Magmatism and Transformation of the Continental Crust of the Active Margins.*), Nedra, Moscow, 261 pp. (in Russian)

Fryer, P., Pearce, J.A., Stokking, L.B. *et al.* (1992) *Proc. ODP, Sci. Results,* 125: College Station, TX (Ocean Drilling Program)

Funerton, S.L. and Barry, A.P. (1984) *Can. J. Earth Sci.,* **21**(5), 615–618

Gaal, G., Bertelsen, A., Gorbatchev, R. *et al.* (1989) *Tectonophys.,* **162,** 1–25

Gaal, G. and Gorbatschev, R. (1987) *Precambrian Res.,* **35,** 15–72

Gast, P.W. (1968) *Geochim. et Cosmochim. Acta,* **10**(10), 1,057–1,068

Geological Atlas of China (1981) Pekin, 140 pp. (in Chinese)

Gerlach *et al.* (1988) *J. Petrol.,* **29**(2), 333–382

Giggenbach, W. and Le Guern, F. (1976) *Geochim. & Cosmochim. Acta,* **40**(1), 57–69

Gibson, I.L. and Gibbs, A.D. (1987) *Tectonophys.,* **133,** 57–64

Gill, J.B. (1981) *Orogenic Andesites and Plate Tectonics.* Springer, New York, 390 pp.

Girnis, A.V. and Ryabchikov, I.D. (1988) Experimentalnaya petrologiya i genezis komatiitov. (Experimental Petrology and Genesis of the Komatiites.) In: O.A. Bogatikov (Ed.) *Early Precambrian Komatiites and High-Magnesian Volcanites of the Baltic Shield,* Nauka, Leningrad. pp. 185–192 (in Russian)

Girnis, A.V., Ryabchikov, I.D. and Bogatikov, O.A. (1987) *Genezis Komatiitov i Komatiitovyh Basaltov.* (*Genesis of Komatiites and Komatiitic Basalts*), Nauka, Moscow, 119 pp. (in Russian)

Gize, P. and Pavlenkova, N.I. (1988) *Izvestiya AN SSSR, Seriya Fizika Zemli,* **10** (in Russian)

Gladyshev, G.P. (1988) *Termodinamika i Makrokinetika Prirodnyh Ierarhicheskih Protsessov.* (*Thermodynamics and Macrokinetics of Natural Hierarchical Processes.*), Nauka, Moscow, 287 pp. (in Russian)

Glebovitsky, V.A. (1997) *The Early Precambrian of Russia.* Harwood Academic Publishers, 261 pp.

Glebovitsky, V.A. and Bushmin, S.A. (1983) *Postmigmatitovyi Metasomatoz.* (*Postmigmatite Metasomatism.*), Nauka, Leningrad, 287 pp. (in Russian)

Glebovitsky, V.A., Zinger, T.F., Kozakov, I.K. *et al.* (1985) *Migmatizatsiya i Granitoobrazovanie v Razlichnyh Termodinamicheskyh Rezhimah.* (*Migmatization and Granite Formation in Various Thermodynamic Regimes.*), Nauka, Leningrad, 310 pp. (in Russian)

Glensdorf, P. and Prigozhin, I.R. (1973) *Termodinamicheskaya Teoriya Struktury, Ustoichivosti i Fluktuatsyi.* (*Thermodynamic Theory of Structure, Tolerance and Fluctuation.*), Mir Publ., Moscow, 256 pp. (in Russian)

Glikson, A.Y. (1984) Significance of Early Archaean mafic–ultramafic xenolith patterns. In: A. Kroner, G.N. Hanson and A.M. Goodwin (Eds.). *Archaean Geochemistry,* Springer, Heidelberg, pp. 262–282

Glikson, A.Y. (1993) *Earth Sci. Rev.,* **35,** 285–319

Goldstein, S.L. and Galer, S.J.G. (1993) 142Nd/144Nd isotopic traces of the first billion years of Earth history. *Terra Abstracts, EUG VII, Strasbourg, France, 4–8 April 1993.* p. 35

Golubev, A.I. and Svetov, A.P. (1983) *Geokhimiya Basaltov Platformennogo Vulkanizma Karelii.* (*Geochemistry of the Platform Basalts of Karelia.*), Karelia, Pertozavodsk, 191 pp. (in Russian)

Gonshakova, V.I. (Ed.) (1973) *Bazit-giperbazitivyi magmatizm i minerageniya yuga Vostochno-Evropeyskoy Platformy.* (*Basic–Ultrabasic Magmatism and Mineralogy of the South of the East European Platform.*), Nedra, Moscow, 294 pp. (in Russian)

Gorbatschev, R. and Bogdanova, S. (1993) *Precambrian Res.,* **64**(1–4), 3–21

Gordienko, I.V. (1987) *Paleozoyskyi Magmatizm i Geodinamika Tsentralno-Aziatskogo Skladchatogo Poyasa.* (*Palaeozoic Magmatism and Geodynamics of the Central Asian Fold Belt.*), Nauka, Moscow, 238 pp. (in Russian)

Gorodnitsky, A.M. and Zonenchain, L.P. (1979) Paleogeodinamika i Dreif Kontinentov. (Palaeogeodynamics and Continental Drift.) In: O.G. Sorokhtin(Ed.), *Geophysics of the Ocean,* Vol. 2, Nauka, Moscow, 327–369 (in Russian)

Gover, C.F. and Owen, V. (1984) *Can. J. Earth Sci.,* **21,** 678–693

Grachev, A.F. (1987) *Riftovye Zony Zemli.* (*Rift Zones of the Earth.*), Nedra, Moscow, 285 pp. (in Russian)

Green, D.H. and Hibberson, W.H. (1970) *Lithos,* **5**(6), 209–222

Green, D.H. and Ringwood, A.E. (1967) *Contrib. Mineral. & Petrol.,* **15,** 103–190

Green, D.H. and Ringwood, A.E. (1968) Genesis of the calc-alkaline igneous rock suite. *Contrib. Mineral. & Petrol.,* **18,** 163–174

Grew, E.S. (1984) *Tectonophysics,* **105**(1–4), 177–191

Griffin, W.L., McGregor, V.R., Nutman, A. *et al.* (1980) *Earth & Planet. Sci. Lett.,* **50**(1), 59–74

Griffin, W.L., O'Reilly, S.Y. and Stabel, A. (1988) *Geochim. et Cosmochim. Acta,* **52**(2), 449–459

Gruau, G., Jahn, B.M., Glikson, A.G., Dany, R., Hickman, A.H. and Chavel, C. (1987) *Earth & Planet. Sci. Lett.,* **85,** 105–116

Gruau, G., Tourpin, S., Jahn, B.M. and Anhaeusser, C.R. (1988) *Chem. Geol.*, **70**, p. 114

Gruenewaldt, G. von and Harmer, R.E. (1992) *Proterozoic Crustal Evolution*. Elsevier, Amsterdam, pp. 181–213

Gruenewaldt, G. Von, Sharpe, M.R. and Hatton, C.Y. (1985) *Econ. Geol.*, **80**, 803–812

Haapala, I. (1977) *Bull. Geol. Surv. Finland*, **286**, 128 pp.

Hale, C.J. (1987) *Nature*, **329**(6136), 233–236

Hall, R.P. and Hughes, D.J. (1987) *Contrib. Mineral. & Petrol.*, **97**(2), 169–172

Hall, R.P. and Hughes, D.J. (1992) Mantle evolution in the Early Proterozoic, *29th IGC, Kyoto, Japan, Abstracts, Vol. 1*, p. 53

Halls, H.C. (1992) *Earth & Planet. Sci. Lett.*, **105**(1/9), 279–292

Halls, H.C. and Fahrig, W.F. (1987) *Mafic Dyke Swarms*. Geol. Assoc. Canada, Spec. Pap. 34, 503 pp.

Hamilton. P.G., O'Nions, R.K., Bridgwater, D. and Nutman, A.P. (1983) *Earth & Planet. Sci. Lett.*, **62**, 263–272

Hanski, E.J. (1992) *Geol. Surv. Finland Bull.*, **367**, 196 pp.

Hanski, E.J., Huhma, H., Smolkin, V.F. and Vaasjoki, M. (1991) *Bull. Geol. Surv. Finland*, **62**, 123–133

Hanski, E.J. and Smolkin, V.F. (1989) *Precambrian. Res.*, **45**, 63–82

Harjono, H., Diament, M., Nouaili, L. and Dubois, J. (1989) *J. Volcanol. & Geotherm. Res.*, **39** (4), 335–348

Harris, N.B.W., Pearce, J.A. and Tindle, A.G. (1986) Geochemical characteristics of collision-zone magmatism. In: M.P.L. Coward and A.S. Ries (Eds.), *Collision Tectonics*. Geol. Soc. Spes. Publ., No. 19, pp. 67–81

Hart, S.R. (1988) *Earth & Planet, Sci. Lett.*, **90**(3), 273–296

Hart, S.R. and Zindler, A. (1986) *Chem. Geol.*, **57**(3/4), 247–267

Heaman, L.M. and Tarney, J. (1989) *Nature*, **340**(6236), 705–708

Hegner, E., Kyser, T.K. and Hulbert, L. (1989) *Can. J. Earth Sci.*, **26**(5), 1,027–1,035

Hekinian, R. (1982) *Petrology of the Oceanic Floor*. Elsevier, Amsterdam, 393 pp.

Helmstaedt, H.H. and Scott, D.J. (1992) The Proterozoic Ophiolite Problem. In: K.S. Condie (Ed.) *Proterozoic Crustal Evolution*. Elsevier, Amsterdam, pp. 55–95

Hermon, R.S. Kempton, P.D., Stosch, H.-G., *et al.* (1987) *Earth Planet. Sci. Lett.*, **81**(2/3), 193–202

Hickey-Vargas *et al.* (1989) *Contrib. Mineral. Petrol.*, **103**, 361–386

Hoffman P.F. (1988) United plates of America, the birth of a craton: Early Proterozoic assembly and growth of Laurentia. In: G.W. Wetherill, A.L. Albee and F.G. Stehly (Eds.) *Annual review Earth & Planetary Sciences*, **16**, 542–603

Huhma, H., Cliff, R.A., Perttunen, V. and Sakko, M. (1990) *Contr. Miner. Petrol.*, **104**, 369–379

Huhma, H. and Merilainen, K. (1991) Provenance of peregneisses from the Lapland granulite Belt. *Joint meeting "Metamorphism, Deformations and structure of the Crust"*. Oulu. abstr., 26 pp.

Hunter, D.R. (1974) *Precambrian Res.*, **1**(4), 259–294

Huppert, H.E. and Sparks, R.S.J. (1985) *Earth & Planet. Sci. Lett.*, **74**, 371–386

Ilupin, I.P., Kaminskiy, F.V. and Frantcesson, E.V. (1978) *Geokhimiya Kimberlitov.* (*Geochemistry of Kimberlites.*), Nedra, Moscow, 352 pp. (in Russian)

Imsland, P. (1983) *Contrib. Mineral. & Petrol.*, **83**(1/2), 31–37

Ionov, D.A. and Jagutz, E. (1988) *Doklady AN SSSR*, **301**(5), 63–79 (in Russian)

Ireland, T.R. and Wlotzka, F. (1992) *Earth & Planet. Sci. Lett.*, **109**(1/2), 1–10

Irvine, T.N. (1974) *Mem. Geol. Soc. Am.*, **138**, 240 pp.

Irving, A.J. (1980) *Am. J. Sci.*, **280-A**(2), 426–683

Isachsen, Y.W. (Ed.) (1969) Origin of Anorthosite and Related Rocks. In: Origin of Anorthosites and Related Rocks. *Mem. N.Y. State Mus. Sci. Serv.*, **18**, 435–445

Ishizuka, H., Kawanobe, Y. and Sakai, H. (1990) *Geochem. J.*, **24**(2), 75–92

Jackson, E.D. (1961) Primary textures and mineral associations in the Ultramafic zone in the Stillwater Complex, Montana. *US Geol. Surv. Prof. Pap.*, **358**, 106 pp.

Jacobsen, S.B. (1988) *Geochim. et Cosmochim. Acta*, **52**, 1,341–1,350

Jacobsen, S.B. and Wasserburg, G.J. (1979) *Proc. Nat. Acad. Sci. USA*, **77**(11), 6,298–6,302

Jagoutz, E., Palme, H., Baddenhausen, H. *et al.*, (1979) The abundance of major, minor and trace elements in the Earth's mantle as derived from primitive ultramafic nodules, *Proc. 10th Lunar Conference*, Vol. 2, 2,031–2,050

Jahn, B.M. and Gruau, G. (1989) Geochemistry and isotopic characteristics of Archaean komatiites and basalts and their inference an Early crust–mantle differentiation. *Ext. Abstr. 29th IGC, Washington*, Vol. 2, p. 112

Jahn, B.M., Shin, C.Y. and Murthy, V.R. (1974) *Geochim. et Cosmochim. Acta*, **38**(4), 873–885

Jahn, B.M., Vidal, P. and Kroner, A. (1984) *Contrib. Mineral. & Petrol.*, **86**(4), 398–408

Jeffries, H. (1929) *The Earth*, 2nd Edn. Cambridge University Press, London

Joplin, G.A. (1968) *J. Geol. Soc. Austral.*, **15**, 275–294

Kadik, A.A. (Ed.) (1991) *Fluidy i Okislitelno-Vosstanovitelnye Reaktsii v Magmaticheskih Sistemah.* (*Fluids and Reduction–Oxidation in Magmatic Systems.*), Nauka, Moscow, 256 pp. (in Russian)

Karig, D.E. (1971) *J. Geophys. Res.*, **76**, 2,542–2,561

Karig, D.E., Barber, A.J., Charlton, T.R., Klemperer, S. and Nuggson, D.M. (1987) *Geol. Soc. Am. Bull.*, **98**(1), 18–32

Kastens, K.A., Mascle, J. *et al.* (1990) *Proc. ODP, Sci. Results*, *107*, Ocean Drilling Program, College Station, TX, 772 pp.

Kato, T., Ringwood, A.E. and Irifune, T. (1988) *Earth & Planet. Sci. Lett.*, **89**, 123–145

Kay, S.M. and Kay, R.W. (1990) Aleutian magmas in space and time. *Decade North Am. Geol.*, **12**, 438–518

Kazansky, V.I. (1988) *Evolutsiya Rudonosnyh Struktur Dokembriya.* (*Evolution of Precambrian Ore-Bearing Structures.*), Nedra, Moscow, 286 pp. (in Russian)

Kempton, P.D., Downes, H., Sharkov, E.V. *et al.* (1995) *Lithos*, **36**, 157–184

Kempton, P.D., Hawkesworth, C.J. and Fowler, M. (1991a). Geochemistry and isotopic composition of gabbro from layer 3 of the Indian Ocean crust, Leg 118, Hole 735B. In: R.P. Von Herzen, P.T. Robinson *et al.*, *Proc. ODP, Sci. Results*, *118*, Oceanic Drilling Program, college Station, TX, pp. 127–144

Kempton, P.D. *et al.* (1991b) *J. Geophys. Res.*, **96**(B8), 13,713–13,735

Kent, G.M., Harding, A.J. and Orcutt, J.A. (1990) *Nature*, **344**(6267), 650–653

Kepezinskas, K.B. and Kepezinskas, V.V. (1985) Geokhimiya vulkanogennyh porod metaofiolitovyh formatsiy Mongolii i usloviya obrazovaniya paleookeanicheskoi litosfery. (Geochemistry of Volcanogenic Rocks of Metaophiolite Formations of Mongolia and Conditions of Palae-Oceanic Lithosphere Formations.) In: *Geochemistry of REE in Basites and Hyperbasites*, Nauka, Novosibirsk, pp. 4–26 (in Russian)

Kepezinskas, P.K., Kepezinskas, K.B. and Pukhtel, I.S. (1990) *Doklady AN SSSR*, **316**(3), 718–721 (in Russian)

Kepezinskas, K.V., Kepezinskas, V.V. and Zaitsev, N.S. (1987) *Evolutsiya Zemnoy Kory Mongolii v Dokembrii. (Precambrian Evolution of the Earth's Crust of Mongolia.)*, Nauka, Moscow, 168 pp. (in Russian)

Khain, V.E. (1984) *Regionalnaya Geotektonika. Alpyiskyi Sredizemnomorskyi Poyas. (Regional Geotectonics. Alpine Mediterranean Belt.)*, Nedra, Moscow, 344 pp. (in Russian)

Khain, V.E. (1985) *Regionalnaya Geotektonika. Okeany. Sintez. (Regional Geotectonics. Oceans: a Synthesis.)*, Nedra, Moscow (in Russian)

Khain, V.E. (1993) *Vestnik MGU*, **6**, 12–25 (in Russian)

Khain, V.E. and Bozhko, N.A. (1988) *Historical Geotectonics. Precambrian*, Nedra, Moscow, 382 pp. (in Russian)

Kinny, P.D. (1986) *Earth & Planet Sci. Lett.*, **79**, 337–347

Kiselyev, A.I., Medvedev, M.E. and Golovko, G.A. (1979) *Vulkanizm Baikalskoi Riftovoi Zony i Problemy Glubinnogo Magmoobrazovaniya. (Volcanism of the Baikal Rift Zone and Problems of Deep Magma Generation.)*, Nauka, Novosibirsk, 197 pp. (in Russian)

Klerkx, I., Deutsch, S. and Pichler, Z.W. (1977) *J. Volcanol. & Geotherm. Res.*, **2**(1), 49–71

Kober, B., Pidgeon, R.T. and Lippolt, H.J. (1989) *Earth & Planet. Sci. Lett*, **91**, 286–296

Kogarko, L.N. (1977) *Problemy Genezisa Agpaitovyh Magm. (Problems of Agpaite Magma Genesis.)*, Nauka, Moscow, 128 pp. (in Russian)

Kogarko, L.N., Kononova, V.A., Orlova, M.P. and Wooley, A.R. (1995) *Alkaline Rocks and Carbonatite of the World. P. 2*: Former USSR. Chapman & Hall. London etc., 226 pp.

Kogarko, L.N., Rudchenko, N.A. and Zakharov, M.V. (1993) *Geochem. Intern.*, **8**, 1,087–1,111

Kogarko, L.N. and Ryabchikov, I.D. (1995) *Geochem. Intern.*, **12**, 1,699–1,709

Konnikov, E.G. (1986) *Differentsirovannye Giperbazit-Bazitovye Kompleksy Docembriya Zabaikalya. (Differentiated Precambrian Ultrabasic–Basic Complexes of Zabaikalye.)*, Nauka, Novosibirsk, 220 pp. (in Russian)

Kononova, V.A. (Ed.) (1984) *Magmaticheskie Gornye Porody. Schelochnye Porody. Magmatic Rocks. Alkaline Rocks.)*, Nauka, Moscow, 415 pp. (in Russian)

Kontinen, A. (1987) *Precambrian Res.*, **35**, 313–341

Kornprobst, J. (Ed.) (1984) Kimberlites and Related Rocks. In: *Proc. III Intern. Kimberlite Conf. Clermont-Ferrand, France*, Vol. 1.

Kornprobst, J. (Ed.) (1987) The Mantle and Crust–Mantle Relationships. Elsevier, Amsterdam, 393 pp.

Korzinsky, M.A., Tkachenko, S.I., Shmulovich, K.I. *et al.* (1994) *Nature*, **369**(6,475), 51–52

Kosygin, V.Yu. (1991) *Gravitatsionnoye Pole i Plotnostnye Modeli Tektonosfery Severo-Zapada Tihogo Okeana. (Gravity Field and Density Models of the Tectonosphere of the North-West Pacific.)*, Nauka, Moscow, 201 pp. (in Russian)

Kotov, A.B., Salnikova, E.B., Morozova, I.N. *et al.* (1993) *Russian Geology and Geophysics*, **2**, 15–21

Kovalenko, V.I. (1977) *Petrologiya i Geokhimiya Redkometalnyh Granitoidov. (Petrology and Geochemistry of Rare-Metal Granitoids.)*, Nauka, Novosibirsk, 250 pp. (in Russian)

Kovalenko, N.I. (1979) *Eksperimentalnoye Issledovaniye Obrazovaniya Redkometalnyh Litiy-Ftoristyh Granitov.* (*Experimental Study of Rare-Metal Lithium–Fluorine Granite Formation.*), Nauka, Moscow, 152 pp. (in Russian)

Kovalenko, V.I. (Ed.) (1982) *Regionalnaya Petrohimiya Mezozoiskih Intruzyi Mongolii.* (*Regional Petrochemistry of Mesozoic Intrusions of Mongolia.*), Nauka, Moscow, 207 pp. (in Russian)

Kovalenko, V.I. and Kovalenko, N.I. (1984) *Physics of the Earth and Plan. Int.*, **35**, 51–62

Kovalenko, V.I. (Ed.) (1987) *Magmaticheskie Gornye Porody. Evolutsiya Magmatizma v Istorii Zemli.* (*Igneous Rocks.* Vol. 6. *Evolution of Magmatism in the Earth's History.*), Nauka, Moscow, 438 pp. (in Russian)

Kovalenko, V.I., Bogatikov, O.A., Tsvetkov, A.A. and Yarmolyuk, V.V. (1987a) Magmatizm Sovremennyh Geodinamicheskyh Obstanovok. (Magmatism of Present-Day Geodynamic Environments.) In: V.I. Kovalenko (Ed.) *Magmatic Rocks. Evolution of Magmatism over the Earth's History*, Nauka, Moscow, pp. 18–85 (in Russian)

Kovalenko, V.I., Goreglyad, A. and Yarmolyuk, V.V. (1985c) *Geol. Zentr. -Bl. -Geol. Carpath.*, **36**(2), 131–166

Kovalenko, V.I., Kuzmin, M.I., Antipin, V.S. and Koval, P.V. (1975) Zonalnost Areala Mezozoiskih Magmaticheskih, Metamorficheskih Porod i Metasomaticheskih Porod Zapadnoi Chasti Mongolo-Ohotskogo Poyasa. (Zonation of the Area of Mesozoic Magmatic, Metamorphic Rocks and Metasomatic Rocks of the Western Part of the Mongolian-Okhotsk Belt.) In: E.I. Popolitov (Ed.) *Geochemistry and Petrology of Metasomatism*, Nauka, Novosibirsk, pp. 75–111 (in Russian)

Kovalenko, V.I., Tsepin, A.I., Ionov, D.A. and Ryabchikov, I.D. (1985b) *Doklady AN SSSR*, **280**(2), 449–453 (in Russian)

Kovalenko, V.I., and Yarmolyuk, V.V. (1990) Evolutsiya Magmatizma v Strukturah Mongolii. (Evolution of the Magmatism in the Structures of Mongolia.) In: N.S. Zaitsev and V.I. Kovalenko (Eds.) *Evolution of Geological Processes and Metallogeny of Mongolia*, Nauka, Moscow, pp. 23–55 (in Russian)

Kovalenko, V.I., Yarmolyuk, V.V., Ionov, D.A. *et al.* (1989) *Geotektonika*, **4**, 3–16 (in Russian)

Kovalenko, V.I. and Yarmolyuk, V.V. (1995) *Econ. Geol.*, **90**(3)

Kovalenko, V.I., Tsaryeva, G.M., Goreglyad, A.V. *et al.* (1995a) *Econ. Geol.*, **90**(3), 530–507

Kovalenko, V.I., Yarmolyuk, V.V. and Bogatikov, O.A. (1995b) *Magmatism, Geodynamics and Metallogeny of Central Asia*, MIKO — Commercial Herald Publishers, 272 pp.

Kovalenko, V.I., Tsaryeva, G.M., Naumov, V.B. *et al.* (1996a) *Petrology*, **4**(3), 277–290

Kovalenko, V.I., Yarmolyuk, V.V., Pukhtel, I.S. *et al.* (1996b) *Petrology*, **4**(5), 420–459

Kovalenko, V.I., Yarmolyuk, V.V., Kovach V.P. *et al.* (1996c) *Geochim. Intern.*, **34**(8), 628–640

Kramers, J.D. and Ridley, J.R. (1989) *Geology*, **17**(5), 442–445

Kramm, U., Kogarko, L.N., Kononova, V.A. and Vartiainen, H. (1993) *Lithos*, **30**, 33–44

Kratz, K.O. (Ed.) (1978) *Geologiya i Petrologiya Arheiskogo Granitno-Zelenokamennogo Kompleksa Tsentralnoy Karelii.* (*Geology and Petrology of the Archaean Granite–Greenstone Complex of Central Karelia.*), Nauka, Leningrad, 262 pp. (in Russian)

Kratz, K.O. (Ed.) (1981) *Drevneyshie Granitoidy SSSR (Kompleks Seryh Gneisov.)* (*The Most Ancient Granitoids of the USSR (Grey Gneisses Complex.)*, Nauka, Leningrad, 152 pp. (in Russian)

Kratz, K.O., Glebovitsky, V.A., Bylinsky, R.V., Duk, V.L., Sharkov, E.V. et al. (1978) *Zemnaya Kora Vostochnoi Chasti Baltyiskogo Shchita.* (*The Earth Crust of the Eastern part of the Baltic Shield.*), Nauka, Leningrad, 230 pp. (in Russian)

Kravtsova, E.I. (1992) Contraints of feldspar miscibility changes on formation of rapakivi granite complexes. *29th IGC, Kyoto, Japan, Abstracts, Vol. 2*, p. 565

Krishnamurthy, P. and Cox, K.G. (1977) *Contrib. Mineral. & Petrol.*, **62**, 53–75

Krogh, T.E., McNutt, R.H. and Danis, G.L. (1982) *Can. J. Earth Sci.*, **19**, 723–728

Kroner, A. (1988) Precambrian Plate Tectonics. In: A. Kroner (Ed.). *Development and Evolution of Continental Crust*. Vol. 4, Elsevier, Amsterdam, pp. 57–90

Kroner, A., Puustinen, K. and Hickman, M. (1981) *Contrib. Mineral. & Petrol.*, **76**(1), 33–41

Krylov, D.P., Meshick, A.P. and Shukolyukov, Yu.A. (1993) *Geochem. J.*, **27**(2), 91–102

Kuehner, S.M. (1989) Petrology and geochemistry of early Proterozoic high-Mg dykes from the Vestfold Hills, Antarctica. In: A.J. Crawford (Ed.) *Boninites*, Unwin Hyman, pp. 208–231

Kulikova, V.V. (1993) *Volotskaya Svita — Stratotip Nizhnego Arheya Baltiyskogo Shchita.* (*Volotsky Suite — the Stratotype of the Early Arcaean in the Baltic Shield.*), Karelian Scientific Centre, Petrozavodsk, 255 pp. (in Russian)

Kuno, H. and Aoki, K. (1970) *Phys. Earth & Planet. Inter.*, **3**, 273–301

Kussmaul, S., Hormann, R.K., Ploskouka, E. and Subieta, T. (1977) *J. Volcanol. & Geotherm. Res.*, **2**(1), 73–111

Kuznetsov, Yu.A. (1964) *Glavnye Tipy Magmaticheskih Formatsyi SSSR.* (*The Major Types of Igneous Formations of the USSR.*), Nedra, Moscow, 387 pp. (in Russian)

Lambert, D.D., Morgan, J.W., Walker, R.J. et al. (1989) *Science*, **244**(4909), 1,169–1,174

Lambert, D.D. and Simmons, E.G. (1988) *Econ. Geol.*, **83**(6), 1,109–1,126

Lambert, D.D., Walker, R.J., Morgan, J.W. et al. (1995) *J. Petrology*, **35**(6), 1717–1753

Larin, A. and Neymark, L. (1992) Trans-Siberian Proterozoic (1.7–1.9 Ga) anorogenic anorthosite–rapakivi-like granite-acid volcanic belt. In: *29th IGC, Kyoto, Japan, Abstracts*, Vol. 2, p. 563

Larson, R.L. and Olsen, P. (1991) *Earth Planet. Sci. Lett.*, **107**, 437–447

Laubscher, H. (1988) *Bull. Geol. Soc. Am.*, **100**(9), 1,313–1,328

Laz'ko, E.E. (1988a) *Ultrabazity of iolitovoi Assotsiatsii.* (*Ultrabasites of the Ophiolitic Assemblage.*) In: E.E. Laz'ko and E.V. Sharkov (Eds.) *Igneous Rocks. Ultramafic Rocks*. Nauka, Moscow, pp. 8–96 (in Russian)

Laz'ko, E.E. (1988b) *Ultrabazity Dunit–Piroksenit–Gabbrovoi Assotsiatsii.* (*Ultrabasites of the Dunite–Pyroxenite–Gabbro Assemblage.*) In: E.E. Laz'ko and E.V. Sharkov (Eds.), *Igneous Rocks. Ultramafic Rocks*, Nauka, Moscow, pp. 96–114 (in Russian)

Laz'ko, E.E. (1988c) *Kimberlity (Kimberlites.)*, In: E.E. Laz'ko and E.V. Sharkov (Eds.) *Igneous Rocks. Ultramafic Rocks*, Nauka, Moscow, pp. 196–217 (in Russian)

Laz'ko, E.E. and Gladkov, N.G. (1991) *Izvestiya AN SSSR*, Ser. Geol., **6**, 47–65

Laz'ko, E.E. and Sharkov, E.V. (Eds.) (1988d) *Magmaticheskie Gornye Porody. Ultramaficheskie Porody.* (*Igneous Rocks. Vol. 5. Ultramafic Rocks.*), Nauka, Moscow, 501 pp. (in Russian)

Laz'ko, E.E., Sharkov, E.V. and Bogatikov, O.A. (1993) *Geochem. Internat.*, **30**(9), 1–24

Le Bas, M.J. (1987) Nephelinites and Carbonatites. In: J.G. Fitton and B.G.J. Upton (Eds.) *Alkaline Igneous Rocks*, Geol. Soc. Spec. Publ., 30, pp. 53–83

Lehtonen, M.I., Manninen, T., Rastas, P. and Rasanen, J. (1992) On the Early Proterozoic Metavolcanic Rocks in Finnish Central Lapland. In: V.V. Balagansky and F.P. Mitrofanov (Eds.) *Correlation of Precambrian Formations of the Kola— Karelian Region and Finland*, Kola Sci. Center RAS Publ., Apatity, pp. 65–85

Le Pichon, X. and Huchon, P. (1984) *Earth Planet. Sci. Lett.*, **67**(1), 123–135

Lichak, I.L. (1983) *Petrologiya Korostenskogo Plutona.* (*Petrology of the Korosten Pluton.*), Naukova Dumka, Kiev, 246 pp. (in Russian)

Lightfood, P.C., Hawkesworth, C.J., Hergt, J. *et al.* (1993) *Contrib. Mineral. Petrol.*, **114**, 171–188

Lightfood, P.C., Naldrett, A.J., Gorbachev, N.S. *et al.* (1990) *Contrib. Mineral. Petrol.*, **104**(6), 631–644

Liotard, I.M., Diphy, C., Dostal, J. and Cornen, G. (1982) *Chem. Geol.*, **35**(1/2), 115–128

Litvinovsky, B.A., Zanvilevich, A.N., Alekshin, A.M. and Pogladchikov, Yu.Yu. (1992) *Angaro-Vitimsky Batolit — Krupneishyi Granitoidnyi Pluton.* (*The Angaro-Vitim Batholith — the Largest Granitoid Pluton.*), Nauka, Novosibirsk, 141 pp. (in Russian)

Lobach-Zhuchenko, S.B. (1979) *Plagiogranity Rannego Dokembriya — geneticheskie svyazi i formatzionnaya Prinadleznost.* (*Plagiogranites of the Early Precambrian — Genetical Relationships and its Formation Position.*). Proc. of Inst. of Geology and Geochemistry, Urals Branch of the USSR Ac. Sci., Sverdlovsk, **155**, 20–31

Lobach-Zhuchenko, S.B., Bibikova, E.V. Drugova, G.M. *et al.* (1993) *Petrology*, **1**(6), 657–677

Lobkovsky L.I. (1988) *Geodinamika Zon Spredinga, Subduktsii i Dvuhyarusnaya Tektonika Plit.* (*Geodynamics of the Spreading and Subduction Zones and Two-Stage Plate Tectonics.*), Nauka, Moscow, 255 pp. (in Russian)

Logachev, N.A. (1977) *Vulkanogennye i Osadochnye Formatsii Riftovyh Zon Vostochnoy Afriki.* (*Volcanogenic and Sedimentary Formations of East African Rift Zones.*), Nauka, Moscow, 183 pp. (in Russian)

Logachev, N.A. and Zorin, Yu.A. (1992) *Tectonophysics*, **208**(1/3), 273–286

Longhi, J.N., Wooden, J.L. and Coopinger, K.D. (1983) *J. Geophys. Res., Suppl.*, **88**, 1,353–1,369

Lowe, D.R. (1994) Archaean Greenstone-Related Sedimentary Rocks. In: K.C. Condie (Ed.) *Archaean Crustal Evolution.* Elsevier, Amsterdam, pp. 121–169

Lucas, S.B., St-Onge, M.R., Parrish, R.R. and Dunphy, J.M. (1992) *Geology*, **20**(2), 113–116

Luchitsky, I.V. (Ed.) (1975) *Granitoidnye i Schelochnye Formatsii v Strukturah Zapadnoi i Severnoi Mongolii.* (*Granitoid and Alkaline Formations in the Structures of Western and Northern Mongolia.*), Nauka, Moscow, 288 pp. (in Russian)

Luchitsky, I.V. (Ed.) (1983) *Kontinentalny Vulkanizm Mongolii.* (*Continental Volcanism of Mongolia.*), Nauka, Moscow, 205 pp. (in Russian)

Lugovic, B., Alther, R., Raczek, I. *et al.* (1991) *Contrib. Mineral. Petrol.*, **106**(2), 201–216

Luth, W.C. (1969) *Am. J. Sci.*, A267, (Schairer Volume), 325–341

Magnitsky, V.A. (1965) *Vnutrenneye Stroeniye i Fizika Zemli.* (*Internal Structure and Physics of the Earth.*), Nedra, Moscow, 379 pp. (in Russian)

Mahoney, J.J., Storey, M., Duncan, R.A. *et al.* (1993) Geochemistry and Geochronology of Leg 130 Basement Lavas: Nature and Origin of the Ontong Java Plateau. In: W.H. Berger, L.W. Kroenke, L.A. Mayer *et al.*, *Proc. ODP, Sci. Results, 130: College Station, TX (Ocean Drilling Program)*, pp. 3–22

Makhotkin, I.L., Sublukov, S.M., Zhuravlev, D.Z. *et al.* (1995) *6th Intern. Kimberlite Conference*. Extended Abstracts. Novosibirsk, Inst. of Geology, Geophys. & Mineralogy Siberian Branch of RAS, pp. 342–344

Makhotkin, I.L. and Zherdev, Yu.P. (1993) *Doklady RAN*, **329**(4), 484–489 (in Russian)

Malov, N.D. (1974) *Sovetskaya Geologiya*, **7**, 1,032–1,039 (in Russian)

Malov, N.D. and Sharkov, E.V. (1978) *Geokhimiya*, **7**, 1,032–1,039 (in Russian)

Mamyrin, B.A. and Tostikhin, I.N. (1981) *Izotopny Sostav Geliya v Prirode*. (*Helium Isotopic Composition in Nature.*), Energoizdat, Moscow, 222 pp. (in Russian)

Mamyrin, B.A., Tolstikhin, I.N., Anufriev, V.S. and Kamensky, I.L. (1968) *Doklady AN SSSR*, **184**, 1,179–1,199 (in Russian)

Manning, D.A.C. (1981) *Contrib. Mineral. Petrol.*, **76**(2), 206–215

March, J.S., Bowen, M.P., Rogers, N.W. and Bowen, T.B. (1989) *Precambrian Res.*, **44**(1), 39–65

Marchenkov, K.I. and Zharkov, V.N. (1989) O Relyefe Granitsy Kora-Mantiya Venery. (On the Relief of the Crust–Mantle Interface on Venus.). In: *Astrochemistry and Comparative Planetology*, Nauka, Moscow, pp. 28–33 (in Russian)

Marchenkov, K.I., Zharkov, V.N. and Nikishin, A. (1992) Investigation of the stresses in the lithosphere of Mars: new tectonic interpretation. *29th IGC. Abstracts. V. 3.* Kyoto, pp. 647–648

Marinov, N.A. (Ed.) (1973) *Geologiya Mongolskoi Narodnoy Respubliki. Magmatizm, Metamorfizm, Tektonika.* (*Geology of the Mongolian People's Republic. Magmatism, Metamorphism and Tectonics.*), Nedra, Moscow, 752 pp. (in Russian)

Marker, M. (1985) Early Proterozoic (c. 2000–1900 Ma) Crustal Structure of the Northeastern Baltic Shield: Tectonic Division and Tectogenesis. *Nor. Geol. Unders. Bull.*, **403**, 55–74

Martin, H. (1994) The Archaen grey gneisses and the genesis of continental crust. In: K.C. Kondie (Ed.) *Archaean crustal evolution*. Elsevier, Amsterdam, 205–259

Martin, H., Auvray, B., Blais, S. *et al.* (1984) *Bull. Geol. Soc. Finland*, **56**(1/2), 135–160

Mason, B. (1958) *Principles of Geochemistry*. Wiley, New York; Chapman and Hall, London, 310 pp.

Mass, R., Kinny, P.D., Froude, D.O., Williams, I.C. and Compston, W. (1990) Geochemical Characteristics of 4100–4200 Ma Crust: the REE Record in the Detrital Zircons from Narryer Gneiss Complex, Western Australia. In: *Abstr. ICOG-7, Canberra, Australia*, p. 62

Master, S. (1992) Early Proterozoic Assemble of "Ubenda" in Proto-West Gondwana. (Equatorial and Southern Africa and Adjacent Parts of South America.). In: *29th IGC, Kyoto, Japan, Abstr.*, Vol. 2, p. 272

Mazarovich, A.O., Frikh-Khar, D.I., Kogarko, L.N., Koporulin, V.I., Rikhter, A.V., Akhmetiev, M.A. and Zolotarev, B.P. (1990) *Tectonism and Magmatism of the Cape Verde Islands*. Nauka, Moscow, 246 pp. (in Russian)

McCulloch, M.T. and Wasserburg, G.J. (1978) *Science*, **200**(4345), 1,003–1,011

McDonough, W.F. (1992) Composition of the Primitive Mantle and Other Mantle Reservoirs. *29th IGC, Kyoto, Japan, Abstracts*, Vol. 1, p. 175

McDonough, W.F. and Frey, F.A. (1989) *Rev. Mineral.*, **21**, 99–145

McKenzie, D.P. (1984) *Earth & Planet. Sci. Lett.*, **74**, 81–91

McKenzie, D.P. and Bickle, M.J. (1988) *J. Petrol.*, **29**(3), 625–679

McLelland, J.M. and Chiarenselli, J. (1990) *J. Geol.*, **98**(1), 19–41

Mehnert, K.R. (1968) *Migmatites and the Origin of Granitic Rocks*, Elsevier, Amsterdam, 393 pp.

Melyahovetsky, A.A., Ashepkov, I.V. and Dobretsov, N.L. (1986) *Doklady AN SSSR*, **286**(5), 1,215–1,219 (in Russian)

Menzies, M.A. and Hawkesworth, C.J. (Eds.) (1987) Mantle metasomatism. *Acad. Press*, London

Merilainen, K. (1976) *Bull. Geol. Surv. Finland*, **281**, 129 pp.

Milanovsky, E.E. and Koronovsky, N.V. (1974) *Orogennyi Vulkamizm i Tektonika Alpiyskogo Poyasa Evrazii. (Orogenic Volcanism and Tectonics of the Alpine Belt of Eurasia.)*, Nedra, Moscow, 279 pp. (in Russian)

Miller Yu.V. and Milkevich R.I. (1995) *Geotectonica*, **6**, 80–92 (in Russian).

Miller, J.D. (Ed.) (1995) Field Trip Guide Book for the Geology and Ore Deposits of the Midcontinent Rift in the Lake Superior Region: Minnesota Geological Survey, Guidebook Series, 20

Mints, M.V., Glasnev, V.N., Konilov, A.I. *et al.* (1996) *Early Precambrian of the north-east of the Baltic Shield: Palaeogeodynamics, Structure and Evolution of Continental Crust.* Moscow, Scientific World, 278 pp. (in Russian, with English abstract)

Mitchel, A.H.G. and Garson, M.S. (1981) *Mineral Deposits and Global Tectonic Settings.* London

Mitrofanov, F.P. (Ed.) (1985) *Migmatizatsiya i Granitoobrazovanie v Razlichnyh Termodinamicheskih Rezhimah. (Migmatization and Granite Formation in Various Thermodynamic Regimes.)*, Nauka, Leningrad, 310 pp. (in Russian)

Mitrofanov, F.P. (Ed.) (1989) *Roi Maficheskih Daek kak Indikatory Endogennyh Rezhimov. (Mafic Dyke Swarms as Indicators of Endogenic Regimes (Kola Peninsula.)*, Kola Sci. Centre RAS Publ., Apatity, 119 pp. (in Russian)

Mitrofanov F.P. (Ed.) (1995) *Geology of the Kola Peninsula (Baltic Shield).* Kola Scientific Centre RAS Publ., Apatity, 144 pp.

Mitrofanov, F.P., Balagansky, V.V., Balashov, Yu.A. *et al.* (1993) *Doklady RAN*, **331**(1), 95–98 (in Russian)

Mitrofanov, F.P. and Torokhov, M. (1994) Kola Belt of Layered Intrusions. *Guide to Field Trip of 7th International Platinum Symposium.* Kola Sci. Centre RAS Publ., Apatity, 110 pp.

Mitrofanov, F.P. and Smolkin, V.F. (Eds.) (1995) *Magmatism, Sedimentogenez i Geodynamica Pechengskoi Paleoriftogennoi Srtuktry. (Magmatism, Sedimentogenesis and Geodynamics of the Pechenga Palaeorift.)*, Apatity, Kola Sci. Centre RAS, 254 pp. (in Russian)

Miyashiro, H. (1974) *Am. J. Sci*, **274**, 321–355

Miyashiro, H., Aki, K. and Sengor, A.M.C. (1982) *Orogeny*, Wiley, Chichester

Molnar, P. and Atwater, T. (1978) *Earth & Planet. Sci. Lett.*, **41**(3), 330–340

Molnar P. and England, P. (1995) *Earth & Planet. Sci. Lett.*, **131**(1/2), 57–70

Molnar, P. and Tapponier, P. (1975) *Science*, **189**, 419–426

Morgan, W.J. (1972) *Bull. Assoc. Petrol. Geol.*, **56**(2), 203–217

Morris, J.D. and Hart, S.R. (1983) *Geochim. et Cosmochim. Acta*, **47**(11), 2015–2030

Morrison, J. and Valley, J.M. (1988) *Contrib. Mineral. & Petrol.*, **98**(1), 97–108

Mossakovsky, A.A. (1975) *Orogennye Struktury i Vulkanizm Paleozoid Evrazii i ih Mesto v Formirovanii Kontinentalnoi Zemnoi Kory*. (*Orogenic Structures and Volcanism of the Palaeozoic of Eurasia and their Role in Continental Crust Formation.*), Nauka, Moscow, 318 pp. (in Russian)

Mueller, P.A., Wooden, J.L. and Nutman, A.P. (1992) *Geology*, **20**(4), 327–330

Murton, B.J., Pette, D.W., Arculus, R.J. *et al.*, Trace Elements Geochemistry of Volcanic Rocks from Site 786: the Izu-Bonin Forearc. In: P. Fryer, J.A. Pearce and L.B. Stokking *et al.*, *Proc. ODP, Sci. Results*, 125: college Station, TX (Ocean Drilling Project), pp. 211–236

Myers, J.S. (1988) *Precambrian Res.*, **38**(4), 309–323

Mysen, B.O. and Kushiro, I. (1977) *Am. Mineral.*, **62**(9/10), p. 843

Nasegawa, A., Zhao, D., Hori, S., Yamamoto, A. and Horiuchi, S. (1991) *Nature*, **352**, 683–689

Navon, O. and Stolper, E. (1987) *J. Geol.*, **95**(3), 285–307

Neal., C.R., Taylor, L.A., Davidson, J.P. *et al.* (1990) *Earth & Planet. Sci. Lett*, **99**(4), 362–379

Nesbitt, R.W., Jahn, B.M. and Purvis, A.C. (1982) *J. Volcanol. & Geotherm. Res.*, **14**, 31–45

Neuman, E.R. and Ramberg, I.B. (Eds.) (1978) *Petrology and Geochemistry of Continental Rifts. Tectonics and Geophysics of Continental Rifts*. NATO Advanced Study Institute Series. Reidel, Dordrecht, Holland

Neymark, L.A., Amelin, J.V. and Larin, A.M. (1994) *Mineral. Petrol.*, **50**, 173–193

Neymark, L.A., Larin, A.M., Ovchinikova, G.V. and Yakovleva, S.Z. (1992) *Doklady RAN*, **323** (3), 514–518 (in Russian)

Nicolas, A. (1989) *Structures of Ophiolites and Dynamics of Oceanic Lithosphere*. Kluwer Academic Press, Boston, Massachusetts, 367 pp.

Nicolas, A., Freydier, Cl., Godard, M. and Vauchez, A. (1993) *Geology*, **21**, 53–56

Nikolayev, V.G. (1986) *Pannonskyi Bassein*. (*Pannonsky Basin.*), Nauka, Moscow, 120 pp. (in Russian)

Nisbett, E.G. and Walker, D. (1982) *Earth & Planet. Sci. Lett.*, **60**, 105–113

Nixon, P.H. (Ed.) (1987) *Mantle Xenoliths*. Wiley, Chichester, 844 pp.

Noble, D.C., Vogel, T.A., Peterson, P.S. *et al.* (1984) *Geology*, **12**(1), 35–39

Nohda, S. and Wasserburg, G.J. (1986) *Earth & Planet. Sci. Lett.*, **78**(2/3), 157–167

Nutman, A.P., Friend, C.R.L., Kinny, P.D. and McGregor, V.R. (1993) *Geology*, **21**, 415–418

Ohtani, E. (1984) *Earth Planet Sci. Lett.*, **67**, 261–272

Ohtani, E. (1990) *Precambrian Res.*, **48**, 195–202

Olsen, P., Silver, P.G. and Carlson, R.W. (1990) *Nature*, **344**, 209–214

O'Nions, R.K., Evensen, N.M. and Hamilton, P.J. (1979) *Geophys. Res.*, **84**(b11), 6,091–6,101

O'Reily, S.Y. and Griffin, W.L. (1985) *Geochim. et Cosmochim. Acta*, **52**(2), 433–447

Orlova, M.P., Zhidov, A.Ya., Orlov, D.M. and Zotova, I.F. (1993) *Geokhim. Int.*, **8**, 1,161–1,182

Orovetsky, Yu.P. (1990) *Mantyinyi Diapirizm*. (*Mantle Diapirism.*), Naukova Dumka, Kiev, 170 pp. (in Russian)

Pactung, D.A. (1987) *Contrib. Mineral. Petrol.*, **97**(3), 289–303

Page, R.W. and McCulloch, M.T. (1984) *Izotopnye Dannye ob Osnovnyh Sobytiyha v Dokembrii Avstralii. (Isotopic Data About Major Events During the Precambrian of Australia.*), 27th IGC, Geology of Precambrian, Nauka, Moscow, Vol. 5, pp. 107–116 (in Russian)

Pallister, J.S., Hobbit, R.P., Crandel, D.R. and Mullineaux, D.R. (1992) *Bull. Volcanol.*, **54**(2), 126–146

Pearce, J.A. (1976) *J. Petrology*, **17**(1), 15–43

Pearce, J.A., Bender, J.F., De Long, S.E. *et al.* (1990) *J. Volcanol. & Geotherm. Res.*, **44**(1–2), 189–229

Pearce, J.A., Harris, N.B.W. and Tindle, A.G. (1984) *J. Petrology*, **25**, 956–983

Pearson, D.G., Carlson, R.W. *et al.* (1995) *Earth & Planet. Sci. Lett.*, **134**, 34–357

Peive, A.V. *et al.* (Ed.) (1978) *Tektonika Evropy i Smezhnyh Oblastey. Varistsidy, Epipaleozoiskie Platformy, Alpidy. (Tectonics of Europe and Adjacent Areas. Variscides, Epipalaeozoic Platforms and Alps.*), Nauka, Moscow, 558 pp. (in Russian)

Peive, V.A., Zonenshain, L.P. and Knipper, A.L. (1980) *Tektonika Severnoi Evrazii. (Tectonics of Northern Eurasia.*), Nauka, Moscow, 221 pp. (in Russian)

Percival, J.A. (1994) Archaean high-grade metamorphism. In: K.C. Condie (Ed.) *Archaean Crustal Evolution*, Elsevier, Amsterdam, 357–410

Phillips, R.J., Grimm, R.E. and Malin, M.C. (1991) *Science*, **252**(5006), 651–658

Pidgeon, R.T., Kober, B. and Lippolt, H.J. (1988) *Chem. Geol.*, **70**, p. 145

Pidgeon, R.T., Wilde, S.A. *et al.* (1990) *Precambrian Res.*, **48**, 309–325

Piper, J.D.A. *et al.* (1983) *Geophys. J. Astron. Soc.*, **74**(2)

Pitcher, W.S. (1974) *Pacific Geol.*, **8**, 51–62

Pitcher, W.S. (1979) *J. Geol. Soc.*, **136**(6), 627–662

Pokhilenko, N.P. (1989) *Mantiynye Paragenezisy v Kimberlitah, in Proishozhdeniye i Poiskovoye Znacheniye. (Mantle Paragenesis in Kimberlites, their Origin and Significance in Exploration.*). D.Sc. Thesis, IGG SO Acad. of Sciences of USSR, Novosibirsk, 40 pp. (in Russian)

Polyak, B.G. (1988) *Teplomassopotok iz Mantii v Glavnyh Strukturah Zemnoi Kory. (Heat-Mass Flow in the Major of the Earth's Crust Structures.*), Nauka, Moscow, 192 pp. (in Russian)

Ponikarov, V.P. (Ed.) (1967) *Siriya. (Syria.*), Nedra, Moscow, 245 pp. (in Russian)

Powell, C.M.A., Roots, S.R. and Veevers, J.J. (1988) *Tectonophysics*, **155**, 261–283

Presnall, D.S., Dixon, S.A., Dixon, J.R. *et al.* (1978) *Contrib. Mineral. Petrol.*, **66**, 203–220

Presvik, T. and Goles, G.G. (1985) *Earth Planet. Sci. Lett.*, **72**, 65–73

Prinzhofer, A. and Allegre, C.J. (1985) *Earth & Planet. Sci. Lett.*, **74**(2/3), 251–265

Prijatkina, L.A. and Sharkov, E.V. (1979) *Geologiya Laplandskogo Glubinnogo Nadviga. (Geology of the Lapland Deep-Seated Thrust.*), Nauka, Leningrad, 128 pp. (in Russian)

Pukhtel, I.S., Bogatikov, O.A. and Simon, A.K. (1993a) *Petrology*, **1**(5), 451–473

Pukhtel, I.S. and Zhuravlev, D.Z. (1992) *Petrology*, **1**(3), 263–299

Pukhtel, I.S., Zhuravlev, D.Z., Ashihmina, N.A. *et al.* (1992) *Doklady RAN*, **326**(4), 706–711 (in Russian)

Pukhtel, I.S., Zhuravlev, D.Z., Kulikov, V.S. and Kulikova, V.V. (1991) *Geokhimiya*, **5**, 625–634 (in Russian)

Pukhtel, I.S., Zhuravlev, D.Z. and Kulikova, V.V. (1993a) *Internat. Geol. Rev.*, **35**, 825–839

Pukhtel, I.S., Zhuravlev, D.Z., Samsonov, A.V. and Arndt, N.T. (1993b) *Precambrian Res.*, **62**, 399–417

Pushcharovsky, Yu.M. and Melanholina, E.N. (1989) *Tektonicheskoe Razvitie Zemli. Tikhyi Okean i ego Obramlenie. (Tectonic Development of the Earth. The Pacific Ocean and its Surroundings.)*, Nauka, Moscow, 264 pp. (in Russian)

Pushkarev, Yu.D. (1990) *Megatsycly v Evolutsii Sistemy Kora-Mantiya. (Megacycles in the Evolution of the Crust–Mantle System.)*, Nauka, Leningrad, 217 pp. (in Russian)

Quick, J.E. and Delinger, R.P. (1993) *J. Geophys. Res.*, **98**(B8), 14,015–14,027

Rampino, M.R. and Self, S. (1992) *Nature*, **359**(6,390), 50–53

Ramo, O.T. (1991) *Geol. Surv. Finland Bull.*, **355**, 161 pp.

Ramo, O.T. and Haapala, I. (1996) Rapakivi granite magmatism: a global review with emphasis on petrogenesis. In: D. Demaiffe (ed.) *Petrology and Geochemistry of Magmatic Suites of Rocks in the Continental and Oceanic Crust.* ULB, Royal Museum for Central Africa, Tervuren, 177–200

Recou, L.E., Dercourt, J., Geyssant, J. *et al.* (1986) *Tectonophysics*, **123**(1–4), 83–122

Reednen D.D., van, Roering, C., Ashwal, L.D. and Wit, M.J. (1992) *Precambrian Res.*, Spec. Iss., **55**(1–4), 587 pp.

Rehault, J.P., Mooussat, E. and Fabri, A. (1987) *Marine Geol.*, **74**(1/2), 123–150

Reisberg, L., Zindler, A. and Jagoutz, E. (1989) *Earth & Planet. Sci. Lett.*, **96**(1/2), 161–180

Renne, P.R. and Basu, A.R. (1991) *Science*, **253**(9184), 176–179

Richardson, S.H., Erlank, A.J. and Hart, S.R. (1985) *Earth & Planet. Sci. Lett.*, **75**(2/3), 116–128

Richardson, S.H., Gurney, J.J., Erlank, A.J. and Harris, J.W. (1984) *Nature*, **310**, 198–202

Ridgen, S.M., Ahrens, T.J. and Stolper, E.M. (1984) *Science*, **226**, 1,071–1,074

Ringwood, A.E. (1975) *Composition and Petrology of the Earth's Mantle.* McGraw-Hill, New York, 618 pp.

Ringwood, A.E. (1978) *Origin of the Earth and the Moon.* Springer Verlag, New York

Ringwood, A.E. and Irifune, T. (1988) *Nature*, **331**, 131–134

Ringwood, A.E. and Kesson, S.E. (1991) *Earth planet. Sci. Lett.*, **114**

Robinson, M.S. and Lucey, P.G. (1997) *Science*, **275**(5297), 197–200

Roden, M.F., Hart, S.R., Frey, F.A. and Melson, W.G. (1984) *Contrib. Mineral. & Petrol.*, **85**(4), 376–390

Roden, M.F., Irving, A.J. and Murthy, V.R. (1988) *Geochim. er Cosmochim. Acta*, **52**(2), 461–473

Ronov, A.B. (1980) *Osadochnaya Obolochka Zemli. (Sedimentary Cover of the Earth.)*, Nauka, Moscow (in Russian)

Ronov, A.B., Yaroshevsky. A.A. and Migdisov, A.A. (1990) *Himicheskoye Stroeniye Zemnoi Kory i Geokhimicheskyi Balans Glavnyh Elementov. (Chemical Structure of the Earth's Crust and Geochemical Balance of the Major Elements.)*, Nauka, Moscow, 182 pp. (in Russian), with English abstract)

Rosen, O.M., Andreev, V.P., Belov, A.N. *et al.* (1988) *The Archaean of the Anabar Shield and the Problems of the Early Evolution of Earth*, Nauka, Moscow, 253 pp. (in Russian)

Royden, L.N. (1989) *Tectonics*, **8**(1), 51–61

Rudnick, R.L. and Fountain, D.M. (1995) *Rew. of Geophys.*, **33**(3), 267–309

Rudnick, R.L. and Taylor S.R. (1987) *J. Geophys. Res.*, **92**(B13), 13,981–14,005

Rundquist D.V. and Mitrofanov F.P. (Eds.) (1993) *Precambrian Geology of the USSR.* Elsevier, Amsterdam, 440 pp.

Ruzhentsev, S.V., Badarch, G., Voznesenskaya, T.A. and Markova, N.G. (1990) Tektonika Yuzhnoi Mongolii. (Tectonics of Southern Mongolia.), In: N.S. Zaitsev and V.I. Kovalenko (Eds.) *Evolution of Geological Processes and Metallogeny of Mongolia.*), Nauka, Moscow, pp. 111–117 (in Russian)

Ryabchikov, I.D. (1987) Protsessy Mantyinogo Magmoobrazovaniya. (Processes of Mantle Magma Generation.), In: V.I. Kovalenko (Ed.) *Igneous Rocks. Evolution of Igneous Rocks*, Nauka, Moscow, 349–371 (in Russian)

Ryabchikov, I.D. (1988) *Geokhimicheskaya Evolutsiya Mantii Zemli.* (*Geochemical Evolution of the Earth's Mantle.*), Nauka, Moscow, 36 pp. (in Russian)

Ryabchikov, I.D. (1995) Different sources of kimberlites and carbonatite parent mag-mas: evidence from high pressure experiments and trace element geochemistry. In: *6th Intern. Kimberlite Conference.* Novosibirsk, Siberian Branch of RAS, 706–707

Ryabchikov, I.D. and Kovalenko, V.I. (1987) *Anateksis Kontinentalnoi Kory.* (*Anatexis of the Continental Crust.*), In: V.I. Kovalenko (Ed.) *Igneous Rocks. Evolution of Igneous Rocks*, Nauka, Moscow, pp. 372–389 (in Russian)

Ryabchikov, I.D., Solovova, I.P., Solovova, I.P., Girnis, A.V. *et al.* (1991) Usloviya Generatsii i i Kristallizatsii Vysokokalievyh Magm. (Generation and Crystallization Conditions of High-Potassium Magmas.), In: O.A. Bogatikov and V.A. Kononova (Eds.) *Lamproites*, Nauka, Moscow, 218–276 (in Russian)

Rybakov, S.I. (1987) *Kolchedannoye Rudoobrazovanie v Rannem Dokembrii Baltiskogo Shchita.* (*Pyrite Ore Formation in the Early Precambrian of the Baltic Shield.*), Nauka, Leningrad, 270 pp. (in Russian)

Safronov, V.S. (1969) *Evolutsiya Doplanetnogo Oblaka i Obrazovanye Zemli i Planet.* (*Evolution of the Pre-Planetary Nebula and the Formation of the Earth and other Planets.*), Nauka, Moscow (in Russian), 256 pp.

Safronov, V.S. (1987) *Proichozhdenie Zemli.* (*Origin of the Earth.*), Znaniye, Moscow, 48 pp. (in Russian)

Safronov, V.S. and Vityazev, A.V. (1983) *Itogi Nauki i Tehniki, Ser. Astronomiya*, **24**, VINITI, Moscow (in Russian)

Sager, W.W. and Han, H.-C. (1993) *Nature*, **364**(6438), 610–612

Samsonov, A.V., Pukhtel, I.S., Zhuravlev, D.Z. and Chernyshev, I.V. (1993) *Petrology*, **1**(1), 23–42

Samsonov, A.V., Puchtel, I.S., Shchipansky, A.A. *et al.* (1997) in: *EUG9, Strausburg-France, 23–27, March 1997. Abstract Suppl. 1, Terra Nova* 9, 363

Scherbak, N.P., Artemenko, G.V., Bartinsky, E.N. *et al.* (1989) *Geochronological Scale of Precambrian of the Ukrainian Shield*, Naukova Dumka, Kiev, 144 pp. (in Russian)

Schiffries, C.M. and Rye, D.N. (1989) *Am. J. Sci.*, **289**(7), 841–873

Semikhatov, M.A., Shurkin, K.A., Aksenov, E.M. *et al.* (1991) *Izvestiya AN SSSR, Ser. Geol.*, **4**, 3–13 (in Russian)

Sergeev, S.A., Bibikova, E.V., Levchenkov, O.A. *et al.* (1990) *Geokhimiya*, **1**, 73–83 (in Russian)

Sergeev, S.A., Bibikova, E.V., Levchenkov, O.A. *et al.* (1991) *Geochem. Internat.*, **8**, 65–74

Shanin, L.L., Arakelyants, M.M., Bogatikov, O.A. *et al.* (1981) *Geokhimiya*, **7**, 970–980 (in Russian)

Sharapov, V.N., Simbireva, I.G. and Bondarenko, P.M. (1984) *Struktura i Geodinamika Seismofokalnyh Zon Kurilo-Kamchatskogo Regiona.* (*Structure and Geodynamics of the Seismofocal Zones of the Kurile-Kamchatka Region.*), Nauka, Novosibirsk, 199 pp. (in Russian)

Sharaskin, A.Ya. (1992) *Tektonika i Magmatizm Okrainnyh Morei v Svyazi s Problemoi Evolyutsii Kory i Mantii.* (*Tectonics and Magmatism of Marginal Seas in Relation to the Problems of Crust and Mantle Evolution.*), Nauka, Moscow, 163 pp. (in Russian)

Sharkov, E.V. (1980) *Russian Geology and Geophysics*, **5**, 80–88

Sharkov, E.V. (1983) *Geokhimiya*, **2**, 283–293 (in Russian)

Sharkov, E.V. (1984) *Geotectonics*, **18**(4), 123–132

Sharkov, E.V. (Ed.) (1985) *Magmaticheskiye Gornye Porody. T. 3. Osnovnye Porody.* (*Igneous Rocks. Vol. 3. Basic Rocks.*), Nauka, Moscow, 487 pp. (in Russian)

Sharkov, E.V. (1992) *Doklady RAN*, **325**(2), 349–353 (in Russian)

Sharkov, E.V. and Bogatikov, O.A. (1987) *Internat. Geol. Rev.*, **29**(10), 1,135–1,149

Sharkov, E.V., Bogatikov, O.A., Krassivskaya, I.S. (1998) *Petrology*, in press.

Sharkov, E.V. and Bindeman, I.N. (1990) *Volcanology & Seismology*, **12**(4), 423–430

Sharkov, E.V., Laz'ko, E.E., Fedosova, S.P. and Hanna, S. (1989) *Geochem. Internat.*, **27**(6), 82–94

Sharkov, E.V. and Ledneva, G.V. (1993) *Zapiski Vserossiyskogo Miner. Obshchestva*, **4**, 35–55 (in Russian)

Sharkov, E.V. Lyakhovich, Vl.V. and Ledneva, G.N. (1994) *Petrology*, **2**(5), 454–475

Sharkov, E.V. and Pukhtel, I.S. (1986) *Izvestiya AN SSSR, Ser. Geol.*, **8**, 32–45 (in Russian)

Sharkov, E.V. and Smolkin, V.F. (1997) *Precambrian Res.*, **82**, 133–151

Sharkov, E.V., Smolkin, V.F. *et al.* (1989) *Doklady AN SSSR*, **309**(6), 137–140

Sharkov, E.V., Smolkin, V.F. and Krassivskaya, I.S. (1997) *Petrology*, **5**(5), 448–465

Sharkov, E.V. and Svalova, V.B. (1991) *Internat. Geol. Rev.*, **33**(12), 1,184–1,198

Sharkov, E.V. and Svalova, V.B. (1993) "Fore-Arc Rifting" in Alpine Belt–the Result of Dynamical Interaction of Lithosphere and Asthenosphere Under Plate Collision. In: *TERRA Abstracts, EUG VII, Strasbourg, France*, pp. 268–269

Sharkov, E.V. and Tsvetkov, A.A. (1990) *Volcanology and Seismology*, **9**(1), 45–66

Sharma, M., Basu, A.R. and Nesterenko, G.V. (1992) *Earth Planet. Sci. Lett.*, **113**, 365–381

Sharma, M., Wasserburg, G.J., Papanastassiou, D.A. *et al.* (1995) *Earth & Planet. Sci. Lett.* **39**, 101–114

Sharpe, M.R. and Hulbert, L.J. (1985) *Econ. Geol.*, **80**, 85–110

Shatsky, V.S., Jagouts E., Kozmenko, O.A. *et al.* (1993) *Geologiya i Geofizika*, **12**, 47–56 (in Russian)

Shaw, H.R. *et al.* (1968) *Am. J. Sci.*, **263**(2), 120–152

Shemyakin, V.M. (1988) *Petrologiya Charnokitoidov Rannego Dokembriya.* (*Petrology of Charnockitoids of the Early Precambrian.*), Nauka, Leningrad, 200 pp. (in Russian)

Shevrais, J.W., Taylor L.A., Lugmair, G.W. *et al.* (1988) *Bull. Geol. Soc. Am.*, **100**(3), 411–423

Schiano, P., Clocchiatti, R., Shimizu, N., Mauri, R.C., Jochum, K.P. and Hofmann, A.W. (1995) *Nature*, **377**(6550), 595–600

Shimizu, M. (1979) *Precambrian. Res.*, 3/4, 311–324

Shimizu, M. and Richardson, S.H. (1987) *Geochim. et Cosmochim Acta,* **51**(3), 755–758

Shirey, S.B. (1990) Constraints on Continental Evolution from Pb and Sm Isotopic Studies of Archaean Crust. In: *Abstr. ICOG-7*, Canberra, Australia, p. 92

Shpunt, B.R. (1987) *Pozdnedokembriuskiy Riftogenez Sibirskoi Platformy.* (Late Precambrian Riftogenesis of the Siberian platform.), Yakutsk, 137 pp. (in Russian)

Shurkin, K.A., Gorlov, N.V., Sal'ye, M.E. *et al.* (1962) *Belomorskiy Kompleks Severnoi Karelii i Yugo-Zapada Kolskogo Poluostrova (Geologiya i Pegmatitonosnosto. (Belomorsky Complex of Northern Karelia and Southwest of the Kola Peninsula (Geology and Pegmatite Occurrence.),* Academy of Sciences, Moscow, Leningrad, 302 pp. (in Russian)

Silver, P.G., Carlson, R.W. and Olson, P. (1988) Deep Slab Geochemical Heterogeneity, and the Large-Scale Structure of Mantle Convection: Investigation and Enduring Paradox. In: G.W. Wetherill and F.G. Stehly (Eds.) *Annual Review of Earth & Planetary Sciences*, **16**, 477–542

Sinitsyn, A.V., Dauyev, Yu.M. and Grib, V.P. (1992) *Geologiya i Geofizika*, **10**, 74–83 (in Russian)

Sivel, W.J. and McCulloch, M.T. (1991) *Nature*, **354**, 384–387

Slabunov, A.I. (1992) Guidebook of the Geological Extrusion on the Archaean of Northern Karelia. Karelian Research Centre, Petrozavodsk, 28–35 (in Russian)

Slezin, Yu.B. (1991) *Vulkanologiya i Seismologiya*, **1**, 21–34 (in Russian)

Smolkin, V.F. (1992) *Komatiitovyi i Pikritovyi Magmatizm Rannego Dokembriya Baltyiskogo Shchita. (Komatiitic and Picritic Magmatism of the Early Precambrian of the Baltic Shield.),* Nauka, St. Petersburg, 263 pp. (in Russian)

Smith, C.B. (1983) *Nature*, **304**, 51–54

Snyder, G.A., Lee, D.-C., Taylor, L.A., Halliday, A.N. and Jerde, E.A. (1994) *Geochim. Cosmochim. Acta*, **58**(21), 4,795–4,808

Snyder, G.A., Taylor, L.A. and Halliday, A.N. (1995) *Geochim. Cosmochim. Acta*, **59**(6), 1,185–1,203

Snyder, G.A., Neal, C.R., Taylor, L.A. and Halliday, A.N. (1996) *J. Geophys. Res.*, **100**(E5), 9365–9388

Sobolev A.V. and Chaussidon M. (1996) *Earth Planet. Sci. Lett.*, **137**(1–4), 45–55

Sobolev, A.V. and Danyushevsky, L.V. (1994) *J. Petrology*, **35**, 1183–1211

Sobolev, A.V. and Shimizu, N. (1993) *Nature*, **363**, 151–154

Sobolev, N.V. (1977) *Deep-seated Inclusions in Kimberlites and the Problem of the Composition of the Upper Mantle.* Am. Geophys. Union, Washington D.C., 279 pp.

Sobolev, N.V., Pohilenko, N.P., Grib, V.P. *et al.* (1992) *Geologiya i Geofizika*, **10**, 84–92 (in Russian)

Sobolev, V.S. (Ed.) (1975) *Glubinnye Ksenolity i Verhnyaya Mantiya Zemli.* (Deep-Seated Xenoliths and the Upper Mantle of the Earth.), Nauka, Novosibirsk, 272 pp. (in Russian)

Sochava, A.B. (1986) *Petrokhimiya Verhnego Arheya i Proterozoya Zapada Vitimo-Aldanskogo Shchita.* (*Petrochemistry of late Archaean and Proterozoic of West Vitimo-Aldan Shield.*), Nauka, Leningrad, 142 pp. (in Russian)

Solovova, I.P., Girnis, A.V., Naumov, V.B. *et al.* (1992) *Doklady RAN*, **324**(4), 861–865 (in Russian)

Sorokhtin, O.G. (Ed.) (1979) *Geofizika Okeanov. T. 2. Geodinamika.* (*Geophysics of the Oceans. Vol. 2. Geodynamics.*), Nauka, Moscow, 415 pp. (in Russian)

Sorokhtin, O.G. and Ushakov, S.A. (1991) *Globalnaya Evolutsiya Zemli.* (*Global Evolution of the Earth.*), Moscow University, 446 pp. (in Russian)

Spadea, P., Beccaluva, L., Conltorti, M. *et al.* (1992) *Ofioliti*, **17**(1), 79–93

Sparks, R.S.J. (1986) *Earth & Planet. Sci. Lett.*, **78**, 211–223

Spiridonov, V.G., Sukhanov, M.K., Karpenko, S.F. and Lyalikov, A.V. (1991) *Doklady AN SSSR*, **319**(5), 1,209–1,212 (in Russian)

Stosch, H.-G. and Lugmaier, G.W. (1986) *Earth Planet. Sci. Lett.*, **80**(3/4), p. 281

Stosch, H.-G., Lugmaier, G.W. and Kovalenko, V.I. (1986) *Geochim. et Cosmochim. Acta*, **50**(12), 2,601–2,614

Stosch, H.-G., and Zeck, H.A. (1980) Geochemistry and Mineralogy of Two Spinel Peridotite Suites from Dreiser Weier, West Germany. *Geochim. et Cosmochim. Acta*, **44**(3), 457–470

Suen, C.J. and Frey, F.A. *et al.* (1987) *Earth & Planet. Sci. Lett.*, **85**(1/3), 183–202

Sukhanov, M.K. (1989) *Geol. Surv. Finland, Spec. Pap.*, **8**, 125

Sukhanov, M.K., Lennikov, A.M. and Zhuravlev, D.Z. (1991) *Doklady AN SSSR*, **320**(1), 187–191 (in Russian)

Sun, S.-S., Nesbitt, R. and McCulloch, M.T. (1989) Geochemistry and Petrogenesis of Archaean and Early Proterozoic Silicoeus High-Magnesium Basalts. In: A.J. Crawford (Ed.), *Boninites*, Unwin Hyman, pp. 148–173

Suominen, V. (1991) *Bull. Geol. Surv. of Finland*, 94–100

Svalova, V.B. and Sharkov, E.V. (1991) *Tikhookaenskaya Geologiya*, **5**, 49–53 (in Russian)

Svalova, V.B. and Sharkov, E.V. (1992a) *Russian Geology and Geophysics*, **33**(5), 15–23

Svalova, V.B. and Sharkov, E.V. (1992b) *Ofioliti*, **17**(1), 165–171

Sviridenko, L.P., Svetov, A.P., Golubev, A.I. and Pavlov, G.M. (1984) *Doklady AN SSSR*, **276**(6), 1,449–1,452 (in Russian)

Sylvester, P.J. (1994) Archaean Granite Plutons. In: K.C. Condie (Ed.) *Archaean Crustal Development*, Elsevier, Amsterdam, pp. 261–314

Tait, S.R. (1985) *Geol. Mag.*, **122**(5), 485–490

Tamaki, K. (1988) *Bull. Geol. Surv. Japan*, **39**, 269–365

Tankard, A.J., Jackson, M.P.A., Eriksson, K.A. *et al.* (1982) *Crustal Evolution of Southern Africa: 3.8 Billion Years of Earth History.* Springer Verlag, Berlin, 523 pp.

Taylor, S.R. (1991) *Nature*, **350**(6317), 376–377

Taylor, S.R. and Bence, A.E. (1975) *Evolution of the Lunar Highland Crust.* Proc. Lunar. Sci. Conf., Houston (Texas), Vol. 1

Taylor, S.R. and McLennon, S.M. (1985) *The Continental Crust: its Composition and Evolution.* Blackwell Scientific Publications, London, 312 pp.

Taylor, L.A., Neal, C.R. *et al.* (1989) *J. Geol.*, **97**(5), 551–567 pp.

Tera, F., Brown, L., Morris, J. *et al.* (1986) *Geochim. et Cosmochim. Acta*, **50**, 535–550

Thorpe, R.S. (Ed.) (1982) *Andesites: Orogenic Andesites and Related Rocks*, Wiley, New York, 730 pp.

Thorpe, R.I., Hickman, A.H., Davis, D.W., Mortensen, J.K. and Tredall, A.F. (1992) *Precambrian Res.*, **56**, 169–189

Timmerman, M.J. and Balagansky, V.V. (1994) *Terra Nova*, 6, Abstract Supplement, **2**, 19–59

Toft, P.B., Hills, D.V. and Haggerty, S.E. (1989) *Tectonophysics*, **161**, 213–231

Tolstikhin, I.N., Dokuchaeva, V.S., Kamensky, I.L. and Amelin, Yu.A. (1992) *Geochim. et Cosmochim. Acta*, **56**, 987–999

Tsaryeva, G.M., Naumov, V.B. and Babansky, A.D. (1992) *Doklady RAN*, **322**(3), 579–583 (in Russian)

Tsvetkov, A.A. (1984) *Izvestiya AN SSSR, Ser. Geol.*, **3**, 24–41 (in Russian)

Tsvetkov, A.A. (1990) *Magmatizm i Geodinamika Komandorsko-Aleutskoi Ostrovnoi Dugi.* (*Magmatism and Geodynamics of the Comandor-Aleutian Island Arc.*), Nauka, Moscow, 325 pp. (in Russian)

Tsvetkov, A.A. and Avdeiko, G.P. (1982) *Doklady AN SSSR*, **267**(15), 985–990 (in Russian)

Tsvetkov, A.A., Volynets, O.N. and Bailey, G. (1993) *Petrology*, **1**(2), 103–127

Tugarinov, A.I. and Bibikova, E.V. (1980) *Geokhronologiya Baltyiskogo Shchita po Opredeleniyam Zirkonov.* (*Geochronology of the Baltic Shield by Zircon Determinations.*), Nauka, Moscow, 130 pp. (in Russian)

Ueda, S. and Kanamori, H. (1979) *J. Geophys. Res.*, **84**, 1,049–1,061

Upton, B.G.J. and Emeleus, C.H. (1987) *Geol. Soc. Spec. Publ.*, **30**, 445–471

Vasil'ev, Yu.R. (1988) *Ultrabazity Schelochno-Ultraosnovnyh Kompleksov.* (*Ultrabasites of Alkaline-Ultrabasic Complexes.*), In: E.E. Laz'ko and E.V. Sharkov (Eds.), Magmatic rocks. Ultrabasic Rocks, Nauka, Moscow, pp. 172–196 (in Russian)

Velikoslavinsky, D.A., Birkis, A.P., Bogatikov, O.A. *et al.* (1978) *Anortozit-Rapakivigranitnaya Formatsiya.* (*Anorthosite–Rapakivi Granite Formation.*), Nauka, Leningrad, 296 pp. (in Russian)

Vidal, P., Dupuy, C., Maury, R. and Richard, M. (1989) *Geology*, **17**, 1,115–1,118

Vinogradov, A.P. (1961) *Geokhimiya*, **1**, 3–29 (in Russian)

Volodichev, O.I. (1990) *Belomorskiy Kompleks Karelii: Geologiya i Petrologiya.* (*Belomorian complex of Karelia: geology and petrology.*). Nauka, Leningrad, 245 pp. (in Russian)

Vorontsov, A.A. (1994) *Geologiya i Geokhimiya Ranne-Srednedevonskih Magmaticheskih Assotsiatsyi so Shchelochnyi Porodami Severo-Zapadnoi Mongolii.* (*Geology and Geochemistry of Early-Middle Devonian Igneous Assemblages with Alkaline Rocks of North-West Mongolia.*). Ph. D. Thesis, Inst. of Geochemistry of SO RAN, Novosibirsk, 25 pp. (in Russian)

Vuollo, J., Nukanen, V.M., Liipo, J.P. and Piirainen, T.A. (1995) Palaeoproterozoic Mafic Dyke Swarms in the Eastern Fennoscandian Shield, Finland. In: Baer G. and Heimann A. (Eds.). Physics and Chemistry of Dykes. Balkema, Rotterdam, 179–192

Wager, L. and Brown, G. (1968) *Layered Igneous Rocks*, Oliver & Boyd, Edinburg and London, 588 pp.

Walker, D. and Heys, J.F. (1977) *Geology*, **5**(7), 425–428

Wasserburg, G.J. (1984) *Some Short-Lived Radioactive Nuclei at the Early Stages of the Solar System Development.* In: 27th IGC, Geochemistry and Astrochemistry, Reports, **11**, 26–31

Weaver, S.D., Storey B., Pankhurst, R.J., Mukasa, S.B., Divenere, V.S. and Breadshow, J.D. (1994) *Geology*, **22**, 811–814

Wedepohl, K.N., Meyer, K. and Muecke, G.K. (1983) Chemical Composition and Genetic Relations of Meta-Volcanic Rocks from the RhenoHercynian Belt of Northern Germany. In: H. Martin and F.W. Edler *Intercontinental Fold Belts*, Springer Verlag, Berlin, 231–256

Weiser, Y. (1976) Evolution of the ^{87}Sr/^{86}Sr Ratio in Sea Water Over the Geological History as an Indicator of Earth's Crust Evolution. In: B.F. Windley (Ed.) *The Early History of the Earth*, London

Whilshire, H.G. and Shervais, J.W. (1975) *Phys. Chem. Earth*, **9**, 257–272

White, R. and McKenzie, D. (1989) *J. Geoph. Res.*, **94**(B6), 7,685–7,729

Wilson, J.T. (1968) *Proc. Am. Phillos. Soc.*, **112**(2), 309–320

Wilson, A.H. (1982) *J. Petrol.*, **23**(2), 240–292

Wilson, M. (1989) *Igneous Petrogenesis, a Global Tectonic Approach*. Umwin Hyman, London, 466 pp.

Windley, B.F. (1992) *The Evolving Continents*. 2nd. Edn, 339 pp. London: John Wiley

Windley, B.F. (1991) *Tectonophysics*, **195**(1), 1–10

Wooden, J.L., Czamanske, G.K. and Zeintek, M.L. (1991) *Contrib. Mineral. & Petrol.*, **107**, 80–93

Wooley, A.R. (1987) Alkaline Rocks and Carbonatites of the World. Part 1: North and South America. British Museum (Natural History), London, 216 pp.

Woster, M.G., Huppert, H.E. and Sparks, R.S.J. (1990) *Earth & Planet. Sci. Lett.*, **101**(1), 78–89

Wyllie, P.J. (1988) *Rev. Geophys.*, **26**, 370–404

Wyllie, P.J., Huang, W.L., Stern, C.R. and Maaloe, S. (1976) *Can. J. Earth. Sci.*, **13**(8), 1007–1019

Yarmolyuk, V.V. (1983) *Pozdnepaleozoyskyi Vulkanizm Kontinentalnyh Riftogennyh Struktur Tsentralnoi Azii*. (*Late Palaeozoic Volcanism of Continental Rift Structures of Central Asia.*), Nauka, Moscow, 200 pp. (in Russian)

Yarmolyuk, V.V. and Kovalenko, V.I. (1987) *Magmaticheskie Gornye Porody. Kislye i Srednie Porody*. (*Magmatic Rocks. Vol. 4. Acid and Intermediate Rocks.*), Nauka, Moscow, 374 pp. (in Russian)

Yarmolyuk, V.V. and Kovalenko, V.I. (1991) *Riftogenny Magmatizm Aktivnyh Kontinentalnyh Okrain i Ego Rudonosnost*. (*Riftogenic Magmatism of Active Continental Margins and its Ore Deposits.*), Nauka, Moscow, 262 pp. (in Russian)

Yarmolyuk, V.V., Kovalenko, V.I. and Bogatikov, O.A. (1992) *Doklady Akademie Nauk SSSR*, **312**, 93–96

Yarmolyuk, V.V., Ivanov, V.G., Kovalenko, V.I. and Samoilov, V.S. (1995) *Geotektonikcs*, **28**(5), 391–407

Yarmolyuk, V.V., Kovalenko, V.I. and Ivanov, V.J. (1996) *Geotectonics*, **29**(5), 395–421

Yarmolyuk, V.V., Kovalenko, V.I. and Samoilov, V.S. (1989) Evolutsiya Riftogennogo magmatizma v Geodinamicheskom Tsykle. (Evolution of Riftogenic Magmatism in the Geodynamic Cycle.). In: O.A. Bogatikov (Ed.) *Petrology and Geodynamics of Riftogenic Magmatism*, Nauka, Moscow, 49–64 (in Russian)

Yarmolyuk, V.V., Kovalenko, V.I. and Samoilov, V.S. (1991) *Geotektonika*, **1**, 69–83 (in Russian)

Yaroshevsky, A.A. (1964) Printsip Zonnoi Plavki i Ego Primenenie pri Resh-
    enii Nekotoryh Geokhimicheskih Voprosov. (The zone Refinement Principle
    and its Application to the Solution of some Geochemical Problems.)
    In: A.P. Vinogradov (Ed.) *Chemistry of the Earth's Crust*, Vol. 2, Nauka, Moscow,
    55–62 (in Russian)
Yashina, R.M. (1982) *Schelochnoi Magmatizm Skladchato-Glybovyh Oblastei. (Alkaline
    Magmatism of Fold-Fault Areas.*), Nauka, Moscow, 273 pp. (in Russian)
Yeroshenko, V.A. and Sharkov, E.V. (1989) *Internat. Geol. Rev.*, **31**(10), 969–985
Yeroshenko, V.A. and Sharkov, E.V. (1993) *Geochem. Internat.*, **30**(10), 53–62
Zaitsev, N.S. and Tauson, L.V. (Eds.) (1971) *Redkometalnye Granitoidy Mongolii. (Rare-
    Metal Granitoids of Mongolia.*), Nauka, Moscow, 221 pp. (in Russian)
Zeil, W. and Pichler, H. (1967) *Geol. Rundsch.*, **57**(1), 48–81
Zharkov, V.N. and Trubitsyn, A.P. (1975) Lunnye Maskony. (Lunar Muskons.). In: *Cosmo-
    chemistry of the Moon and Planets*, Nauka, Moscow, 311–313 (in Russian)
Zhang, Y.-S. and Tanimoto, T. (1993) *J. Geophys. Res.*, **98**(B6), 9798–9823
Zhuravlev, A.Z., Laz'ko, E.E. and Ponomarenko, A.I. (1991) *Geokhimiya*, **7**, 982–995
    (in Russian)
Zhuravlev, D.Z., Tsvetkov, A.A., Zhuravlev, A.Z. et al. (1985) *Izotopniy Sostav Sr i Nd v
    Chetvertichnyh Vulkanitah Kurilshoi Ostrovnoi Dugi v Svyasi s Problemoi
    Genezisa Ostrovoduzhnyh Magm. (Isotope Composition of Sr and Nd in
    Quaternary Volcanics of Kurile Island Arc.*), Nauka, Moscow (in Russian)
Zimin, S.S. (1973) *Formatsii Nikelenosnyh Rogovoobmankovyh Bazitov Dalnego
    Vostoka. (Formations of Nickel-Bearing Hornblende Basic Rocks of the Far East.*),
    Nauka, Novosibirsk, 90 pp. (in Russian)
Zindler, A. and Jagoutz, E. (1988) *Geochim. et Cosmochim. Acta*, **52**(2), 319–333
Zinger, T.F., Gotze, J., Levchenkov, O.A. et al. (1996) Zircon in polydeformed and meta-
    morphosed Precambrian granitoids from the White Sea tectonic zone, Russia: mor-
    phology, cathodoluminiscence, and U-Pb chronology, *Intern. Geol. Rev.*, **38**, 57–73
Zonenshain, L.P. and Kuzmin, M.I. (1978) *Geotektonika*, **1**, 19–42
Zonenshain, L.P., Kuzmin, M.I. and Bocharova, N.Yu. (1990), *Tectonophysics*, **199**,
    165–192
Zonenshain L.P., Kuzmin M.I. and Natapov L.M. (1990) *Geology of the USSR: A plate
    tectonic synthesis.* Am. Geophys. Union, 21, Washington D.C., 242 pp.
Zonenshain, L.P. and Le Pichon, X. (1986) *Tectonophysics*, **123**(1–4), 181–212
Zonenshain, L.P. and Tomurtogoo, O. (1979) *Ofiolity i Osnovnye Zakonomernosti
    formirovaniya Zemnoi Kory Mongolii. (Ophiolites and Main Regulaties of
    Formation of the Earth's Crust of Mongolia.*), Nauka, Moscow, 67–76 (in Russian)
Zorin, J.A., Kozhevnikov, V.M., Novoselova, M.R. and Turutanov, E.K. (1989)
    Tectonophysics, **168**(4), 327–337

INDEX

Milton Keynes UK
Ingram Content Group UK Ltd.
UKHW020007071024
449327UK00031B/2687